AF806155

# Fascinating Life Sciences

This interdisciplinary series brings together the most essential and captivating topics in the life sciences. They range from the plant sciences to zoology, from the microbiome to macrobiome, and from basic biology to biotechnology. The series not only highlights fascinating research; it also discusses major challenges associated with the life sciences and related disciplines and outlines future research directions. Individual volumes provide in-depth information, are richly illustrated with photographs, illustrations, and maps, and feature suggestions for further reading or glossaries where appropriate.

Interested researchers in all areas of the life sciences, as well as biology enthusiasts, will find the series' interdisciplinary focus and highly readable volumes especially appealing.

David Costantini

# The Role of Organismal Oxidative Stress in the Ecology and Life-History Evolution of Animals

Springer

David Costantini
Department of Ecological and Biological Sciences
University of Tuscia
Viterbo, Italy

UMR 7221 CNRS - Muséum National d'Histoire Naturelle, Laboratoire Physiologie Moléculaire et Adaptation, Département Adaptation du vivant
Paris, France

ISSN 2509-6745 ISSN 2509-6753 (electronic)
Fascinating Life Sciences
ISBN 978-3-031-65182-3 ISBN 978-3-031-65183-0 (eBook)
https://doi.org/10.1007/978-3-031-65183-0

© The Editor(s) (if applicable) and The Author(s), under exclusive license to Springer Nature Switzerland AG 2024
This work is subject to copyright. All rights are solely and exclusively licensed by the Publisher, whether the whole or part of the material is concerned, specifically the rights of translation, reprinting, reuse of illustrations, recitation, broadcasting, reproduction on microfilms or in any other physical way, and transmission or information storage and retrieval, electronic adaptation, computer software, or by similar or dissimilar methodology now known or hereafter developed.
The use of general descriptive names, registered names, trademarks, service marks, etc. in this publication does not imply, even in the absence of a specific statement, that such names are exempt from the relevant protective laws and regulations and therefore free for general use.
The publisher, the authors and the editors are safe to assume that the advice and information in this book are believed to be true and accurate at the date of publication. Neither the publisher nor the authors or the editors give a warranty, expressed or implied, with respect to the material contained herein or for any errors or omissions that may have been made. The publisher remains neutral with regard to jurisdictional claims in published maps and institutional affiliations.

This Springer imprint is published by the registered company Springer Nature Switzerland AG
The registered company address is: Gewerbestrasse 11, 6330 Cham, Switzerland

If disposing of this product, please recycle the paper.

*To the natural world*

# Preface

Life on Earth is astonishingly rich. It comes in many shapes, sizes, and colourations, but also countless behaviours and molecules. Why is our planet full of life? What are the sources of this massive biological diversity? And how does biological diversity evolve? These questions alone can mean many different things to many people. And so, they can have different answers. Many scientists are particularly interested in understanding the proximate causes of this biological diversity, and what this means for the resilience and adaptability of animals to natural and anthropogenic environmental changes. From an eco-evolutionary perspective, measures of physiological functions inform predictions on how these functions vary at individual, population, or species level and drive the generation of biological diversity. Intense research over the last century showed that hormonal control of development or behaviour, regulation of energy intake and expenditure, or immune function are vitally important. In recent decades, scientists have increasingly come to appreciate that the interplay between molecules called reactive oxygen species and antioxidants can have dramatic effects on the generation and evolution of biological diversity. We have recent evidence that variation in organismal oxidative status poses significant consequences for growth, reproduction, or ageing, and may therefore evolve in response to selection. However, there is an intriguing amount of unexplained variation within and among species in antioxidant types and levels, or in the generation of reactive oxygen species. Thus, it remains unclear when and how oxidative status evolves, partly because there is not a single metric or tissue that gives us a comprehensive framework of an individual oxidative status. Relying on an eco-evolutionary perspective of the biology of reactive oxygen species and antioxidants, this book explores and discusses how the need to manage the organismal oxidative status in an optimal way to maintain homeostasis is an important proximate mechanism driving the resilience and adaptability of individuals or species to environmental stressors, and the outcome of many life-history trade-offs. The book provides many examples of the types of questions that scientists have addressed in studies of oxidative stress physiology in non-model species, highlighting the value of studying this topic outside of a biomedical context.

The first six chapters of the book focus more on the proximate mechanisms that mould the organismal oxidative status, and the cross-talk between oxidative status and other processes, including immunity, nutrition, endocrine regulation, metabolic rate, and (epi)genetics under different environmental circumstances. With Chap. 1 "*Oxidative Status and Oxidative Stress: Definitions and Mechanisms*" the book kicks off focusing on the toolkit that organisms have evolved to regulate their cellular oxidative status and combat the damaging consequences of reactive oxygen species. Staying on the subject of endogenous mechanisms, Chap. 2 "*The Immuno-Oxidative Ecology*" deals with the role of oxidative stress in generating variation in immunological performance in animal populations. Chapter 3 "*Oxidative Costs and Antioxidant Rewards of Nutrition*" focusses on the effects of diet on the organismal resistance to oxidative stress, and how the links between diet and oxidative status have influenced the evolution of foraging behaviour, food choice, and use of nutrients to program the oxidative status of offspring. In Chap. 4 "*Diversity in Oxidative Status: Roles of Stress Response, Metabolism, Rhythmicity, and (Epi)Genetics*", it is illustrated the roles of some proximate mechanisms (e.g. types of hormones, stress response, and epi(genetic) factors) in controlling the individual, spatial, temporal, and species variation in levels of reactive oxygen species and antioxidants. Chapter 5 "*Exposome and Regulation of Oxidative Status Across Different Environments*" shows how the need to maintain the oxidative status homeostasis is key in the responses of organisms to changes in temperature, ultraviolet radiation, partial pressure of oxygen, acidification, salinity stress, availability of water, and habitat. Next, in Chap. 6 "*From Strenuous Migration to Dormancy and Arousal: How Physical (In)Activity Affects Oxidative Status*", it is discussed the cross-talk between metabolism and oxidative status in the contexts of particular biological phenomena that are characterised by strenuous metabolic activity (e.g. migration) or reduced metabolic rate (e.g. hibernation).

In the last four chapters, the book examines the ways in which balancing investment in defences against the consequences of sustaining oxidative damage has influenced trade-offs in the allocation of resources to the key life-history traits of growth and development, sexual signalling, reproduction, and ageing. Chapter 7 "*For Better or Worse: How Early Life Oxidative Stress Moulds the Phenotype*" illustrates the fundamental roles of early in life exposure to oxidative stress in shaping schedules of growth and development and in orchestrating the life of individuals from conception to death. In Chaps. 8 "*The Oxidative Costs of a Colourful Life*" and 9 "*Oxidative Stress and Reproductive Trade-offs: From Courtship to Parental Care*", the book shows the many routes through which oxidative stress can impact on reproductive fitness, from the production of sexual signals and tactics of mate acquisition to the reproductive decisions and parental care. The book ends with Chap. 10 "*The Long and The Short of It: How Oxidative Stress Drives Cellular Senescence, Ageing, and Lifespan of Animals*" discussing the controversial literature about the role of oxidative stress as a driver of ageing.

In bringing together issues in life-history evolution to the physiology of oxidative stress, this book provides a fascinating examination of how very small-scale processes, like those that occur within our cells, can affect large-scale processes that generate biological diversity. It also offers opportunities for the discipline to yield considerable insight into the ecological and evolutionary significance of variation in oxidative status and oxidative stress. The complexity and diversity of the natural world is a powerful experimental resource. Recognising and valuing this complexity will be instructive and illuminating.

Viterbo, Italy David Costantini

# Competing Interests

I declare that I have no competing financial and/or non-financial interests in relation to the content of the book.

# Contents

# Chapter 1
# Oxidative Status and Oxidative Stress: Definitions and Mechanisms

**Abstract** The origin of aerobic life is one of the most fascinating and elusive topics in biology. Multiple lines of evidence build a compelling case for a central role of oxygen in driving the evolution of complex multicellular life. With the transition from a reducing to an oxidising chemistry in the atmosphere and in the oceans, organisms expanded their biochemical and metabolic capacities to use oxygen to generate energy more efficiently. At the same time, organisms have evolved the so-called antioxidant mechanisms to protect themselves against the toxic effects of oxygen derivatives, such as free radicals and non-radical reactive oxygen species. There is now consensus that reactive oxygen species and antioxidants are of profound biological relevance across a wide range of organisms. In this chapter, I have summarised the definitions and fundamentals of oxidative status physiology to provide the basic background information for practitioners of the field. Specifically, I have provided (1) a general historical background and definitions of reactive oxygen species, antioxidants, and oxidative stress and (2) an overview of the main antioxidants that have so far been discovered. Finally, I have examined the circumstances under which oxidative stress can surprisingly increase phenotypic stress resistance or tolerance.

## 1.1 Introduction: Living with Oxygen

We are so used to the presence of oxygen on our planet that we cannot even imagine that it has not always been the case. When the earth was formed around 4.5 billion years ago, the primordial atmosphere was very different from what we observe nowadays. Oxygen was absent from the earth's atmosphere for close to half of its lifespan. The primordial atmosphere was indeed mildly reducing, consisting of large proportions of methane, ammonia, carbon dioxide, and hydrogen (Schopf and Klein 1992; Sessions et al. 2009). There is consensus that the evolution of cyanobacteria had a major impact on the composition of the atmosphere. These microbes had the remarkable capacity to perform photosynthesis, and oxygen was a major by-product

© The Author(s), under exclusive license to Springer Nature Switzerland AG 2024
D. Costantini, *The Role of Organismal Oxidative Stress in the Ecology and Life-History Evolution of Animals*, Fascinating Life Sciences,
https://doi.org/10.1007/978-3-031-65183-0_1

of it. As a result, around a couple billion years ago, oxygen became the major component of the atmosphere, an event known as the Great Oxidation Event (Sessions et al. 2009). A second significant increase in the concentration of oxygen in the atmosphere occurred around 600–800 million years ago, and this second increase was accompanied by the oxygenation of the deep oceans and the emergence of multicellular animals (Sessions et al. 2009).

The transition from a reducing to an oxidising atmosphere has determined the evolution of metabolic networks of increasing complexity (Falkowski 2006; Raymond and Segrè 2006; Weaver 2019). Although oxygen was likely poisonous for anaerobic organisms, because of its high redox potential, this molecule was also an ideal terminal acceptor of electrons in the cellular processes of energy generation. The stimulatory effect of oxygen on the evolution of metabolic networks was not a dinner party. Harnessing oxygen in aerobic metabolism to generate energy exposes cells to molecules called reactive oxygen species. If uncontrolled, these highly reactive by-products can wreak havoc, attacking all of the main building blocks from which bodies are made, including DNA, resulting in oxidative damage. As a consequence, the need to evolve adequate defences against the generation or accumulation of such damage arose in parallel with the use of oxygen by cells to generate energy. This also raises the obvious chicken-and-egg question: what came first, the oxygen-producing photosynthesis or the compounds that protect cells against oxygen-induced damage? Current evidence suggests that at least some compounds with antioxidant properties (e.g. ergothioneine, superoxide dismutase) might have already been present in microbes before Earth's oxygenation (Fig. 1.1; Boden et al. 2021; Burn et al. 2017; Case 2017; Inupakutika et al. 2016; Kornas et al. 2010; Lane 2002; Ruszczycky and Liu 2017; Weaver 2019). This is reasonable to expect because (i) pro-oxidants could have been formed even in absence of the high concentrations of oxygen that occur in our atmosphere, (ii) the generation of pro-oxidants in the initial stages of the evolution of oxygen metabolism would have been probably too detrimental if organisms would not have already had some antioxidant compounds to counteract them, and (iii) some antioxidant enzymes occur even in anaerobic organisms (Borda et al. 2001; Brioukhanov and Netrusov 2004; Kornas et al. 2010; Weaver 2019). The nature of, and the interplay between, the benefits and costs involved in balancing offspring or energy generation against damage mitigation is a major area of research that cuts across many biological disciplines and levels of enquiry. We think that oxidative stress has played a central role as driver of the evolution of animal diversity.

With this first chapter that kicks off the book, I aim to summarise the definitions and fundamentals of the oxidative status physiology in order to provide the basic background information for practitioners that are new to the field. Specifically, I have provided (i) a general historical background and definitions of reactive oxygen species, antioxidants, and oxidative stress, and (ii) an overview of the main antioxidants that have so far been discovered. I have also examined the circumstances under which exposure to oxidative stress is not necessarily detrimental, but can surprisingly increase phenotypic stress resistance or tolerance.

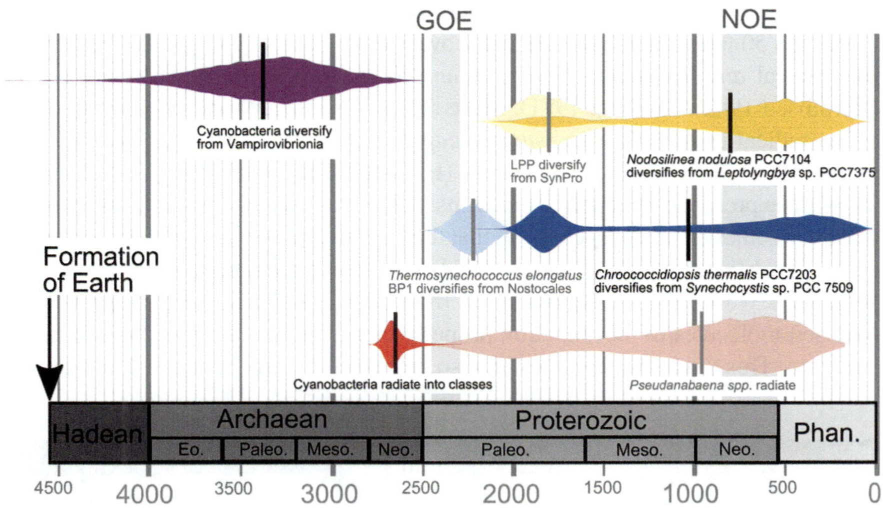


**Fig. 1.1** Emergence of cyanobacteria (purple) and their copper/zinc superoxide dismutase (CuZnSODs; red), nickel superoxide dismutases (NiSODs; orange), and manganese and/or iron superoxide dismutases (Mn- and/or Fe-utilising SODs; blue) inferred through geological time. Phylogenetic evidence suggests that ancestral cyanobacteria used CuZnSODs during the Archaean. Coloured bands represent the distribution of age posterior estimates for key nodes in a Bayesian molecular clock calibrated with the first radiation of cyanobacteria between 2.32 and 2.7 billion years ago. Vertical black lines represent the mean in each instance. Dark-coloured bands represent the earliest cyanobacteria predicted to have genes encoding the relevant SOD isoform based on a hypothesis of vertical inheritance, whereas paler nodes require horizontal gene transfer or other patterns of reticulated evolution. Phan = Phanerozoic Aeon; Meso = Mesoproterozoic Era; Neo = Neoproterozoic Era; GOE = Great Oxidation Event; NOE = Neoproterozoic Oxygenation Event. Reprinted from Boden et al. (2021). © 2021, Crown; distributed under the terms of the Creative Commons Attribution License (CC BY)

## 1.2 The Odd Couple: Reactive Oxygen Species and Antioxidants

### *1.2.1 Reactive Oxygen Species: Definitions, Properties, Sources, and Targets*

Many important scientific discoveries were achieved accidently because they were not initially planned to seek. Serendipity is the term used to categorise these unplanned discoveries. Penicillin is perhaps the most famous example of serendipity. But the discovery of free radicals is not so least. At the end of the 1800s, whilst attempting to prepare a compound called hydrocarbon hexaphenylethane, the scientist Moses Gomberg (1900) at the University of Michigan came across a novel molecule that was identified as the triphenylmethyl radical. This molecule represents the first free radical ever described. For this discovery, Gomberg has been recognised as the founder of radical chemistry.

The first proof of free radical generation in biological materials was published more than 50 years later Gomberg's discovery (Commoner et al. 1954). We now call free radical *any transient (chemically unstable) species (atom, molecule, or ion)* (Bernthsen 1942; Herzberg 1971). More recently, some authors proposed to define a free radical as *any chemical species capable of independent existence that contains one or more unpaired electrons* (Halliwell and Gutteridge 2015; Li et al. 2016). The presence of unpaired electrons makes free radicals unstable and prone to react with other chemical species, so that they are able to oxidise cellular macromolecules, such as proteins, lipids, or DNA, and cause damage to cells and tissues (Fig. 1.2). We refer to these oxidation by-products as oxidative damage because oxidised molecules do not function properly due to an alteration of their chemical structure. This oxidative damage can also be caused by other chemical species that

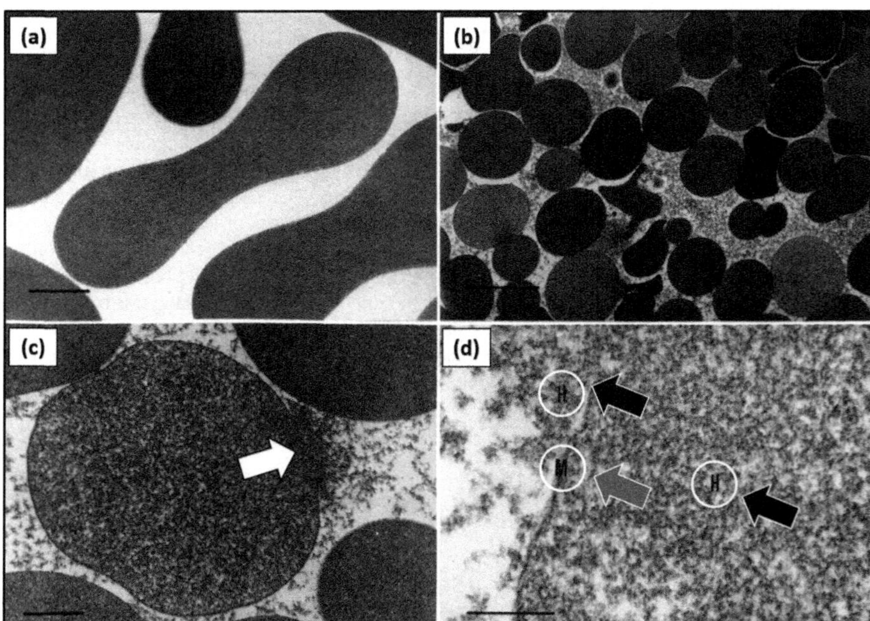


**Fig. 1.2** Transmission electron micrographs of erythrocytes during the oxidation induced by free radicals generated from 2,2′-azobis(2-amidinopropane) dihydrochloride. (**a**) untreated control erythrocytes from a rabbit. The scale mark represents 1 μm. (**b**) erythrocytes incubated in a medium containing 2,2′-azobis(2-amidinopropane) dihydrochloride for 100 min. The scale mark represents 5 μm. (**c**) an arrow indicates the site of haemolysis, i.e. rupture of erythrocytes. The scale mark represents 1 μm. (**d**) part of panel c. A strand of haemoglobin (H) leaked out from the site of haemolysis (black arrow). M is a remnant of plasma membrane of the erythrocytes (grey arrow). The scale mark represents 0.5 μm. These in vitro studies revealed that free radicals induced lipid peroxidation to membrane phospholipids of erythrocytes with consequent haemolysis, which was suppressed by vitamin E (Miki et al. 1987; Niki 2014; Niki et al. 1988; Yamamoto et al. 1986). Reprinted from Niki et al. (1988) with minor modifications. © 1988 ASBMB; Niki, Komuro, Takahashi, Urano, Ito, Terao; distributed under the terms of the Creative Commons Attribution License (CC BY)

lack radical properties, such as hydrogen peroxide, hypochlorous acid, peroxynitrite, or singlet oxygen. Halliwell and Gutteridge (2015) proposed to use the term *reactive species* to refer to this wide variety of chemical species with pro-oxidant properties, including both radical or non-radical chemicals, as well as chemicals that are derivatives of oxygen (reactive oxygen species) or of other elements (e.g. reactive nitrogen species).

Reactive oxygen species can originate from a variety of sources (Halliwell and Gutteridge 2015). The major sites of their generation in animal cells include mitochondria, peroxisomes, endoplasmic reticulum, phagocytes, and plasma membrane (Asada 1999; Halliwell and Gutteridge 2015; Zhang and Wong 2021). In particular, during the oxidative phosphorylation in mitochondria, some electrons leak out of the electron transport chain and reduce oxygen to superoxide free radical (Fig. 1.3). Free radicals can arise from the loss of a single electron from a non-radical chemical species:

$$A - e^{-} \rightarrow A^{\bullet +} \left(\text{radical cation}\right)$$

or from the gain of a single electron from a non-radical chemical species:

$$A + e^{-} \rightarrow A^{\bullet -} \left(\text{radical anion}\right)$$

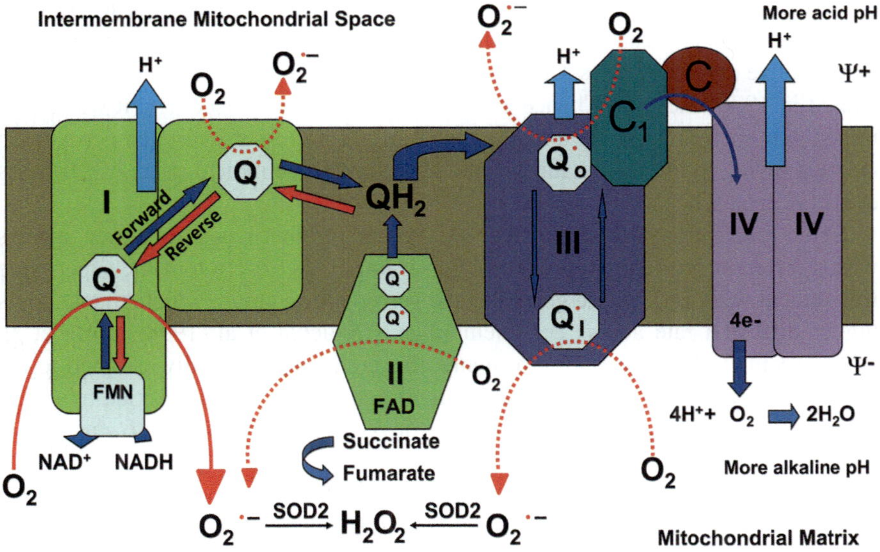


**Fig. 1.3** Schematic representation of the mitochondrial electron transport chain and generation of the mitochondrial superoxide free radical ($O_2^{\bullet -}$). At several sites of the respiratory chain, electrons leak from ubiquinone (Q) to oxygen creating the free radical superoxide. The main sources of mitochondrial reactive oxygen species are complexes I, II, and III, which produce superoxide mainly on the matrix side, where it is rapidly dismutated to hydrogen peroxide ($H_2O_2$) by mitochondrial manganese superoxide dismutase (SOD2). Reprinted from Dikalov (2011) with permission of Elsevier

A free radical can also originate from homolytic fission of a molecule: a covalent bond is broken, and one electron from the bonding electron pair remains on each atom:

$$A:B \rightarrow A^{\bullet} + B^{\bullet}$$

By contrast, in the heterolytic fission, one atom receives both bonding electrons from the other atom, so that it is given a negative charge, whilst the other atom is given a positive charge; neither of these two atoms is a free radical:

$$A:B \rightarrow A^{-} + B^{+}$$

Reactive oxygen species differ largely in their chemical properties and damaging potential. For example, at neutral pH in water, the rate constant in the reaction with methionine is $7 \times 10^9$ $M^{-1}$ $s^{-1}$ for the hydroxyl radical ($^{\bullet}OH$), $3.8 \times 10^7$ $M^{-1}$ $s^{-1}$ for hypochlorous acid (HOCl), $3.6 \times 10^2$ $M^{-1}$ $s^{-1}$ for peroxynitrous acid (ONOOH), and $2 \times 10^{-2}$ $M^{-1}$ $s^{-1}$ for hydrogen peroxide ($H_2O_2$), respectively (Buxton et al. 1988; Davies 2005, 2016). The most reactive compounds are expected to have negligible capability to migrate across cells because, owing to their strong reactivity, they would rapidly react with any molecule around causing oxidation locally (e.g. hydroxyl radical). By contrast, hydrogen peroxide, which is much less reactive than the hydroxyl radical, can work as a Trojan horse, diffusing away from sites of generation towards other cellular locations or cells (e.g. through aquaporins, Bestetti et al. 2018), where it can propagate the oxidative cascade through the generation of new reactive oxygen species. Cellular generation of hydrogen peroxide was demonstrated in 1970 (Sies and Chance 1970). Shortly after this discovery, it was determined that mitochondria are a major source of generation of hydrogen peroxide in animals owing to the dismutation of the superoxide free radical (Fig. 1.3; Boveris et al. 1972; Loschen et al. 1971).

Some reactive oxygen species can also generate other reactive oxygen species with a different pro-oxidant potential. For example, the reaction of the superoxide free radical with nitric oxide generates the peroxynitrite anion, which, thanks to its higher diffusion rate across cell membranes (Denicola et al. 1998; Marla et al. 1997), seems to be responsible for many effects that were initially ascribed to the superoxide free radical (Halliwell and Gutteridge 2015). The superoxide free radical can also give rise to the more reactive hydroxyl radical through the *Haber–Weiss reaction*:

$$O_2^{\bullet -} + H_2O_2 \rightarrow O_2 + {}^{-}OH + {}^{\bullet}OH$$

In the 1950s and 1960s, research on free radicals gained momentum because it was demonstrated that their occurrence in living organisms, leading to the hypothesis that they might have a role as promoters of cellular senescence (Commoner et al. 1954; Gershman et al. 1954; Harman 1956; McCord and Fridovich 1968, 1969). Since then, research on the biological meaning of reactive oxygen species has acquired a prominent position in medicine, biochemistry, physiology, pharmacology, environmental toxicology, behavioural ecology, evolutionary ecology, and conservation physiology (Beaulieu and Costantini 2014; Costantini 2008, 2014, 2019; Fridovich 1978; Halliwell and Gutteridge 2015; McGraw et al. 2010).

## *1.2.2 The Other Face of the Coin: The Antioxidant Machinery*

In the nineteenth century, engineers discovered that certain substances could prevent metals from corroding by shutting off the oxidation processes. Although the reasons behind this discovery were unknown, scientists found that the addition of molecules with antioxidant properties to foods rich in unsaturated fatty acids prevented the onset of rancidity (Halliwell and Gutteridge 2015). But it was in the 1960s that antioxidant research took off. McCord and Fridovich (1968, 1969) purified an enzyme from bovine erythrocytes that (i) was capable of catalysing the dismutation of superoxide free radicals and (ii) was widespread across tissues. This enzyme was called superoxide dismutase (SOD). We now know that living organisms have evolved a number of ways for coping with reactive oxygen species that we refer to as antioxidants (Pamplona and Costantini 2011). These molecules vary widely in function, chemical properties, and concentrations across tissues (Fig. 1.4).

In both the fourth and fifth edition of their book, Halliwell and Gutteridge (fifth ed. Halliwell and Gutteridge 2015; p. 77) defined an antioxidant as *any substance that delays, prevents, or removes oxidative damage to a target molecule*. Pamplona and Costantini (2011) proposed a broader definition of antioxidant, even encompassing organismal structures (e.g. cell membrane composition) because they can influence the resistance or vulnerability of cells to pro-oxidants: *any mechanism, structure, and/or substance that prevents, delays, removes, or protects against oxidative non-enzymatic chemical modification (damage) to a target molecule*. In this definition, Pamplona and Costantini also pointed out that oxidative damage includes only molecules that underwent non-enzymatic modifications, in order to distinguish them from those molecules that participate in redox reactions catalysed by enzymes.

### 1.2.2.1 Enzymatic and Non-enzymatic Molecular Antioxidants

Antioxidant enzymes play an important role in the removal of reactive oxygen species (including free radicals and intermediate derivatives of oxidative damage, such as organic hydroperoxides) or in their transformation into less reactive molecules (Halliwell and Gutteridge 2015). The organism also relies on non-enzymatic antioxidants, which may work as co-factors of antioxidant enzymes, sequester metal ions, or quench reactive oxygen species (Halliwell and Gutteridge 2015). Maintaining a variety of components in the antioxidant machinery could provide a 'fail-safe' to avoid the case that the failure of one antioxidant compromises the whole system. This is particularly relevant for those antioxidant mechanisms that rely on a delicate equilibrium among molecules involved. For example, deficiency

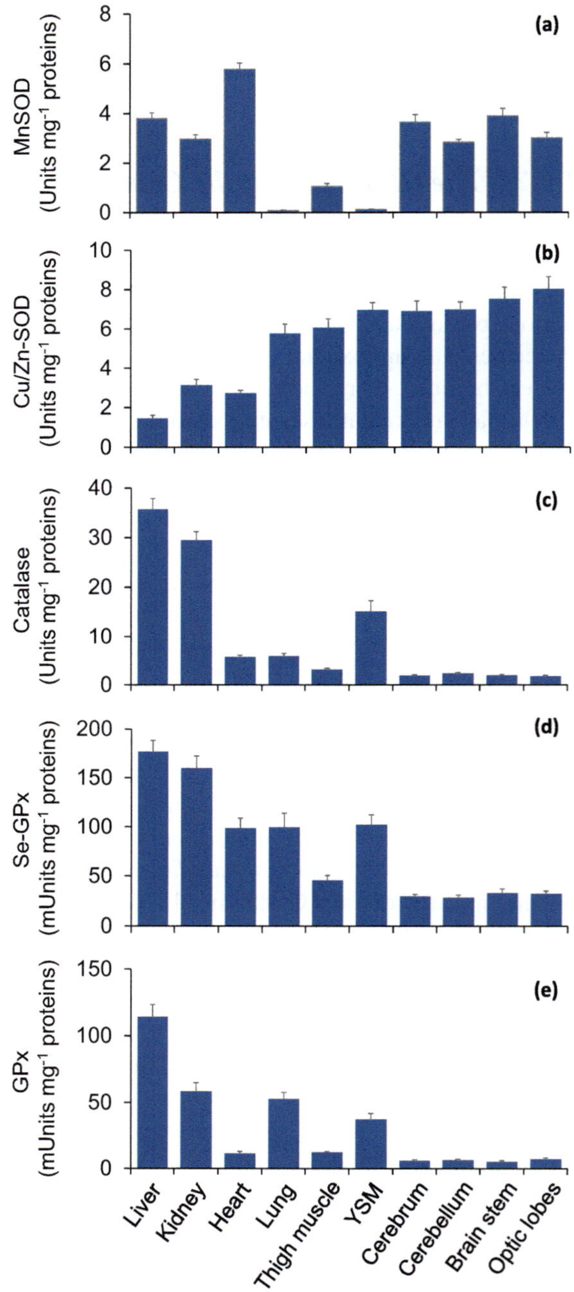


**Fig. 1.4** Activities of (**a**) Mn-superoxide dismutase, (**b**) Cu/Zn-superoxide dismutase, (**c**) catalase, (**d**) selenium-dependent glutathione peroxidase, and (**e**) non-selenium-dependent glutathione peroxidase across 10 tissues of newly hatched chicks *Gallus gallus domesticus*. Values are shown as mean and standard error. Data were collected from Surai et al. (1999)

in certain antioxidants may compromise the activity of other antioxidants, jeopardising the organismal protection against oxidative stress (e.g. Lebold et al. 2013). In the following paragraphs, I have illustrated the main categories of enzymatic and non-enzymatic antioxidants.

In animals, there are three main isoforms of the enzyme superoxide dismutase: the copper-zinc superoxide dismutase (SOD1), located in the cell cytoplasm, nucleus, and in the lumen between outer and inner membranes of mitochondria; the manganese superoxide dismutase (SOD2), located in the matrix of mitochondria; and the extracellular copper-zinc superoxide dismutase (SOD3), located extracellularly in plasma, lymph, or synovial fluid. All three of them catalyse the dismutation of the superoxide free radical into oxygen and hydrogen peroxide using protons of their metal co-factor (Halliwell and Gutteridge 2015):

$$2O_2^{\bullet -} + 2H^+ \xrightarrow{SOD} O_2 + H_2O_2$$

These three types of superoxide dismutases can also be found in organisms other than animals together with other isoforms, such as the iron-containing superoxide dismutase of bacteria, algae, and some plants. A particular family of superoxide dismutases includes the superoxide dismutase with iron or manganese. These ancient superoxide dismutases lack metal-binding selectivity because they can be active with either or both iron and manganese (i.e. cambialistic superoxide dismutase; Valenti et al. 2022). In animals, concentrations of superoxide dismutases vary among tissues. For example, in humans, levels of copper-zinc superoxide dismutase were particularly high in liver, cerebral grey matter, or testis compared to thyroid gland or lung (Halliwell and Gutteridge 2015). In newly hatched chicks *Gallus gallus domesticus*, levels of copper-zinc superoxide dismutase were particularly high in brain, skeletal muscle, or yolk sac membrane compared to liver (Fig. 1.4; Surai et al. 1999). Analyses of structure and molecular evolution of the superoxide dismutase gene family indicated that specific amino acid changes influenced the capacity of the enzyme to eliminate reactive oxygen species (Tian et al. 2021a).

Catalase is another important antioxidant enzyme, which is specialised in the removal of hydrogen peroxide. Most animal cells have a form of catalase that has four haem molecules embedded in its core; some microbes have manganese instead of haem. In human tissues, the activity of catalase was particularly high in liver or erythrocytes compared to brain, heart, skeletal muscle, or spleen (Halliwell and Gutteridge 2015). In newly hatched chicks, the activity of catalase was particularly high in liver or kidney compared to brain, thigh muscle, or heart (Fig. 1.4; Surai et al. 1999). The action of catalase is very important for removing hydrogen peroxide when its concentration is high. This is because the catalase has a low affinity for hydrogen peroxide, so that a high concentration of it makes easier for catalase to catch two molecules of hydrogen peroxide. The catalytic mechanism through which catalase performs its action is a two-step process: (i) it binds a molecule of hydrogen peroxide and breaks it into an atom of oxygen that remains bound to the haem iron (generating a porphyrin cation radical called compound I) and a molecule of water, and then (ii) it binds another molecule of hydrogen peroxide to compound I, breaking it again into an atom of oxygen and a molecule of water, respectively. The two atoms of oxygen are combined with each other by catalase producing oxygen:

$$2H_2O_2 \overset{CAT}{\rightarrow} 2H_2O + O_2$$

Glutathione peroxidases (GPx) of vertebrates are a group of antioxidant enzymes that preferentially use the electrons from the molecule glutathione (GSH) to reduce peroxides and organic hydroperoxides to water and alcohols, respectively (Maiorino et al. 2015). By contrast, bacteria, invertebrates, plants, and fungi use mostly electrons from thioredoxins; thioredoxin peroxidases are widespread and ancient homologs of glutathione peroxidases (Maiorino et al. 2015). Glutathione peroxidase is more effective than catalase at controlling levels of hydrogen peroxide when this reactive oxygen species occurs in low concentrations. The activity of glutathione peroxidase varies among tissues. In humans, it was more active in liver than in skeletal muscle or lung (Halliwell and Gutteridge 2015). In newly hatched chicks, the activity of selenium-dependent glutathione peroxidase was particularly high in liver or kidney compared to brain or thigh muscle (Fig. 1.4; Surai et al. 1999). Glutathione peroxidases occur as selenium- or non-selenium-dependent forms, and a number of isozymes have been identified across many diverse organisms (e.g. Behne and Kyriakopoulos 2001; Margis et al. 2008). For example, in mammals, there are four major selenium-dependent isozymes: (i) classical glutathione peroxidase (GPx1), which is found in red blood cells, liver, lung, and kidney, (ii) gastrointestinal glutathione peroxidase (GPx2), (iii) plasma glutathione peroxidase (GPx3), which is extracellular, but it is also present in different organs, such as kidney, lung, epididymis, vas deferens, placenta, seminal vesicle, heart, and muscle, and (iv) phospholipid glutathione peroxidase (PHGPx4 or GPx4), which is also broadly distributed in different tissues (Margis et al. 2008; Tian et al. 2021b). A phylogenetic analysis of the glutathione peroxidase family (including plants, fungi, bacteria, invertebrates, and vertebrates) suggested that basal classes of glutathione peroxidases originated from independent evolutionary events, such as gene duplication, gene losses, and lateral gene transfer (Margis et al. 2008). The analysis also identified two potential evolutionary patterns in animals, one leading to the origin of most isozymes and another one leading to the glutathione peroxide 4 isozyme, which would also include plants and fungi. This second pattern would suggest evolutionary convergence mediated by functional pressures (Margis et al. 2008). Further work on mammals found a correlation between oxidative stress tolerance and the divergent selection and gene copy number of GPx subclasses, suggesting that the evolution of the GPx gene family contributed to the adaptation of mammals to stressful environments (Tian et al. 2021b).

Thiols are organosulfur compounds of the form R–SH. The antioxidant activity of thiols is owing to their reduced sulfhydryl (SH) group. Glutathione (GSH) is a tripeptide thiol (γ-L-glutamyl-L-cysteinyl glycine) that occurs at high concentrations in its reduced protective form in most tissues (Halliwell and Gutteridge 2015).

In both animals and plants, glutathione is synthesised in two ATP-dependent steps from amino acid precursors (Fig. 1.5): in the first step, the enzyme γ-glutamylcysteine synthetase (γ-GCS) joins the two amino acids glutamic acid and cysteine; in the second step, the enzyme glutathione synthetase adds glycine (Halliwell and Gutteridge 2015; Noctor et al. 2012). The antioxidant activity of glutathione is due to the SH group of its cysteine residue. This residue is oxidised to disulfide, generating oxidised glutathione (GSSG). Once oxidised, the cysteine residue can be transformed back to its reduced form through the action of the enzyme glutathione reductase, which takes hydrogen from nicotinamide adenine dinucleotide phosphate (NADPH) (Fig. 1.5). Concentration of glutathione varies among tissues. For example, in rat and human tissues, levels of glutathione were higher in liver than in erythrocytes, blood plasma, or kidney (Halliwell and Gutteridge 2015). In the scallop *Adamussium colbecki*, levels of glutathione were higher in mantle than in gills (Philipp et al. 2006).

Another thiol-related redox-active molecule is the protein thioredoxin (Mitsui et al. 2002). Thioredoxin can regulate the activity of transcription factors (e.g. nuclear factor-kappaB, activated protein-1) and can scavenge reactive oxygen species in cooperation with the peroxiredoxin/thioredoxin-dependent peroxidase (Mitsui et al. 2002). Ovothiols are another group of thiols with important antioxidant properties. They occur at very high levels in the eggs of several marine invertebrates and fish, where they scavenge hydrogen peroxide released during fertilisation (Turner et al. 1987, 1988). They were also identified in some intracellular pathogens (Spies and Steenkamp 1994).

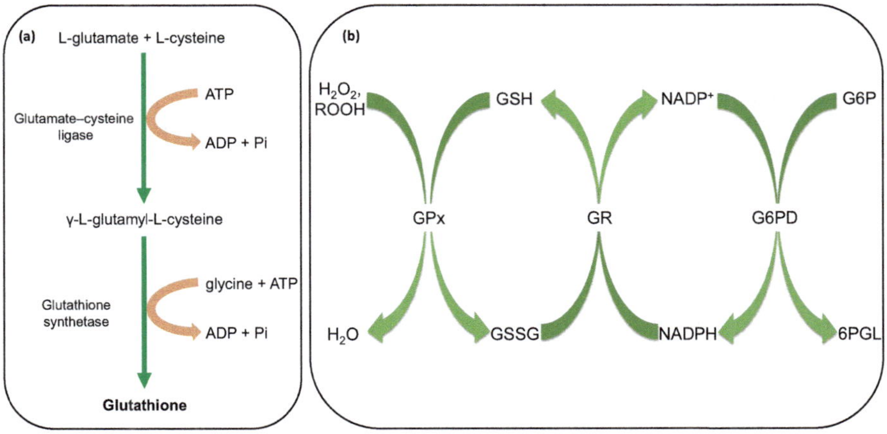


**Fig. 1.5** Metabolism of glutathione. (**a**) biosynthetic pathway of glutathione; it involves two ATP-requiring enzymatic steps. (**b**) glutathione cycle. GSH = reduced glutathione; GSSG = oxidised glutathione; GPx = glutathione peroxidase; GR = glutathione reductase; G6PD = glucose-6-phosphate dehydrogenase; G6P = glucose 6-phosphate; 6PGL = 6-phosphogluconolactone; NADPH = reduced nicotinamide adenine dinucleotide phosphate; $NADP^+$ = oxidised nicotinamide adenine dinucleotide phosphate; ROOH = organic hydroperoxide; $H_2O_2$ = hydrogen peroxide; $H_2O$ = water

Some studies suggested that haemoglobin might have a potential protective role against reactive oxygen species because it contains thiol groups (Reischl et al. 2007). For example, Hausladen et al. (1998) found that haemoglobin protects *Escherichia coli* against nitrosative stress. Similarly, Nishi et al. (2008) found that the overexpression of haemoglobin reduced the generation of hydrogen peroxide and enhanced cell viability against exposure to hydrogen peroxide in rats. Erythrocytes face high partial pressures of oxygen, high oxygen flow, and high concentrations of iron. Hence, selective pressures may have favoured the evolution of thiol-rich haemoglobins in order to make red blood cells buffered against oxidation. However, the propensity to (auto)oxidation of haemoglobin might fuel a state of oxidative stress with a consequent loss in red blood cell function (Katsu et al. 2010; Misra and Fridovich 1972; Tappel 1955). For example, the loss of iron-centred oxygen-binding proteins (haemoglobin and myoglobin) in Antarctic fishes does not correspond with an overall reduction across all tissues in levels of oxidised macromolecules, antioxidants, or rates of protein degradation (O'Brien et al. 2018). Goodchild and DuRant (2020) found that reactive oxygen species may indirectly impair red blood cell membrane integrity via haemoglobin degradation products that associate with the membrane itself. Taken together, it may be hypothesised that natural selection has operated in order to optimise a molecular compromise among benefits and costs of having or not having haemoglobin, or of having evolved more or less potent antioxidant defences that protect haemoglobin from oxidation, respectively (e.g. Reinke and O'Brien 2006).

We have also antioxidants that are not endogenously synthesised. The so-called dietary antioxidants include molecules, such as vitamin E, carotenoids, and polyphenols that animals obtain from food. These antioxidants can interact with each other to eliminate reactive oxygen species more efficiently (ascorbate may also be involved): when a given antioxidant is oxidised, another antioxidant can transform it back to its reduced form (see also Chap. 3). For example, the reaction between a carotenoid and a reactive oxygen species generates a carotenoid radical, which can be converted back to its reduced form using the electrons provided by tocopherols:

$$Car^{\bullet +} + \alpha TOH \rightarrow Car + \alpha TO^{\bullet} + H^{+}$$

It is, therefore, increasingly recognised that, in order to achieve significant protection against oxidation, it is needed a balance in the concentrations of the different dietary antioxidants. Some examples of these dietary antioxidants are reported in the following paragraphs.

Vitamin E is a chain-breaker dietary antioxidant (i.e. it is able of shortening the oxidation chain length as a result of its reaction with peroxyl radicals) that protects fatty acids against oxidation (Esterbauer et al. 1991; Niki 2014; Niki et al. 1988). The term vitamin E refers to a class of lipophilic molecules that includes tocopherols (d-α, d-β, d-γ, and d-δ tocopherol) and their corresponding tocotrienols (Surai 2002; Wang and Quinn 1999). Tocopherols also promote repair of oxidative damage in cell membranes of myoblasts (Howard et al. 2011) and protect telomeres from oxidation (Pérez-Rodríguez et al. 2019; Pineda-Pampliega et al. 2019). The reaction between vitamin E and free radicals generates the α-tocopherol radical, which is reconverted into vitamin E using the electrons provided by the ascorbic acid.

Carotenoids are lipophilic dietary antioxidants that occur in animal cells, but also in bacteria or in plants, where they protect tissues against the pro-oxidant effects of ultraviolet light. Hundreds of different types of carotenoids have been described so far, but only a few of them (e.g. β-carotene, lutein, zeaxanthin) occur at relevant concentrations in animal tissues (Krinski 1993; Surai 2002). Carotenoids are synthesised by photosynthetic organisms; by contrast, animals generally assume carotenoids from their diet, and accumulate them as they are or after transformation in novel carotenoids. However, evidence of laterally transferred genes for carotenoid biosynthesis in some arthropods challenged the view that all animal species are unable of synthesising carotenoids (Altincicek et al. 2011; Moran and Jarvik 2010). The antioxidant role of carotenoids is highly debated and appears to vary widely across organisms. For example, carotenoids play an important role in preventing protein oxidation in *Deinococcus radiodurans* (e.g. Krisko and Radman 2010; Tian et al. 2009). However, despite their in vitro scavenging properties, empirical evidence for carotenoids being important antioxidants in vivo is generally weak in other taxa, such as birds (Costantini and Møller 2008; Simons et al. 2012; Weaver et al. 2018; see also Chap. 3). Moreover, some supplementation trials of carotenoids (administered alone or in combination with other dietary antioxidants) even found toxic effects on the organism (e.g. Lee et al. 1999; Muntwyler et al. 2002; The Age-Related Eye Disease Research Group 2001). One mechanistic reason for these inconsistent antioxidant effects of carotenoids across organisms might lie with the type of carotenoid. For example, deinoxanthin, which is the major product of the carotenoid synthesis pathway in *Deinococcus radiodurans* (Lemee et al. 1997), has a stronger in vitro scavenging capability against hydrogen peroxide and singlet oxygen than carotenoids like β-carotene, zeaxanthin, and lutein, which are the main carotenoids of birds (Tian et al. 2007).

Another major class of dietary antioxidants are polyphenols. Numbering several thousands of distinct species, polyphenols represent a group of hydrophilic secondary metabolites, mostly derivatives and/or isomers of flavones, isoflavones, flavonols, catechins, and phenolic acids (Halliwell and Gutteridge 2015; Han et al. 2007). Polyphenols have diverse properties, such as antioxidant, anti-apoptosis, anti-ageing, or anti-inflammation protection, improvement of the endothelial function, inhibition of angiogenesis, and cell proliferation activity (Halliwell and Gutteridge 2015; Han et al. 2007). Polyphenols can also have pro-oxidant effects (Rietjens et al. 2002). In contrast to lipophilic antioxidants, polyphenols cannot be stored in body tissues, so that they are rapidly metabolised.

Ascorbic acid (ascorbate or vitamin C) can be synthesised endogenously in kidney or liver. Many animal taxa (e.g. some primate species including humans, some bats, many rodent and passerine species, teleost fish) lack ascorbate synthesis capacity owing to functional mutations in the L-gulono-γ-lactone oxidase (GLO) gene, which codes for the enzyme responsible for catalysing the last step of ascorbate synthesis (e.g. Birney et al. 1976, 1980; Chaudhuri and Chatterjee 1969; Cui et al. 2011; Drouin et al. 2011; Jenness et al. 1980; Martínez del Rio 1997; Roy and

Guha 1958; Smirnoff 2018). Although this loss of the capability to synthetise ascorbate may appear counterproductive, we know that the synthesis of ascorbate may lead to the formation of hydrogen peroxide and the depletion of glutathione (Bánhegyi et al. 1996; Drouin et al. 2011). Thus, high rates of ascorbate synthesis might generate large amounts of hydrogen peroxide, favouring ironically pro-oxidant side-effects (Halliwell and Gutteridge 2015). Moreover, ascorbate can act as a pro-oxidant due to its interaction with iron metabolism. Ascorbate-like compounds were also identified in several taxa (Smirnoff 2018). Ascorbic acid detoxifies cells from free radicals (e.g. superoxide, tocopheroxyl radicals, and alkoxyl/peroxyl radicals) via one-electron reduction, with the formation of the weakly reactive monodehydroascorbate radical (Smirnoff 2018). Thus, on the one side, ascorbic acid blocks free-radical chain reactions, and, on the other side, the generated radical does not perpetuate the chain reaction. In photosynthetic eukaryotes, ascorbic acid also works as a coenzyme of the enzyme ascorbate peroxidase in the removal of hydrogen peroxide (Maruta et al. 2016). Once oxidised, the ascorbic acid is restored to its non-oxidised form by the enzyme dehydroascorbate reductase, which takes electrons from glutathione (e.g. Maellaro et al. 1994; Navas and Villalba 1994; Rose and Bode 1993; Smirnoff 2018; Wells and Xu 1994; Zhang et al. 2015).

We have also molecules with antioxidant properties that are generated from metabolic pathways. Uric acid is a by-product of nitrogen metabolism; it seems to have important antioxidant properties in the organism, being able of reducing many reactive oxygen species (Iqbal et al. 1999; Klandorf et al. 1999; Muraoka and Miura 2003). The interpretation of the antioxidant role of uric acid is not straightforward as changes in its concentration in blood are possibly influenced by several factors, including excretory pathways, protein turnover (proteins are substrates of oxidation), and antioxidant demands. Multiple selective pressures may have, therefore, operated to generate variation among species in concentration and dynamics of uric acid (Monaghan and Costantini 2014).

Luciferins are substrates of the luminous reactions that occur in many marine bioluminescent species. Luciferins have also antioxidant properties, as they are highly reactive with molecules, such as the superoxide free radical or peroxides. Rees et al. (1998) suggested that the primary function of luciferins was originally the detoxification of cells from oxygen derivatives. The functional shift from antioxidant to light-emitting function might have occurred when the strength of selection for antioxidant defence mechanisms decreased. Rees et al. (1998) suggested that this might have been made possible when marine organisms began colonising deeper layers of the oceans, where exposure to oxidative stress is considerably reduced because of the lower light irradiance and oxygen levels that occur in deep waters.

Proteins synthesised during inflammation can also play an indirect role in antioxidant protection. Free metal ions promote the pro-oxidant activity of hydrogen peroxide through its one-electron reduction to the more reactive hydroxyl radical (Fenton reaction):

$$Mn^{+} + H_2O_2 \rightarrow M^{(n+1)+} + {}^{\bullet}OH + {}^{-}OH$$

($Mn^{+}$ and $M^{(n+1)+}$ are transition metal ions; $Mn^{+}$ stands for ions, such as $Cu^{+}$ and $Fe^{2+}$; $M^{(n+1)+}$ stands for ions, such as $Cu^{2+}$ and $Fe^{3+}$). Free metal ions can also cleave organic hydroperoxides into highly reactive free radicals, such as alkoxyl (R-O•) and alkylperoxyl (R-OO•) radicals, through the Fenton reaction. However, living organisms have evolved specific proteins (e.g. ceruloplasmin, ferritin, haptoglobin) that can sequester free metal ions, thus indirectly contributing to protection against oxidation of biomolecules (Halliwell and Gutteridge 2015).

Finally, some studies isolated a number of peptides with antioxidant properties (e.g. antioxidin-I, antioxidin-RP1) from the skin of several amphibian species (Barbosa et al. 2018; Plácido et al. 2022; Yang et al. 2009). These peptides appear to offer protection against several environmental stressors, such as elevated ultraviolet radiation, that can potentially cause increased generation of reactive oxygen species (Yang et al. 2016).

#### 1.2.2.2 Structural Components of Cells

Macromolecules differ in their susceptibility to oxidative damage, a property intrinsically related to the chemical characteristics of the molecules themselves. Thus, molecular composition of the cell membrane influences indirectly its resistance against oxidative stress, so that this function might be under natural selection (Bielski et al. 1983; Hulbert et al. 2007; O'Connor et al. 2002; Pamplona and Barja 2007; Pamplona and Costantini 2011; Porter et al. 1995). This hypothesis is based on the following evidence.

Polyunsaturated fatty acids are macromolecules highly susceptible to oxidative damage. They are less abundant in species that live longer (Hulbert et al. 2007; Pamplona 2008). However, the relative abundance of polyunsaturated fatty acids in cell membranes varies across animal taxa because it also affects other import cell functions, such as the fluidity. Guanine is the least abundant nitrogenous base in mitochondrial DNA (Samuels 2005) and is generally more easily oxidised as compared to the other nucleotides (Bjelland and Seeberg 2003; Kovacic and Wakelin 2001). Cell proteins have a low content of the amino acid methionine (Aledo et al. 2011; Pamplona et al. 2005; Portero-Otín et al. 2004; Ruiz et al. 2005), which is one of the amino acids most susceptible to oxidation (Stadtman et al. 2003). Several studies found a connection between fatty acid composition of the cell membrane, guanine abundance, or amino acid profile and lifespan across a diverse group of vertebrates, suggesting that structural resistance against reactive oxygen species has a relevant role in ageing (see Chap. 10).

#### 1.2.2.3 Regulators of Reactive Oxygen Species Generation

In animal cells, mitochondria are the major sites of reactive oxygen species generation (Fig. 1.3). In particular, respiratory chain complexes I and III, and to a lower degree the complex II, are the main producers of superoxide free radicals and derived reactive oxygen species within the mitochondrial respiratory chain, owing to the reaction of electrons (that leak out of the electron transport chain) with oxygen (Bleier and Dröse 2013; Brand et al. 2004; Dikalov 2011; Hulbert et al. 2007; Pamplona and Barja 2007). We have identified several factors that influence to some degree the activity of these respiratory complexes and, consequently, their contribution to the generation of reactive oxygen species.

The concentration of respiratory chain complexes varies among taxa that differ in the rate of reactive oxygen species generation. For example, birds have lower concentrations of both complex I and reactive oxygen species than mammals (Lambert et al. 2007; Pamplona and Barja 2007; St-Pierre et al. 2002). Therefore, regulation of the expression of the mitochondrial complexes could be part of an adaptive mechanism to control the generation of reactive oxygen species (Pamplona and Costantini 2011). The degree of reduction of these respiratory complexes is also relevant, as they produce more reactive oxygen species when they are more reduced (e.g. Barja 2007; Kushnareva et al. 2002; Pamplona and Barja 2006, 2007). Chemical modifications of respiratory complexes can influence their activity (Pamplona and Costantini 2011). For example, the glutathionylation of complex I increases the generation of the superoxide free radical (Han et al. 2003; Taylor et al. 2003).

Uncoupling proteins are important regulators of reactive oxygen species generation. Respiratory complexes use energy released by electrons to pump protons from the matrix of the mitochondrion into its intermembrane space, creating a proton motive force. Protons move back into the matrix through the adenosine-5′-triphosphate (ATP) synthases, following an electrochemical gradient; adenosine-5′-triphosphate synthases use this energy to produce ATP. Under these conditions, there is a coupling between the electron transport chain and the oxidative phosphorylation. However, the reaction of the superoxide free radical with membrane phospholipids generates the hydroxynonenal, an end product of lipid peroxidation, which is capable of activating uncoupling proteins. Uncoupling proteins provide a path for protons to diffuse from the intermembrane space into the matrix without ATP synthesis. Therefore, energy is released as heat. In this way, electron transport is uncoupled from the oxidative phosphorylation. This mild uncoupling lowers the proton motive force, causing the oxidation of respiratory complexes and lowering the local concentration of oxygen. These changes translate into a reduced generation of reactive oxygen species (feedback response to overproduction of superoxide; Echtay et al. 2003; Brand et al. 2004).

Organisms can also regulate the generation of reactive oxygen species modifying the partial pressure of oxygen into the mitochondria. In so doing, there is a reduction of ATP production, but also of the superoxide free radical because there is less oxygen that can react with electrons (Halliwell and Gutteridge 2015).

Finally, the mitochondrial generation of reactive oxygen species can be influenced by the cardiolipin concentration. Cardiolipin is a phospholipid located almost exclusively within the inner mitochondrial membrane (Hoch 1992; Portero-Otin et al. 2001). Cardiolipin is a potential target of reactive oxygen species because it contains unsaturated fatty acids. Peroxidation of cardiolipin influences the activity of mitochondrial complexes I, III, and IV (Paradies et al. 2000, 2001, 2002). However, cardiolipin has a low unsaturation degree, which makes it more resistant to peroxidation than other phospholipids. This low unsaturation of cardiolipin might explain why this particular phospholipid has been positively selected as a regulator of mitochondrial bioenergetics (Pamplona 2008; Pamplona and Costantini 2011).

#### 1.2.2.4 Repair and Detoxifying Systems

Oxidative damage is an unavoidable consequence of oxidative metabolism. The primary antioxidant mechanisms are unable of avoiding completely the generation of oxidative damage. However, some of this damage can be repaired or removed from the organism.

Although most oxidative modifications to proteins are irreversible (and, therefore, need to be catabolised; e.g. most of the protein carbonyls), certain oxidative modifications can be repaired. For example, methionine sulfoxide reductases are thioredoxin-linked enzymes involved in the enzymatic conversion of methionine sulfoxide to methionine. Overexpression of this enzyme increases resistance against oxidative stress, while a decrease of its expression increases cell sensitivity to oxidative stress and reduced cell survival, respectively (Moskovitz et al. 2001; Ruan et al. 2002). Degradation of oxidised proteins is needed to avoid aggregation and the build-up of modified proteins. This is done through the activity of various systems, including lysosomal proteases, calcium-dependent proteases, proteasomal system, and caspases (Pamplona and Costantini 2011).

Damage to lipids can also be repaired. Peroxidised phospholipids can be repaired by the enzymes phospholipase A2 and acyltransferase. Phospholipase A2 is an enzyme that removes acyl chains from cell membrane phospholipids, while acyltransferase and transacylase enzymes are responsible for the reacylation of phospholipids (Farooqui et al. 2000; Lauritzen et al. 2001). Organisms have also evolved mechanisms to transfer polyunsaturated fatty acids across tissues and incorporate them in cell membranes to compensate for the metabolic loss of phospholipids, such as the transfer to the nervous system and incorporation in the membranes of neurons (Rapoport et al. 2001).

Finally, repair of DNA damage is essential to limit the generation of mutations. Base excision repair is a widespread mechanism that has been evolved to remove a damaged base directly from DNA. This mechanism relies on the activity of DNA glycosylase enzymes, which remove the damaged base by hydrolysing the bond between the base itself and the sugar-phosphate backbone. Nucleotide excision repair is another widespread mechanism that enables the removal of a damaged nucleotide. The 8-hydroxy-2′-deoxyguanosine (8-OHdG) and the 8-oxo-7,8-dihydro-2′-deoxyguanosine (8-oxodG) are two of the major forms of free radical-induced oxidative lesions to the base guanine, which may cause mutations (David et al. 2007; Lu et al. 2001; Valavanidis et al. 2009).

#### 1.2.2.5 Redox Signalling

Reactive oxygen species also work as specific second messengers in the intracellular signal transduction pathway (*redox signalling*) in a variety of cellular processes, including cell proliferation and differentiation, cell reprogramming, and inflammation (Fig. 1.6) (Dröge 2002; Flohé 2016; Giles 2009; Sauer et al. 2001; Sinenko et al. 2021; Thannickal and Fanburg 2000). For example, hydrogen peroxide

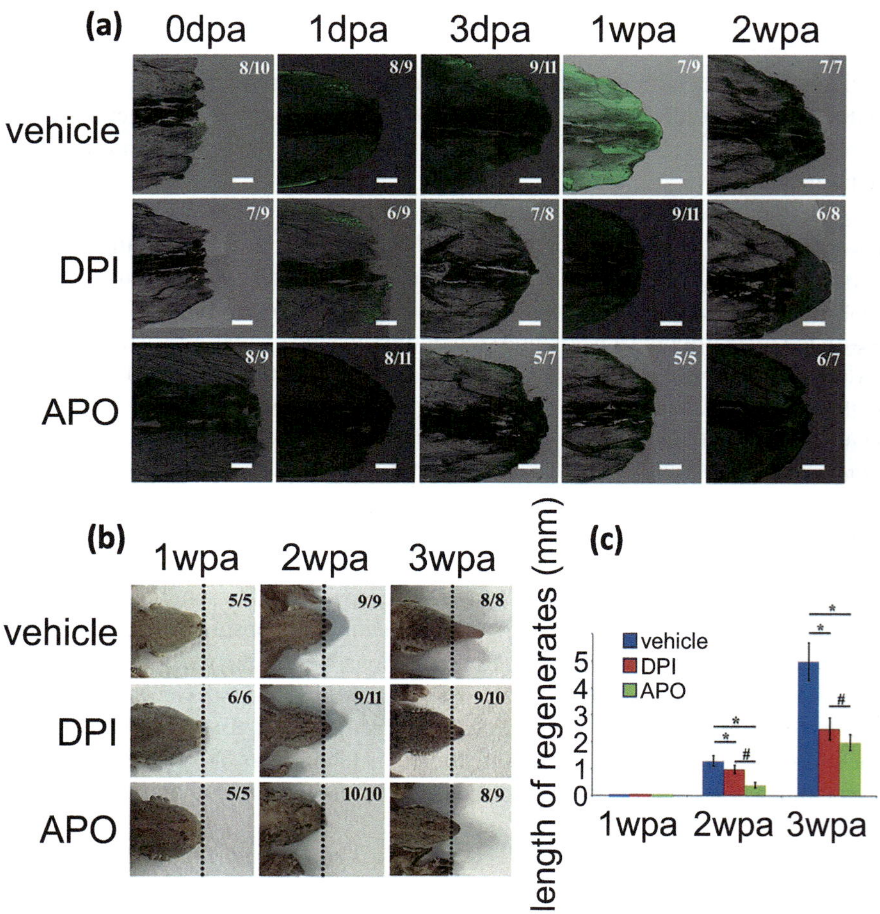


**Fig. 1.6** Generation of reactive oxygen species during tail regeneration of geckos and the effects of inhibitors on the regenerates. (**a**) amputation-induced reactive oxygen species were detected with a fluorescent probe (H2DCFDA) before or after treatment with reactive oxygen species inhibitors DPI and APO, respectively. (**b**) effects of reactive oxygen species inhibitors DPI and APO on the tail regeneration. (**c**) length of regenerating tails. Values are expressed as mean and standard error. The ratio in the right upper of images indicates the statistical samples out of the whole numbers. Asterisks or # indicate significant differences between groups: * $P < 0.05$ versus vehicle; # $P < 0.05$ between different inhibitors of reactive oxygen species. dpa = days after amputation; wpa = weeks after amputation. Scale bars represent 400 μm in (**a**). Reprinted from Zhang et al. (2016) with minor modifications. © 2016, The Author(s); distributed under the terms of the Creative Commons Attribution License (CC BY)

stimulates the initiation of tissue regeneration for wound healing (Cordeiro and Jacinto 2013), and the generation of reactive oxygen species is important for neurogenesis in red-spotted newts *Notophthalmus viridescens* (Hameed et al. 2015) or tail regeneration in Schlegel's Japanese geckos *Gekko japonicus* (Zhang et al. 2016). Reactive oxygen species can also alter the epigenetic programme through the modulation of the activity of enzymes responsible for demethylation and deacetylation of histones (Niu et al. 2015). In particular, reactive oxygen species can also activate and orchestrate the antioxidant response.

Reactive oxygen species can determine a change of the expression of many antioxidant genes (e.g. glutathione peroxidase, catalase, superoxide dismutase) through the activation of specific transcription factors (Copple et al. 2008; Kaspar et al. 2009; Pamplona and Costantini 2011; Turpaev 2002). The pathway involving the transcription factors Nrf2 (nuclear factor-E2-related factor 2) and Keap1 (Kelch-like ECH-associated protein 1) plays a central role in the redox control of physiological functions (Fig. 1.7). Nrf2 activates the expression of antioxidant genes,

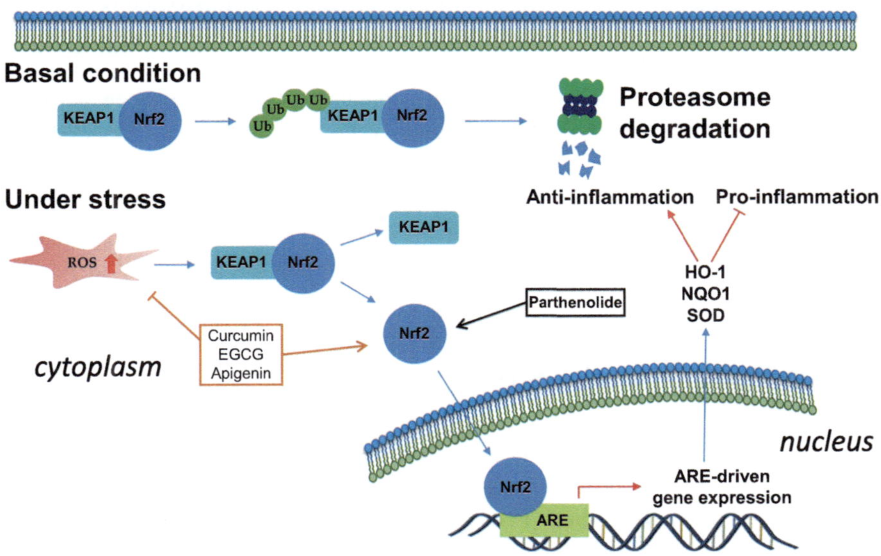


**Fig. 1.7** Schematic representation of the pathway involving the transcription factors Nrf2 (nuclear factor-E2-related factor 2) and Keap1 (Kelch-like ECH-associated protein 1). Nrf2 activates the expression of antioxidant genes, while Keap1 suppresses the transcriptional activity of Nrf2 under unstressed conditions, respectively. Under unstressed conditions, KEAP1 binds Nrf2 and targets it for rapid proteasomal degradation through ubiquitination (Ub). Under conditions of oxidative stress, reactive oxygen species (ROS) can oxidise the cysteine residues (SH) on KEAP1, leading to structural changes that enable nuclear translocation and accumulation of Nrf2. In the nucleus, Nrf2 binds with one of small Maf proteins and then binds to the antioxidant response element (ARE), initiating the transcription of antioxidant genes. Several phytochemicals (e.g. curcumin, EGCG, apigenin, parthenolide) can also directly upregulate Nrf2, resulting in increased expression of cytoprotective genes or antioxidant enzymes, such as haem oxygenase 1 (HO-1), NAD(P)H quinone oxidoreductase 1 (NQO1), or superoxide dismutase (SOD). Reprinted from Shin et al. (2020) with minor modifications. © 2020 by the authors, licensee MDPI, Basel, Switzerland; distributed under the terms of the Creative Commons Attribution License (CC BY)

while Keap1 suppresses the transcriptional activity of Nrf2 under unstressed conditions (Itoh et al. 1997; Sies et al. 2017). Nrf2 works predominantly through binding to an antioxidant response element (ARE) in order to active the expression of an extensive set of genes that regulate antioxidant levels, cytoprotective mechanisms, inflammation, and metabolism (Castiglione et al. 2020; Yamamoto et al. 2018). Under unstressed conditions, KEAP1 binds Nrf2 and targets it for rapid proteasomal degradation through ubiquitination, thereby maintaining Nrf2 at relatively low levels in the cell (Castiglione et al. 2020; Yamamoto et al. 2018). Under conditions of oxidative stress, reactive oxygen species can oxidise cysteine residues on KEAP1, leading to structural changes that enable nuclear translocation and accumulation of Nrf2, thus activating its wide-ranging cytoprotective transcriptional programme (Castiglione et al. 2020; Yamamoto et al. 2018). Activation of the nuclear factor κB (NF-κB) owing to its oxidation by hydrogen peroxide can also rapidly active the expression of several antioxidants (Schreck et al. 1991).

Reactive oxygen species can also induce the activation of heat-shock transcription factors with consequent elevated expression of heat shock proteins and molecular chaperones that protect cells against the accumulation of misfolded or oxidised proteins (e.g. Feder and Hofmann 1999; Kristensen et al. 2003; Morimoto 1998; Parsell and Lindquist 1993; Sun and MacRae 2005).

Finally, the redox signalling and control mediated by the hydrogen peroxide involves the oxidative modification of some types of amino acids (e.g. cysteine, methionine, proline, histidine, and tryptophan) in proteins (Sies 1986). Thiol peroxidases are also major regulators of global gene expression in response to hydrogen peroxide; they sense and transfer oxidative signals to the signalling proteins and regulate transcription (Fomenko et al. 2011).

## 1.3 Defining Oxidative Stress

There is considerable evidence across a wide range of taxa that oxidative status (defined by the levels of pro-oxidants and antioxidants that occur in cells or tissues) plays a central role in biological processes. The redox system (defined as a group of interacting atoms or molecules that can either oxidise biomolecules or protect biomolecules from oxidation caused by reactive oxygen species) remains highly conserved across living organisms, but there are also significant differences among taxa. Cells live in a dynamic redox environment, whose status is determined by a complex balance between pro-oxidant and antioxidant molecules (Fig. 1.8). Oxidative stress is complicated to define because (i) many molecules and biochemical or genetic pathways are involved, (ii) it is a latent variable because we cannot directly measure it, so that we have to rely on proxy variables, (iii) there is no single marker of oxidative stress, and (iv) scientists have different opinions about what is more relevant to measure. The first definitions of oxidative stress were focussed on the biochemical aspects of the phenomenon without explicit consideration of the fitness or health consequences for the organism. Helmut Sies (1985, 1991) has been

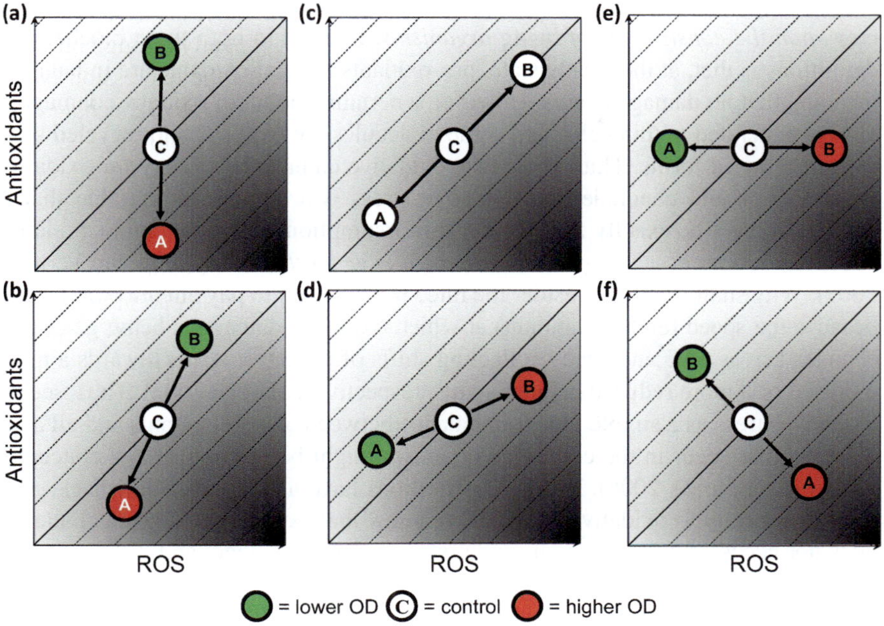


**Fig. 1.8** Relationships among reactive oxygen species (ROS), antioxidants, and oxidative stress. Oxidative damage (and so oxidative stress) increases when antioxidant levels decrease for a given level of pro-oxidants. Therefore, oxidative stress increases from top left corner to bottom right corner as indicated by background shading (darker indicates higher oxidative stress level). Dashed lines are "iso-oxidative stress" lines, in that all combinations of antioxidants and ROS levels on the line yield the same level of oxidative stress. The solid line represents combinations of antioxidants and ROS with an equal level of oxidative stress as the control group. Panels from (**a**) to (**f**) illustrate different hypothetical situations that are compared with respect to possible ways that two (experimental) groups A and B could differ from a control (C) group. In scenario (**a**), A and B do not differ in generation of ROS, but B has more antioxidants per unit of ROS and, consequently, attains a lower level of oxidative stress than A (e.g. Gaál et al. 2006). In scenario (**b**), A has lower levels of ROS but also of antioxidants than B, hence A suffers more oxidative stress than B (e.g. Gaál et al. 2006; Lin et al. 2004). In scenario (**c**), antioxidant levels of both A and B are matched to the different ROS levels, and consequently, there is no difference in oxidative damage between the two groups (e.g. López-Torres et al. 1993; Pérez-Campo et al. 1993). In scenario (**d**), B is exposed to higher oxidative damage than A because the antioxidant response of B to a ROS challenge, while still operating, is not sufficient to maintain oxidative damage at the same level of A (e.g. Gianni et al. 2004; Lin et al. 2008). In scenario (**e**), A and B do not differ in antioxidant levels, however, B suffers higher oxidative damage because its generation of ROS is higher than A (e.g. Costantini et al. 2008). Finally, in scenario (**f**), B has lower ROS generation and higher antioxidant levels than A, hence suffering lower oxidative damage than A (e.g. Costantini and Dell'Omo 2006). OD = oxidative damage. Reprinted from Costantini and Verhulst (2009) with permission of John Wiley and Sons

the first to provide a definition of oxidative stress: *a disturbance in the pro-oxidant–antioxidant balance in favour of the former, leading to potential damage*. Halliwell and Gutteridge (2015) proposed a revised version of Sies' definition: *oxidative stress is a disturbance in the pro-oxidant–antioxidant balance in favour of the former, leading to potential biomolecular damage caused by attack of reactive species upon the constituents of living organisms.* Implicit in both these biochemical definitions is that, at the steady state, pro-oxidants and antioxidants are in balance, meaning that no damage occurs. However, a certain amount of oxidative damage is always generated within cells, and some molecules are oxidised and recycled back to their reduced forms. Thus, if we take oxidative damage as a proxy for oxidative stress, we should conclude that oxidative stress is a continuous variable that is unlikely to ever be exactly zero. Under this assumption, we could define oxidative stress as *the rate at which oxidative damage is generated* (Costantini and Verhulst 2009). Why shall we also consider the rate? It is relevant to pay attention at the rate because the speed (and the amount) at which oxidative damage is being generated changes across time and space, reflecting shifts of oxidative status towards a more oxidised or a more reduced cellular status, respectively (Dröge 2002). Thus, seeing oxidative stress as a simple global imbalance between all pro-oxidants and all antioxidants that occur in the cell/tissue/organism might be inadequate and conceptually limiting (Jones 2006). All these definitions have also another drawback: they refer generically to oxidative damage. We know that exposure to a given challenge does not necessarily increase all markers of oxidative damage, and different types of damage seem to differ in their functional consequences for the organism (Dotan et al. 2004). For example, we do not know whether oxidation of DNA involves active genes or the so-called junk DNA, which does not code for any protein. The levels of the various antioxidants also do not change in the same way in response to an increased generation of reactive oxygen species, because antioxidants vary in function, have different substrates, and differ in concentration and dynamics (Halliwell and Gutteridge 2015).

Damage-centred definitions of oxidative stress also ignore the fact that oxidative stress may result from the disruption of signalling molecules with antioxidant properties, such as thiol molecules (e.g. glutathione) without concomitant or proportional oxidative damage (Jones 2006, 2008; Sies and Jones 2007; Sohal and Orr 2012). For example, pro-oxidant peroxides may disrupt signalling without accumulation of oxidative damage in the cell because oxidation of some protein antioxidants stimulates the release of the signal-regulating kinase 1, which activates apoptosis (Saitoh et al. 1998; Zhang et al. 2004). Thus, Jones (2006) defined oxidative stress as a *disruption of redox signalling and control*, which involves those mechanisms that regulate the redox balance and the antioxidant response to oxidative insults. According to the Jones' definition of oxidative stress, Sohal and Orr (2012) proposed the *redox stress hypothesis of ageing*, which states that ageing is caused by a pro-oxidising shift in the redox system of cells, which leads to the over-oxidation of redox-sensitive protein thiols and the consequent disruption of the redox-regulated signalling mechanisms. Thus, a trade-off between redox signalling and oxidative stress might also underlie the regulation of reactive oxygen species generation, and this trade-off would have implications for life-history theory (*redox signalling hypothesis of life history* that complements the *oxidative stress hypothesis of life history*; Costantini 2019; see paragraph 1.2.2.5). This trade-off might also be important in other contexts other than life history, such as tissue development

(Love et al. 2013), regulation of immune function (Nathan and Cunningham-Bussel 2013), and spermatogenesis (Guerriero et al. 2014), where reactive oxygen species play fundamental roles as cellular signals.

Biochemical definitions of oxidative stress have also other limitations. Although the term stress is commonly used in the field, it is important to be aware of the effects that its use has on our thinking. The negative meaning given to the word stress is actually misleading because biochemical definitions of oxidative stress do not make any predictions about the consequences for organismal fitness. Moreover, there is growing awareness that the intensity of stress is very relevant in terms of consequences for organismal function; exposure to stressors of low to mild intensity can potentially stimulate the organismal performance (see paragraph 1.4).

In biochemical terms, variation in the levels of reactive oxygen species, increased generation of oxidative damage, decreased GSH:GSSG ratio, or up-/down-regulation of antioxidant enzymes all reflect a perturbation of the oxidative status, with potential significant consequences for the organismal fitness (Table 1.1). Therefore, we cannot neglect the functional relevance of antioxidants in favour of a

**Table 1.1** Examples of experimental manipulations of specific components of the redox system that caused a significant effect on fitness-related metrics

| Experiment | Species | Effect | Article |
|---|---|---|---|
| Increased ROS | *Caenorhabditis elegans* | Reduced longevity | Larsen (1993) and de Castro et al. (2004) |
| Increased ROS | *Passer domesticus* | Reduced sperm quality | Rojas Mora et al. (2017) |
| Inhibition of GSH synthesis | *Serinus canaria* | Delayed laying date and reduced clutch size | Costantini et al. (2016) |
| Inhibition of GSH synthesis | *Taeniopygia guttata* | Increased reproductive effort | Romero-Haro et al. (2016) |
| Inhibition of GSH synthesis | *Sturnus vulgaris* | Reduced song rate | Messina et al. (2017) |
| Increased GSH availability | *Oryctolagus cuniculus* | Reduced activity of herpesvirus | Nucci et al. (2000) |
| Knock out of antioxidant genes | Laboratory mouse | Accelerated pathology and reduced healthy lifespan | Melov et al. (1998, 1999), de Haan et al. (1998) and Salmon et al. (2010) |
| Knock out of Cu/Zn-SOD | *Salmonella typhimurium* | Reduced survival | De Groote et al. (1997) |
| Knock out of Cu/Zn-SOD | *Caenorhabditis elegans* | Increased generation of superoxide and reduced lifespan | Yanase et al. (2009) |
| Overexpression of Cu/Zn-SOD and catalase | *Drosophila melanogaster* | Increased lifespan | Orr and Sohal (1994) |
| Overexpression of Cu/Zn-superoxide dismutase | *Drosophila melanogaster* | Increased mean and maximum lifespan | Sun and Tower (1999) |

*ROS* = reactive oxygen species; *GSH* = reduced glutathione; *Cu/Zn-SOD* = copper-zinc superoxide dismutase damage-centred view. It is, however, difficult to determine the severity of oxidative stress based on observed changes in the components of the redox system. The analysis of the different types of cell death, in combination with the quantification of oxidative status markers, might provide a valuable tool to infer about the severity of oxidative stress (Fig. 1.9). For example, increased cellular necrosis and apoptosis might reflect high oxidative stress at the biochemical level, while upregulation of antioxidant enzymes might reveal a condition of low to mild oxidative stress in which enzymes are not being inactivated by reactive oxygen species, or it is physiologically possible for the organism to upregulate their expression (Lushchak 2014). Another important distinction to make is between acute (or transient) and chronic oxidative stress, because their consequences for organismal fitness are likely to

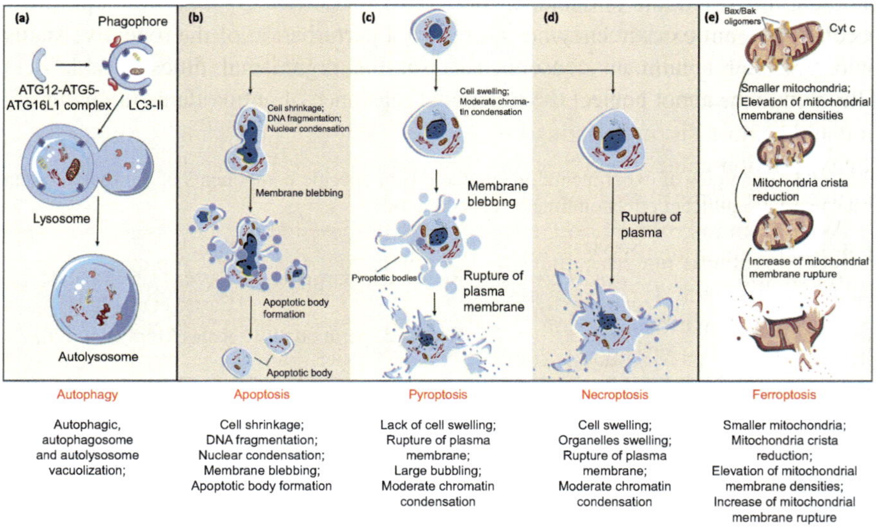


**Fig. 1.9** Morphological changes of regulated cell death. Different types of cell death that are induced with various pathways present diverse morphological changes. (**a**) the formation of autophagosomes is the typical characteristic of autophagy. When the autophagosome is completely formed, the protein LC3-II will separate from the outer membrane, and the autophagosome fuses with the lysosome so that an autolysosome finally emerged. (**b**) in the early stage of apoptosis, the nuclear condensation starts with cell shrinkage. During later stages, nucleus breaks up and the plasma membrane bubbles with no rupture, naturally associating with no inflammation. Finally, it forms apoptotic bodies. (**c**) in contrast to apoptosis, after cell swelling and large bubbling, pyroptosis has plasma membrane rupture in the final stage. (**d**) in the necroptosis, the appearance of cell swelling is often followed by organelles dilation, and the nucleus disintegrates late. In some cases, chromatin condensation also occurs. Finally, with the rupture of plasma membrane, a massive inflammation in the tissue is triggered. (**e**) the morphological characteristics of ferroptosis mainly reflect on the changes of mitochondria. Compared with organelles swelling in necroptosis, in the first stage of ferroptosis, the mitochondria become smaller, and membrane densities are elevated, followed by mitochondria crista reduction and plasma membrane rupture. Reprinted with minor modifications from Hu et al. (2021). © 2021 Hu, Li, Lin, Shan, Yu, Wang, Liao, Yan, Wang, Shang, Huang, Zhang and Xiong; distributed under the terms of the Creative Commons Attribution License (CC BY)

differ (Dröge 2002; Lushchak 2014). Finally, a review article proposed to distinguish between a biochemical definition and a biologically oriented definition of oxidative stress; this second definition points to the effects of oxidative stress on fitness (Costantini 2019). In other words, we should refer to *biochemical oxidative stress* when the study did not measure any fitness outcomes; by contrast, we could refer to *biological oxidative stress* if we establish a cause–effect relationship between markers of oxidative status and fitness metrics.

## 1.4 The Paradox of Oxidative Stress

It is increasingly recognised that a wide range of environmental stressors or molecules, including reactive oxygen species or oxidised biomolecules, can have stimulatory or even beneficial effects on organismal function at low to mild concentrations (or intensities), but toxic at high concentrations (Calabrese 2010; Costantini 2014; Costantini and Borremans 2019; Costantini et al. 2010a; Mattson and Calabrese 2010). To some extent, this is not surprising because life on earth appeared and diversified in challenging environments, so that selection has favoured evolution of many molecular mechanisms to deal with stressful conditions that organisms retain at present. Consequently, we would expect that any detrimental effects of stress on fitness should show up only when the organism is being exposed above certain levels.

This kind of biphasic responses with low-dose stimulation and high-dose inhibition was named hormesis by Southam and Ehrlich (1943). It is only since the 1980s that there has been increasing interest in biological meaning of hormesis. We now know that hormetic responses occur across a wide range of organisms (from plants to bacteria and vertebrates) exposed to a diverse range of over 1000 chemical and environmental stressors (Calabrese 2010; Calabrese and Baldwin 2003; Calabrese and Blain 2005, 2009; Costantini et al. 2010a; Mattson and Calabrese 2010).

The term hormesis is currently used in at least two or three different, albeit related, ways. In its original use, hormesis describes a biphasic response to an environmental stressor, with a mild-dose stimulation and a high-dose inhibition of a certain trait (e.g. antioxidant activity, longevity) compared to a control group, respectively. Let us take the relationship between a pro-oxidant molecule (e.g. heavy metal) and oxidative damage as an example. We might expect that the negative effects of exposure to the pro-oxidant molecule become more severe (and oxidative damage to therefore increase) as exposure increases (*linear no-threshold model*, Fig. 1.10a). The assumptions underlying the linear no-threshold model are, however, irreconcilable with current evolutionary theory (see Costantini and Borremans 2019). Another possibility is that there is no noticeable effect until the exposure to the pro-oxidant molecule reaches a threshold value (termed the *no observed adverse effect level* in toxicology), above which oxidative damage increases linearly or nonlinearly with concentration of the pro-oxidant (Fig. 1.10b). In a third scenario, we can observe a biphasic change in the generation of oxidative damage (Fig. 1.10c): a low dose of a pro-oxidant molecule would elicit the

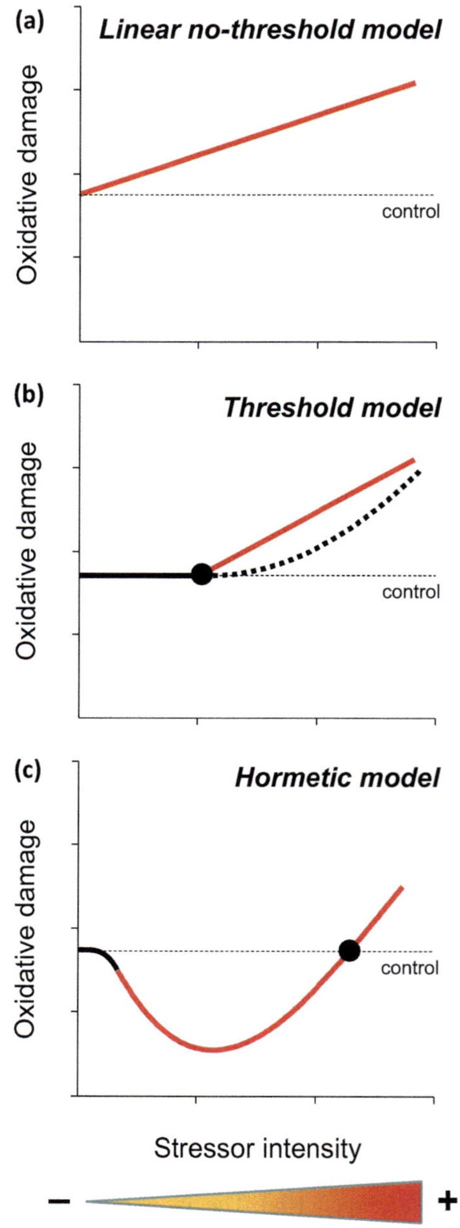


**Fig. 1.10** Examples of common dose–response models; oxidative damage is used as an end-point. (**a**) in the linear no-threshold model, oxidative damage increases linearly with the intensity of a given stressor, with no threshold for a physiological effect to be observed. This model is based on the postulate that any stressor intensity is harmful, regardless of how low the dose is. (**b**) in the threshold model, the stressor has no biological effect until a threshold (indicated by the black circle) is reached, above which oxidative damage can increase linearly or nonlinearly. (**c**) in the hormetic model, the response to the environmental stressor is biphasic, with oxidative damage decreasing and increasing at mild and high intensities of the stressor, respectively. The zero-equivalent point (black circle) refers to the area of the hormetic curve where oxidative damage equals that of the control (i.e. zero exposure). Reprinted with minor modifications from Costantini (2013) with permission of Springer Nature

organismal stress response (possibly beneficial), reducing the generation and accumulation of oxidative damage as a consequence; on the other hand, exposure to a high dose of a pro-oxidant molecule would cause inhibition of stress response pathways and toxicity, leading to an increase in the generation and accumulation of oxidative damage as a consequence (Costantini 2013; Costantini et al. 2010a;

Mattson and Calabrese 2010). This last kind of dose–response relationship is what we call hormesis.

In this definition of hormesis, we have to distinguish between stimulatory and beneficial. Stimulatory refers to an upregulation of a given physiological mechanism that favours the maintenance of homeostasis. However, this stimulation is not necessarily beneficial in terms of reproduction or longevity in case, for example, a trade-off occurs between these two traits because they cannot be increased simultaneously (e.g. increased reproduction results in reduced longevity and vice versa). Moreover, the stimulatory response can be induced without or after a transient disruption of cell homeostasis, followed by an overcompensatory response (Calabrese 2010; Calabrese and Baldwin 2002). The biphasic nature of hormesis might also indicate that, at mild doses of stress, the organism can upregulate self-maintenance mechanisms because it requires less resources compared to the exposure to high doses.

In most experimental investigations of this process, various treatment groups are exposed to differing levels of a stressor, and the average response of each group is then measured and compared to a control group. The timescale over which the response is measured can vary from the immediate (as is typical with biochemical measurements, such as concentration of oxidative damage) to the long term (as in the case of longevity; Costantini et al. 2010a). For example, the viability of *Saccharomyces cerevisiae* increases with the concentration of hydrogen peroxide at which they are exposed (peak of stimulation at around 0.45 mmol $L^{-1}$) until a threshold value (0.8 mmol $L^{-1}$) beyond which the viability declines below the control level, indicating that hydrogen peroxide has biphasic effects on the organismal performance of *Saccharomyces cerevisiae* (Davies et al. 1995).

A second use of the term hormesis relies on the phenomenon of conditioning or priming, whereby exposure to a mild level of a stressor results in the organism being better able to cope with exposure to higher levels of the stressor when encountered on subsequent occasions (Costantini et al. 2010a). This type of hormetic response is called *conditioning hormesis* according to the terminology proposed by Calabrese et al. (2007) or *priming hormesis*. This hormetic priming model predicts that individuals previously exposed to mild levels of a certain stressor would withstand the subsequent exposure to high levels of the stressor better than individuals that did not experience any stress exposure or were previously exposed to a high level of the stressor. For example, conditioning of PC12 cells (a rat cell line derived from pheochromocytoma cells) with low concentrations of hydrogen peroxide made them more resistant to subsequent exposure to high levels of hydrogen peroxide compared to cells that were not previously exposed to or were exposed to high concentrations of hydrogen peroxide, respectively (Tang et al. 2005). Specifically, cells were conditioned with 0, 5, 10, 20, or 30 μmol $L^{-1}$ of hydrogen peroxide for 90 min, followed by a recovery period of 24 h, and subsequent exposure to 50 or 100 μmol $L^{-1}$ hydrogen peroxide for 24 h. Apoptosis was significantly lower in those cells that were conditioned with 10 or 20 μmol $L^{-1}$ hydrogen peroxide in both experiments (Tang et al. 2005). Moreover, conditioning with 10 μmol $L^{-1}$ hydrogen peroxide had a protective effect on hydrogen peroxide-induced dissipation of the mitochondrial membrane potential, inhibited hydrogen peroxide-induced intracellular accumulation of reactive oxygen species, and induced overexpression of the antiapoptotic protein Bcl-2 (Tang et al. 2005).

Finally, a third emerging hormetic model relates to a prolonged exposure of the organism to a given stressor that elicits a stimulatory or beneficial response only if the dose of such stressor is moderate (e.g. Ayyanath et al. 2013; Costantini 2022). However, this type of hormesis needs to be distinguished from acclimation, which, conversely from hormesis, is a reversible process, i.e. the organism's status bounces back to that expressed prior to exposure to the environmental stimulus. This potential third type of hormesis needs more work in order to be validated.

Calabrese (2013) provided an extensive documentation of mechanisms that drive the hormetic responses to a number of chemical substances or abiotic factors. The mechanisms selected were principally those mediated via receptor and/or cell signalling pathways. Calabrese (2013) proposed that there are several types of hormetic responses that can arise through both mechanisms. In receptor-mediated hormetic mechanisms, both the stimulatory and the inhibitory responses can be mediated by the same or different receptor/s. In signalling-mediated hormetic mechanisms, the same signalling pathway can affect both stimulation and inhibition, or there might be distinct mechanisms for stimulation and inhibition, respectively. Additionally, it is increasingly suggested that stress priming might induce a molecular memory (e.g. epigenetic memory in Ding et al. 2012; Sani et al. 2013) that would prepare organisms to cope with stressors later in life.

Environmental stressors have been key driving forces of evolution. It has long been recognised by evolutionary biologists that exposure to moderate amounts of stress may facilitate adaptation by inducing a number of genetic and phenotypic responses (Badyaev 2005; Bijlsma and Loeschcke 1997; Gonzalez et al. 2013; Steinberg 2012). Although hormesis might come at a cost for certain life-history traits, a number of studies found that hormesis may increase both longevity and fecundity, or only fecundity without any costs for survival (Calabrese and Baldwin 2003; Calabrese and Blain 2005; Cardoso et al. 2002; Cutler et al. 2009; Giesy et al. 2000; Parkhurst et al. 1981; Qu et al. 2015). Thus, there are conditions whereby hormetic mechanisms might boost fitness to some degree, or buffer fitness against any detrimental effects, indicating that hormesis might provide selective advantages. Importantly, we have evidence for genetic background underlying some types of hormetic responses (e.g. Rodriguez et al. 2012; Sarup and Loeschcke 2011). In conclusion, exposure to mild oxidative stress might be one engine of adaptation because it would elicit genetic and phenotypic hormetic responses of organisms to stressful events, promoting maintenance or even boosting of their fitness.

## 1.5 Conclusions and Future Directions

Elucidating the physiological mechanisms underlying variation in organismal functions has a central role for our understanding of how biodiversity is generated (e.g. Costantini 2014; Flatt and Heyland 2011; Ricklefs and Wikelski 2002; Spicer et al.

2019; Zera and Harshman 2001). This is particularly relevant for life-history theory, an analytical framework designed to study the diversity of life-history traits (e.g. growth pattern, age at sexual maturity, reproduction, longevity) (Flatt and Heyland 2011; Roff 1992; Stearns 1992).

Over the last 20 years or so, we have been witnessing an increasing interest in the role of oxidative stress as a physiological mediator of (i) short- and long-term life-history trade-offs and (ii) acclimatisation (or adaptability) of organisms to natural and anthropogenic environmental changes (Beaulieu and Costantini 2014; Costantini 2008, 2010, 2014, 2019; Costantini et al. 2010b; Dowling and Simmons 2009; Isakkson et al. 2011; McGraw et al. 2010; Metcalfe and Alonso-Alvarez 2010; Monaghan et al. 2009; Selman et al. 2012). However, research on the role of oxidative stress as a constraint on and cost of life-history strategies has proven problematic because, in many cases, physiological data are not linked to functional or fitness outcomes. This is a critical missing information given that the individual performance may display a hormetic, rather than a linear response, to biochemical oxidative stress. Our predictions are indeed often based on the assumption that organismal oxidative status varies linearly with environmental factors. We need a shift of theoretical focus and an introduction of non-linearity into our models if we aim at better understanding the connections between organismal oxidative status and life-history evolution.

Smith et al. (2014) also suggested that traditional trade-offs might not be necessary to invoke oxidative stress as a mediator of the relationships between life-history traits because increasing oxidative insults may uncouple phenotypic correlations, particularly between reproduction and lifespan. Tüzün et al. (2020) also suggested that ecological differences between species may cause a decoupling of the covariation between species differences in life-history pace and their oxidative status. Thus, running simple bivariate models between a marker of oxidative status and, for example, longevity might not be appropriate to test the role of oxidative stress in generating among species differences in lifespan.

By investigating the roles of reactive oxygen species, antioxidants, oxidative damage, and oxidative stress with the same rigour that ecophysiologists have used in evaluating other physiological traits (e.g. hormones, metabolic rate), we will be able to better elucidate the significance of the variation in oxidative stress, and the causes and consequences of life-history variation at the individual, population, and species levels. The definitions and fundamentals of oxidative status physiology presented in this chapter provide the basic background information for practitioners, essential to make robust inferences about the proximate routes through which variation in morphological, behavioural, or life-history traits is generated and evolve, and to understand how and why oxidative status itself evolves.

## References

Aledo JC, Li Y, De Magalhães JP, Ruíz-Camacho M, Pérez-Claros JA (2011) Mitochondrially encoded methionine is inversely related to longevity in mammals. Aging Cell 10:198–207

Altincicek B, Kovacs JL, Gerardo NM (2011) Horizontally transferred fungal carotenoid genes in the two-spotted spider mite *Tetranychus urticae*. Biol Lett 8:253–257

Asada K (1999) The water-water cycle in chloroplasts: scavenging of active oxygen and dissipation of excess photons. Annu Rev Plant Physiol Plant Mol Biol 50:601–639
Ayyanath M-M, Cutler GC, Scott-Dupree CD, Sibley PK (2013) Transgenerational shifts in reproduction hormesis in green peach aphid exposed to low concentrations of imidacloprid. PLoS One 8:e74532
Badyaev AV (2005) Stress-induced variation in evolution: from behavioural plasticity to genetic assimilation. Proc R Soc Lond B 272:877–886
Bánhegyi G, Csala M, Braun L, Garzó T, Mandl J (1996) Ascorbate synthesis-dependent glutathione consumption in mouse liver. FEBS Lett 381:39–41
Barbosa EA, Oliveira A, Plácido A, Socodato R, Portugal CC, Mafud AC, Ombredane AS, Moreira DC, Vale N, Bessa LJ, Joanitti GA, Alves C, Gomes P, Delerue-Matos C, Mascarenhas YP, Marani MM, Relvas JB, Pintado M, Leite JRSA (2018) Structure and function of a novel antioxidant peptide from the skin of tropical frogs. Free Radic Biol Med 115:68–79
Barja G (2007) Mitochondrial oxygen consumption and reactive oxygen species production are independently modulated: implications for aging studies. Rejuv Res 10:215–224
Beaulieu M, Costantini D (2014) Biomarkers of oxidative status: missing tools in conservation physiology. Conserv Physiol 2:cou014
Behne D, Kyriakopoulos A (2001) Mammalian selenium-containing proteins. Annu Rev Nutr 21:453–473
Bernthsen A (1942) A textbook of organic chemistry. Blackie and Sons, Cadiz
Bestetti S, Medraño-Fernandez I, Galli M, Ghitti M, Bienert GP, Musco G, Orsi A, Rubartelli A, Sitia R (2018) A persulfidation-based mechanism controls aquaporin-8 conductance. Sci Adv 4:eaar5770
Bielski BHJ, Arudi RL, Sutherland MW (1983) A study of the reactivity of HO2/O2 with unsaturated fatty acids. J Biol Chem 258:4759–4761
Bijlsma R, Loeschcke V (1997) Environmental stress, adaptation and evolution. Birkhauser, Basel
Birney EC, Jenness R, Ayaz KM (1976) Inability of bats to synthesise L-ascorbic acid. Nature 260:626–628
Birney EC, Jenness R, Hume ID (1980) Evolution of an enzyme system: ascorbic acid biosynthesis in monotremes and marsupials. Evolution 34:230–239
Bjelland S, Seeberg E (2003) Mutagenicity, toxicity and repair of DNA base damage induced by oxidation. Mutat Res 531:37–80
Bleier L, Dröse S (2013) Superoxide generation by complex III: from mechanistic rationales to functional consequences. Biochim Biophys Acta 1827:1320–1331
Boden JS, Konhauser KO, Robbins LJ, Sánchez-Baracaldo P (2021) Timing the evolution of antioxidant enzymes in cyanobacteria. Nat Commun 12:4742
Borda MJ, Elsetinow AR, Schooen MA, Strongin DR (2001) Pyrite-induced hydrogen peroxide formation as a driving force in the evolution of photosynthetic organisms on an early Earth. Astrobiology 1:283–288
Boveris A, Oshino N, Chance B (1972) The cellular production of hydrogen peroxide. Biochem J 128:617–630
Brand MD, Affourtit C, Esteves TC, Green K, Lambert A, Miwa S, Pakay JL, Parker N (2004) Mitochondrial superoxide: production, biological effects, and activation of uncoupling proteins. Free Radic Biol Med 37:755–767
Brioukhanov AL, Netrusov AI (2004) Catalase and superoxide dismutase: distribution, properties, and physiological role in cells of strict anaerobes. Biochemistry 69:949–962
Burn R, Misson L, Meury M, Seebeck FP (2017) Anaerobic origin of ergothioneine. Angew Chem Int Ed Engl 56:12508–12511
Buxton GV, Greenstock CL, Helman WP, Ross AB (1988) Critical review of rate constants for reactions of hydrated electrons, hydrogen atoms, and hydroxyl radicals ($^{\bullet}OH/^{\bullet}O^-$) in aqueous solution. J Phys Chem Ref Data 17:513–886
Calabrese EJ (2010) Hormesis: once marginalized, evidence now supports hormesis as the most fundamental dose response. In: Mattson MP, Calabrese EJ (eds) Hormesis: a revolution in biology, toxicology and medicine. Springer, New York, pp 15–56

Calabrese EJ (2013) Hormetic mechanisms. Crit Rev Toxicol 43:580–606
Calabrese EJ, Baldwin LA (2002) Defining hormesis. Human Exp Toxicol 21:91–97
Calabrese EJ, Baldwin LA (2003) The hormetic dose response model is more common than the threshold model in toxicology. Toxicol Sci 71:414–428
Calabrese EJ, Blain R (2005) The occurrence of hormetic dose responses in the toxicological literature, the hormesis database: an overview. Toxicol Appl Pharmacol 202:289–301
Calabrese EJ, Blain R (2009) Hormesis and plant biology. Environm Poll 157:42–48
Calabrese EJ, Bachmann KA, Bailer AJ, Bolger PM, Borak J, Cai L, Cedergreen N, Cherian MG, Chlueh CC, Clarkson TW, Cook RR, Diamond DM, Doolittle DJ, Dorato MA, Duke SO, Feinendegen L, Gardner DE, Hart RW, Hastings KL, Hayes AW, Hoffmann GR, Ives JA, Jaworowski Z, Johnson TE, Jonas WB, Kaminski NE, Keller JG, Klaunig JE, Knudsen TB, Kozumbo WJ, Lettleri T, Liu SZ, Maisseu A, Maynard KI, Masoro EJ, McClellan RO, Mehendale HM, Mothersill C, Newlin DB, Nigg HN, Oehme FW, Phalen RF, Philbert MA, Rattan SIS, Riviere JE, Rodricks J, Sapolsky RM, Scott BR, Seymour C, Sinclair DA, Smith-Sonneborn J, Snow ET, Spear L, Stevenson DE, Thomas Y, Tubiana M, Williams GM, Mattson MP (2007) Biological stress response terminology: integrating the concepts of adaptive response and preconditioning stress within a hormetic dose-response framework. Toxicol Appl Pharmacol 222:122–128
Cardoso VV, Ferreira MP, Montagner JM, Fernandez CG, Moreira JC, Oliveira AK (2002) The effects of constant and alternating temperatures on the reproductive potential, life span, and life expectancy of *Anastrepha fraterculus* (Wiedemann) (Diptera: Tephritidae). Braz J Biol 62:775–786
Case AJ (2017) On the origin of superoxide dismutase: an evolutionary perspective of superoxide-mediated redox signaling. Antioxidants 6:82
Castiglione GM, Xu Z, Zhou L, Duh EJ (2020) Adaptation of the master antioxidant response connects metabolism, lifespan and feather development pathways in birds. Nat Commun 11:2476
Chaudhuri CR, Chatterjee IB (1969) Ascorbic acid synthesis in birds: phylogenetic trend. Science 164:435–436
Commoner B, Townsend J, Pake GE (1954) Free radicals in biological materials. Nature 174:689–691
Copple IM, Goldring CE, Kitteringham NR, Park BK (2008) The Nrf2-Keap1 defence pathway: role in protection against drug-induced toxicity. Toxicology 246:24–33
Cordeiro JV, Jacinto A (2013) The role of transcription-independent damage signals in the initiation of epithelial wound healing. Nat Rev Mol Cell Biol 14:249–262
Costantini D (2008) Oxidative stress in ecology and evolution: lessons from avian studies. Ecol Lett 11:1238–1251
Costantini D (2010) Redox physiology in animal function: the struggle of living in an oxidant environment. Curr Zool 56:687–702
Costantini D (2013) Oxidative stress and hormetic responses in the early life of birds. In: Laviola G, Macrì S (eds) Adaptive and maladaptive aspects of developmental stress. Current topics in neurotoxicity 3. Springer Science and Business Media, New York, pp 257–273
Costantini D (2014) Oxidative stress and hormesis in evolutionary ecology and physiology: a marriage between mechanistic and evolutionary approaches. Springer-Verlag, Berlin, p 348
Costantini D (2019) Understanding diversity in oxidative status and oxidative stress: the opportunities and challenges ahead. J Exp Biol 222:jeb194688
Costantini D (2022) For better or worse: benefits and costs of transgenerational plasticity and the transhormesis hypothesis. In: Costantini D, Marasco V (eds) Development strategies and biodiversity: Darwinian fitness and evolution in the Anthropocene. Springer, Cham, pp 37–49
Costantini D, Borremans B (2019) The linear no-threshold model is less realistic than threshold or hormesis-based models: an evolutionary perspective. Chem Biol Inter 301:26–33
Costantini D, Dell'Omo G (2006) Effects of T-cell-mediated immune response on avian oxidative stress. Comp Biochem Physiol Part A 145:137–142

Costantini D, Møller AP (2008) Carotenoids are minor antioxidants for birds. Funct Ecol 22:367–370
Costantini D, Verhulst S (2009) Does high antioxidant capacity indicate low oxidative stress? Funct Ecol 23:506–509
Costantini D, Fanfani A, Dell'Omo G (2008) Effects of corticosteroids on oxidative damage and circulating carotenoids in captive adult kestrels (*Falco tinnunculus*). J Comp Physiol B 178:829–835
Costantini D, Metcalfe NB, Monaghan P (2010a) Ecological processes in a hormetic framework. Ecol Lett 13:1435–1447
Costantini D, Rowe M, Butler MW, McGraw KJ (2010b) From molecules to living systems: historical and contemporary issues in oxidative stress and antioxidant ecology. Funct Ecol 24:950–959
Costantini D, Casasole G, AbdElgawad H, Asard H, Eens M (2016) Experimental evidence that oxidative stress influences reproductive decisions. Funct Ecol 30:1169–1174
Cui J, Yuan X, Wang L, Jones G, Zhang S (2011) Recent loss of vitamin C biosynthesis ability in bats. PLoS One 6:e27114
Cutler GC, Ramanaidu K, Astatkie T, Isman MB (2009) Green peach aphid, *Myzus persicae* (Hemiptera: Aphididae), reproduction during exposure to sublethal concentrations of imidacloprid and azadirachtin. Pest Manag Sci 65:205–209
David S, O'Shea V, Kundu S (2007) Base-excision repair of oxidative DNA damage. Nature 447:941–950
Davies MJ (2005) The oxidative environment and protein damage. Biochim Biophys Acta 1703:93–109
Davies MJ (2016) Protein oxidation and peroxidation. Biochem J 473:805–825
Davies JMS, Lowry CV, Davies KJA (1995) Transient adaptation to oxidative stress in yeast. Arch Biochem Biophys 317:1–6
de Castro E, Hegi de Castro S, Johnson TE (2004) Isolation of long-lived mutants in *Caenorhabditis elegans* using selection for resistance to juglone. Free Radic Biol Med 37:139–145
De Groote MA, Ochsner UA, Shiloh MU, Nathan C, McCord JM, Dinauer MC, Libby SJ, Vazquez-Torres A, Xu Y, Fang FC (1997) Periplasmic superoxide dismutase protects salmonella from products of phagocyte NADPH-oxidase and nitric oxide synthase. Proc Natl Acad Sci USA 94:13997–14001
de Haan JB, Bladier C, Griffiths P, Kelner M, O'Shea RD, Cheung NS, Bronson RT, Silvestro MJ, Wild S, Zheng SS, Beart PM, Hertzog PJ, Kola I (1998) Mice with a homozygous null mutation for the most abundant glutathione peroxidase, Gpx1, show increased susceptibility to the oxidative stress-inducing agents paraquat and hydrogen peroxide. J Biol Chem 273:22528–22536
Denicola A, Souza JM, Radi R (1998) Diffusion of peroxynitrite across erythrocyte membranes. Proc Natl Acad Sci USA 95:3566–3571
Dikalov S (2011) Cross talk between mitochondria and NADPH oxidases. Free Radic Biol Med 51:1289–1301
Ding Y, Fromm M, Avramova Z (2012) Multiple exposures to drought 'train' transcriptional responses in Arabidopsis. Nat Commun 3:740
Dotan Y, Lichtenberg D, Pinchuk I (2004) Lipid peroxidation cannot be used as a universal criterion of oxidative stress. Progr Lipid Res 43:200–227
Dowling DK, Simmons LW (2009) Reactive oxygen species as universal constraints in life-history evolution. Proc R Soc Lond B 276:1737–1745
Dröge W (2002) Free radicals in the physiological control of cell function. Physiol Rev 82:47–95
Drouin G, Godin J-R, Pagé B (2011) The genetics of vitamin C loss in vertebrates. Curr Genomics 12:371–378
Echtay KS, Esteves TC, Pakay JL, Jekabsons MB, Lambert AJ, Portero-Otín M, Pamplona R, Vidal-Puig AJ, Wang S, Roebuck SJ, Brand MD (2003) A signalling role for 4-hydroxy-2-nonenal in regulation of mitochondrial uncoupling. EMBO J 22:4103–4110
Esterbauer H, Schaur RJ, Zollner H (1991) Chemistry and biochemistry of 4-hydroxynonenal, malonaldehyde and related aldehydes. Free Radic Biol Med 11:81–128

Falkowski PG (2006) Tracing oxygen's imprint on Earth's metabolic evolution. Science 311:1724–1725
Farooqui AA, Horrocks LA, Farooqui T (2000) Deacylation and reacylation of neural membrane glycerophospholipids - a matter of life and death. J Mol Neurosci 14:123–135
Feder ME, Hofmann GE (1999) Heat-shock proteins, molecular chaperones, and the stress response: evolutionary and ecological physiology. Annu Rev Physiol 61:243–282
Flatt T, Heyland A (2011) Mechanisms of life history evolution: the genetics and physiology of life history traits and trade-offs, 1st edn. Oxford University Press, Oxford
Flohé L (2016) The impact of thiol peroxidases on redox regulation. Free Radic Res 50:126–142
Fomenko DE, Koc A, Agisheva N, Jacobsen M, Kaya A, Malinouski M, Rutherford JC, Siu KL, Jin DY, Winge DR, Gladyshev VN (2011) Thiol peroxidases mediate specific genome-wide regulation of gene expression in response to hydrogen peroxide. Proc Natl Acad Sci USA 108:2729–2734
Fridovich I (1978) The biology of oxygen radicals. Science 201:875–880
Gaál T, Ribiczeyné-Szabó P, Stadler K, Jakus J, Reiczigel J, Kövér P, Mézes M, Sümeghy L (2006) Free radicals, lipid peroxidation and the antioxidant system in the blood of cows and newborn calves around calving. Comp Biochem Physiol Part B 143:391–396
Gershman R, Gilbert DL, Nye F, Dwyer P, Fenw MV (1954) Oxygen poisoning and X-irradiation: a mechanism in common. Science 119:623–626
Gianni P, Jan KJ, Douglas MJ, Stuart PM, Tarnopolsky MA (2004) Oxidative stress and the mitochondrial theory of aging in human skeletal muscle. Exp Geront 39:1391–1400
Giesy JP, Pierens SL, Snyder EM, Miles-Richardson S, Kramer VJ, Snyder SA, Nichols KM, Villeneuve DA (2000) Effects of 4-nonylphenol on fecundity and biomarkers of estrogenicity in fathead minnows (*Pimephales promelas*). Environ Toxicol Chem 19:1368–1377
Giles GI (2009) Redox-controlled transcription factors and gene expression. In: Jacob C, Winyard PG (eds) Redox signalling and regulation in biology and medicine. Wiley-VCH Verlag, Weinheim, pp 245–270
Gomberg M (1900) An instance of trivalent carbon: triphenyl-methyl. J Am Chem Soc 22:757–771
Gonzalez A, Ronce O, Ferriere R, Hochberg ME (2013) Evolutionary rescue: an emerging focus at the intersection between ecology and evolution. Phil Trans R Soc Lond B 368:20120404
Goodchild CG, DuRant SE (2020) Fluorescent heme degradation products are biomarkers of oxidative stress and linked to impaired membrane integrity in avian red blood cells. Physiol Biochem Zool 93:129–139
Guerriero G, Trocchia S, Abdel-Gawad FK, Ciarcia G (2014) Roles of reactive oxygen species in the spermatogenesis regulation. Front Endocrinol 5:56
Halliwell BH, Gutteridge JMC (2015) Free radicals in biology and medicine, 5th edn. Oxford University Press, Oxford
Hameed LS, Berg DA, Belnoue L, Jensen LD, Cao Y, Simon A (2015) Environmental changes in oxygen tension reveal ROS-dependent neurogenesis and regeneration in the adult newt brain. elife 4:e08422
Han D, Canali E, Rettori D, Kaplowitz N (2003) Effect of glutathione depletion on sites and topology of superoxide and hydrogen peroxide production in mitochondria. Mol Pharmacol 64:1136–1144
Han X, Shen T, Lou H (2007) Dietary polyphenols and their biological significance. Int J Mol Sci 8:950–988
Harman D (1956) Aging: a theory based on free radical and radiation chemistry. J Geront A 11:298–300
Hausladen A, Gow AJ, Stamler JS (1998) Nitrosative stress: metabolic pathway involving the flavohemoglobin. Proc Natl Acad Sci USA 95:14100–14105
Herzberg G (1971) The spectra and structures of simple free radicals: an introduction to molecular spectroscopy. Dover Phoenix Editions, New York
Hoch FL (1992) Cardiolipins and biomembrane function. Biochim Biophys Acta 1113:71–133
Howard AC, McNeil AK, McNeil PL (2011) Promotion of plasma membrane repair by vitamin E. Nat Commun 2:597

Hu XM, Li ZX, Lin RH, Shan JQ, Yu QW, Wang RX, Liao LS, Yan WT, Wang Z, Shang L, Huang Y, Zhang Q, Xiong K (2021) Guidelines for regulated cell death assays: a systematic summary, a categorical comparison, a prospective. Front Cell Dev Biol 9:634690

Hulbert AJ, Pamplona R, Buffenstein R, Buttemer WA (2007) Life and death: metabolic rate, membrane composition, and life span of animals. Physiol Rev 87:1175–1213

Inupakutika MA, Sengupta S, Devireddy AR, Azad RK, Mittler R (2016) The evolution of reactive oxygen species metabolism. J Exp Bot 67:5933–5943

Iqbal M, Probert LL, Alhumadi ND, Klandorf H (1999) Protein glycosylation and advanced glycosylated endproducts (AGEs) accumulation: an avian solution? J Geront A 54:B171–B176

Isakkson C, Sheldon BC, Uller T (2011) The challenges of integrating oxidative stress into life-history biology. Bioscience 61:194–202

Itoh K, Chiba T, Takahashi S, Ishii T, Igarashi K, Katoh Y, Oyake T, Hayashi N, Satoh K, Hatayama I, Yamamoto M, Nabeshima Y (1997) An Nrf2/small Maf heterodimer mediates the induction of phase II detoxifying enzyme genes through antioxidant response elements. Biochem Biophys Res Commun 236:313–322

Jenness R, Birney E, Ayaz K (1980) Variation of L-gulonolactone oxidase activity in placental mammals. Comp Biochem Physiol Part B 67:195–204

Jones DP (2006) Redefining oxidative stress. Antioxid Redox. Signal 8:1865–1879

Jones DP (2008) Radical-free biology of oxidative stress. Am J Phys 295:C849–C868

Kaspar JW, Niture SK, Jaiswal AK (2009) Nrf2:Inrf2 (Keap1) signaling in oxidative stress. Free Radic Biol Med 47:1304–1309

Katsu M, Niizuma K, Yoshioka H, Okami N, Sakata H, Chan PH (2010) Hemoglobin-induced oxidative stress contributes to matrix metalloproteinase activation and blood-brain barrier dysfunction in vivo. J Cereb Blood Flow Metab 30:1939–1950

Klandorf H, Probert LL, Iqbal M (1999) In the defense against hyperglycaemia: an avian strategy. World's Poul Sci J 55:251–268

Kornas A, Kuźniak E, Slesak I, Miszalski Z (2010) The key role of the redox status in regulation of metabolism in photosynthesizing organisms. Acta Biochim Pol 57:143–151

Kovacic P, Wakelin LPG (2001) DNA molecular electrostatic potential: novel perspectives for the mechanism of action of anticancer drugs involving electron transfer and oxidative stress. Anticancer Drug Des 16:175–184

Krinski I (1993) Micronutrients and their influence on mutagenicity and malignant transformation. Ann N Y Acad Sci 686:229–242

Krisko A, Radman M (2010) Protein damage and death by radiation in Escherichia coli and Deinococcus radiodurans. Proc Natl Acad Sci USA 107:14373–14377

Kristensen TN, Sørensen JG, Loeschcke V (2003) Mild heat stress at a young age in *Drosophila melanogaster* leads to increased Hsp70 synthesis after stress exposure later in life. J Genet 82:89–94

Kushnareva Y, Murphy AN, Andreyev A (2002) Complex I-mediated reactive oxygen species generation: modulation by cytochrome c and NAD(P)+ oxidation reduction state. Biochem J 368:545–553

Lambert AJ, Boysen HM, Buckingham JA, Yang T, Podlutsky A, Austad SN, Kunz TH, Buffenstein R, Brand MD (2007) Low rates of hydrogen peroxide production by isolated heart mitochondria associate with long maximum lifespan in vertebrate homeotherms. Aging Cell 6:607–618

Lane L (2002) Oxygen: the molecule that made the world. Oxford University Press, Oxford

Larsen PL (1993) Aging and resistance to oxidative damage in *Caenorhabditis elegans*. Proc Natl Acad Sci USA 90:8905–8909

Lauritzen L, Hansen HS, Jorgensen MH, Michaelsen KF (2001) The essentiality of long-chain n-3 fatty acids in relation to development and function of the brain and retina. Prog Lipid Res 40:1–94

Lebold KM, Löhr CV, Barton CL, Miller GW, Labut EM, Tanguay RL, Traber MG (2013) Chronic vitamin E deficiency promotes vitamin C deficiency in zebrafish leading to degenerative myopathy and impaired swimming behavior. Comp Biochem Physiol Part C 157:382–389

Lee IM, Cook NR, Manson JE, Buring JE, Hennekens CH (1999) Betacarotene supplementation and incidence of cancer and cardiovascular disease: the Women's health study. J Natl Cancer Inst 91:2102–2106

Lemee L, Peuchant E, Clerc M, Brunner M, Pfander H (1997) Deinoxanthin: a new carotenoid isolated from Deinococcus radiodurans. Tetrahedron 53:919–926

Li R, Jia Z, Trush MA (2016) Defining ROS in biology and medicine. Reactive Oxygen Species 1:9–21

Lin H, Decuypere E, Buyse J (2004) Oxidative stress induced by corticosterone administration in broiler chickens (*Gallus gallus domesticus*) 2. Short-term effect. Comp Biochem Physiol Part B 139:745–751

Lin H, De Vos D, Decuypere E, Buyse J (2008) Dynamic changes in parameters of redox balance after mild heat stress in aged laying hens (*Gallus gallus domesticus*). Comp Biochem Physiol Part C 147:30–35

López-Torres M, Pérez-Campo R, Cadenas S, Rojas C, Barja G (1993) A comparative study of free radicals in vertebrates—II. Non-enzymatic antioxidants and oxidative stress. Comp Biochem Physiol Part B 105:757–763

Loschen G, Flohé L, Chance B (1971) Respiratory chain linked H2O2 production in pigeon heart mitochondria. FEBS Lett 18:261–264

Love NR, Chen Y, Ishibashi S, Kritsiligkou P, Lea R, Koh Y, Gallop JL, Dorey K, Amaya E (2013) Amputation-induced reactive oxygen species are required for successful Xenopus tadpole tail regeneration. Nat Cell Biol 15:222–228

Lu AL, Li X, Gu Y, Wright PM, Chang DY (2001) Repair of oxidative DNA damage: mechanisms and functions. Cell Biochem Biophys 35:141–170

Lushchak VI (2014) Free radicals, reactive oxygen species, oxidative stress and its classification. Chem-Biol Inter 224:164–175

Maellaro E, Del Bello B, Sugherini L, Santucci A, Comporti M, Casini AF (1994) Purification and characterization of glutathione-dependent dehydroascorbate reductase from rat liver. Biochem J 301:471–476

Maiorino M, Bosello-Travain V, Cozza G, Miotto G, Roveri A, Toppo S, Zaccarin M, Ursini F (2015) Understanding mammalian glutathione peroxidase 7 in the light of its homologs. Free Radic Biol Med 83:352–360

Margis R, Dunand C, Teixeira FK, Margis-Pinheiro M (2008) Glutathione peroxidase family—an evolutionary overview. FEBS J 275:3959–3970

Marla SS, Lee J, Groves JT (1997) Peroxynitrite rapidly permeates phospholipid membranes. Proc Natl Acad Sci USA 94:14243–14248

Martínez del Rio C (1997) Can Passerines synthesize vitamin C? Auk 114:513–516

Maruta T, Sawa Y, Shigeoka S, Ishikawa T (2016) Diversity and evolution of ascorbate peroxidase functions in chloroplasts: more than just a classical antioxidant enzyme? Plant Cell Physiol 57:1377–1386

Mattson MP, Calabrese EJ (2010) Hormesis: a revolution in biology, toxicology and medicine. Springer-Humana Press, New York

McCord JM, Fridovich I (1968) The reduction of cytochrome c by milk xanthine oxidase. J Biol Chem 243:5753–5760

McCord JM, Fridovich I (1969) Superoxide dismutase: an enzymic function for erythrocuprein (hemocuprein). J Biol Chem 244:6049–6055

McGraw KJ, Cohen AA, Costantini D, Hõrak P (2010) The ecological significance of antioxidants and oxidative stress: a marriage between mechanistic and functional perspectives. Funct Ecol 24:947–949

Melov S, Schneider JA, Day BJ, Hinerfeld D, Coskun P, Mirra SS, Crapo JD, Wallace DC (1998) A novel neurological phenotype in mice lacking mitochondrial manganese superoxide dismutase. Nat Genet 18:159–163

Melov S, Coskun P, Patel M, Tuinstra R, Cottrell B, Jun AS, Zastawny TH, Dizdaroglu M, Goodman SI, Huang TT, Miziorko H, Epstein CJ, Wallace DC (1999) Mitochondrial disease in superoxide dismutase 2 mutant mice. Proc Natl Acad Sci USA 96:846–851

Messina S, Eens M, Casasole G, Ablelgawad H, Asard H, Pinxten R, Costantini D (2017) Experimental inhibition of a key cellular antioxidant affects vocal communication. Funct Ecol 31:1101–1110
Metcalfe NB, Alonso-Alvarez C (2010) Oxidative stress as a life-history constraint: the role of reactive oxygen species in shaping phenotypes from conception to death. Funct Ecol 24:984–996
Miki M, Tamai H, Mino M, Yamamoto Y, Niki E (1987) Free-radical chain oxidation of rat red blood cells by molecular oxygen and its inhibition by alpha-tocopherol. Arch Biochem Biophys 258:373–380
Misra HP, Fridovich I (1972) The generation of superoxide radical during the autoxidation of haemoglobin. J Biol Chem 247:6960–6962
Mitsui A, Hamuro J, Nakamura H, Kondo N, Hirabayashi Y, Ishizaki-Koizumi S, Hirakawa T, Inoue T, Yodoi J (2002) Overexpression of human thioredoxin in transgenic mice controls oxidative stress and life span. Antiox Redox Sign 4:693–696
Monaghan P, Costantini D (2014) Free radicals: an evolutionary perspective. In: Laher I (ed) Systems biology of free radicals and anti-oxidants. Springer-Verlag, Berlin, pp 39–64
Monaghan P, Metcalfe NB, Torres R (2009) Oxidative stress as a mediator of life history trade-offs: mechanisms, measurements and interpretation. Ecol Lett 12:75–92
Moran NA, Jarvik T (2010) Lateral transfer of genes from fungi underlies carotenoid production in aphids. Science 328:624–627
Morimoto RI (1998) Regulation of the heat shock transcriptional response: crosstalk between a family of heat shock factors, molecular chaperones, and negative regulators. Genes Dev 12:3788–3796
Moskovitz J, Bar-Noy S, Williams WM, Requena J, Berlett BS, Stadtman ER (2001) Methionine sulfoxide reductase (MsrA) is a regulator of antioxidant defense and lifespan in mammals. Proc Natl Acad Sci USA 98:12920–12925
Muntwyler J, Hennekens CH, Manson JE, Buring JE, Gaziano JM (2002) Vitamin supplement use in a low-risk population of US male physicians and subsequent cardiovascular mortality. Arch Intern Med 62:1472–1476
Muraoka S, Miura T (2003) Inhibition by uric acid of free radicals that damage biological molecules. Pharmacol Toxicol 93:284–289
Nathan C, Cunningham-Bussel A (2013) Beyond oxidative stress: an immunologist's guide to reactive oxygen species. Nature Rev Immunol 13:349–361
Navas P, Villalba JM (1994) Cordoba F. Ascorbate function at the plasma membrane. Biochim Biophys Acta 1197:1–13
Niki E (2014) Role of vitamin E as a lipid-soluble peroxyl radical scavenger: in vitro and in vivo evidence. Free Radic Biol Med 66:3–12
Niki E, Komuro E, Takahashi M, Urano S, Ito E, Terao K (1988) Oxidative hemolysis of erythrocytes and its inhibition by free radical scavengers. J Biol Chem 263:19809–19814
Nishi H, Inagi R, Kato H, Tanemoto M, Kojima I, Son D, Fujita T, Nangaku M (2008) Hemoglobin is expressed by mesangial cells and reduces oxidant stress. J Am Soc Nephrol 19:1500–1508
Niu Y, DesMarais TL, Tong Z, Yao Y, Costa M (2015) Oxidative stress alters global histone modification and DNA methylation. Free Radic Biol Med 82:22–28
Noctor G, Mhamdi A, Chaouch S, Han Y, Neukermans J, Marquez-Garcia B, Queval G, Foyer CH (2012) Glutathione in plants: an integrated overview. Plant Cell Environ 35:454–484
Nucci C, Palamara AT, Ciriolo MR, Nencioni L, Savini P, D'Agostini C, Rotilio G, Cerulli L, Garaci E (2000) Imbalance in corneal redox state during herpes simplex virus 1-induced keratitis in rabbits: effectiveness of exogenous glutathione supply. Exp Eye Res 70:215–220
O'Brien KM, Crockett EL, Philip J, Oldham CA, Hoffman M, Kuhn DE, Barry R, McLaughlin J (2018) The loss of hemoglobin and myoglobin does not minimize oxidative stress in Antarctic icefishes. J Exp Biol 221:jeb162503
O'Connor TP, Lee A, Jarvis JU, Buffenstein R (2002) Prolonged longevity in naked mole-rats: age-related changes in metabolism, body composition and gastrointestinal function. Comp Biochem Physiol Part A 133:835–842

Orr WC, Sohal RS (1994) Extension of life-span by overexpression of superoxide dismutase and catalase in Drosophila melanogaster. Science 263:1128–1130

Pamplona R (2008) Membrane phospholipids, lipoxidative damage and molecular integrity: a causal role in aging and longevity. Biochim Biophys Acta-Bioenerget 1777:1249–1262

Pamplona R, Barja G (2006) Mitochondrial oxidative stress, aging and caloric restriction: the protein and methionine connection. Biochim Biophys Acta-Bioenerget 1757:496–508

Pamplona R, Barja G (2007) Highly resistant macromolecular components and low rate of generation of endogenous damage: two key traits of longevity. Ageing Res Rev 6:189–210

Pamplona R, Costantini D (2011) Molecular and structural antioxidant defences against oxidative stress in animals. Am J Phys 301:R843–R863

Pamplona R, Portero-Otín M, Sanz A, Ayala V, Ilieva EV, Barja G (2005) Protein and lipid oxidative damage and complex I content are lower in the brain of budgerigards and canaries than in mice. Relation to aging rate Age 27:267–280

Paradies G, Petrosillo G, Pistolese M, Ruggiero FM (2000) The effect of reactive oxygen species generated from mitochondrial electron transport chain on the cytochrome c oxidase activity and on the cardiolipin content in bovine heart submitochondrial particles. FEBS Lett 466:323–326

Paradies G, Petrosillo G, Pistolese M, Ruggiero FM (2001) Reactive oxygen species generated by the mitochondrial respiratory chain affect the complex III activity via cardiolipin peroxidation in beefheart submitochondrial particles. Mitochondrion 1:151–159

Paradies G, Petrosillo G, Pistolese M, Ruggiero FM (2002) Reactive oxygen species affect mitochondrial electron transport complex I activity through oxidative cardiolipin damage. Gene 286:135–141

Parkhurst BR, Bradshaw AS, Forte JL, Wright GP (1981) The chronic toxicity to *Daphnia magna* of acridine, a representative azaarene present in synthetic fossil fuel products and wastewaters. Environ Pollut 24:21–30

Parsell DA, Lindquist S (1993) The function of heat-shock proteins in stress tolerance: degradation and reactivation of damaged proteins. Annu Rev Genet 27:437–496

Pérez-Campo R, López-Torres M, Rojas C, Cadenas S, Barja G (1993) A comparative study of free radicals in vertebrates—I. Antioxidants enzymes. Comp Biochem Physiol Part B 105:749–755

Pérez-Rodríguez L, Redondo T, Ruiz-Mata R, Camacho C, Moreno-Rueda G, Potti J (2019) Vitamin E supplementation—but not induced oxidative stress—influences telomere dynamics during early development in wild passerines. Front Ecol Evol 7:173

Philipp E, Brey T, Heilmayer O, Abele D, Pörtner HO (2006) Physiological ageing in a temperate and a polar swimming scallop. Mar Ecol Progr Ser 307:187–198

Pineda-Pampliega J, Herrera-Dueñas A, Mulder E, Aguirre JI, Höfle U, Verhulst S (2019) Antioxidant supplementation slows telomere shortening in free-living white stork chicks. Proc R Soc Lond B 286:20191917

Plácido A, do Pais do Amaral C, Teixeira C, Nogueira A, Brango-Vanegas J, Alves Barbosa E, Moreira D, Silva-Carvalho AÉ, da Silva MDG, do Nascimento Dias J, Albuquerque P, Saldanha-Araújo F, Lima F, Batagin-Neto A, Kuckelhaus S, Bessa LJ, Freitas J, Dotto Brand G, Santos N, Relvas J, Gomes P, Leite JR, Eaton P (2022) Neuroprotective effects on microglia and insights into the structure-activity relationship of an antioxidant peptide isolated from Pelophylax perezi. J Cell Mol Med 26:2793–2807

Porter NA, Caldwell SE, Mills KA (1995) Mechanisms of free radical oxidation of unsaturated lipids. Lipids 30:277–290

Portero-Otin M, Bellmunt MJ, Ruiz MC, Barja G, Pamplona R (2001) Correlation of fatty acid unsaturation of the major liver mitochondrial phospholipid classes in mammals to their maximum life span potential. Lipids 36:491–498

Portero-Otín M, Requena JR, Bellmunt MJ, Ayala V, Pamplona R (2004) Protein nonenzymatic modifications and proteasome activity in skeletal muscle from the short-lived rat and long-lived pigeon. Exp Gerontol 39:1527–1535

Qu Y, Xiao D, Li J, Chen Z, Biondi A, Desneux N, Gao X, Song D (2015) Sublethal and hormesis effects of imidacloprid on the soybean aphid Aphis glycines. Ecotoxicology 24:479–487

Rapoport SI, Chang MCJ, Spector AA (2001) Delivery and turnover of plasma-derived essential PUFAs in mammalian brain. J Lipid Res 42:678–685
Raymond J, Segrè D (2006) The effect of oxygen on biochemical networks and the evolution of complex life. Science 311:1764–1767
Rees JF, de Wergifosse B, Noiset O, Dubuisson M, Janssens B, Thompson EM (1998) The origins of marine bioluminescence: turning oxygen defence mechanisms into deep-sea communication tools. J Exp Biol 201:1211–1221
Reinke NB, O'Brien GM (2006) High activity antioxidant enzymes protect flying-fox haemoglobin against damage: an evolutionary adaptation for flight? J Comp Physiol B 176:729–737
Reischl E, Dafre AL, Franco JL, Filho DW (2007) Distribution, adaptation and physiological meaning of thiols from vertebrate hemoglobins. Comp Biochem Physiol Part C 146:22–53
Ricklefs RE, Wikelski M (2002) The physiology/life-history nexus. Trends Ecol Evol 17:462–468
Rietjens IMCM, Boersma MG, de Haan L, Spenkelink B, Awad HM, Cnubben NHP, van Zanden JJ, van der Woude H, Alink GM, Koeman JH (2002) The pro-oxidant chemistry of the natural antioxidants vitamin C, vitamin E, carotenoids and flavonoids. Environ Toxicol Pharmacol 11:321–333
Rodriguez M, Basten Snoek L, Riksen JAG, Bevers RP, Kammenga JE (2012) Genetic variation for stress-response hormesis in *C. elegans* lifespan. Exp Gerontol 47:581–587
Roff DA (1992) The evolution of life histories: theory and analysis. Chapman & Hall, Atlanta
Rojas Mora A, Firth A, Blareau S, Helfenstein F (2017) Oxidative stress affects sperm performance and ejaculate redox status in subordinate house sparrows. J Exp Biol 220:2577–2588
Romero-Haro AA, Sorci G, Alonso-Alvarez C (2016) The oxidative cost of reproduction depends on early development oxidative stress and sex in a bird species. Proc R Soc Lond B 283:20160842
Rose RC, Bode AM (1993) Biology of free radical scavengers: an evaluation of ascorbate. FASEB J 7:1135–1142
Roy RN, Guha BC (1958) Species difference in regard to the biosynthesis of ascorbic acid. Nature 182:319–320
Ruan H, Tang XD, Chen ML, Joiner MA, Sun G, Brot N, Weissbach H, Heinemann SH, Iverson L, Wu CF, Hoshi T (2002) High-quality life extension by the enzyme peptide methionine sulfoxide reductase. Proc Natl Acad Sci USA 99:2748–2753
Ruiz MC, Ayala V, Portero-Otin M, Requena JR, Barja G, Pamplona R (2005) Protein methionine content and MDA-lysine adducts are inversely related to maximum life span in the heart of mammals. Mech Ageing Dev 126:1106–1114
Ruszczycky MW, Liu HW (2017) The surprising history of an antioxidant. Nature 551:37–38
Saitoh M, Nishitoh H, Fujii M, Takeda K, Tobiume K, Sawada Y, Kawabata M, Miyazono K, Ichijo H (1998) Mammalian thioredoxin is a direct inhibitor of apoptosis signal-regulating kinase (ASK) 1. EMBO J 17:2596–2606
Salmon AB, Richardson A, Perez VI (2010) Update on the oxidative stress theory of aging: does oxidative stress play a role in aging or healthy aging? Free Radic Biol Med 48:642–655
Samuels DC (2005) Life span is related to the free energy of mitochondrial DNA. Mech Ageing Dev 126:1123–1129
Sani E, Herzyk P, Perrella G, Colot V, Amtmann A (2013) Hyperosmotic priming of Arabidopsis seedlings establishes a long-term somatic memory accompanied by specific changes of the epigenome. Genome Biol 14:R59
Sarup P, Loeschcke V (2011) Life extension and the position of the hormetic zone depends on sex and genetic background in Drosophila melanogaster. Biogerontology 12:109–117
Sauer H, Wartenberg M, Hescheler J (2001) Reactive oxygen species as intracellular messengers during cell growth and differentiation. Cell Physiol Biochem 11:173–186
Schopf JW, Klein C (1992) The Proterozoic Biosphere. Cambridge University Press, Cambridge
Schreck R, Rieber P, Baeuerle PA (1991) Reactive oxygen intermediates as apparently widely used messengers in the activation of the NF-κB transcription factor and HIV-1. EMBO J 10:2247–2258

Selman C, Blount D, Nussey DH, Speakman JR (2012) Oxidative damage, ageing and life history evolution: where now? Trends Ecol Evol 27:570–577

Sessions AL, Doughty DM, Welander PV, Summons RE, Newman DK (2009) The continuing puzzle of the great oxidation event. Curr Biol 19:567–574

Shin SA, Joo BJ, Lee JS, Ryu G, Han M, Kim WY, Park HH, Lee JH, Lee CS (2020) Phytochemicals as anti-inflammatory agents in animal models of prevalent inflammatory diseases. Molecules 25:5932

Sies H (1985) Oxidative stress: introductory remarks. In: Sies H (ed) Oxidative stress. London Academic Press, London, pp 1–8

Sies H (1986) Biochemistry of oxidative stress. Angew Chem Int Ed 25:1058–1071

Sies H (1991) Oxidative stress: oxidants and antioxidants. Academic Press, London

Sies H, Chance B (1970) The steady state level of catalase compound I in isolated hemoglobin-free perfused rat liver. FEBS Lett 11:172–176

Sies H, Jones D (2007) Oxidative stress. In: Fink G (ed) Encyclopedia of stress, vol 3, 2nd edn. Elsevier, Amsterdam, pp 45–48

Sies H, Berndt C, Jones DP (2017) Oxidative stress. Annu Rev Biochem 86:715–748

Simons MJP, Cohen AA, Verhulst S (2012) What does carotenoid-dependent coloration tell? Plasma carotenoid level signals immunocompetence and oxidative stress state in birds–a meta-analysis. PLoS One 7:e43088

Sinenko SA, Starkova TY, Kuzmin AA, Tomilin AN (2021) Physiological signaling functions of reactive oxygen species in stem cells: from flies to man. Front Cell Dev Biol 9:714370

Smirnoff N (2018) Ascorbic acid metabolism and functions: a comparison of plants and mammals. Free Radic Biol Med 122:116–129

Smith SW, Latta LC, Denver DR, Estes S (2014) Endogenous ROS levels in C. Elegans under exogenous stress support revision of oxidative stress theory of life-history tradeoffs. BMC Evol Biol 14:161

Sohal RS, Orr WC (2012) The redox stress hypothesis of aging. Free Radic Biol Med 52:539–555

Sørensen JG, Sarup P, Kristensen TN, Loeschcke V (2008) Temperature-induced hormesis in *Drosophila*. In: Le Bourg E, Rattan SIS (eds) Mild stress and healthy aging. Springer, Cham, pp 65–79

Southam CM, Ehrlich J (1943) Effects of extracts of western red cedar heartwood on certain wood-decaying fungi in culture. Phytopathology 33:517–524

Spicer JI, Morley SA, Bozinovic F (2019) Physiological diversity, biodiversity patterns and global climate change: testing key hypotheses involving temperature and oxygen. Phil Trans R Soc Lond B 374:20190032

Spies HS, Steenkamp DJ (1994) Thiols of intracellular pathogens. Identification of ovothiol a in *Leishmania donovani* and structural analysis of a novel thiol from *Mycobacterium bovis*. Eur J Biochem 224:203–213

Stadtman ER, Moskovitz J, Levine RL (2003) Oxidation of methionine residues of proteins: biological consequences. Antiox Redox Sign 5:577–582

Stearns SC (1992) The evolution of life histories. Oxford University Press, Oxford

Steinberg CEW (2012) Stress ecology: environmental stress as ecological driving force and key player in evolution. Springer Science & Business Media, Dordrecht

St-Pierre J, Buckingham JA, Roebuck SJ, Brand MD (2002) Topology of superoxide production from different sites in the respiratory chain. J Biol Chem 277:44784–44490

Sun Y, MacRae TH (2005) Small heat shock proteins: molecular structure and chaperone function. Cell Mol Life Sci 62:2460–2476

Sun J, Tower J (1999) FLP recombinase-mediated induction of cu/Zn-superoxide dismutase transgene expression can extend the lifespan of adult Drosophila melanogaster flies. Mol Cell Biol 19:216–228

Surai P (2002) Natural antioxidants in avian nutrition and reproduction. Nottingham University Press, Nottingham

Surai PF, Speake BK, Noble RC, Sparks NH (1999) Tissue-specific antioxidant profiles and susceptibility to lipid peroxidation of the newly hatched chick. Biol Trace Elem Res 68:63–78

Tang X-Q, Feng J-Q, Chen J, Chen P-X, Zhi J-L, Cui Y, Guo R-X, Yum H-M (2005) Protection of oxidative preconditioning against apoptosis induced by $H_2O_2$ in PC 12 cells: mechanisms via MMP, ROS, and Bcl-2. Brain Res 1057:57–64
Tappel AL (1955) Unsaturated lipide oxidation catalyzed by hematin compounds. J Biol Chem 217:721–733
Taylor ER, Hurrell F, Shannon RJ, Lin TK, Hirst J, Murphy MP (2003) Reversible glutathionylation of complex I increases mitochondrial superoxide formation. J Biol Chem 278:19603–19610
Thannickal VJ, Fanburg BL (2000) Reactive oxygen species in cell signaling. Am J Phys 279:L1005–L1028
The Age-Related Eye Disease Research Group (2001) A randomized, placebocontrolled, clinical trial of high-dose supplementation with vitamins C and E, beta carotene, and zinc for age-related cataract and vision loss. Arch Ophthalmol 119:1439–1452
Tian B, Xu Z, Sun Z, Lin J, Hua Y (2007) Evaluation of the antioxidant effects of carotenoids from Deinococcus radiodurans through targeted mutagenesis, chemiluminescence, and DNA damage analyses. Biochim Biophys Acta 1770:902–911
Tian B, Sun Z, Shen S, Wang H, Jiao J, Wang L, Hu Y, Hua Y (2009) Effects of carotenoids from *Deinococcus radiodurans* on protein oxidation. Lett Appl Microbiol 49:689–694
Tian R, Geng Y, Guo H, Yang C, Seim I, Yang G (2021a) Comparative analysis of the superoxide dismutase gene family in Cetartiodactyla. J Evol Biol 34:1046–1060
Tian R, Geng Y, Yang Y, Seim I, Yang G (2021b) Oxidative stress drives divergent evolution of the glutathione peroxidase (GPX) gene family in mammals. Integr Zool 16:696–711
Turner E, Klevit R, Hager LJ, Shapiro BM (1987) Ovothiols, a family of redox-active mercaptohistidine compounds from marine invertebrate eggs. Biochemistry 26:4028–4036
Turner E, Hager LJ, Shapiro BM (1988) Ovothiol replaces glutathione-peroxidase as a hydrogen-peroxide scavenger in sea-urchin eggs. Science 242:939–941
Turpaev KT (2002) Reactive oxygen species and regulation of gene expression. Biochemistry 67:281–292
Tüzün N, Debecker S, Stoks R (2020) Strong species differences in life history do not predict oxidative stress physiology or sensitivity to an environmental oxidant. J Anim Ecol 89:1711–1721
Valavanidis A, Vlachogianni T, Fiotakis C (2009) 8-hydroxy-2′-deoxyguanosine (8-OHdG): a critical biomarker of oxidative stress and carcinogenesis. J Environ Sci Health C Environ Carcinog Ecotoxicol Rev 27:120–139
Valenti R, Jabłońska J, Tawfik DS (2022) Characterization of ancestral Fe/Mn superoxide dismutases indicates their cambialistic origin. Protein Sci 31:e4423
Wang X, Quinn PJ (1999) Vitamin E and its function in membranes. Progr Lipid Res 38:309–336
Weaver RJ (2019) Hypothesized evolutionary consequences of the alternative oxidase (AOX) in animal mitochondria. Integr Comp Biol 59:994–1004
Weaver RJ, Santos ESA, Tucker AM, Wilson AE, Hill GE (2018) Carotenoid metabolism strengthens the link between feather coloration and individual quality. Nat Commun 2018:73
Wells WW, Xu DP (1994) Dehydroascorbate reduction J Bioenerget Biomembr 26:369–377
Yamamoto Y, Niki E, Kamiya Y, Miki M, Tamai H, Mino M (1986) Free radical chain oxidation and hemolysis of erythrocytes by molecular oxygen and their inhibition by vitamin E. J Nutr Sci Vitaminol 32:475–479
Yamamoto M, Kensler TW, Motohashi H (2018) The KEAP1-NRF2 system: a thiol-based sensor-effector apparatus for maintaining redox homeostasis. Physiol Rev 98:1169–1203
Yanase S, Onodera A, Tedesco P, Johnson TE, Ishii N (2009) SOD-1 deletions in Caenorhabditis elegans alter the localization of intracellular reactive oxygen species and show molecular compensation. J Gerontol A Biol Sci Med Sci 64:530–539
Yang H, Wang X, Liu X, Wu J, Liu C, Gong W, Zhao Z, Hong J, Lin D, Wang Y, Lai R (2009) Antioxidant peptidomics reveals novel skin antioxidant system. Mol Cell Proteomics 8:571–583
Yang X, Wang Y, Zhang Y, Lee WH, Zhang Y (2016) Rich diversity and potency of skin antioxidant peptides revealed a novel molecular basis for high-altitude adaptation of amphibians. Sci Rep 6:19866

Zera AJ, Harshman LG (2001) The physiology of life history trade-offs in animals. Annu Rev Ecol Syst 32:95–126

Zhang Y, Wong HS (2021) Are mitochondria the main contributor of reactive oxygen species in cells? J Exp Biol 224:jeb221606

Zhang R, Al-Lamki R, Bai L, Streb JW, Miano JM, Bradley J, Min W (2004) Thioredoxin-2 inhibits mitochondria-located ASK1-mediated apoptosis in a JNK-independent manner. Circ Res 94:1483–1491

Zhang Y-J, Wang W, Yang H-L, Li Y, Kang X-Y, Wang X-R, Yang Z-L (2015) Molecular properties and functional divergence of the dehydroascorbate reductase gene family in lower and higher plants. PLoS One 10:e0145038

Zhang Q, Wang Y, Man L, Zhu Z, Bai X, Wei S, Liu Y, Liu M, Wang X, Gu X, Wang Y (2016) Reactive oxygen species generated from skeletal muscles are required for gecko tail regeneration. Sci Rep 6:20752

# Chapter 2
# The Immuno-Oxidative Ecology

**Abstract** Immune function plays a fundamental role in the maintenance of organismal homeostasis. It protects the organism against parasites, pathogens, toxic substances, and cancerous cells. To this end, immune function relies on a complex system, involving molecules, cells, and tissues. However, immune responses against harmful agents carry different types of costs for the host that can manifest themselves in diverse ways, such as increased metabolic rate or trade-offs with other life processes (e.g. reduced growth or fertility). The emergence of these costs would explain why we observe a wide individual variation in immunological function. But how are these costs incurred? A pillar of the immuno-oxidative ecology is that oxidative stress is a major physiological cost of immune function that mediates the trade-off between the expression of life-history traits and the need to mount an immune response. In this chapter, I have reviewed (i) the main molecular and cellular mechanisms through which immune function generates oxidative stress, (ii) the studies that tested the effects of antigen injection and parasite exposure on oxidative damage and antioxidant levels in a wide range of vertebrates, and (iii) the ways by which pathogens respond to or evade the pro-oxidant attack generated by the host to kill them.

## 2.1 Introduction to the Immuno-Oxidative Ecology

Immunological function plays a crucial role in the maintenance of physiological homeostasis of living organisms, recognising and protecting against infectious agents, toxic substances, and cancerous cells. Parasites and pathogens can have a number of detrimental effects on organisms, such as reductions in host energetic stores, harm to tissues, alterations of behaviour, or impaired reproduction (e.g. Reddy et al. 2006; Brecchia et al. 2010; Graham et al. 2011; Losdat et al. 2011; Watson 2013; Asghar et al. 2015; Sanchez et al. 2018). Consequently, we should expect individuals with poorly-functioning immune responses be counterselected. Surprisingly, we observe much variation among individuals in immunological

© The Author(s), under exclusive license to Springer Nature Switzerland AG 2024

D. Costantini, *The Role of Organismal Oxidative Stress in the Ecology and Life-History Evolution of Animals*, Fascinating Life Sciences,
https://doi.org/10.1007/978-3-031-65183-0_2

function, such as lack of antibody production in some individuals, or low to high expression levels of inflammatory molecules. The immunoecological theory states that one explanation for this wide individual variation lies in the factors that shape the immune function, and the benefits and costs of an immune response, which in turn influence host–parasite relationships and disease outcomes (Sheldon and Verhulst 1996; Martin et al. 2011; Sorci et al. 2013, 2017; Ohmer et al. 2021). Physiological costs might favour the maintenance of immunological variation because these costs are likely to vary with individual condition, environment, sex, age, and diversity of pathogens the host is exposed to (e.g. Sheldon and Verhulst 1996; Klasing 2004; Martin et al. 2008; Seppälä and Jokela 2010; Maizels and Nussey 2013). The immune function requires several resources, including energy and nutrients. For example, Iseri and Klasing (2014) estimated that (i) the systemic immune system at maintenance in adult male chickens *Gallus gallus domesticus* has the same lysine content as 332 average-sized feathers, 16% of an egg, or 5.4% of a pectoralis muscle from an adult chicken, respectively, and (ii) during the acute-phase response to *Escherichia coli*, the additional lysine needed would equal 355 feathers, 17% of an egg, or 5.5% of a pectoralis muscle, respectively.

Inferring from patterns observed in biomedical research, ecoimmunological theory predicts that the immune response also carries costs for the organism through the generation of oxidative stress, which has detrimental effects on a number of organismal functions (Costantini 2008, 2014). Several lines of evidence provide support for a mechanistic link between immune response (including inflammation) and oxidative stress in free-living animals (Costantini 2008, 2014, 2022; Sorci and Faivre 2009). In this chapter, I have reviewed (i) the main molecular and cellular mechanisms through which immune function generates oxidative stress, (ii) the studies that tested the effects of antigen injection and parasite exposure on oxidative damage and antioxidant levels in a wide range of vertebrates, and (iii) the ways by which pathogens respond to or evade the pro-oxidant attack generated by the host to kill them.

## 2.2 Role of Reactive Oxygen Species in Immunity

In the 1930s, Baldridge and Gerrard carried out the first studies on the oxygen metabolism of phagocytes. They showed that neutrophils of dogs underwent a marked increase in oxygen consumption while phagocytosing bacteria. Some 30 years later, Iyer and collaborators (Iyer et al. 1961) hypothesised that the augmented oxygen uptake of neutrophils observed during phagocytosis is necessary for the generation of hydrogen peroxide. A few years later, Paul and Sbarra (1968) measured directly the generation of hydrogen peroxide during phagocytosis, formally supporting the hypothesis proposed by Iyer and collaborators. At that time, it was well known that mammalian cells can generate hydrogen peroxide from the superoxide free radical through the action of the superoxide dismutase (see Chap. 1).

In 1973, Babior and collaborators demonstrated that activated phagocytes generate indeed the superoxide free radical (Babior et al. 1973). It was later discovered that immune cells generate different kinds of reactive oxygen species to control or eliminate parasites and pathogens (e.g. Kozlov et al. 2003; Babior 2004; Levy 2004; Segal 2005; Halliwell 2006; Üllen et al. 2013).

Phagocytes contain a multi-component enzyme complex, the nicotinamide adenine dinucleotide phosphate (NADPH) oxidase (Hampton et al. 1998; Babior 1999, 2004). This enzyme complex is responsible for the generation of reactive oxygen species after immune stimulation. To do so, phagocytes increase their oxygen uptake as much as 10–20 times the resting values (*respiratory burst*) (Fig. 2.1). The NADPH is oxidised into NADP⁺, releasing two electrons, which are used to reduce oxygen to superoxide free radical:

$$NADPH + 2O_2 \rightarrow NADP^+ + H^+ + 2O_2^{\cdot -}$$

The superoxide free radical is highly reactive so that it cannot pass through the cell wall or membrane of the pathogen (e.g. Hassan and Fridovich 1979; Halliwell 2006). The superoxide free radical serves as the starting material for the generation of other less reactive molecules, such as hydrogen peroxide, hypochlorous acid, or peroxynitrite (Halliwell 2006). For example, two superoxide radicals are converted into one molecule of hydrogen peroxide:

$$O_2^{\cdot -} + e^- + 2H^+ \rightarrow H_2O_2$$

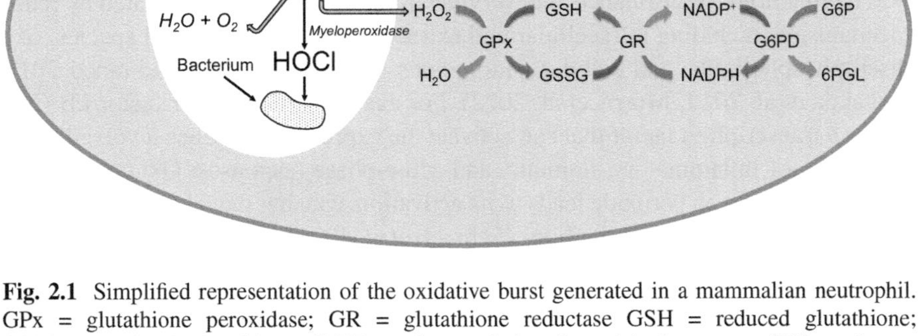


**Fig. 2.1** Simplified representation of the oxidative burst generated in a mammalian neutrophil. GPx = glutathione peroxidase; GR = glutathione reductase GSH = reduced glutathione; GSSG = oxidised glutathione; G6PD = glucose-6-phosphate dehydrogenase; G6P = glucose 6-phosphate; 6PGL = 6-phosphogluconolactone; NADPH = reduced nicotinamide adenine dinucleotide phosphate; $NADP^+$ = oxidised nicotinamide adenine dinucleotide phosphate; $H_2O_2$ = hydrogen peroxide; HOCl = hypochlorous acid; $O_2$ = oxygen; $O_2^{\bullet -}$ = superoxide free radical; $H_2O$ = water

In contrast to the superoxide radical, hydrogen peroxide is less reactive, so that it can easily enter into pathogens and kill them, often working as material for the generation of the hydroxyl radical (Repine et al. 1981; Halliwell 2006).

The hydrogen peroxide is also precursor of the hypochlorous acid (HOCl), a powerful antibacterial and antifungal reactive oxygen species (Winterbourn 2002; Halliwell 2006). The conversion of hydrogen peroxide to hypochlorous acid is catalysed by the enzyme myeloperoxidase; this enzyme is, however, absent in some animal groups (Brune et al. 1972; Penniall and Spitznagel 1975; Stabler et al. 1994):

$$Cl^- + H_2O_2 \rightarrow OCl^- + H_2O$$

$$OCl^- + H^+ \leftrightarrow HOCl$$

$$HOCl + H^+ + Cl^- \leftrightarrow H_2O + Cl_2$$

In turn, the hypochlorous acid can generate the highly reactive hydroxyl radical through its reaction with either $Fe^{2+}$ (Dukan and Touati (1996) or the superoxide free radical (Shen et al. 2000):

$$Fe^{2+} + HOCl \rightarrow Fe^{3+} + OH^{"} + Cl^-$$

$$HOCl + O_2^{"-} \rightarrow O_2 + OH^{"} + Cl^-$$

The generation of reactive nitrogen species is another important route through which immune cells can kill pathogens (Halliwell 2006). They can do so, for example, by inhibiting respiration mechanisms (Shi et al. 2005) or by reacting with the superoxide free radical to generate peroxynitrite (Halliwell 2006):

$$O_2^{"-} + NO^{"} \rightarrow ONOO^-$$

Increased generation of the superoxide free radical and of the peroxynitrite following an immune challenge (lipopolysaccharide) was confirmed using spin probes and electron paramagnetic resonance spectroscopy (Kozlov et al. 2003).

The functions, performance, and survival of immune cells are regulated by redox mechanisms, including intracellular and extracellular reactive oxygen species, oxidised phospholipids, and cellular antioxidants (e.g. Wheeler and DeFranco 2012; Angajala et al. 2018; Morris et al. 2022). For example, the nuclear factor κB (NF-κB) is a transcription factor that can activate the expression of genes involved in the regulation of inflammatory, immune, and acute-phase responses. Oxidation of the NF-κB by hydrogen peroxide leads to its activation, causing the release of the inhibitory subunit of the NF-κB inhibitor (Schreck et al. 1991). The superoxide free radical also appears to increase the release of pro-inflammatory cytokines by neutrophils, whereas elevated intracellular concentrations of hydrogen peroxide diminish lipopolysaccharide-induced neutrophil activation (e.g. Strassheim et al. 2004; Brigelius-Flohe 2006; Zmijewski et al. 2007; Lorne et al. 2008). The anti-inflammatory cytokine IL10 may also decrease the inflammation through the inhibition of the capacity of macrophages to generate hydroxyl free radicals following their induction by lipopolysaccharide exposure (Dokka et al. 2001). Finally, reactive oxygen species generated in the mitochondria also play a primary role in the regulation of immune response (Angajala et al. 2018).

There is also evidence that antioxidants affect the activity of immune cells. For example, glutathione supplementation improved macrophage functions in individuals with HIV infection resulting in improved control of *Mycobacterium tuberculosis* (Morris et al. 2013). On the other hand, depletion of glutathione compromised phagocytosis and macrophage survival (e.g. Fitzpatrick et al. 2011) and antiviral responses (e.g. Diotallevi et al. 2017). Regulation of antioxidant enzymes is also likely important during the oxidative burst. For example, increased synthesis of those antioxidant enzymes (catalase and glutathione peroxidase) that reduce hydrogen peroxide to water might compromise the phagocytic response. However, these enzymes are also needed to avoid that these reactive oxygen species are generated in excess, jeopardising cell homeostasis. We know little about how organisms handle this complex trade-off between the need to kill pathogens and to protect their own tissues from oxidation.

## 2.3 Immune Function and Metabolic Activity

Beyond oxidative burst, the activity of mitochondria increases during an immune response, raising oxygen consumption of cells as a consequence (e.g. Lochmiller and Deerenberg 2000; Martin et al. 2003). Cellular oxygen consumption generates reactive oxygen species, so that we can expect that increased metabolic rate also contributes to some extent to increase oxidative stress during an immune response, such as in the case of fever (Demas et al. 1997; Riedel and Maulik 1999; Ots et al. 2001).

Estimated energetic costs (percentage of increment in resting metabolic rate compared to controls) of mounting an immune response during vaccination or sepsis may vary from 10 to 57% in humans or laboratory rodents compared to controls (Lochmiller and Deerenberg 2000). In mammals, serum levels of antibodies IgG against the antigen keyhole limpet haemocyanin in rodents that received previously the removal of their stored energetic reserves (epididymal white adipose tissue or inguinal white adipose tissue) were lower than those measured in control rodents (Demas et al. 2003). In birds, Martin et al. (2003) estimated that the total cost of an immune response to an antigen injected in house sparrows *Passer domesticus* was ca. 29% of their resting metabolic rate, an amount equivalent to the cost of production of half of an egg (8.23 kJ $egg^{-1}$) in this species. Importantly, it has been found that mitochondrial reactive oxygen species can upregulate the expression of the hypoxia-inducible factor 1-α (HIF-1α), which in turn activates a response to the hypoxia generated in inflamed tissues, improving the inflammatory response itself (Wang et al. 2010). Thus, the mitochondrial activity can also modulate indirectly the immune response relying on the signalling role of reactive oxygen species.

In conclusion, the contribution of increased metabolic rate to the generation of reactive oxygen species during an immune response cannot be neglected. However, we lack direct estimates of the amount of reactive oxygen species that are generated from cellular metabolism and oxidative burst in immune-challenged organisms under different ecologically-relevant conditions, respectively.

## 2.4 Immune Response and Organismal Oxidative Stress

We have evidence in full organisms that the oxidative burst plays an important role in protecting against potential pathogens, but also that alterations of host oxidative status are a widespread consequence of immune function in animals. Several studies found that individuals that were made experimentally unable to synthesise certain enzymes were less resistant to severe bacterial or fungal infections (Nathan and Shiloh 2000). For example, all wild-type mice that were experimentally infected with *Salmonella typhimurium* were still alive 20 days later; by contrast, knockout mice that did not express *phox* (phagocyte oxidase, i.e. NADPH-based respiratory burst) or iNOS (nitric oxide synthase that generates nitric oxide) died during the course of the experiment (Mastroeni et al. 2000). Moreover, *phox* mice died more quickly (100% mortality at day 5 post-infection) than iNOS mice (100% mortality at day 19 post-infection) (Mastroeni et al. 2000). Further studies showed that the experimental inhibition of nitric oxide generation made mice less resistant to infections with larvae of the nematode *Brugia malayi* (Rajan et al. 1996). By contrast, experimental stimulation of nitric oxide generation made mice more resistant to infections (Rajan et al. 1996). It has also been shown that eggs of a solitary digger wasp, the European beewolf *Philanthus triangulum*, emit large amounts of gaseous nitric oxide to protect themselves and their provisions, paralysed honeybees, against mould fungi (Strohm et al. 2019). Finally, studies on ants *Formica fusca* found that, when on a diet enriched with hydrogen peroxide, they were significantly more resistant to fungal infections than those that received control food (Bos et al. 2015).

Reactive oxygen species play a major role in the defence of organisms against pathogens. However, they do not target specifically any pathogens, so that they can also oxidise host biomolecules, causing cell and tissue damages (*immunopathological effects*; Sorci and Faivre 2009; Costantini and Møller 2009). Reactive oxygen species can also damage immune cells themselves, potentially impairing the capacity of the host to cope with pathogens. For example, it has been found that an experimental increase of mitochondrial superoxide generation (by knockout of the antioxidant enzyme manganese superoxide dismutase) in the thymus can disrupt normal T cell development and impair the response of the mammalian adaptive immune system to the influenza A virus H1N1 (Case et al. 2011). High generation of reactive oxygen species may compromise the oxidative burst, the production of neutrophil extracellular traps, or the migration of neutrophils towards sites of infection (e.g. Araźna et al. 2015; Sônego et al. 2016; Fokam 2020). Excessive generation of reactive oxygen species may also inhibit the activation of B cells and their differentiation into antibody-producing plasma cells (Jang et al. 2015). Moreover, immune cells with low antioxidant levels are more rapidly inactivated during phagocytosis than immune cells with high antioxidant levels (Halliwell and Gutteridge 2015). Do these research results translate from these mechanistic models to organismal outcomes? Tobler et al. (2011, 2015) found that pre-immune challenge levels of the superoxide free radical were negatively correlated with the strength of the subsequent immune response towards exposure to either phytohaemagglutinin or lipopolysaccharide in painted dragon lizards *Ctenophorus pictus*, suggesting that oxidative stress may have constrained the immune function of lizards. Similarly, Ridgway et al. (2014) found that exposure to a pro-oxidant molecule (tert-butyl hydroperoxide) reduced haemocyte counts in four bivalve species. Finally, Gu et al. (2022) found that polyphenols increased the survival rate, mitigated oxidative damage, inflammation, tissue lesions, and improved the immunity of gibel carps *Carassius auratus gibelio* challenged with *Aeromonas hydrophila*.

Recent phylogenetic meta-analyses showed that changes of oxidative status markers owing to either injection of an antigen or infection are a widespread consequence of immune response across vertebrates (Fig. 2.2; Costantini 2022). These meta-analyses, involving 141 studies, 1262 effect sizes, and 97 species, identified significant effects of taxonomic class and experimental environment (captivity vs. wild). By contrast, age category of animals (young vs. adult), study design (correlational vs. experimental), and proxies of pace of life (clutch size, litter size, and body mass) were negligible in this dataset (Costantini 2022). These meta-analyses also identified that (i) the strength and the direction of the effect may be affected by methodological aspects, such as type of immunostimulant, laboratory assay, or tissue analysed, and (ii) there was large heterogeneity in strength and direction of effect sizes, which suggests that immune response does not necessarily result in strong oxidative stress. Finally, this work identified a significant taxonomic bias, because the dataset was dominated by studies on fish, birds, and mammals, which provided 95.8% of effect sizes (Costantini 2022).

A closer look at single studies included in the meta-analyses by Costantini et al. (2022) provides further relevant information about the interplay between oxidative status and immune function across a diverse range of organisms. The immune challenges caused changes in oxidative status markers under very different experimental conditions: (i) in both the short (Fig. 2.3) (e.g. 1 day after antigen exposure, Costantini and Dell'Omo 2006; Baylor and Butler 2019) and long term (e.g. 6 days after antigen exposure, Casagrande et al. 2012); (ii) in response to different doses of immunostimulant injected into the animal (Armour et al. 2020; Butler et al. 2021; Hudson et al. 2021); and (iii) in response to both artificial antigens (e.g. phytohaemagglutinin in Costantini and Dell'Omo 2006; lipopolysaccharide in Casasole et al. 2016) and real pathogens (e.g. *Trichostrongylus tenuis* larvae in Mougeot et al. 2009; *Isospora* coccidians in Sepp et al. 2012).

This heterogeneity in strength and direction of effect sizes is particularly intriguing. Studies on mammals found that the stimulation of immune cells enhanced the levels of oxidative damage in blood or liver (Sirak et al. 1991; Cadenas et al. 1998; Schneeberger et al. 2013). However, amongst free-living adult female sheep *Ovis*

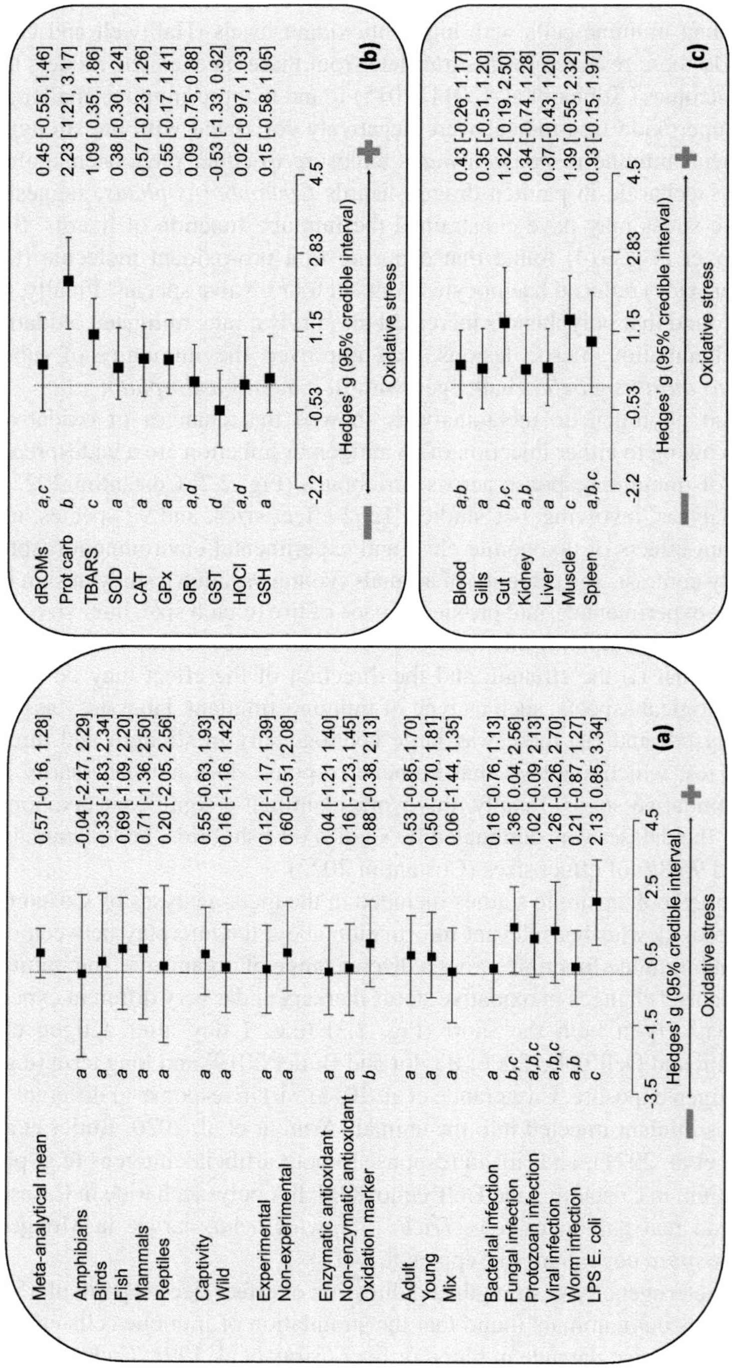
Meta-analytical mean a,c 0.57 [-0.16, 1.28]
Amphibians a -0.04 [-2.47, 2.29]
Birds a 0.33 [-1.83, 2.34]
Fish a 0.69 [-1.09, 2.20]
Mammals a 0.71 [-1.36, 2.50]
Reptiles a 0.20 [-2.05, 2.56]
Captivity a 0.55 [-0.63, 1.93]
Wild a 0.16 [-1.16, 1.42]
Experimental a 0.12 [-1.17, 1.39]
Non-experimental a 0.60 [-0.51, 2.08]
Enzymatic antioxidant a 0.04 [-1.21, 1.40]
Non-enzymatic antioxidant a 0.16 [-1.23, 1.45]
Oxidation marker b 0.88 [-0.38, 2.13]
Adult a 0.53 [-0.85, 1.70]
Young a 0.50 [-0.70, 1.81]
Mix a 0.06 [-1.44, 1.35]
Bacterial infection a 0.16 [-0.76, 1.13]
Fungal infection b,c 1.36 [0.04, 2.56]
Protozoan infection a,b,c 1.02 [-0.09, 2.13]
Viral infection a,b,c 1.26 [-0.26, 2.70]
Worm infection a,b 0.21 [-0.87, 1.27]
LPS E. coli c 2.13 [0.85, 3.34]
-3.5 -1.5 0.5 2.5 4.5
Hedges' g (95% credible interval)
Oxidative stress
(a)
dROMs a,c 0.45 [-0.55, 1.46]
Prot carb b 2.23 [1.21, 3.17]
TBARS c 1.09 [0.35, 1.86]
SOD a 0.38 [-0.30, 1.24]
CAT a 0.57 [-0.23, 1.26]
GPx a 0.55 [-0.17, 1.41]
GR a,d 0.09 [-0.75, 0.88]
GST d -0.53 [-1.33, 0.32]
HOCl a,d 0.02 [-0.97, 1.03]
GSH a 0.15 [-0.58, 1.05]
-2.2 -0.53 1.15 2.83 4.5
Hedges' g (95% credible interval)
Oxidative stress
(b)
Blood a,b 0.43 [-0.26, 1.17]
Gills a 0.35 [-0.51, 1.20]
Gut b,c 1.32 [0.19, 2.50]
Kidney a,b 0.34 [-0.74, 1.28]
Liver a 0.37 [-0.43, 1.20]
Muscle c 1.39 [0.55, 2.32]
Spleen a,b,c 0.93 [-0.15, 1.97]
-2.2 -0.53 1.15 2.83 4.5
Hedges' g (95% credible interval)
Oxidative stress
(c)

**Fig. 2.2** Summary of the evidence for changes in oxidative status markers in individuals challenged with an immunostimulant (antigen or live parasite/pathogen). Forest plot show predictions of models for the overall meta-analytical mean with 95% credible intervals and for the effects of moderators on this mean; effect size estimates (Hedges'g) are statistically significant when their own credible intervals do not include zero. (**a**) taxonomic class, experimental environment (captivity or wild), study design (experimental or non-experimental), marker category (oxidation, antioxidant enzyme, or non-enzymatic antioxidant), age category (young, adult, or mix for studies that used a pool of young and adults), and type of immunostimulant, (**b**) laboratory assay, and (**c**) tissue. Markers are categorised as follows: (i) oxidation: dROMs (reactive oxygen metabolites), prot carb (protein carbonyls), TBARS (thiobarbituric acid reactive substances); (ii) enzymatic antioxidants: SOD (superoxide dismutase), CAT (catalase), GPx (glutathione peroxidase), GR (glutathione reductase), GST (glutathione-S-transferase); (iii) non-enzymatic antioxidants: HOCl (non-enzymatic antioxidant capacity), GSH (reduced glutathione). Outcomes of contrasts between levels are shown as letters: shared letters (e.g. both levels with **a**) indicate that the two levels of a given moderator did not differ significantly. The dataset included 141 studies that involved 1262 effect sizes from 97 species: (Srivastava et al. 1992; Belló et al. 2000; Ito et al. 2000; Dautremepuits et al. 2003; Alonso-Alvarez et al. 2004; Marcogliese et al. 2005, 2010, 2021; Bertrand et al. 2006; Costantini and Dell'Omo 2006; Hõrak et al. 2006, 2007; Torres and Velando 2007; Perez-Rodriguez et al. 2008; Kim et al. 2009; Kurhalyuk et al. 2009, 2011; Mougeot et al. 2009, 2010; Saleh et al. 2009; Devevey et al. 2010; Espinal-Carrion and Lopez-Lopez 2010; Ali et al. 2011; Charoensuk et al. 2011; Garcia et al. 2011; Kurhalyuk and Tkachenko 2011; Losdat et al. 2011; Pap et al. 2011, 2014; Tkachenko et al. 2011, 2014, 2016; Vitula et al. 2011; Casagrande et al. 2012; van de Crommenacker et al. 2012; De Coster et al. 2012; Heidarpour et al. 2012; Hoogenboom et al. 2012; Orledge et al. 2012; Sepp et al. 2012; Serra et al. 2012; Stumbo et al. 2012; Dkhil et al. 2013; Hanssen et al. 2013; Isaksson et al. 2013; Moore et al. 2013; Schneeberger et al. 2013; Zahran and Risha 2013; Adeyemi 2014; Allaam et al. 2014; Giraudeau et al. 2014; Guzmán-Villanueva et al. 2014; Jolly et al. 2014; Lilley et al. 2014; Rodriguez et al. 2014; Sobroza et al. 2014; Tonin et al. 2014; Azma et al. 2015; Costantini et al. 2015a, b; Cram et al. 2015; Du et al. 2015; Hau et al. 2015; Lopez-Arrabe et al. 2015; Marri and Richner 2015; Skuratovskaya et al. 2015, 2020; Sweazea et al. 2015; Tancredo et al. 2015; Tobler et al. 2015; Wegmann et al. 2015; Broggi et al. 2016; Casasole et al. 2016; da Silva et al. 2016; Delhaye et al. 2016, 2018; Gasso et al. 2016; Le Guernic et al. 2016; Razavi et al. 2016; Schneebauer et al. 2016; Azimzadeh and Amniattalab 2017; Baldissera et al. 2017, 2019; De Angeli et al. 2017; Jiang et al. 2017; Kumar et al. 2017; Peixoto et al. 2017; Rodríguez-Quiroga et al. 2017; Şahan et al. 2017; Schoenle et al. 2017; Sebastiano et al. 2017, 2018; Stauffer et al. 2017; Abasubong et al. 2018; Costantini et al. 2018; Ghirardi et al. 2018, 2020; Maronde et al. 2018; Norte et al. 2018; Pacios-Palma et al. 2018; Zahran et al. 2018, 2019; Balasch et al. 2019; Baylor and Butler 2019; Charpentier et al. 2019; da Rosa et al. 2019; Fritze et al. 2019, 2021; Geng et al. 2019; Neuman-Lee et al. 2019; Ramirez-Otarola et al. 2019, 2021; Roth et al. 2019; Roy et al. 2019; Yun et al. 2019; Zhao et al. 2019; Abdelkhalek et al. 2020; Akinsanya et al. 2020; Amin et al. 2020; Armour et al. 2020; Chen et al. 2020; El-Son et al. 2020; Giri et al. 2020; Harikrishnan et al. 2020; Hernández-Arciga et al. 2020; Kim et al. 2020; Li et al. 2020; Liu et al. 2020; Molbert et al. 2020; Pal et al. 2020; Ren et al. 2020; Sun et al. 2020; Voigt et al. 2020; Xu et al. 2020; Zhu et al. 2020; Alaasam et al. 2021; Petitjean et al. 2021a, b; Zhang et al. 2021; Messina et al. 2022; Casagrande et al., unpublished results). Reprinted with minor modifications from Costantini et al. (2022). © 2022 by the author; published by Oxford University Press and the Society for Experimental Biology; distributed under the terms of the Creative Commons Attribution License (CC BY)

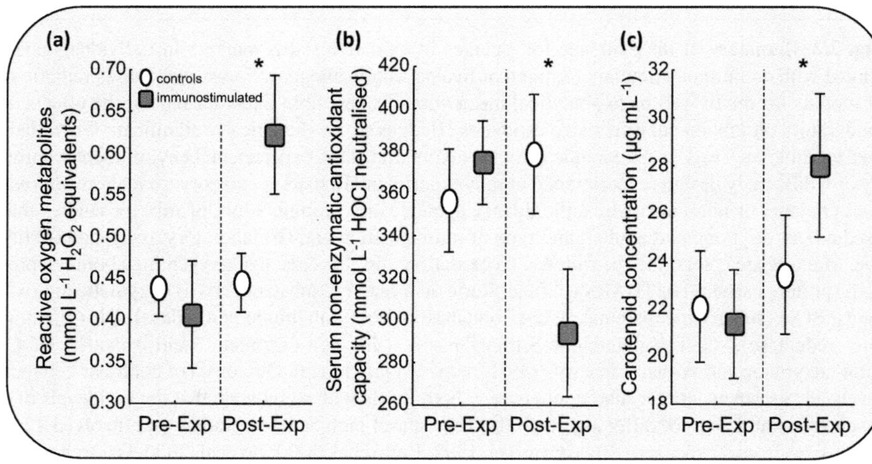


**Fig. 2.3** Intradermal injection of the antigen phytohaemagglutinin increased (**a**) serum levels of reactive oxygen metabolites (marker of oxidative damage) and decreased (**b**) the serum non-enzymatic antioxidant capacity and (**c**) the serum concentration of carotenoids in nestling common kestrels *Falco tinnunculus*. Pre-Exp = values of physiological markers recorded before immunostimulation; Post-Exp = values of physiological markers recorded 24 h after immunostimulation. Values are shown as mean and standard error. Asterisks indicate significant differences between groups: * $P < 0.05$. Reprinted with minimal modifications from Costantini and Dell'Omo (2006) with permission of Elsevier

*aries*, there was no evidence that current parasite burden (strongyle faecal egg count) significantly explained individual variation in oxidative damage levels (Nussey et al. 2009). Experimental increase of parasites did not affect markers of oxidative status in Columbian ground squirrels *Urocitellus columbianus* (Roth et al. 2019). Chronic coccidian infections were also not associated with seasonal changes of oxidative status markers in house sparrows *Passer domesticus* (Pap et al. 2014, 2015). By contrast, it has been found that blood oxidative stress may be higher in dogs with *Leishmaniasis* infection (Paltrinieri et al. 2010; Almeida et al. 2013) and dogs affected by lymphoma (Winter et al. 2009) or mast cell (cells involved in wound healing and defence against pathogens) tumour (Finotello et al. 2014). In fish, three-spined sticklebacks *Gasterosteus aculeatus* with high levels of expression at major histocompatibility complex class IIB loci had elevated levels of acrolein, a marker of oxidative protein damage (Kurtz et al. 2006). At spawning, prioritisation of investment in egg production may make female brown trout *Salmo trutta* especially vulnerable to ulcerative dermal necrosis and fungal infection (*Saprolegnia* spp.) with associated increased oxidative damage level and decreased antioxidant protection, respectively (Kurhalyuk et al. 2009, 2010; Hoogenboom et al. 2012; Zahran and Risha 2013). Correlative studies also suggested that, when the fungal infection becomes very intense, immune activity might be suppressed through the action of stress hormones, and the generation of early derivatives of oxidative damage decreases as a consequence (Fig. 2.4; Hoogenboom et al. 2012).

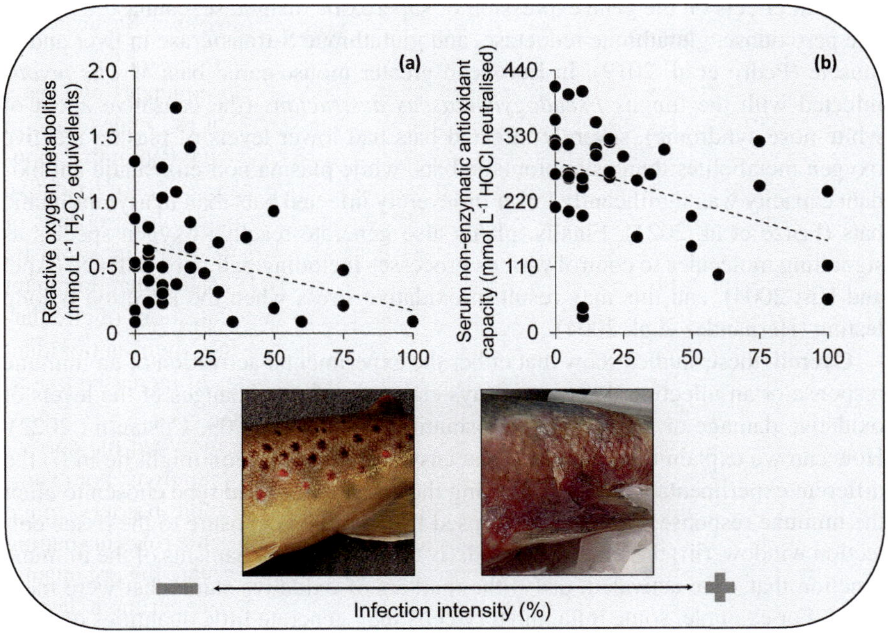


**Fig. 2.4** Relationship between intensity of *Saprolegnia* fungal infection (% cover of fins) and (**a**) plasma levels of reactive oxygen metabolites (marker of oxidative damage) and (**b**) plasma non-enzymatic antioxidant capacity. Data points represent individual trout females sampled at the time of spawning. Reprinted with minor modifications from Hoogenboom et al. (2012) with permission of Elsevier and Springer Nature. Photograph of healthy fish courtesy of Mia Hoogenboom; photograph of fish infected with *Saprolegnia* fungus courtesy of David Bruno

High levels of glucocorticoids can, for example, suppress the activity of antioxidant enzymes and the generation of hydrogen peroxide in macrophages (Pereira et al. 1995). Similarly, levels of reactive oxygen metabolites were significantly lower in chub *Squalius cephalus* infected by acanthocephalan parasites *Pomphorhynchus* sp. compared to uninfected fish, while the non-enzymatic antioxidant capacity of plasma was not related to their infection status (Molbert et al. 2020). In brown trout *Salmo trutta* infected with *Tetracapsuloides bryosalmonae*, levels of several antioxidant enzymes and of glutathione were not correlated with parasite load (Stauffer et al. 2017). However, infected animals may reduce their swimming activity, and this reduction in active metabolic rate might buffer any increased generation of reactive oxygen species due to the immune response (Petitjean et al. 2021a). Mourning doves *Zenaida macroura* naturally infected with the blood trematode *Gigantobilharzia huronensis* had higher levels of thiobarbituric acid reactive substances than those that were not infected (Sweazea et al. 2015). Experimental infection by the parasite tick-borne *Borrelia burgdorferi* affected the oxidative status in its avian reservoir host the blackbird *Turdus merula* (Norte et al. 2018). Experimental infection of Atlantic salmons with the bacterium *Piscirickettsia salmonis* had transient effects on the gene expression of superoxide dismutase, catalase, glutathione peroxidase, glutathione reductase, and glutathione S-transferase in liver and in muscle (Pedro et al. 2019). In European greater mouse-eared bats *Myotis myotis* infected with the fungus *Pseudogymnoascus destructans* (the causative agent of white-nose syndrome), severely infected bats had lower levels of plasma reactive oxygen metabolites than asymptomatic bats, while plasma non-enzymatic antioxidant capacity was significantly lower in severely infected bats than in asymptomatic bats (Fritze et al. 2021). Finally, plants also generate reactive oxygen species as signalling molecules to control various processes including pathogen defence (Apel and Hirt 2004), and this may result in oxidative stress when the infection is long lasting (Hernández et al. 2004).

Overall, these studies show that either the experimental activation of an immune response or an infection does not always cause significant changes of the levels of oxidative damage or antioxidants (Costantini and Møller 2009; Costantini 2022). How can we explain these results? One answer to this question might lie in (i) the different experimental designs, including the antigen dose and type chosen to elicit the immune response, (ii) the time elapsed from antigen exposure to the tissue collection window, (iii) the tissue analysed, (iv) the type of mechanisms of the immune function that were activated, or (v) the markers of oxidative status that were measured. For example, some inflammatory cells may generate little quantities of reactive oxygen species (Conlon et al. 1991). Therefore, it might be relevant to quantify the total and relative numbers of different leucocytes that are activated following experimental stimulation. Intra-individual repeated measurements over time would also help overcome any issue related to the time lag of the response.

The results of the meta-analysis did not suggest a relevant role of the pace of life in explaining the effects of an immune response on the cellular oxidative status. However, we cannot rule out that some of this heterogeneity in effect sizes might be explained by the life history of the species, the phase of the life cycle, the intensity of sexual selection, the history of parasite-mediated selection, or the feeding habits. If so, we might predict that species have evolved different mechanisms to handle immune response-induced oxidative stress threats. Avian malaria is an interesting example of a globally widespread disease, caused by haemosporidian parasites of the genus *Plasmodium*, that show considerable variation in both prevalence and parasitaemia at small spatial scales (Wood et al. 2007). The malaria parasite is transmitted by insect vectors like mosquitoes (*Culex*, *Aedes*, and *Culiseta*), and chronic infection may have detrimental effects on reproductive activity and survival of avian hosts (Knowles et al. 2010). An increase of oxidative damage might be implicated in the reduction of the reproductive fitness, especially when the malaria infection is chronic. van de Crommenacker et al. (2012) found that malaria was associated with increased susceptibility to oxidative stress in the Seychelles warbler *Acrocephalus sechellensis*, but this effect depended on the breeding stage: only during the energetically demanding provisioning stage did infected birds have higher oxidative stress susceptibility than non-infected birds. The alteration of the oxidative status was caused by a marked increase of oxidative damage levels observed only in infected birds during provisioning, and by an overall reduction of the non-enzymatic antioxidant capacity observed in all birds across the breeding cycle. These findings suggested that an intense workload while dealing with an infection could reduce the capacity of the host to maintain the homeostasis of oxidative status (van de Crommenacker et al. 2012), likely because individuals are more constrained in the optimisation of trade-offs. In a study on great tits *Parus major*, the reproductive effort decreased levels of blood antioxidants and increased malaria load, respectively; however, antioxidants and parasite load were not related to each other (Christie et al. 2012). In great tits, Delhaye et al. (2016) also found that infected individuals had higher generation of the superoxide free radical in red blood cells, but the intensity of this generation of superoxide was not related to the reproductive effort of the birds. Isaksson et al. (2013) found that the oxidative status of the host showed species-specific patterns in relation to infection, and that oxidative damage increased with the level of *Plasmodium* parasitaemia in breeding free-living great tits. The interplay between parasitaemia and oxidative status also seems to depend on the phase of infection (acute vs. chronic, Delhaye et al. 2018). Conversely to birds, *Plasmodium* infection was not associated with two plasma markers of oxidative status in free-living mandrills *Mandrillus sphinx* (Charpentier et al. 2019).

Bats are also particularly special with regard to their immunological function and consequences in terms of oxidative stress. Hibernating greater mouse-eared bats mounted an inflammatory response to a fungal antigen (zymosan), with only mild to negligible consequences for their energy budget and their plasma oxidative status (Fritze et al. 2019). However, outside the hibernating season, bacterial (lipopolysaccharide) and fungal (zymosan) antigens induced a significant increase of plasma oxidative damage in greater mouse-eared bats, whereas exposure to a viral antigen (polyinosinic:polycytidylic acid) did not cause any increase of plasma oxidative damage at all albeit the mild immune response (Costantini et al. 2023). Egyptian fruit bats *Rousettus aegyptiacus*, that were experimentally challenged with the antigen lipopolysaccharide (which simulates a bacterial infection), mounted an inflammatory response (elevation in haptoglobin and lysozyme), but did not show any changes in blood levels of protein carbonyls, reactive oxygen metabolites, glutathione peroxidase, superoxide dismutase, total thiols, and non-enzymatic antioxidant capacity (Fig. 2.5; Costantini et al. 2022). Circumstances under which the immune response is activated are also important for other taxa, like fish. For example, it has been found that the immune response, and so the capacity to degrade pathogens, may be suppressed by a combination of decreased oxygen and pH (Boleza et al. 2001) and increased carbon dioxide, or may be reduced at certain water temperatures than at others (Ndong et al. 2006).

The effect of parasites on the oxidative stress of the host may also be buffered by parental care. De Coster et al. (2012) assessed how mothers mediate the link between parasite infection and oxidative stress in nestling great tits. They carried out an ectoparasite treatment before egg-laying combined with a partial cross-fostering experiment between broods of infested and uninfested nests. The ectoparasite treatment consisted of a first phase in which nest parasites were killed and a second

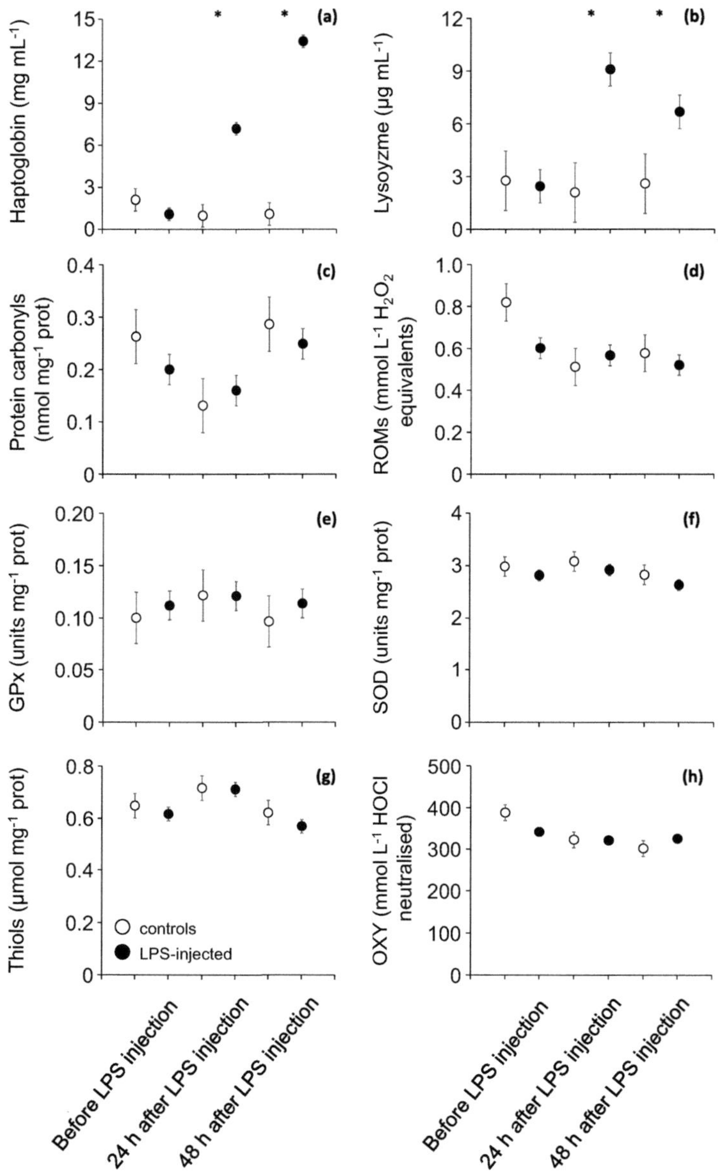


**Fig. 2.5** Effects of lipopolysaccharide injection on inflammatory and oxidative status markers in Egyptian fruit bats *Rousettus aegyptiacus*. Inflammation: (**a**) serum concentration of haptoglobin, (**b**) serum concentration of lysozyme. Oxidative damage: (**c**) concentration of protein carbonyls in red blood cells, (**d**) serum levels of reactive oxygen metabolites (ROMs). Antioxidants: (**e**) GPx, activity of glutathione peroxidase in red blood cells, (**f**) SOD, activity of superoxide dismutase in red blood cells, (**g**) concentration of total thiols in red blood cells, (**h**) OXY, serum non-enzymatic antioxidant capacity. Values are shown as least square mean and standard error. Asterisks indicate significant differences between groups. Reprinted with minor modifications from Costantini et al. (2022). © 2022 by the authors; published by Oxford University Press and the Society for Experimental Biology; distributed under the terms of the Creative Commons Attribution License (CC BY)

phase in which half of nests was inoculated with hen fleas *Ceratophyllus gallinae*, which are blood-sucking ectoparasites of great tits that induce immune responses (De Coster et al. 2010) and various deleterious effects, especially in males (Heeb et al. 1998; Tschirren et al. 2003). Then, half broods were reciprocally swapped between pairs of infested and uninfested nests. De Coster and collaborators found that, when pre- and post-hatching environments were similar (control-control, infested-infested), (i) daughters had significantly lower plasma levels of oxidative damage than sons in infested nests, but not in uninfested nests, and (ii) daughters had significantly higher and lower plasma non-enzymatic antioxidant capacity than sons in infested and uninfested nests, respectively. Maronde et al. (2018) also found that neither plasma non-enzymatic antioxidant capacity nor reactive oxygen metabolites of nestling great tits were significantly influenced by the flea infestation. They also observed that parents of flea-infested nests fed nestlings with a higher proportion of caterpillars, the main source of antioxidants, suggesting that parents can compensate for negative effects of parasitism by modifying the food composition.

Another reason for the lack of changes of oxidative status during an immune response may also lie in the strategies that parasites adopt to invade the host. For example, various strains or species of bacteria, protozoa, and worms have evolved mechanisms to suppress the generation of reactive oxygen species of their hosts. Thus, we should not overlook the biology of the parasite or pathogen when interpreting the effects of infection on the oxidative status of the host (see paragraph 2.5). In line with this, it is also largely overlooked the possibility that any increases of oxidative damage levels observed during an infection are not entirely due to the physiological response of the host, but are also influenced to some extent by the release of reactive oxygen species from the invading microorganism. For example, *Mycoplasma* species are highly contagious bacteria that cause severe and chronic respiratory diseases. It has been reported that various *Mycoplasma* species can release hydrogen peroxide, superoxide free radical, and hydroxyl radical, which are used as a primary virulence mechanism, inflicting damage to host tissues (Lynch and Cole 1980; Baumann 1989).

There is general consensus that oxidative stress is implicated in the pathogenesis of a number of viral infections, and that certain viruses can take advantage of a cellular state of oxidative stress because it favours their replication (e.g. Peterhans 1979; Peterhans et al. 1987; Schwarz 1996; Reshi et al. 2014; Camini et al. 2017; Costantini 2022). Viral interactions with mitochondrial membranes and other mitochondria-associated components lead to increased generation of reactive oxygen species. In turn, these reactive oxygen species benefit viral replication through a number of mechanisms, such as modulation of host signalling pathways and covalent changes in viral components (Foo et al. 2022). Herpesviruses are one of the most common viral agents in wild animals. They establish a latent long-life infection, with reactivation under phases of stress and immunosuppression of the host. Active herpesvirus infections cause a number of clinical signs, including depression of normal activity, respiratory distress, extremity paralysis, head-shaking, haemorrhagic lesions, twisting of the neck, and even sudden mortality (Kaleta 1990).

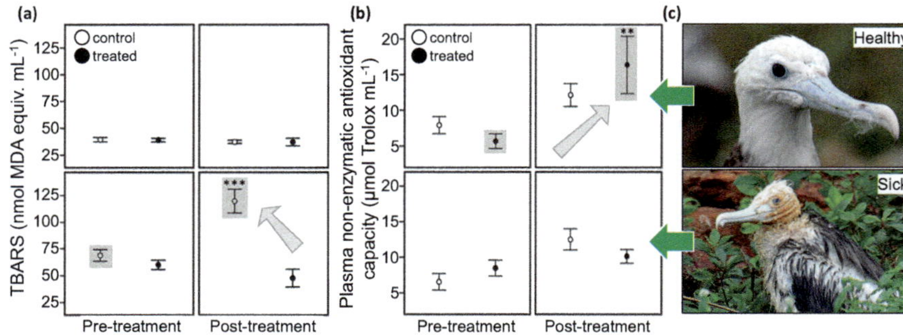


**Fig. 2.6** Effect of resveratrol administration on the (**a**) plasma concentration of thiobarbituric acid reactive substances (TBARS) and (**b**) plasma non-enzymatic antioxidant capacity of (**c**) healthy and sick (viral disease) chicks of the Magnificent frigatebird *Fregata magnificens*. Asterisks indicate significant differences between groups: * $P < 0.05$; ** $P < 0.01$; *** $P < 0.001$. Values are shown as mean and standard error. Reprinted with minor modifications from Sebastiano et al. (2018) with permission of John Wiley and Sons. Photographs of the healthy and of the sick chicks by David Costantini

There is meta-analytical, correlative, and experimental evidence that oxidative stress is strongly associated with herpesvirus infection, and supplementation of specific antioxidants may inhibit virus replication (e.g. Valyi-Nagy et al. 2000; Faith et al. 2006; Kavouras et al. 2007; Sheridan and Beck 2008; Mathew et al. 2010; De Luca et al. 2012; Xu et al. 2013; da Silva et al. 2016; Sebastiano et al. 2016, 2017, 2018; Costantini et al. 2018). For example, both acute and latent experimental infections with herpes simplex virus type 1 in the murine nervous system were associated with increased levels of two oxidative damage markers, 4-hydroxy-2-nonenal and 8-hydroxyguanosine (Valyi-Nagy et al. 2000). da Silva et al. (2016) found that female green sea turtles *Chelonia mydas* affected by fibropapillomatosis had higher levels of thiobarbituric acid reactive substances than those that did not show any visible clinical signs of the disease. Finally, Sebastiano et al. (2017, 2018) found that chicks of Magnificent frigatebirds *Fregata magnificens* with visible clinical signs of a disease associated with herpesvirus activity had strong alterations of oxidative status markers (particularly lipid peroxidation), and that the administration of the antioxidant resveratrol to the birds strongly protected the chicks against oxidative damage to lipids (Fig. 2.6).

Other viruses can also generate strong oxidative stress or even exploit a state of slight cellular oxidative stress to increase their replication. For example, deficiency of dietary antioxidants (vitamin E) and co-factors of glutathione peroxidase (selenium) can cause transition of avirulent to virulent and virulent to extremely virulent mutation of coxsackievirus in mice (Beck 1997). However, a link between viral infection and oxidative stress is not always demonstrated. For example, poxvirus infection was not associated with plasma levels of thiobarbituric acid reactive substances in male house finches *Haemorhous mexicanus* (Giraudeau et al. 2014). Similarly, the oxidative status of wild European rabbits naturally infected

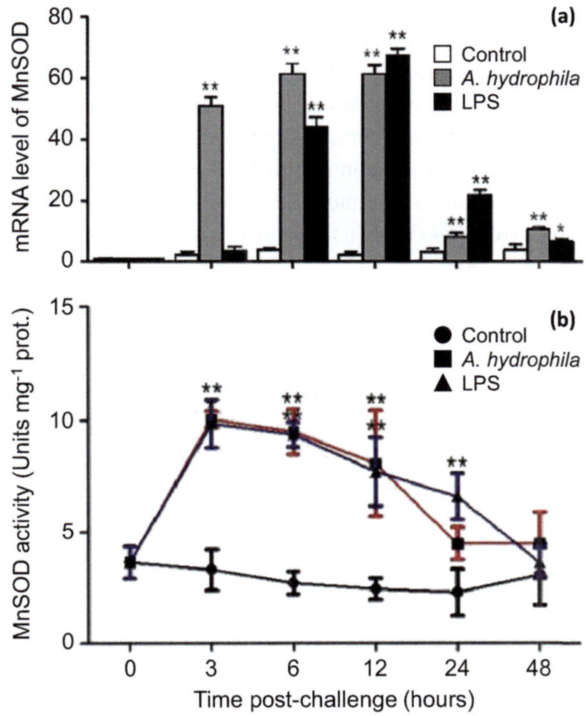


**Fig. 2.7** Temporal changes of gene expression and enzymatic activity after immunostimulation. (**a**) expression of the manganese superoxide dismutase (MnSOD) gene and (**b**) activity of the antioxidant enzyme MnSOD in gill of Qihe crucian carps *Carassius auratus* after either infection with the gram-negative bacterium *Aeromonas hydrophila* or injection of lipopolysaccharide (LPS, wall component of gram-negative bacteria). Asterisks indicate significant differences between groups: * $P < 0.05$; ** $P < 0.01$. Values are shown as mean and standard deviation. Reprinted with minor modifications from Kong et al. (2017) with permission of Elsevier

with myxoma virus and rabbit haemorrhagic disease virus did not differ from the oxidative status of rabbits that were not infected by either virus (Pacios-Palma et al. 2018).

We have also a number of studies that found significant changes of levels of certain antioxidants in response to an immune challenge. Many studies found a significant upregulation of the expression of antioxidant enzymes (Fig. 2.7). Other studies found significant reductions of antioxidants. In this second case, it is not always clear if these reductions reflect a depletion of antioxidants owing to their oxidation or a downregulation of antioxidant genes. There is, however, good consensus that some antioxidants, like carotenoids and vitamin E, can boost the immune response. For example, the administration of vitamin E into the egg increased the phagocytic potential and the generation of nitric oxide in macrophages isolated from birds when they were 4 weeks old (Gore and Qureshi 1997). However, Marri and Richner

(2015) and Maronde et al. (2018) did not find any evidence for either an immunostimulatory and/or an antioxidant role of vitamin E in nestling great tits. Similarly, Possenti et al. (2017) found that an experimental increase of vitamin E in yolk did not affect the oxidative status of embryos of yellow legged gulls *Larus michahellis*. Male zebra finches *Taeniopygia guttata* supplemented with carotenoids produced stronger immune responses than control birds to an injection with (i) the antigen phytohaemagglutinin, which stimulates the activity of immune cells (Blount et al. 2003; McGraw and Ardia 2003) or (ii) the antigen sheep red blood cells, which also stimulate a humoral response to T-dependent antigens (McGraw and Ardia 2003). Similarly, supplementation of β-carotene to tadpoles of wood frogs *Lithobates sylvaticus* stimulated their immune response to infection with the trematode parasite *Echinostoma trivolvis* (Szuroczki et al. 2019). However, many studies found that carotenoids have low to negligible antioxidant effects in birds (Costantini and Møller 2008; Simons et al. 2012) and do not influence the oxidative burst of phagocytes (Sild et al. 2011). These results might be reconciled with each other if carotenoids act mainly as molecules that stimulate and modulate the immune response (e.g. Bendich 1989; Chew 1993; Kim et al. 2000), while other compounds, like vitamins A and E, other fat-soluble antioxidants, or vitamin C, act mainly as free-radical scavengers (e.g. Cadenas et al. 1998; Surai 2002; Hartley and Kennedy 2004). In this scenario, the pro-oxidant effects of high levels of immuno-stimulation or immuno-modulation mediated by carotenoids would be neutralised to some degree by the protective effects of other antioxidants, so that the host can prevent immuno-pathological damages (Costantini and Møller 2009). However, the antioxidant effects of carotenoids might be circumstantial. For example, supplementation of carotenoids, but not of vitamin E, reduced oxidative damage in greenfinches *Carduelis chloris* (Hõrak et al. 2007).

Although the immunological machinery of invertebrates differs in many respects from that of vertebrates, it also relies on the oxidative burst in order to degrade pathogens (Philipp et al. 2012). For example, studies on bivalves found that their immune cells increase significantly the generation of reactive oxygen species in the first hour after infection, but at much lower concentrations than those recorded in vertebrates (Pipe et al. 1997; Hégaret et al. 2003; Donaghy et al. 2009; Philipp et al. 2012). Antioxidant defences of bivalves may also be upregulated during infection. For example, *Crassostrea gigas* larvae, that were experimentally infected with the bacterium *Vibrio coralliilyticus*, upregulated antioxidant defences within 24 h from infection, as revealed by the higher transcript abundance of the antioxidant enzymes glutathione peroxidase, glutathione reductase, superoxide dismutase, catalase, and peroxiredoxines 4 and 5 (Genard et al. 2013). Similarly, after injection of giant freshwater prawns *Macrobrachium rosenbergii* with the yeast *Debaryomyces hansenii*, the respiratory burst increased significantly, stimulating the upregulation of the antioxidant enzymes superoxide dismutase and glutathione peroxidase (Yeh et al. 2009). Similar results were also found in giant tiger shrimps *Penaeus monodon* after injection with either the bacterium *Photobacterium damselae* or the white spot syndrome virus *Whispovirus* (Liu et al. 2010).

## 2.5 Arms Race: How Pathogens and Parasites Evade or Counterattack the Host's Oxidative Weapons

Pathogens and parasites have evolved a number of strategies to evade or cope with the host's inflammatory and immune responses (e.g. Hornef et al. 2002; Sacks and Sher 2002; Sorci and Faivre 2009; Li et al. 2023), or even to mitigate the pro-oxidant consequences of feeding on the host, as in the case of blood-feeding arthropods (e.g. Toh et al. 2010; Whiten et al. 2017; Hernandez et al. 2022). Pathogens and parasites may rely on mechanisms that enable them to avoid recognition from the host immune system or to interfere with the secretion of inflammatory cytokines that the host uses in the chemical communication between immune cells. Pathogens and parasites may also rely on antioxidant defences to resist the oxidative burst generated by the host (Fig. 2.8).

Several studies found that pathogens with impaired antioxidant defences are more sensitive to phagocytic killing, providing an indirect proof for an important role of reactive oxygen species as antimicrobial agents (Halliwell and Gutteridge 2015). Pathogens and parasites may (i) increase quickly their antioxidant levels in response to the oxidative burst of the host, (ii) maintain high constitutive levels of antioxidant enzymes, (iii) express a number of proteins that repair oxidative damage, and (iv) reduce or prevent the activation of reactive oxygen species generation system during phagocytosis (e.g. Liu et al. 2005; Ishibe et al. 2008; Bosch et al.

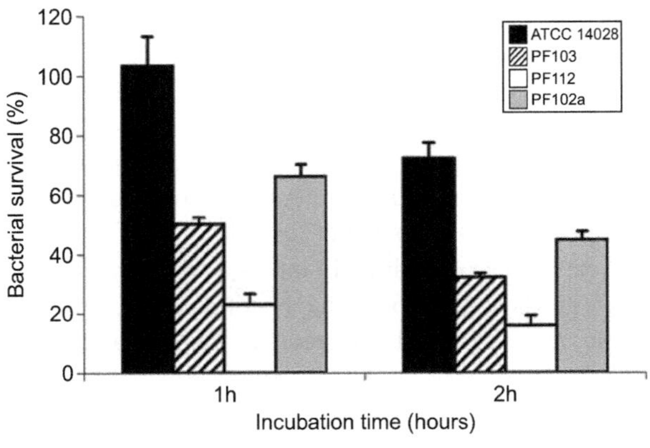


**Fig. 2.8** Susceptibility of wild-type strains and mutant strains of *Salmonella enterica* serovar Typhimurium to hydrogen peroxide. Results are expressed as percent survival after 1 h and 2 h of incubation at 37 °C. The percent survival was calculated for each strain by dividing the number of CFU/ml obtained from incubation in PBS alone by the number of CFU/ml obtained from incubation in 250 μmol $L^{-1}$ of hydrogen peroxide. ATCC 14028 = wild type; PF103 = mutant lacking Dps protein (DNA-binding proteins from starved cells); PF112 = mutant lacking Dps, sodC1, and sodC2 (both refer to superoxide dismutase); PF102a = mutant lacking sodC1 and sodC2. Values are shown as mean and standard error. Reprinted with minor modifications from Pacello et al. (2008) with permission of Elsevier

2015; Gennaris et al. 2015; Ezraty et al. 2017). For example, Radovanović et al. (2015) found that the seasonal patterns of antioxidants in the intestinal parasite *Pomphorhynchus laevis* are closely correlated with those recorded in its host, the barbel *Barbus barbus*, possibly reflecting the physiological adjustments of the parasite in response to the changes of the antioxidant system of the host. Microorganisms, such as *Toxoplasma gondii*, *Pseudomonas aeruginosa*, and *Babesia* spp., have evolved a complex antioxidant defence system for surviving inside host cells (Bosch et al. 2015; Alegria et al. 2017). Glutathione-deficient *Plasmodium berghei* parasites exhibited growth delay and increased nuclear DNA damage (Padín-Irizarry et al. 2016). Moreover, antioxidant mechanisms may increase virulence of certain pathogens (Brenot et al. 2004), or even facilitate the successful migration of some pathogens (e.g. the bacterium *Borrelia*) from the mammalian host to the arthropod vector (Narasimhan et al. 2007). The bacterium *Staphylococcus aureus* uses the pigment staphyloxanthin to prevent damage from peroxides released by neutrophils, promoting its virulence (Liu et al. 2005). The protein SOK (surface factor promoting resistance to oxidative killing), that is expressed on the bacteria surface, can block the pro-oxidant effects of reactive oxygen species (Malachowa et al. 2011). The bacterium *Listeria monocytogenes* uses glutathione as a critical signalling molecule to activate virulence gene expression; in so doing it might also regulate its biphasic lifestyle switching from saprophyte to pathogen (Reniere et al. 2015). Similarly, changes in some components of the intracellular oxidative status work as physiological triggers of gametocytogenesis for *Plasmodium* (Beri et al. 2017).

The key role of antioxidants in the defence of parasites against ROS-dependent immune response is also well supported by those studies showing that parasites, whose antioxidant defences were experimentally impaired, are more sensitive to phagocytic killing. For example, *Salmonella typhimurium* strains deficient in superoxide dismutase were less resistant to reactive oxygen species generated by the host (De Groote et al. 1997). Similarly, strains of *Staphylococcus aureus* unable to synthesise carotenoids were less resistant to neutrophil oxidative burst and were less pathogenic compared to wild strains (Liu et al. 2005).

Some parasite species can also directly interfere with the generation of reactive oxygen species of the host (Lambert and Nicolas 1998; Broeg and Steinhagen 2012). For example, *Helicobacter pylori* reduces the availability of substrates that the host uses for the generation of nitric oxide (Gobert et al. 2001). Similarly, *Leishmania* protozoans inhibit the generation of reactive oxygen species by phagocytes. This capability enables parasites to survive and infect macrophages within which they can replicate (Panaro et al. 1998). As a response, the organism recruits immune cells into inflammatory sites, where they can release larger amounts of reactive oxygen species than in other inflammatory responses (Paltrinieri et al. 2010; Almeida et al. 2013).

Viruses have evolved sophisticated mechanisms to exploit reactive oxygen species, in particular those generated in mitochondria, to thrive in their hosts (Fig. 2.9). These mechanisms involve an increase in the antioxidant cell capacity and control of lipid peroxidation, modulation of reactive oxygen species generation through interaction between electron transport chain and viral proteins or metabolic

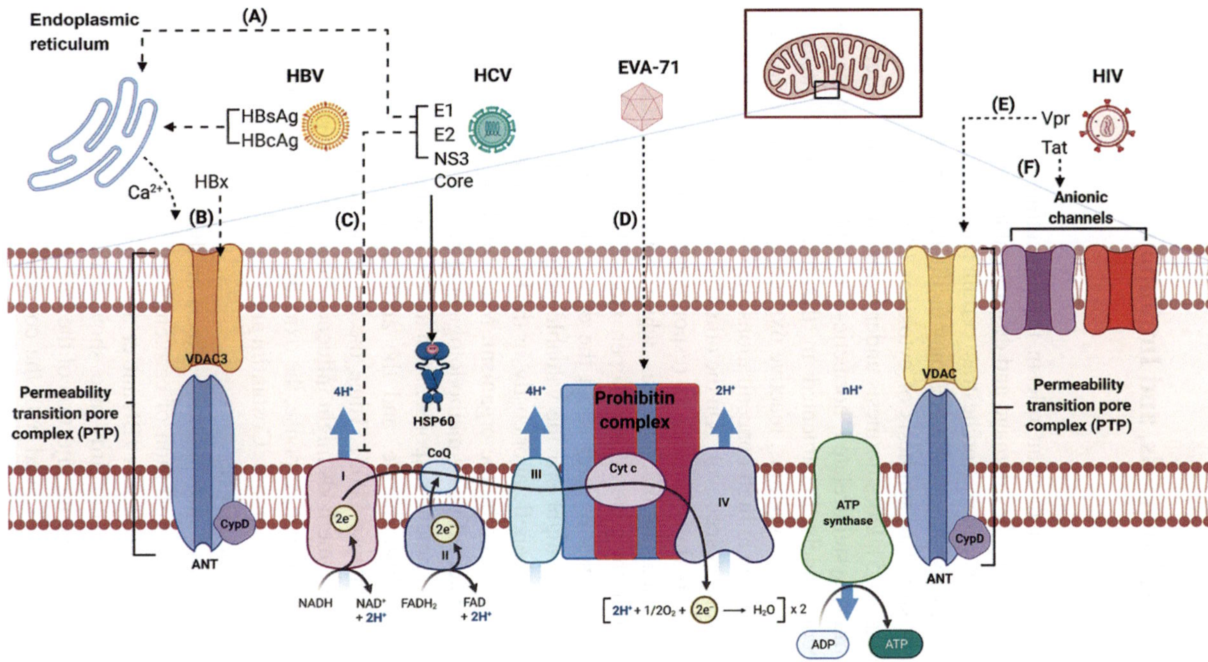


**Fig. 2.9** Interactions between viral proteins and components of the electron transport chain. (**a**) viral proteins from hepatitis B virus (HBV) induce stress in the endoplasmic reticulum and calcium efflux; the calcium can be taken up by mitochondria, leading to the generation of reactive oxygen species; (**b**) HBx protein interacts with voltage-dependent anion channel 3 (VDAC3) and promotes influx of mitochondrial calcium; (**c**) hepatitis C virus (HCV) proteins inhibit complex I of the electron transport chain, resulting in the generation of reactive oxygen species; (**d**) enterovirus A71 (EV-A71) replication complexes interact with mitochondrial prohibitin, which may alter mitochondria integrity and function; (**e**) human immunodeficiency virus (HIV) Vpr protein increases mitochondrial membrane permeabilisation by interacting with adenine nucleotide translocator (ANT) of the mitochondrial permeability transition pore (mPTP), leading to efflux of apoptogenic factors; (**f**) HIV Tat protein dysregulates anionic channels, leading to loss of mitochondrial transmembrane potential. Reprinted from Foo et al. (2022) with permission of Elsevier

reprogramming, ROS-mediated modulation of host signalling pathways, or oxidative modifications of viral proteins, resulting in loss or gain of function (Foo et al. 2022).

## 2.6 Oxidative Stress, Hormesis, and Immunology: A Ménage à Trois?

Reactive oxygen species play a relevant role in the protection of the organism against pathogens. Not only do they damage and destroy pathogens owing to their pro-oxidant potential, but they can also orchestrate certain pathways of the immune response. For example, reactive oxygen species may facilitate the phagocyte adherence to endothelium or stimulate synthesis of cytokines (Sorci and Faivre 2009; Halliwell and Gutteridge 2015). Does hormesis underlie this dual nature of reactive oxygen species? Immune-system-related hormetic-like biphasic dose–response relationships have been widely documented in the immunological literature (Calabrese 2005). At low concentrations, reactive oxygen species would modulate cellular communication, while at high concentrations they would damage host tissues including immune cells, respectively. On the other hand, low concentrations of certain molecules involved in the inflammatory response may stimulate the generation of superoxide free radicals by macrophages, while high concentrations of the same molecules may inhibit the generation of superoxide (Khan and Rai 2007).

Relying on hormetic models to infer about the connections between immune function and oxidative stress might provide valuable insight about the conditions that make immune responses beneficial or costly, or the intensities of pathogen (co) infections that could be tolerated by the organism. Many of these immunological hormetic responses were detected using methodological approaches that ecophysiologists use to stimulate an immune response, such as the activating agents phytohaemagglutinin and lipopolysaccharide, and the antibody-forming response to sheep red blood cells (Calabrese 2005). Thus, the extent to which an hormetic response is triggered by these methods might influence the generation rate of reactive oxygen species, the antioxidant response, the oxidative damage, and the potential trade-offs with the immune response (Costantini and Dell'Omo 2006; Costantini and Møller 2009; McClure et al. 2014).

Dietary antioxidants can also show biphasic effects on the organismal immune function. For example, the administration of carotenoids to captive zebra finches increased their immune responsiveness (Blount et al. 2003; McGraw and Ardia 2003). Meta-analytical evidence on birds also showed that the intensity of the immune response to the subcutaneous injection of the antigen phytohaemagglutinin is significantly and positively associated with the concentration of carotenoids in blood (Simons et al. 2012). Other studies, however, found that carotenoids may also have toxic effects at high concentrations (Costantini et al. 2007; Huggins et al. 2010). Similarly, polyphenols can cause stimulating or detrimental effects on the organism. For example, they had hormetic effects on human immune cell function (Fig. 2.10) (Falchetti et al. 2001), ameliorated neuronal markers of inflammation

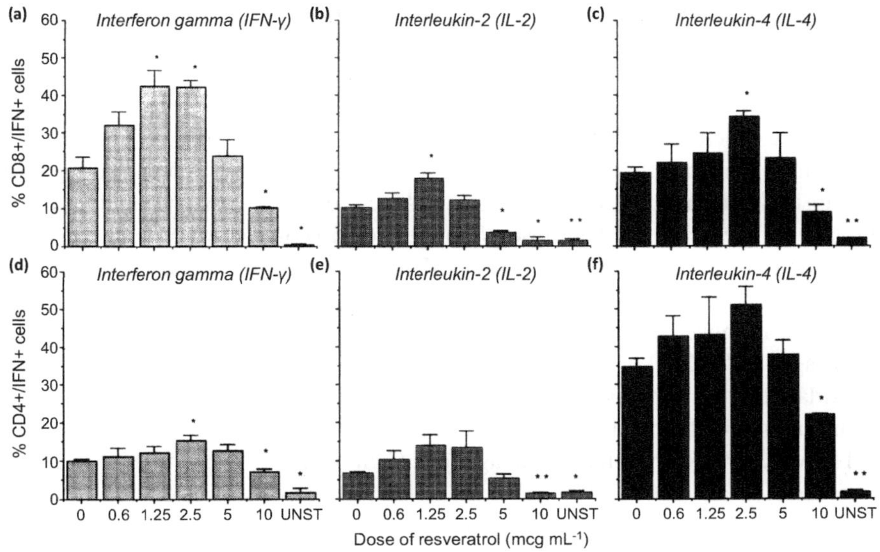


**Fig. 2.10** Effects of resveratrol on the frequency of CD8+ T cells expressing (**a**) interferon gamma (IFN-γ), (**b**) interleukin-2 (IL2), or (**c**) interleukin-4 (IL4), and on the frequency of CD4+ T cells expressing (**d**) IFN-γ, (e) IL2, or (**f**) IL4 in anti-CD3/anti-CD28-stimulated peripheral blood mononuclear cells, respectively. The threshold for detection of positive cytokine-producing cells was set at 1% above the negative control consisting of cells treated with an isotype-specific antibody. UNST = unstimulated controls. Asterisks indicate significant differences between groups: * $P < 0.05$ vs. UNST as reference group; ** $P < 0.01$ vs. UNST as reference group. Values are shown as mean and standard error. Reprinted with minor modifications from Falchetti et al. (2001) with permission of Elsevier

and oxidative stress in rats (Shukitt-Hale et al. 2005), and increased the likelihood of blackcaps *Sylvia atricapilla* to mount a humoral immune response after being injected with sheep red blood cells compared to control blackcaps (Catoni et al. 2008).

We have also evidence that certain environmental abiotic factors can induce hormetic effects on the immune function. For example, Carter et al. (2013) found an hormetic effect of cold stress on wound healing in ducklings of the wild wood duck *Aix sponsa*. Finally, Ramegowda et al. (2013) found that exposure to drought may enhance tolerance to subsequent pathogen infection in some organisms, and this priming effect seems to be connected to an increase of the generation rate of reactive oxygen species during the drought period.

## 2.7 Conclusions and Future Directions

Studies on humans and laboratory rodents provide robust evidence for a connection between immune function and oxidative status. Inferring from the patterns observed in these studies, it is reasonable to predict that the activation of the immune response would generally cause alterations of the cellular oxidative status in animals. This book chapter showed that an immune response owing to either antigen injection into the body or exposure to a pathogenic agent (e.g. virus, fungus) cause changes in oxidative status in animals. However, there is substantial variation in the direction and strength of effects across studies, indicating that an immune response does not always cause detectable changes to the organismal oxidative status.

We have evidence that some of the heterogeneity in the direction and strength of effects is determined by methodological aspects of study designs, including (i) immune marker, (ii) type and dose of antigen, or (iii) mechanism of action of the pathogen (Keil et al. 2001; Adamo 2004; Martin et al. 2011; Pedersen and Babayan 2011; Costantini 2022), (iv) intensity of the generation of reactive oxygen species during an immune response (e.g. higher in mammals than in birds; Penniall and Spitznagel 1975; Conlon et al. 1991; Stabler et al. 1994; Harmon 1998), (v) marker of oxidative status and tissue analysed (Costantini 2022), (vi) time elapsed between the initiation of inflammatory response and the occurrence of detectable costs (Sorci and Faivre 2009), or (vii) individual age (e.g. Cichoń et al. 2003; Gruver et al. 2007; Torres and Velando 2007). For example, Jurk et al. (2014) suggested that systemic chronic inflammation can accelerate senescence via ROS-mediated exacerbation of telomere dysfunction. Thus, it may take time before a significant fitness cost of the immune response can be recorded.

We need more studies that address the role of hormones in the interplay between immune function and oxidative status. For example, many studies found that sexual hormones and glucocorticoids have significant effects on both the immune function and the cellular oxidative status (e.g. Kurtz et al. 2007; Torres and Velando 2007; Costantini et al. 2011; Casagrande et al. 2012). Glucocorticoids are also known for their capacity to suppress the inflammatory response; however, we have also evidence for stimulatory effects, such as enhancement of thymocyte proliferation (Whitfield et al. 1970) and cytokine production (Stanulis et al. 1997a, b). Some studies also found biphasic (hormetic) effects of glucocorticoids on organismal resistance to infections (see review in Calabrese 2005). Glucocorticoids can also regulate the activities of certain antioxidant enzymes (e.g. superoxide dismutase, glutathione peroxidase) and the generation of hydrogen peroxide of macrophages (Pereira et al. 1995).

As more studies take place, across a broader range of taxa, it will be possible to identify the factors that mediate the cross-talk between oxidative status and immune function. This information will provide novel insight into the role of immune response-induced oxidative stress in driving life-history strategies, and the ultimate factors that would determine the heterogeneity among individuals and species in the oxidative consequences of an immune response.

## References

Abasubong KP, Liu WB, Zhang DD, Yuan XY, Xia SL, Xu C, Li XF (2018) Fishmeal replacement by rice protein concentrate with xylooligosaccharides supplement benefits the growth performance, antioxidant capability and immune responses against *Aeromonas hydrophila* in blunt snout bream (*Megalobrama amblycephala*). Fish Shellfish Immunol 78:177–186

Abdelkhalek NK, Risha E, El-Adl MA, Salama MF, Dawood MAO (2020) Antibacterial and antioxidant activity of clove oil against *Streptococcus iniae* infection in Nile tilapia (*Oreochromis niloticus*) and its effect on hepatic hepcidin expression. Fish Shellfish Immunol 104:478–488

Adamo SA (2004) How should behavioural ecologists interpret measurements of immunity? Anim Behav 68:1443–1449

Adeyemi JA (2014) Oxidative stress and antioxidant enzymes activities in the African catfish, *Clarias gariepinus*, experimentally challenged with *Escherichia coli* and *Vibrio fischeri*. Fish Physiol Biochem 40:347–354

Akinsanya B, Isibor PO, Onadeko B, Tinuade AA (2020) Impacts of trace metals on African common toad, Amietophrynus regularis (Reuss, 1833) and depuration effects of the toad's enteric parasite, *Amplicaecum africanum* (Taylor, 1924) sampled within Lagos metropolis, Nigeria. Heliyon 6:e03570

Alaasam VJ, Keehn JE, Durso AM, French SS, Feldman CR (2021) Ectoparasite load is reduced in side-blotched lizards (*Uta stansburiana*) at wind farms: implications for oxidative stress. Physiol Biochem Zool 94:35–49

Alegria TGP, Meireles DA, Cussiol JRR, Hugo M, Trujillo M, de Oliveira MA, Miyamoto S, Queiroz RF, Valadares NF, Garratt RC, Radi R, Di Mascio P, Augusto O, Netto LES (2017) Ohr plays a central role in bacterial responses against fatty acid hydroperoxides and peroxynitrite. Proc Natl Acad Sci USA 114:E132–E141

Ali EH, Hashem M, Al-Salahy MB (2011) Pathogenicity and oxidative stress in Nile tilapia caused by *Aphanomyces laevis* and *Phoma herbarum* isolated from farmed fish. Diseas Aquat Organis 94:17–28

Allaam MA, Allam TS, Elkhatam AO (2014) Biochemical and circulating oxidative stress biomarkers in Egyptian buffaloes (*Bubalus bubalis*) infested by sarcoptic mange. Global Veter 13:656–661

Almeida BFM, Narciso LG, Melo LM, Preve PP, Bosco AM, Lima VMF, Ciarlini PC (2013) Leishmaniasis causes oxidative stress and alteration of oxidative metabolism and viability of neutrophils in dogs. Vet J 198:599–605

Alonso-Alvarez C, Bertrand S, Devevey G, Gaillard M, Prost J, Faivre B, Sorci G (2004) An experimental test of the dose-dependent effect of carotenoids and immune activation on sexual signals and antioxidant activity. Am Nat 164:651–659

Amin YA, Noseer EA, Fouad SS, Ali RA, Mahmoud H (2020) Changes of reproductive indices of the testis due to *Trypanosoma evansi* infection in dromedary bulls (*Camelus dromedarius*): semen picture, hormonal profile, histopathology, oxidative parameters, and hematobiochemical profile. J Advan Vet Anim Res 7:537–545

Angajala A, Lim S, Phillips JB, Kim JH, Yates C, You Z, Tan M (2018) Diverse roles of mitochondria in immune responses: novel insights into immuno-metabolism. Front Immunol 9:1605

Apel K, Hirt H (2004) Reactive oxygen species: metabolism, oxidative stress, and signal transduction. Annu Rev Plant Biol 55:373–399

Araźna M, Pruchniak MP, Demkow U (2015) Reactive oxygen species, granulocytes, and NETosis. Adv Exp Med Biol 836:1–7

Armour EM, Bruner TL, Hines JK, Butler MW (2020) Low-dose immune challenges result in detectable levels of oxidative damage. J Exp Biol 223:jeb220095

Asghar M, Hasselquist D, Zehtindjiev P, Westerdahl H, Bensch S (2015) Hidden costs of infection: chronic malaria accelerates telomere degradation and senescence in wild birds. Science 347:436–438

Azimzadeh K, Amniattalab A (2017) Total sialic acid, oxidative stress and histopathological changes in rainbow trout saprolegniasis (*Oncorhynchus mykiss*). Kafkas Universitesi Veteriner Fakultesi Dergisi 23:55–62

Azma F, Razavi SM, Nazifi S, Rakhshandehroo E, Sanati AR (2015) A study on the status of inflammatory systems in camels naturally infected with *Toxoplasma gondii*. Trop Anim Health Prod 47:909–914

Babior BM (1999) NADPH oxidase: an update. Blood 93:1464–1476

Babior BM (2004) NADPH oxidase. Curr Opin Immunol 16:42–47

Babior BM, Kipnes RS, Curnutte JT (1973) Biological defense mechanisms. The production by leukocytes of superoxide, a potential bactericidal agent. J Clin Invest 52:741–744

Balasch JC, Vargas R, Brandts I, Tvarijonaviciute A, Reyes-López F, Tort L, Teles M (2019) Divergent personalities influence the myogenic regulatory genes myostatin, myogenin and ghr2 transcript responses to *Vibrio anguillarum* vaccination in fish fingerlings (*Sparus aurata*). Physiol Behav 212:112697

Baldissera MD, Souza CF, Moreira KLS, da Rocha M, da Veiga ML, Baldisserotto B (2017) *Melaleuca alternifolia* essential oil prevents oxidative stress and ameliorates the antioxidant system in the liver of silver catfish (*Rhamdia quelen*) naturally infected with *Ichthyophthirius multifiliis*. Aquacult 480:11–16

Baldissera MD, Souza CF, Descovi SN, Verdi CM, Zeppenfeld CC, da Silva AS, Santos RCV, Baldisserotto B (2019) Grape pomace flour ameliorates *Pseudomonas aeruginosa*-induced bioenergetic dysfunction in gills of grass carp. Aquacult 506:359–366

Baumann L (1989) The capacity of *Ureaplasma urealyticum*, *Mycoplasma hominis* and seven other *Mycoplasma* species for hemadsorption, sperm adsorption, hemolysis and peroxide formation. Arch Exp Vetmed 43:789–800

Baylor JL, Butler MW (2019) Immune challenge-induced oxidative damage may be mitigated by biliverdin. J Exp Biol 222:jeb200055

Beck MA (1997) Rapid genomic evolution of a non-virulent coxsackievirus B3 in mice due to vitamin E or selenium-deficient mice. Biomed Environ Sci 10:307–315

Belló ARR, Fortes E, Bello-Klein A, Bello AA, Llesuy SF, Robaldo RB, Bianchini A (2000) Lipid peroxidation induced by *Clinostomum detruncatum* in muscle of the freshwater fish *Rhamdia quelen*. Dis Aquatic Organ 42:233–236

Bendich A (1989) Carotenoids and the immune response. J Nutr 119:112–115

Beri D, Balan B, Chaubey S, Subramaniam S, Surendra B, Tatu U (2017) A disrupted transsulphuration pathway results in accumulation of redox metabolites and induction of gametocytogenesis in malaria. Sci Rep 7:40213

Bertrand S, Criscuolo F, Faivre B, Sorci G (2006) Immune activation increases susceptibility to oxidative tissue damage in zebra finches. Funct Ecol 20:1022–1027

Blount JD, Metcalfe NB, Birkhead TR, Surai PF (2003) Carotenoid modulation of immune function and sexual attractiveness in zebra finches. Science 300:125–127

Boleza KA, Burnett LE, Burnett KG (2001) Hypercapnic hypoxia compromises bactericidal activity of fish anterior kidney cells against opportunistic environmental pathogens. Fish Shellfish Immunol 11:593–610

Bos N, Sundström L, Fuchs S, Freitak D (2015) Ants medicate to fight disease. Evol 69:2979–2984

Bosch SS, Kronenberger T, Meissner KA, Zimbres FM, Stegehake D, Izui NM, Schettert I, Liebau E, Wrenger C (2015) Oxidative stress control by apicomplexan parasites. Biomed Res Int 2015:351289

Brecchia G, Cardinali R, Mourvaki E, Collodel G, Moretti E, Dal Bosco A, Castellini C (2010) Short- and long-term effects of lipopolysaccharide-induced inflammation on rabbit sperm quality. Anim Reprod Sci 118:310–316

Brenot A, King KY, Janowiak B, Griffith O, Caparon MG (2004) Contribution of glutathione peroxidase to the virulence of *Streptococcus pyogenes*. Infect Immun 72:408–413

Brigelius-Flohe R (2006) Glutathione peroxidases and redox-regulated transcription factors. Biol Chem 387:1329–1335

Broeg K, Steinhagen D (2012) Attack and defense: reactive oxygen and nitrogen species in teleost fish immune response and the coevolved evasion of microbes and parasites. In: Abele D, Vázquez-Medina JP, Zenteno-Savín T (eds) Oxidative stress in aquatic ecosystems. Wiley-Blackwell, Hoboken, pp 247–260

Broggi J, Soriguer RC, Figuerola J (2016) Transgenerational effects enhance specific immune response in a wild passerine. Peer J 4:e1766

Brune K, Leffell MS, Spitznagel JK (1972) Microbicidal activity of peroxidaseless chicken heterophile leukocytes. Infect Immun 5:283–287

Butler MW, Stierhoff EN, Carpenetti JM, Bertone MA, Addesso AM, Knutie SA (2021) Oxidative damage increases with degree of simulated bacterial infection, but not ectoparasitism, in tree swallow nestlings. J Exp Biol 224:jeb243116

Cadenas S, Rojas C, Barja G (1998) Endotoxin increases oxidative injury to proteins in Guinea pig liver: protection by dietary vitamin C. Pharmacol Toxicol 82:11–18

Calabrese EJ (2005) Hormetic dose-response relationships in immunology: occurrence, quantitative features of the dose response, mechanistic foundations, and clinical implications. Crit Rev Toxic 35:89–295

Camini FC, da Silva Caetano CC, Almeida LT, de Brito Magalhães CL (2017) Implications of oxidative stress on viral pathogenesis. Arch Virol 162:907–917

Carter AW, DuRant SE, Hepp GR, Hopkins WA (2013) Thermal challenge severity differentially influences wound healing in Wood duck (*Aix sponsa*) ducklings. J Exp Zool 319:422–429

Casagrande S, Costantini D, Groothuis TGG (2012) Interaction between sexual steroids and immune response in affecting oxidative status of birds. Comp Biochem Physiol Part A 163:296–301

Casasole G, Costantini D, Cichón M, Rutkowska J (2016) The effect of maternal immunization on female oxidative status, yolk antioxidants and offspring survival in a songbird. Comp Biochem Physiol Part A 194:56–61

Case AJ, McGill JL, Tygrett LT, Shirasawa T, Spitz DR, Waldschmidt TJ, Legge KL, Domann FE (2011) Elevated mitochondrial superoxide disrupts normal T cell development, impairing adaptive immune responses to an influenza challenge. Free Radic Biol Med 50:448–458

Catoni C, Schaffer HM, Peters A (2008) Fruit for health: the effect of flavonoids on humoral immune response and food selection in a frugivorous bird. Funct Ecol 22:649–645

Charoensuk L, Pinlaor P, Prakobwong S, Hiraku Y, Laothong U, Ruangjirachuporn W, Yongvanit P, Pinlaor S (2011) Curcumin induces a nuclear factor-erythroid 2-related factor 2-driven response against oxidative and nitrative stress after praziquantel treatment in liver fluke-infected hamsters. Inter J Parasit 41:615–626

Charpentier MJE, Boundenga L, Beaulieu M, Dibakou SE, Arnathau C, Sidobre C, Willaume E, Mercier-Delarue S, Simon F, Rougeron V, Prugnolle FA (2019) A longitudinal molecular study of the ecology of malaria infections in free-ranging mandrills. Inter J Parasit 10:241–251

Chen JJ, Liu NN, Zhang HJ, Zhao YD, Cao XL (2020) The effects of Aeromonas hydrophila infection on oxidative stress, nonspecific immunity, autophagy, and apoptosis in the common carp. Devel Comp Immunol 105:103587

Chew BP (1993) Role of carotenoids in the immune response. J Dairy Sci 76:2804–2811

Christie P, Glaizot O, Strepparava N, Devevey G, Fumagalli L (2012) Twofold cost of reproduction: an increase in parental effort leads to higher malarial parasitaemia and to a decrease in resistance to oxidative stress. Proc R Soc Lond B 279:1142–1149

Cichoń M, Sendecka J, Gustafsson L (2003) Age-related decline in humoral immune function in collared flycatchers. J Evol Biol 16:1205–1210

Conlon P, Smith D, Gowlett T (1991) Oxygen radical production by avian leukocytes. Can J Vet Res 55:193–195

Costantini D (2008) Oxidative stress in ecology and evolution: lessons from avian studies. Ecol Lett 11:1238–1251

Costantini D (2014) Oxidative stress and hormesis in evolutionary ecology and physiology: a marriage between mechanistic and evolutionary approaches. Springer-Verlag, Berlin, p 348

Costantini D (2022) A meta-analysis of impacts of immune response and infection on oxidative status in vertebrates. Cons Physiol 10:coac018

Costantini D, Dell'Omo G (2006) Effects of T-cell-mediated immune response on avian oxidative stress. Comp Biochem Physiol Part A 145:137–142

Costantini D, Møller AP (2008) Carotenoids are minor antioxidants for birds. Funct Ecol 22:367–370

Costantini D, Møller AP (2009) Does immune response cause oxidative stress? A meta-analysis. Comp Biochem Physiol Part A 153:339–344

Costantini D, Coluzza C, Fanfani A, Dell'Omo G (2007) Effects of carotenoid supplementation on colour expression, oxidative stress and body mass in rehabilitated captive adult kestrels (*Falco tinnunculus*). J Comp Physiol B 177:723–731

Costantini D, Marasco V, Møller AP (2011) A meta-analysis of glucocorticoids as modulators of oxidative stress in vertebrates. J Comp Physiol B 181:447–456

Costantini D, Casagrande S, Casasole G, AbdElgawad H, Asard H, Pinxten R, Eens M (2015a) Immunization reduces vocal communication but does not increase oxidative stress in a songbird species. Behav Ecol Sociobiol 69:829–839

Costantini D, Goutte A, Barbraud C, Faivre B, Sorci G, Weimerskirch H, Delord K, Chastel O (2015b) Demographic responses to oxidative stress and inflammation in the wandering albatross (*Diomedea exulans*). PLoS One 10:e0133967

Costantini D, Seeber PA, Soilemetzidou SE, Azab W, Bohner J, Buuveibaatar B, Czirják GÁ, East ML, Greunz EM, Kaczensky P, Lamglait B, Melzheimer J, Uiseb K, Ortega A, Osterrieder N, Sandgreen DM, Simon M, Walzer C, Greenwood AD (2018) Physiological costs of infection: herpesvirus replication is linked to blood oxidative stress in equids. Sci Rep 8:10347

Costantini D, Weinberg M, Jordán L, Moreno KR, Yovel Y, Czirják GÁ (2022) Induced bacterial sickness causes inflammation but not blood oxidative stress in Egyptian fruit bats (*Rousettus aegyptiacus*). Conserv Physiol 10:coac028

Costantini D, Schad J, Czirjak GA, Voigt CC (2023) Oxidative damage varies in response to bacterial, fungal and viral antigen challenges in bats. J Exp Biol 226:jeb246332

Cram DL, Blount JD, York JE, Young AJ (2015) Immune response in a wild bird is predicted by oxidative status, but does not cause oxidative stress. PLoS One 10:e0122421

da Rosa VM, Ariotti K, Bressan CA, da Silva EG, Dallaporta M, Junior GB, da Costa ST, de Vargas AC, Baldisserotto B, Finamor IA, Pavanato MA (2019) Dietary addition of rutin impairs inflammatory response and protects muscle of silver catfish (*Rhamdia quelen*) from apoptosis and oxidative stress in Aeromonas hydrophila-induced infection. Comp Biochem Physiol Part C 226:108611

da Silva CC, Klein RD, Barcarolli IF, Bianchini A (2016) Metal contamination as a possible etiology of fibropapillomatosis in juvenile female green sea turtles *Chelonia mydas* from the southern Atlantic Ocean. Aquat Toxicol 170:42–51

Dautremepuits C, Betoulle S, Vernet G (2003) Stimulation of antioxidant enzymes levels in carp (*Cyprinus carpio* L.) infected by *Ptychobothrium* sp. (Cestoda). Fish Shellfish Immunol 15:467–471

De Angeli DD, Silveira P, Ramos JA, Sousa JP, Braga ÉM, Norte AC (2017) Haemosporidian infections affect antioxidant defences in great tits *Parus major* but are not related to exposure to aerial pollutants. Parasitol Open 3:e4

De Coster G, De Neve L, Martín-Gálvez D, Therry L, Lens L (2010) Variation in innate immunity in relation to ectoparasite load, age and season: a field experiment in great tits (*Parus major*). J Exp Biol 231:3012–3018

De Coster G, De Neve L, Verhulst S, Lens L (2012) Maternal effects reduce oxidative stress in female nestlings under high parasite load. J Avian Biol 43:177–185

De Groote MA, Ochsner UA, Shiloh MU, Nathan C, McCord JM, Dinauer MC, Libby SJ, Vazquez-Torres A, Xu Y, Fang FC (1997) Periplasmic superoxide dismutase protects *Salmonella* from products of phagocyte NADPH-oxidase and nitric oxide synthase. Proc Natl Acad Sci USA 94:13997–14001

De Luca C, Kharaeva Z, Raskovic D, Pastore P, Luci A, Korkina L (2012) Coenzyme Q10, vitamin E, selenium, and methionine in the treatment of chronic recurrent viral mucocutaneous infections. Nutrition 28:509–514

Delhaye J, Jenkins T, Christe P (2016) *Plasmodium* infection and oxidative status in breeding great tits, *Parus major*. Malar J 15:531

Delhaye J, Glaizot O, Christe P (2018) The effect of dietary antioxidant supplementation in a vertebrate host on the infection dynamics and transmission of avian malaria to the vector. Parasitol Res 117:2043–2052

Demas GE, Chefer V, Talan MI, Nelson RJ (1997) Metabolic costs of mounting an antigen-stimulated immune response in adult and aged C57BL/6J mice. Am J Phys 273:1631–1637

Demas GE, Drazen DL, Nelson RJ (2003) Reductions in total body fat decrease humoral immunity. Proc R Soc Lond B 270:905–911

Devevey G, Bize P, Fournier S, Person E, Christe P (2010) Testing the predictive adaptive response in a host-parasite system. Funct Ecol 24:178–185

Diotallevi M, Checconi P, Palamara AT, Celestino I, Coppo L, Holmgren A, Abbas K, Peyrot F, Mengozzi M, Ghezzi P (2017) Glutathione fine-tunes the innate immune response toward antiviral pathways in a macrophage cell line independently of its antioxidant properties. Front Immunol 8:1239

Dkhil MA, Abdel-Baki AS, Al-Quraishy S, Abdel-Moneim AE (2013) Hepatic oxidative stress in Mongolian gerbils experimentally infected with *Babesia divergens*. Ticks Tick-Borne Dis 4:346–351

Dokka S, Shi X, Leonard S, Wang L, Castranova V, Rojanasakul Y (2001) Interleukin-10-mediated inhibition of free radical generation in macrophages. Am J Physiol Lung Cell Mol Physiol 280:L1196–L1202

Donaghy L, Lambert C, Choi K-S, Soudant P (2009) Hemocytes of the carpet shell clam (*Ruditapes decussatus*) and the Manila clam (*Ruditapes philippinarum*): current knowledge and future prospects. Aquacult 297:10–24

Du YS, Yi MM, Xiao P, Meng LJ, Li X, Sun GX, Liu Y (2015) The impact of *Aeromonas salmonicida* infection on innate immune parameters of Atlantic salmon (*Salmo salar*). Fish Shellfish Immunol 44:307–315

Dukan S, Touati D (1996) Hypochlorous acid stress in *Escherichia coli*: resistance, DNA damage, and comparison with hydrogen peroxide stress. J Bacteriol 178:6145–6150

El-Son MAM, Elbahnaswy S, Ibrahim I (2020) Molecular and histopathological characterization of *Photobacterium damselae* in naturally and experimentally infected Nile tilapia (*Oreochromis niloticus*). J Fish Dis 43:1505–1517

Espinal-Carrion T, Lopez-Lopez E (2010) Helminths and lipid peroxidation in *Astyanax aeneus* (Pisces: Characidae) from a river in the humid subtropics of southeastern Mexico. Dis Aquat Org 88:215–224

Ezraty B, Gennaris A, Barras F, Collet JF (2017) Oxidative stress, protein damage and repair in bacteria. Nat Rev Microbiol 15:385–396

Faith SA, Sweet TJ, Bailey E, Booth T, Docherty JJ (2006) Resveratrol suppresses nuclear factor-[kappa] B in herpes simplex virus infected cells. Antivir Res 72:242–251

Falchetti R, Fuggetta MP, Lanzilli G, Tricarico M, Ravagnan G (2001) Effects of resveratrol on human immune cell function. Life Sci 70:81–96

Finotello R, Pasquini A, Meucci V, Lippi I, Rota A, Guidi G, Marchetti V (2014) Redox status evaluation in dogs affected by mast cell tumour. Vet Comp Oncol 12:120–129

Fitzpatrick AM, Teague WG, Burwell L, Brown MS, Brown LA (2011) Glutathione oxidation is associated with airway macrophage functional impairment in children with severe asthma. Pediatr Res 69:154–159

Fokam D (2020) Instrumental role for reactive oxygen species in the inflammatory response. Front Biosci 25:1110–1119

Foo J, Bellot G, Pervaiz S, Alonso S (2022) Mitochondria-mediated oxidative stress during viral infection. Trends Microbiol 30:679–692

Fritze M, Costantini D, Fickel J, Wehner D, Czirjak GA, Voigt C (2019) Immune response of hibernating European bats to a fungal challenge. Biol Open 8:bio046078

Fritze M, Puechmaille SJ, Costantini D, Fickel J, Voigt CC, Czirjak GA (2021) Determinants of defence strategies of a hibernating European bat species towards the fungal pathogen *Pseudogymnoascus destructans*. Devel Comp Immunol 119:104017

Garcia LO, Becker AG, Bertuzzi T, Cunha MA, Kochhann D, Finamor IA, Riffel APK, Llesuy S, Pavanato MA, Baldisserotto B (2011) Oxidative stress parameters in silver catfish (*Rhamdia quelen*) juveniles infected with *Ichthyophthirius multifiliis* and maintained at different levels of water pH. Vet Parasitol 178:15–21

Gasso D, Vicente J, Mentaberre G, Soriguer R, Rodriguez RJ, Navarro-Gonzalez N, Tvarijonaviciute A, Lavin S, Fernandez-Llario P, Segales J, Serrano E (2016) Oxidative stress in wild boars naturally and experimentally infected with *Mycobacterium bovis*. PLoS One 11:e0163971

Genard B, Miner P, Nicolas J-L, Moraga D, Boudry P, Pernet F, Tremblay R (2013) Integrative study of physiological changes associated with bacterial infection in pacific oyster larvae. PLoS One 8:e64534

Geng RJ, Jia YY, Chi ML, Wang ZG, Liu H, Wang WM (2019) Rnase1 alleviates the *Aeromonas hydrophila*-induced oxidative stress in blunt snout bream. Develop Comp Immunol 91:8–16

Gennaris A, Ezraty B, Henry C, Agrebi R, Vergnes A, Oheix E, Bos J, Leverrier P, Espinosa L, Szewczyk J, Vertommen D, Iranzo O, Collet J-F, Barras F (2015) Repairing oxidized proteins in the bacterial envelope using respiratory chain electrons. Nature 528:409–412

Ghirardi R, Cazenave J, Lopez JA, Antoniazzi CE, Perotti MG (2018) Evaluation of stress responses to water molds in embryos of *Physalaemus albonotatus* (Anura: Leptodactylidae). Herpetol Conserv Biol 13:216–223

Ghirardi R, Cazenave J, Lopez JA, Antoniazzi CE, Perotti MG (2020) Water mould exposure induces enzymatic antioxidant defences in embryos of the two-colored oval frog (*Elachistocleis bicolor*) (Anura: Microhylidae). Can J Zool 98:411–416

Giraudeau M, Mousel M, Earl S, McGraw K (2014) Parasites in the city: degree of urbanization predicts poxvirus and coccidian infections in house finches (*Haemorhous mexicanus*). PLoS One 9:e86747

Giri SS, Kim HJ, Kim SG, Kim SW, Kwon J, Lee SB, Sukumaran V, Park SC (2020) Effectiveness of the guava leaf extracts against lipopolysaccharide-induced oxidative stress and immune responses in *Cyprinus carpio*. Fish Shellfish Immunol 105:164–176

Gobert AP, McGee DJ, Akhtar M, Mendz GL, Newton JC, Cheng Y, Mobley HLT, Wilson KT (2001) *Helicobacter pylori* arginase inhibits nitric oxide production by eukaryotic cells: a strategy for bacterial survival. Proc Natl Acad Sci USA 98:13844–13849

Gore AB, Qureshi MA (1997) Enhancement of humoral and cellular immunity by vitamin E after embryonic exposure. Poult Sci 76:984–991

Graham AL, Shuker DM, Pollitt LC, Auld S, Wilson AJ, Little TJ (2011) Fitness consequences of immune responses: strengthening the empirical framework for ecoimmunology. Funct Ecol 25:5–17

Gruver AL, Hudson LL, Sempowski GD (2007) Immunosenescence of ageing. J Pathol 211:144–156

Gu Y, Chen K, Xi B, Xie J, Bing X (2022) Paeonol increases the antioxidant and anti-inflammatory capacity of gibel carp (*Carassius auratus gibelio*) challenged with *Aeromonas hydrophila*. Fish Shellfish Immunol 123:479–488

Guzmán-Villanueva LT, Ascencio-Valle F, Macías-Rodríguez ME, Tovar-Ramírez D (2014) Effects of dietary β-1,3/1,6-glucan on the antioxidant and digestive enzyme activities of pacific red snapper (*Lutjanus peru*) after exposure to lipopolysaccharides. Fish Physiol Biochem 40:827–837

Halliwell BH (2006) Phagocyte-derived reactive species: salvation or suicide? Trends Biochem Sci 31:509–515

Halliwell BH, Gutteridge JMC (2015) Free radicals in biology and medicine, 5th edn. Oxford University Press, Oxford

Hampton MB, Kettle AJ, Winterboum CC (1998) Inside the neutrophil phagosome: oxidants, myeloperoxidase, and bacterial killing. Blood 92:3007–3017

Hanssen SA, Bustnes JO, Schnug L, Bourgeon S, Johnsen TV, Ballesteros M, Sonne C, Herzke D, Eulaers I, Jaspers VL, Covaci A, Eens M, Halley DJ, Moum T, Ims RA, Erikstad KE (2013) Antiparasite treatments reduce humoral immunity and impact oxidative status in raptor nestlings. Ecol Evol 3:5157–5166

Harikrishnan R, Thamizharasan S, Devi G, Van Doan H, Kumar TTA, Hoseinifar SH, Balasundaram C (2020) Dried lemon peel enriched diet improves antioxidant activity, immune response and modulates immuno-antioxidant genes in *Labeo rohita* against *Aeromonas sorbia*. Fish Shellfish Immunol 106:675–684

Harmon BG (1998) Avian heterophils in inflammation and disease resistance. Poultry Sci 77:972–977

Hartley RC, Kennedy MW (2004) Are carotenoids a red herring in sexual display? Trends Ecol Evol 19:353–354

Hassan HM, Fridovich I (1979) Paraquat and *Escherichia coli*. Mechanism of production of extracellular superoxide radical. J Biol Chem 254:10846–10852

Hau M, Haussmann MF, Greives TJ, Matlack C, Costantini D, Quetting M, Adelman JS, Miranda AC, Partecke J (2015) Repeated stressors in adulthood increase the rate of biological ageing. Front Zool 12:4

Heeb P, Werner I, Kölliker M, Richner H (1998) Benefits of induced host responses against an ectoparasite. Proc R Soc Lond B 265:51–56

Hégaret H, Wikfors GH, Soudant P (2003) Flow cytometric analysis of haemocytes from eastern oysters, *Crassostrea virginica*, subjected to a sudden temperature elevation: II. Haemocyte functions: aggregation, viability, phagocytosis, and respiratory burst. J Exp Mar Biol Ecol 293:249–265

Heidarpour M, Mohri M, Borji H, Moghdass E (2012) Oxidative stress and trace elements in camel (*Camelus dromedarius*) with liver cystic echinococcosis. Vet Parasitol 187:459–463

Hernández JA, Rubio M, Olmos E, Ros-Barceló A, Martínez-Gómez P (2004) Oxidative stress induced by long-term plum pox virus infection in peach (*Prunus persica*). Physiol Plant 122:486–495

Hernandez EP, Anisuzzaman AMA, Kawada H, Kwofie KD, Ladzekpo D, Koike Y, Inoue T, Sasaki S, Mikami F, Matsubayashi M, Tanaka T, Tsuji N, Hatta T (2022) Ambivalent roles of oxidative stress in triangular relationships among arthropod vectors, pathogens and hosts. Antioxidants 11:1254

Hernández-Arciga U, Herrera LG, Konigsberg M, Valdez RA, Flores-Martinez JJ, Romano MC (2020) Synergetic effects of immune challenge and stress depress cortisol, inflammatory response and antioxidant activity in fish-eating Myotis. J Exp Biol 223:jeb234914

Hoogenboom MO, Metcalfe NB, Groothuis TGG, de Vries B, Costantini D (2012) Relationship between oxidative stress and circulating testosterone and cortisol in pre-spawning female brown trout. Comp Biochem Physiol Part A 163:379–387

Hõrak P, Zilmer M, Saks L, Ots I, Karu U, Zilmer K (2006) Antioxidant protection, carotenoids and the costs of immune challenge in greenfinches. J Exp Biol 209:4329–4338

Hõrak P, Saks L, Zilmer M, Karu U, Zilmer K (2007) Do dietary antioxidants alleviate the cost of immune activation? An experiment with greenfinches. Am Nat 170:625–635

Hornef MW, Wick MJ, Rhen M, Normark S (2002) Bacterial strategies for overcoming host innate and adaptive immune responses. Nat Immunol 3:1033–1040

Hudson SB, Virgin EE, Kepas ME, French SS (2021) Energy expenditure across immune challenge severities in a lizard: consequences for innate immunity, locomotor performance and oxidative status. J Exp Biol 224:jeb242608

Huggins KA, Navara KJ, Mendonça MT, Hill GE (2010) Detrimental effects of carotenoid pigments: the dark side of bright coloration. Naturwissenschaften 97:637–644

Isaksson C, Sepil I, Baramidze V, Sheldon BC (2013) Explaining variance of avian malaria infection in the wild: the importance of host density, habitat, individual life-history and oxidative stress. BMC Ecol 13:15

Iseri VJ, Klasing KC (2014) Changes in the amount of lysine in protective proteins and immune cells after a systemic response to dead Escherichia coli: implications for the nutritional costs of immunity. Integr Comp Biol 54:922–930

Ishibe K, Osatomi K, Hara K, Kanai K, Yamaguchi K, Oda T (2008) Comparison of the responses of peritoneal macrophages from Japanese flounder (*Paralichthys olivaceus*) against high virulent and low virulent strains of Edwardsiella tarda. Fish Shellfish Immunol 24:243–251

Ito T, Kera A, Murata H, Yoshida T, Sakai T, Yamauchi K, Yamasaki Y, Yamaguchi T, Ukawa M (2000) Experimentally induced bacterial hemolytic jaundice of yellowtail and oxidative stress. Nippon Suisan Gakkaishi 66:50–54

Iyer GYN, Islam DMF, Quastel JH (1961) Biochemical aspects of phagocytosis. Nature 192:535–542

Jang KJ, Mano H, Aoki K, Hayashi T, Muto A, Nambu Y, Takahashi K, Itoh K, Taketani S, Nutt SL, Igarashi K, Shimizu A, Sugai M (2015) Mitochondrial function provides instructive signals for activation-induced B-cell fates. Nat Commun 6:6750

Jiang B, Wang J, Luo HL, Lu GL, Li YW, Li AX (2017) L-amino acid oxidase expression profile and biochemical responses of rabbitfish (*Siganus oramin*) after exposure to a high dose of *Cryptocaryon irritans*. Fish Shellfish Immunol 69:85–89

Jolly S, Jaffal A, Delahaut L, Palluel O, Porcher JM, Geffard A, Sanchez W, Betoulle S (2014) Effects of aluminium and bacterial lipopolysaccharide on oxidative stress and immune parameters in roach, *Rutilus rutilus* L. Environ Sci Pollut Res 21:13103–13117

Jurk D, Wilson C, Passos JF, Oakley F, Correia-Melo C, Greaves L, Saretzki G, Fox C, Lawless C, Anderson R, Hewitt G, Pender SL, Fullard N, Nelson G, Mann J, van de Sluis B, Mann DA, von Zglinicki T (2014) Chronic inflammation induces telomere dysfunction and accelerates ageing in mice. Nat Commun 2:4172

Kaleta EF (1990) Herpesviruses of birds—a review. Avian Pathol 19:193–211

Kavouras JH, Prandovszky E, Valyi-Nagy K, Krisztian Kovacs S, Tiwari V, Kovacs M, Shukla D, Valyi-Nagy T (2007) Herpes simplex virus type 1 infection induces oxidative stress and the release of bioactive lipid peroxidation by-products in mouse P19N neural cell cultures. J Neurovirol 13:416–425

Keil D, Luebke RW, Pruett SB (2001) Quantifying the relationship between multiple immunological and host resistance: probing the limits of reductionism. J Immunol 167:4543–4552

Khan UW, Rai U (2007) Differential effects of histamine on Leydig cell and testicular macrophage activities in wall lizards: precise role of H1/H2 receptor subtypes. J Endocrinol 194:441–448

Kim HW, Chew BP, Wong TS, Park JS, Weng BBC, Byrne KM, Hayek MG, Reinhart GA (2000) Modulation of humoral and cell-mediated immune responses by dietary lutein in cats. Vet Immunol Immunopathol 73:331–341

Kim YS, Ke F, Zhang QY (2009) Effect of beta-glucan on activity of antioxidant enzymes and Mx gene expression in virus infected grass carp. Fish Shellfish Immunol 27:336–340

Kim JH, Sohn S, Kim SK, Hur YB (2020) Effects on hematological parameters, antioxidant and immune responses, ache, and stress indicators of olive flounders, *Paralichthys olivaceus*, raised in bio-floc and seawater challenged by *Edwardsiella tarda*. Fish Shellfish Immunol 97:194–203

Klasing KC (2004) The costs of immunity. Acta Zool Sin 50:961–969

Knowles SCL, Palinauskas V, Sheldon BC (2010) Chronic malaria infections increase family inequalities and reduce parental fitness: experimental evidence from wild bird population. J Evol Biol 23:557–569

Kong X, Qiao D, Zhao X, Wang L, Zhang J, Liu D, Zhang H (2017) The molecular characterizations of Cu/ZnSOD and MnSOD and its responses of mRNA expression and enzyme activity to *Aeromonas hydrophila* or lipopolysaccharide challenge in Qihe crucian carp *Carassius auratus*. Fish Shellfish Immunol 67:429–440

Kozlov AV, Szalay L, Umar F, Fink B, Kropik K, Nohl H, Redl H, Bahrami S (2003) Epr analysis reveals three tissues responding to endotoxin by increased formation of reactive oxygen and nitrogen species. Free Radic Biol Med 34:1555–1562

Kumar S, Raman RP, Prasad KP, Srivastava PP, Kumar S, Rajendran KV (2017) Modulation of innate immune responses and induction of oxidative stress biomarkers in *Pangasianodon hypophthalmus* following an experimental infection with Dactylogyrid monogeneans. Fish Shellfish Immunol 63:334–343

Kurhalyuk N, Tkachenko H (2011) Induction of oxidative stress and antioxidant defenses in the livers of sea trout, *Salmo trutta* L., with ulcerative dermal necrosis. Fish Aquat Life 19:229–240

Kurhalyuk N, Tkachenko H, Pałczyńska K (2009) Antioxidant enzymes profile in the brown trout (*Salmo trutta trutta*) with ulcerative dermal necrosis. Bull Vet Inst Pulawy 53:813–818

Kurhalyuk N, Tkachenko H, Palczyńska K (2010) Lipid peroxidation and antioxidant defense system in spawn of brown trout (*Salmo trutta* m. *trutta* L.) affected by ulcerative dermal necrosis. Arch Pol Fish 18:115–122

Kurhalyuk N, Tkachenko H, Palczynska K (2011) Resistance of erythrocytes from brown trout (*Salmo trutta* m. *trutta* L.) affected by ulcerative dermal necrosis syndrome. Polish J Vet Sci 14:443–448

Kurtz J, Wegner MK, Kalbe M, Reusch TBH, Schaschl H, Hasselquist D, Milinski M (2006) MHC genes and oxidative stress in sticklebacks: an immuno-ecological approach. Proc R Soc Lond B 273:1407–1414

Kurtz J, Kalbe M, Langefors Å, Mayer I, Milinski M, Hasselquist D (2007) An experimental test of the immunocompetence handicap hypothesis in a teleost fish: 11-ketotestosterone suppresses innate immunity in three-spined sticklebacks. Am Nat 170:509–519
Lambert C, Nicolas JL (1998) Specific inhibition of chemiluminescent activity by pathogenic vibrios in hemocytes of two marine bivalves: *Pecten maximus* and *Crassostrea gigas*. J Invert Pathol 71:53–63
Le Guernic A, Sanchez W, Palluel O, Bado-Nilles A, Floriani M, Turies C, Chadili E, Vedova CD, Cavalié I, Adam-Guillermin C, Porcher JM, Geffard A, Betoulle S, Gagnaire B (2016) Acclimation capacity of the three-spined stickleback (*Gasterosteus aculeatus*, L.) to a sudden biological stress following a polymetallic exposure. Ecotoxicology 25:1478–1499
Levy O (2004) Antimicrobial proteins and peptides: anti-infective molecules of mammalian leukocytes. J Leukoc Biol 76:909–925
Li MY, Guo WQ, Guo GL, Zhu XM, Niu XT, Shan XF, Tian JX, Wang GQ, Zhang DM (2020) Effects of dietary astaxanthin on lipopolysaccharide-induced oxidative stress, immune responses and glucocorticoid receptor (gr)-related gene expression in *Channa argus*. Aquacul 517:734816
Li W, Ren Q, Ni T, Zhao Y, Sang Z, Luo R, Li Z, Li S (2023) Strategies adopted by Salmonella to survive in host: a review. Arch Microbiol 205:362
Lilley TM, Stauffer J, Kanerva M, Eeva T (2014) Interspecific variation in redox status regulation and immune defence in five bat species: the role of ectoparasites. Oecologia 175:811–823
Liu GY, Essex A, Buchanan JT, Datta V, Hoffman HM, Bastian JF, Fierer J, Nizet V (2005) *Staphylococcus aureus* golden pigment impairs neutrophil killing and promotes virulence through its antioxidant activity. J Exp Med 202:209–215
Liu KF, Yeh MS, Kou GH, Cheng W, Lo CF (2010) Identification and cloning of a selenium-dependent glutathione peroxidase from tiger shrimp, Penaeus monodon, and its transcription following pathogen infection and related to the molt stages. Dev Comp Immunol 34:935–944
Liu C, Liu HK, Han D, Xie SQ, Jin JY, Yang YX, Zhu XM (2020) Effects of dietary *Arthrospira platensis* supplementation on the growth performance, antioxidation and immune related-gene expression in yellow catfish (*Pelteobagrus fulvidraco*). Aquacul Rep 17:100297
Lochmiller RL, Deerenberg C (2000) Trade-offs in evolutionary immunology: just what is the cost of immunity? Oikos 88:87–98
Lopez-Arrabe J, Cantarero A, Perez-Rodriguez L, Palma A, Alonso-Alvarez C, Gonzalez-Braojos S, Moreno J (2015) Nest-dwelling ectoparasites reduce antioxidant defences in females and nestlings of a passerine: a field experiment. Oecologia 179:29–41
Lorne E, Zmijewski JW, Zhao X, Liu G, Tsuruta Y, Park YJ, Dupont H, Abraham E (2008) Role of extracellular superoxide in neutrophil activation: interactions between xanthine oxidase and TLR4 induce proinflammatory cytokine production. Am J Physiol Cell Physiol 294:C985–C993
Losdat S, Richner H, Blount JD, Helfenstein F (2011) Immune activation reduces sperm quality in the great tit. PLoS One 6:e22221
Lynch RE, Cole BC (1980) *Mycoplasma pneumoniae*: a prokaryote which consumes oxygen and generates superoxide but which lacks superoxide dismutase. Biochem Biophys Res Commun 96:98–105
Maizels RM, Nussey DH (2013) Into the wild: digging at immunology's evolutionary roots. Nature Immunol 14:879–883
Malachowa N, Kohler PL, Schlievert PM, Chuang ON, Dunny GM, Kobayashi SD, Miedzobrodzki J, Bohach GA, Seo KS (2011) Characterization of a *Staphylococcus aureus* surface virulence factor that promotes resistance to oxidative killing and infectious endocarditis. Infect Immun 79:342–352
Marcogliese DJ, Brambilla LG, Gagne F, Gendron AD (2005) Joint effects of parasitism and pollution on oxidative stress biomarkers in yellow perch *Perca flavescens*. Dis Aquat Org 63:77–84
Marcogliese DJ, Dautremepuits C, Gendron AD, Fournier M (2010) Interactions between parasites and pollutants in yellow perch (*Perca flavescens*) in the st. Lawrence river, Canada: implications for resistance and tolerance to parasites. Can J Zool 88:247–258
Marcogliese DJ, King KC, Bates KA (2021) Effects of multiple stressors on northern leopard frogs in agricultural wetlands. Parasitology 148:827–834

Maronde L, Losdat S, Richner H (2018) Do parasites and antioxidant availability affect begging behaviour, growth rate and resistance to oxidative stress? J Evol Biol 31:904–913
Marri V, Richner H (2015) Immune response, oxidative stress and dietary antioxidants in great tit nestlings. Comp Biochem Physiol Part A 179:192–196
Martin LB, Scheuerlein A, Wikelski M (2003) Immune activity elevates energy expenditure of house sparrows: a link between direct and indirect costs? Proc R Soc Lond B 270:153–158
Martin LB, Weil ZM, Nelson RJ (2008) Seasonal changes in vertebrate immune activity: mediation by physiological trade-offs. Philos Trans R Soc B 363:321–339
Martin LB, Hawley DM, Ardia DR (2011) An introduction to ecological immunology. Funct Ecol 25:1–4
Mastroeni P, Vazquez-Torres A, Fang FC, Xu Y, Khan S, Hormaeche CE, Dougan G (2000) Antimicrobial actions of the NADPH phagocyte oxidase and inducible nitric oxide synthase in experimental salmonellosis. II. Effects on microbial proliferation and host survival *in vivo*. J Exp Med 192:237–247
Mathew SS, Bryant PW, Burch AD (2010) Accumulation of oxidized proteins in Herpesvirus infected cells. Free Radic Biol Med 49:383–391
McClure CD, Zhong W, Hunt VL, Chapman FM, Hill FV, Priest NK (2014) Hormesis results in trade-offs with immunity. Evolution 68:2225–2233
McGraw KJ, Ardia DR (2003) Carotenoids, immunocompetence, and the information content of sexual colors: an experimental test. Am Nat 162:704–712
Messina S, Edwards DP, Van Houtte N, Tomassi S, Benedick S, Eens M, Costantini D (2022) Impacts of selective logging on haemosporidian infection and physiological correlates in tropical birds. Int J Parasit 52:87–96
Molbert N, Alliot F, Leroux-Coyau M, Médoc V, Biard C, Meylan S, Jacquin SR, Goutte A (2020) Potential benefits of acanthocephalan parasites for chub hosts in polluted environments. Environ Sci Technol 54:5540–5549
Moore MS, Reichard JD, Murtha TD, Nabhan ML, Pian RE, Ferreira JS, Kunz TH (2013) Hibernating little brown myotis (*Myotis lucifugus*) show variable immunological responses to white-nose syndrome. PLoS One 8:e58976
Morris D, Guerra C, Khurasany M, Guilford F, Saviola B, Huang Y, Venketaraman V (2013) Glutathione supplementation improves macrophage functions in HIV. J Interf Cytokine Res 33:270–279
Morris G, Gevezova M, Sarafian V, Maes M (2022) Redox regulation of the immune response. Cell Mol Immunol 19:1079–1101
Mougeot F, Martínez-Padilla J, Webster LMI, Blount JD, Pérez-Rodríguez L, Piertney SB (2009) Honest sexual signalling mediated by parasite and testosterone effects on oxidative balance. Proc R Soc Lond B 276:1093–1100
Mougeot F, Martinez-Padilla J, Blount JD, Perez-Rodriguez L, Webster LMI, Piertney SB (2010) Oxidative stress and the effect of parasites on a carotenoid-based ornament. J Exp Biol 213:400–407
Narasimhan S, Sukumaran B, Bozdogan U, Thomas V, Liang X, DePonte K, Marcantonio N, Koski RA, Anderson JF, Kantor F, Fikrig E (2007) A tick antioxidant facilitates the Lyme disease agent's successful migration from the mammalian host to the arthropod vector. Cell Host Microbe 2:7–18
Nathan C, Shiloh MU (2000) Reactive oxygen and nitrogen intermediates in the relationship between mammalian hosts and microbial pathogens. Proc Natl Acad Sci USA 97:8841–8848
Ndong D, Chen Y-Y, Lin Y-H, Vaseeharan B, Chen J-C (2006) The immune response of tilapia *Oreochromis mossambicus* and its susceptibility to *Streptococcus iniae* under stress in low and high temperatures. Fish Shellfish Immunol 22:686–694
Neuman-Lee LA, Van Wettere AJ, French SS (2019) Interrelations among multiple metrics of immune and physiological function in a squamate, the common garter snake (*Thamnophis sirtalis*). Physiol Biochem Zool 92:12–23
Norte AC, Costantini D, Araújo PM, Eens M, Ramos JA, Heylen D (2018) Experimental infection by microparasites affects the oxidative balance in their avian reservoir host the blackbird *Turdus merula*. Ticks Tick-Borne Dis 9:720–729

Nussey DH, Pemberton JM, Pilkington JG, Blount JD (2009) Life history correlates of oxidative damage in a free-living mammal population. Funct Ecol 23:809–817

Ohmer MEB, Costantini D, Czirják G, Downs CJ, Ferguson LV, Flies A, Franklin CE, Kayigwe AN, Knutie S, Richards-Zawacki CL, Cramp RL (2021) Applied ecoimmunology: using immunological tools to improve conservation efforts in a changing world. Conserv Physiol 9:coab074

Orledge JM, Blount JD, Hoodless AN, Royle NJ (2012) Antioxidant supplementation during early development reduces parasite load but does not affect sexual ornament expression in adult ring-necked pheasants. Funct Ecol 26:688–700

Ots I, Kerimov AB, Ivankina EV, Ilyna TA, Hõrak P (2001) Immune challenge affects basal metabolic activity in wintering great tits. Proc R Soc Lond B 268:1175–1181

Pacello F, Ceci P, Ammendola S, Pasquali P, Chiancone E, Battistoni A (2008) Periplasmic Cu,Zn superoxide dismutase and cytoplasmic Dps concur in protecting *Salmonella enterica* serovar Typhimurium from extracellular reactive oxygen species. Biochim Biophys Acta 1780:226–232

Pacios-Palma I, Moreno S, Selman C, Rouco C (2018) Oxidative stress in wild European rabbits naturally infected with myxoma virus and rabbit haemorrhagic disease virus. Eur J Wildl Res 64:47

Padín-Irizarry V, Colón-Lorenzo EE, Vega-Rodríguez J, Castro Mdel R, González-Méndez R, Ayala-Peña S, Serrano AE (2016) Glutathione-deficient *Plasmodium berghei* parasites exhibit growth delay and nuclear DNA damage. Free Radic Biol Med 95:43–54

Pal S, Roy D, Ray S, Homechaudhuri S (2020) *Aeromonas hydrophila* induced mitochondrial dysfunction and apoptosis in liver and spleen of Labeo rohita mediated by calcium and reactive oxygen species. Turk J Fish Aquat Sci 20:255–266

Paul B, Sbarra AJ (1968) The role of the phagocyte in host-parasite interactions. 13. The direct quantitative estimation of H2O2 in phagocytizing cells. Biochimica et biophysica acta, 156(1):168–178

Paltrinieri S, Ravicini S, Rossi G, Roura X (2010) Serum concentrations of the derivatives of reactive oxygen metabolites (d-ROMs) in dogs with leishmaniosis. Veter J 186:393–395

Panaro MA, Lisi S, Mitolo V, Acquafredda A, Fasanella A, Carelli MG, Brandonisio O (1998) Evaluation of killing, superoxide anion and nitric oxide production by *Leishmania infantum*-infected dog monocytes. Cytobios 95:151–160

Pap PL, Vagasi CI, Czirjak GA, Titilincu A, Pintea A, Osvath G, Fulop A, Barta Z (2011) The effect of coccidians on the condition and immune profile of molting house sparrows (*Passer domesticus*). Auk 128:330–339

Pap PL, Sesarman A, Vágási CI, Buehler DM, Pătraş L, Versteegh MA, Banciu M (2014) No evidence for parasitism-linked changes in immune function or oxidative physiology over the annual cycle of an avian species. Physiol Biochem Zool 87:729–739

Pap PL, Pătraş L, Osváth G, Buehler DM, Versteegh MA, Sesarman A, Banciu M, Vágási CI (2015) Seasonal patterns and relationships among coccidian infestations, measures of oxidative physiology, and immune function in free-living house sparrows over an annual cycle. Physiol Biochem Zool 88:395–405

Pedersen AB, Babayan SA (2011) Wild immunology. Mol Ecol 20:872–880

Pedro AVF, Martínez D, Pontigo JP, Vargas-Lagos C, Hawes C, Wadsworth S, Morera FJ, Vargas-Chacoff L, Yáñez AJ (2019) Transcriptional activation of genes involved in oxidative stress in *Salmo salar* challenged with *Piscirickettsia salmonis*. Comp Biochem Physiol Part B 229:18–25

Peixoto MJ, Salas-Leiton E, Brito F, Pereira L, Svendsen J, Baptista T, Pereira R, Abreu H, Reis P, Gonçalves JFM, Ozório ROA (2017) Effects of dietary *Gracilaria* sp. and *Alaria* sp. supplementation on growth performance, metabolic rates and health in meagre (*Argyrosomus regius*) subjected to pathogen infection. J Appl Phycol 29:433–447

Penniall R, Spitznagel JK (1975) Chicken neutrophils: oxidative metabolism in phagocytic cells devoid of myeloperoxidase. Proc Natl Acad Sci USA 72:5012–5015

Pereira B, Rosa LF, Safi DA, Bechara EJ, Curi R (1995) Hormonal regulation of superoxide dismutase, catalase, and glutathione peroxidase activities in rat macrophages. Biochem Pharmacol 50:2093–2098

Perez-Rodriguez L, Fo M, Alonso-Alvarez C, Blas J, Viñuela J, Bortolotti GR (2008) Cell-mediated immune activation rapidly decreases plasma carotenoids but does not affect oxidative stress in red-legged partridges (*Alectoris rufa*). J Exp Biol 211:2155–2161

Peterhans E (1979) Sendai virus stimulates chemiluminescence in mouse spleen cells. Biochem Biophys Res Commun 91:383–392

Peterhans E, Grob M, Burge T, Zanoni R (1987) Virus-induced formation of reactive oxygen intermediates in phagocytic cells. Free Radic Res Commun 3:39–46

Petitjean Q, Jacquin L, Le Henaff M, Perrault A, Cousseau M, Laffaille P, Jean S (2021a) Dose- and time- dependent effects of an immune challenge on fish across biological levels. J Exp Zool Part A 335:250–264

Petitjean Q, Jacquin L, Riem L, Pitout M, Perrault A, Cousseau M, Laffaille P, Jean S (2021b) Intraspecific variability of responses to combined metal contamination and immune challenge among wild fish populations. Environ Pollut 272:116042

Philipp EER, Lipinski S, Rast J, Rosenstiel P (2012) Immune defense of marine invertebrates: the role of reactive oxygen and nitrogen species. In: Abele D, Vázquez-Medina JP, Zenteno-Savín T (eds) Oxidative stress in aquatic ecosystems. Wiley-Blackwell, Hoboken, pp 236–246

Pipe RK, Farley SR, Coles JA (1997) The separation and characterisation of haemocytes from the mussel *Mytilus edulis*. Cell Tissue Res 289:537–545

Possenti CD, Karadas F, Colombo G, Caprioli M, Rubolini D, Milzani A, Donne ID, Saino N, Parolini M (2017) Antioxidants and embryo phenotype: is there experimental evidence for strong integration of the antioxidant system? J Exp Biol 220:615–624

Radovanović TB, Prokić MD, Gavric J, Despotović SG, Gavrilović BR, Borković-Mitić S, Pavlović S, Saicic Z (2015) Glutathione-dependent enzyme activities and concentrations of glutathione, vitamin E and sulfhydryl groups in barbel (*Barbus barbus*) and its intestinal parasite *Pomphorhynchus laevis* (Acanthocephala). Ecol Indic 54:31–38

Rajan TV, Porte P, Yates JA, Keefer L, Shultz LD (1996) Role of nitric oxide in host defense against an extracellular, metazoanparasite, *Brugia malayi*. Infect Immun 64:3351–3353

Ramegowda V, Senthil-Kumar M, Ishiga Y, Kaundal A, Udayakumar M, Mysore KS (2013) Drought stress acclimation imparts tolerance to *Sclerotinia sclerotiorum* and *Pseudomonas syringae* in *Nicotiana benthamiana*. Int J Mol Sci 14:9497–9513

Ramirez-Otarola N, Sarria M, Rivera DS, Sabat P, Bozinovic F (2019) Ecoimmunology in degus: interplay among diet, immune response, and oxidative stress. J Comp Physiol B 189:143–152

Ramirez-Otarola N, Maldonado K, Cavieres G, Bozinovic F, Sabat P (2021) Nutritional ecology and ecological immunology in degus: does early nutrition affect the postnatal development of the immune function? J Exp Zool Part A 335:239–249

Razavi SM, Nazifi S, Afsar M, Yazdanpanah Z, Rakhshandehroo E (2016) Evaluation of the blood oxidant-antioxidant interactions in pigeons naturally infected with *Haemoproteus columbae*. Veterin Arhiv 86:395–405

Reddy MM, Mahipal SVK, Subhashini J, Reddy MC, Roy KR, Reddy GV, Reddy PR, Reddanna P (2006) Bacterial lipopolysaccharide-induced oxidative stress in the impairment of steroidogenesis and spermatogenesis in rats. Reprod Toxicol 22:493–500

Ren ZL, Wang SF, Cai Y, Wu Y, Tian LJ, Liao JQ, Wang SQ, Jiang L, Guo WL, Zhou YC (2020) Antioxidant capacity, non-specific immunity, histopathological analysis and immune-related genes expression in Nile tilapia *Oreochromis niloticus* infected with *Aeromonas schubertii*. Aquaculture 529:735642

Reniere ML, Whiteley AT, Hamilton KL, John SM, Lauer P, Brennan RG, Portnoy DA (2015) Glutathione activates virulence gene expression of an intracellular pathogen. Nature 517:170–173

Repine JE, Fox RB, Berger EM (1981) Hydrogen peroxide kills *Staphylococcus aureus* by reacting with staphylococcal iron to form hydroxyl radical. J Biol Chem 256:7094–7096

Reshi ML, Su Y-C, Hong J-R (2014) RNA viruses: ROS-mediated cell death. Int J Cell Biol 2014:467452

Ridgway I, Bowden TJ, Roman-Gonzalez A, Richardson CA (2014) Resistance to oxidative stress is not associated with the exceptional longevity of the freshwater pearl mussel, *Margaritifera margaritifera* nor three unionid species. Aquat Sci 76:259–267

Riedel W, Maulik G (1999) Fever: an integrated response of the central nervous system to oxidative stress. Mol Cell Biochem 196:125–132

Rodriguez A, Broggi J, Alcaide M, Negro JJ, Figuerola J (2014) Determinants and short-term physiological consequences of PHA immune response in lesser kestrel nestlings. J Exp Zool Part A 321:376–386

Rodríguez-Quiroga JJ, Otero-Rodino C, Suarez P, Nieto TP, Estevez JMG, San Juan F, Soengas JL (2017) Differential effects of exposure to parasites and bacteria on stress response in turbot *Scophthalmus maximus* simultaneously stressed by low water depth. J Fish Biol 91:242–259

Roth JD, Dobson FS, Criscuolo F, Uhlrich P, Zahariev A, Bergouignan A, Viblanc VA (2019) Subtle short-term physiological costs of an experimental augmentation of fleas in wild Columbian ground squirrels. J Exp Biol 222:jeb203588

Roy D, Pal S, Datta Ray S, Homechaudhuri S (2019) Evaluating oxidative stress in *Labeo rohita*, infected asypmtomatically with native and invasive aeromonads using biochemical indices. Proc Nat Acad Sci India Section B 89:973–978

Sacks D, Sher A (2002) Evasion of innate immunity by parasitic protozoa. Nat Immunol 3:1041–1047

Şahan A, Duman S, Colak SO, Cinar E, Bilgin R (2017) Determination of some hematological and non-specific immune defences, oxidative stress and histopathological status in rainbow trout (*Oncorhynchus mykiss*) fed rosehip (*Rosa canina*) to *Yersinia ruckeri*. Turk J Fish Aquat Sci 17:91–100

Saleh MA, Al-Salahy MB, Sanousi SA (2009) Oxidative stress in blood of camels (*Camelus dromedarius*) naturally infected with *Trypanosoma evansi*. Vet Parasitol 162:192–199

Sanchez CA, Becker DJ, Teitelbaum CS, Barriga P, Brown LM, Majewska AA, Hall RJ, Altizer S (2018) On the relationship between body condition and parasite infection in wildlife: a review and meta-analysis. Ecol Lett 21:1869–1884

Schneebauer G, Hanel R, Pelster B (2016) *Anguillicola crassus* impairs the silvering-related enhancements of the ROS defense capacity in swimbladder tissue of the european eel (Anguilla anguilla). J Comp Physiol B 186:867–877

Schneeberger K, Czirják GA, Voigt CC (2013) Inflammatory challenge increases measures of oxidative stress in a free-ranging, long-lived mammal. J Exp Biol 216:4514–4519

Schoenle LA, Kernbach M, Haussmann MF, Bonier F, Moore IT (2017) An experimental test of the physiological consequences of avian malaria infection. J Anim Ecol 86:1483–1496

Schreck R, Rieber P, Baeuerle PA (1991) Reactive oxygen intermediates as apparently widely used messengers in the activation of the NF-κB transcription factor and HIV-1. EMBO J 10:2247–2258

Schwarz KB (1996) Oxidative stress during viral infection: a review. Free Radic Biol Med 21:641–649

Sebastiano M, Chastel O, de Thoisy B, Eens M, Costantini D (2016) Oxidative stress favours herpes virus infection in vertebrates: a meta-analysis. Curr Zool 62:325–332

Sebastiano M, Eens M, Ablelgawad H, de Thoisy B, Lacoste V, Pineau K, Asard H, Chastel O, Costantini D (2017) Oxidative stress biomarkers are associated with visible clinical signs of a disease in frigatebird nestlings. Sci Rep 7:1599

Sebastiano M, Eens M, Messina S, AbdElgawad H, Pineau K, Gerrit B, Chastel O, Costantini D (2018) Resveratrol supplementation reduces oxidative stress and modulates the immune response in free-living animals during a viral infection. Funct Ecol 32:2509–2519

Segal AW (2005) How neutrophils kill microbes. Annu Rev Immunol 23:197–223

Sepp T, Karu U, Blount JD, Sild E, Männiste M, Hõrak P (2012) Coccidian infection causes oxidative damage in greenfinches. PLoS One 7:e36495

Seppälä O, Jokela J (2010) Maintenance of genetic variation in immune defense of a freshwater snail: role of environmental heterogeneity. Evolution 64:2397–2407

Serra L, Pirrello S, Caprioli M, Griggio M, Andreotti A, Romano A, Pilastro A, Saino N, Sacchi R, Galeotti P, Fasola M, Spina F, Rubolini R (2012) Seasonal decline of offspring quality in the European starling *Sturnus vulgaris*: an immune challenge experiment. Behav Ecol Sociobiol 66:697–709

Sheldon BC, Verhulst S (1996) Ecological immunology: costly parasite defences and trade-offs in evolutionary ecology. Trends Ecol Evol 11:317–321
Shen Z, Wu W, Hazen SL (2000) Activated leukocytes oxidatively damage DNA, RNA, and the nucleotide pool through halide-dependent formation of hydroxyl radical. Biochemist 39:5474–5482
Sheridan PA, Beck MA (2008) The immune response to herpes simplex virus encephalitis in mice is modulated by dietary vitamin E. J Nutr 138:130–137
Shi L, Sohaskey CD, Kana BD, Dawes S, North RJ, Mizrahi V, Gennaro ML (2005) Changes in energy metabolism of *Mycobacterium tuberculosis* in mouse lung and under in vitro conditions affecting aerobic respiration. Proc Natl Acad Sci USA 102:15629–15634
Shukitt-Hale B, Galli RL, Meterko V, Carey A, Bielinski DF, McGhie T, Joseph JA (2005) Dietary supplementation with fruit polyphenolics ameliorates age-related deficits in behavior and neuronal markers of inflammation and oxidative stress. Age 27:49–57
Sild E, Sepp T, Männiste M, Hõrak P (2011) Carotenoid intake does not affect immune-stimulated oxidative burst in greenfinches. J Exp Biol 214:3467–3473
Simons MJP, Cohen AA, Verhulst S (2012) What does carotenoid-dependent coloration tell? Plasma carotenoid level signals immunocompetence and oxidative stress state in birds–a meta-analysis. PLoS One 7:e43088
Sirak AA, Beavis AJ, Robertson FM (1991) Enhanced hydroperoxide production by peripheral blood leukocytes following exposure of murine epidermis to 12-Otetradecanoylphorbol-13-acetate. Carcinogenesis 12:91–95
Skuratovskaya EN, Zavyalov AV, Rudneva II (2015) Black Sea whiting *Merlangus merlangus euxinus* (Nordmann, 1840) to parasitic nematode *Hysterothylacium aduncum* (Rudolphi, 1802) infection. Bull Eur Assoc Fish Pathol 35:170–176
Skuratovskaya EN, Yurakhno VM, Chesnokova II (2020) Biochemical response of two fish species of Gobiidae (Gobiiformes) to Cryptocotyle (Opisthorchiida, Heterophyidae) metacercariae infection from the mouth of the river Chernaya (Black Sea). Ecolog Montenegri 38:158–165
Sobroza AO, Tonin AA, Da Silva AS, Dornelles GL, Wolkmer P, Duarte M, Hausen BS, Sangoi MB, Moresco RN, Stefani LM, Mazzantti CM, Lopes ST, Leal ML (2014) Iron metabolism in hamsters experimentally infected with *Leptospira interrogans* serovar pomona: influence on disease pathogenesis. Comp Immunol Microbiol Infect Dis 37:299–304
Sônego F, Castanheira FVES, Ferreira RG, Kanashiro A, Leite CAVG, Nascimento DC, Colón DF, Borges Vde F, Alves-Filho JC, Cunha FQ (2016) Paradoxical roles of the neutrophil in sepsis: protective and deleterious. Front Immunol 7:155
Sorci G, Faivre B (2009) Inflammation and oxidative stress in vertebrate host–parasite systems. Philos Trans R Soc B 364:71–83
Sorci G, Cornet S, Faivre B (2013) Immune evasion, immunopathology and the regulation of the immune system. Pathogens 2:71–91
Sorci G, Lippens C, Lechenault C, Faivre B (2017) Benefits of immune protection versus immunopathology costs: a synthesis from cytokine KO models. Infect Genet Evol 54:491–495
Srivastava P, Puri SK, Dutta GP, Pandey VC (1992) Status of oxidative stress and antioxidant defenses during *Plasmodium knowlesi* infection and chloroquine treatment in *Macaca mulatta*. Inter J Parasitol 22:243–245
Stabler JG, McCormick TW, Powell KC, Kogut MH (1994) Avian heterophils and monocytes: phagocytic and bactericidal activities against *Salmonella enteritidis*. Vet Microbiol 38:293–305
Stanulis ED, Jordan SD, Rosecrans JA, Holsapple MP (1997a) Disruption of Th1/Th2 cytokine balance by cocaine is mediated by corticosterone. Immunopharmacology 37:25–33
Stanulis ED, Matulka RA, Jordan SD, Rosecrans JA, Holsapple MP (1997b) Role of corticosterone in the enhancement of the antibody response after acute cocaine administration. J Pharmacol Exp Ther 280:284–291
Stauffer J, Bruneaux M, Panda B, Visse M, Vasemägi A, Ilmonen P (2017) Telomere length and antioxidant defense associate with parasite-induced retarded growth in wild brown trout. Oecologia 185:365–374
Strassheim D, Asehnoune K, Park JS, Kim JY, He Q, Richter D, Mitra S, Arcaroli J, Kuhn K, Abraham E (2004) Modulation of bone marrow-derived neutrophil signaling by $H_2O_2$: dis-

parate effects on kinases, NF-kappaB, and cytokine expression. Am J Physiol Cell Physiol 286:C683–C692
Strohm E, Herzner G, Ruther J, Kaltenpoth M, Engl T (2019) Nitric oxide radicals are emitted by wasp eggs to kill mold fungi. elife 8:e43718
Stumbo AD, Goater CP, Hontela A (2012) Parasite-induced oxidative stress in liver tissue of fathead minnows exposed to trematode cercariae. Parasitology 139:1666–1671
Sun LL, Sun Y, Jiang M, Luo LJ, Yu XB, Yao WZ, Wu ZL (2020) Isolation, identification and pathogenicity of *Proteus vulgaris* from moribund common carp (*Cyprinus carpio*) farmed in China. Aquacult 525:735294
Surai PF (2002) Natural antioxidants in avian nutrition and reproduction. Nottingham University Press, Nottingham
Sweazea KL, Simperova A, Juan T, Gadau A, Brant SV, Deviche P, Jarrett C (2015) Pathophysiological responses to a schistosome infection in a wild population of mourning doves (*Zenaida macroura*). Zoology 118:386–393
Szuroczki D, Koprivnikar J, Baker RL (2019) Effects of dietary antioxidants and environmental stressors on immune function and condition in *Lithobates* (*Rana*) *sylvaticus*. Comp Biochem Physiol Part A 229:25–32
Tancredo KR, Goncalves ELT, Brum A, Acchile M, Hashimoto GSO, Pereira SA, Martins ML (2015) Hemato-immunological and biochemical parameters of silver catfish *Rhamdia quelen* immunized with live theronts of *Ichthyophthirius multifiliis*. Fish Shellf Immunol 45:689–694
Tkachenko H, Kurhalyuk N, Palczynska K (2011) Responses of antioxidant status in the gills of brown trout (*Salmo trutta* m. *trutta* L.) with ulcerative dermal necrosis. Baltic Coastal Zone. J Ecol Prot Coastl 15:145–158
Tkachenko H, Kurhaluk N, Andriichuk A, Gasiuk E, Beschasniu S (2014) Oxidative stress biomarkers in liver of sea trout (*Salmo trutta* m. *trutta* L.) affected by ulcerative dermal necrosis syndrome. Turk J Fish Aquat Sci 14:391–402
Tkachenko H, Grudniewska J, Pekala A, Terech-Majewska E (2016) Oxidative stress and antioxidant defence markers in muscle tissue of rainbow trout (*Oncorhynchus mykiss*) after vaccination against *Yersinia ruckeri*. J Veter Res 60:25–33
Tobler M, Healey M, Wilson M, Olsson M (2011) Basal superoxide as a sex-specific immune constraint. Biol Lett 7:906–908
Tobler M, Ballen C, Healey M, Wilson M, Olsson M (2015) Oxidant trade-offs in immunity: an experimental test in a lizard. PLoS One 10:e0126155
Toh SQ, Glanfield A, Gobert GN, Jones MK (2010) Heme and blood-feeding parasites: friends or foes? Parasit Vectors 3:108
Tonin AA, da Silva AS, Thome GR, Bochi GV, Schetinger MRC, Moresco RN, Camillo G, Toscan G, Vogel FF, Lopes STA (2014) Oxidative stress in brain tissue of gerbils experimentally infected with *Neospora caninum*. J Parasitol 100:154–156
Torres R, Velando A (2007) Male reproductive senescence: the price of immune-induced oxidative damage on sexual attractiveness in the blue-footed booby. J Anim Ecol 76:1161–1168
Tschirren B, Fitze PS, Richner H (2003) Sexual dimorphism in susceptibility to parasites and cell-mediated immunity in great tit nestlings. J Anim Ecol 72:839–845
Üllen A, Singewald E, Konya V, Fauler G, Reicher H, Nusshold C, Hammer A, Kratky D, Heinemann A, Holzer P, Malle E, Sattler W (2013) Myeloperoxidase-derived oxidants induce blood-brain barrier dysfunction *in vitro* and *in vivo*. PLoS One 8:e64034
Valyi-Nagy T, Olson SJ, Valyi-Nagy K, Montine TJ, Dermody TS (2000) Herpes simplex virus type 1 latency in the murine nervous system is associated with oxidative damage to neurons. Virology 278:309–321
van de Crommenacker J, Richardson DS, Koltz AM, Hutchings K, Komdeur J (2012) Parasitic infection and oxidative stress are associated and vary with breeding activity in the Seychelles warbler. Proc R Soc Lond B 279:1466–1476
Vitula F, Peckova L, Bandouchova H, Pohanka M, Novotny L, Jira D, Kral J, Ondracek K, Osickova J, Zendulkova D, Rosenbergova K, Treml F, Pikula J (2011) *Mycoplasma gallisepticum* infection in the grey partridge *Perdix perdix:* outbreak description, histopathology, biochemistry and antioxidant parameters. BMC Vet Res 7:34

Voigt CC, Fritze M, Lindecke O, Costantini D, Petersons G, Czirjak GA (2020) The immune response of bats differs between pre-migration and migration seasons. Sci Rep 10:17384

Wang D, Malo D, Hekimi S (2010) Elevated mitochondrial reactive oxygen species generation affects the immune response via hypoxia-inducible factor-1α in long-lived Mclk1$^{+/-}$ mouse mutants. J Immunol 184:582–590

Watson MJ (2013) What drives population-level effects of parasites? Meta-analysis meets life-history. Int J Parasit 2:190–196

Wegmann M, Voegeli B, Richner H (2015) Physiological responses to increased brood size and ectoparasite infestation: adult great tits favour self-maintenance. Physiol Behav 141:127–134

Wheeler ML, DeFranco AL (2012) Prolonged production of reactive oxygen species in response to B cell receptor stimulation promotes B cell activation and proliferation. J Immunol 189:4405–4416

Whiten SR, Eggleston H, Adelman ZN (2017) Ironing out the details: exploring the role of iron and heme in blood-sucking arthropods. Front Physiol 8:1134

Whitfield JF, MacManus JP, Rixon RH (1970) Cyclic AMP mediated stimulation of thymocyte proliferation by low concentrations of cortisol. Proc Soc Exp Biol Med 134:1170–1174

Winter JL, Barber LG, Freeman L, Griessmayr PC, Milbury PE, Blumberg JB (2009) Antioxidant status and biomarkers of oxidative stress in dogs with lymphoma. J Vet Inter Med 23:311–316

Winterbourn CC (2002) Biological reactivity and biomarkers of the neutrophil oxidant, hypochlorous acid. Toxicology 181(182):223–227

Wood MJ, Cosgrove CL, Wilkin TA, Knowles SCL, Day KP, Sheldon BC (2007) Within-population variation in prevalence and lineage distribution of avian malaria in blue tits, *Cyanistes caeruleus*. Mol Ecol 16:3263–3273

Xu J, Yin Z, Li L, Cheng A, Jia R, Song X, Lu H, Dai S, Lv C, Liang X, He C, Zhao L, Su G, Ye G, Shi F (2013) Inhibitory effect of resveratrol against duck enteritis virus *in vitro*. PLoS One 8:e65213

Xu WJ, Li HY, Wu LY, Jin JY, Zhu XM, Han D, Liu HK, Yang YX, Xu XD, Xie SQ (2020) Dietary *Scenedesmus ovalternus* improves disease resistance of overwintering gibel carp (*Carassius gibelio*) by alleviating toll-like receptor signaling activation. Fish Shellf Immunol 97:351–358

Yeh SP, Liu KF, Chiu ST, Jian SJ, Cheng W, Liu CH (2009) Identification and cloning of a selenium dependent glutathione peroxidase from giant freshwater prawn, *Macrobrachium rosenbergii*. Fish Shellfish Immunol 27:181–191

Yun S, Giri SS, Kim HJ, Kim SG, Kim SW, Kang JW, Han SJ, Kwon J, Oh WT, Chi C, Jun JW, Chang Park S (2019) Enhanced bath immersion vaccination through microbubble treatment in the cyprinid loach. Fish Shellf Immunol 91:12–18

Zahran E, Risha E (2013) Protective role of adjuvant and potassium permanganate on oxidative stress response of Nile tilapia (*Oreochromis niloticus*) challenged with *Saprolegnia ferax*. Springerplus 2:94

Zahran E, El-Gawad EAA, Risha E (2018) Dietary Withania somnifera root confers protective and immunotherapeutic effects against *Aeromonas hydrophila* infection in Nile tilapia (*Oreochromis niloticus*). Fish Shellf Immunol 80:641–650

Zahran E, Mahgoub HA, Abdelhamid F, Sadeyen JR, Risha E (2019) Experimental pathogenesis and host immune responses of Enterococcus faecalis infection in Nile tilapia (*Oreochromis niloticus*). Aquacult 512:734319

Zhang Y, Hill GE, Ge Z, Park NR, Taylor HA, Andreasen V, Tardy L, Kavazis AN, Bonneaud C, Hood WR (2021) Effects of a Bacterial Infection on Mitochondrial Function and Oxidative Stress in a Songbird. Physiological and biochemical zoology 94:71–82

Zhao H, Luo YE, Wu ZX, Zhou Y, Guo DY, Wang H, Chen XX (2019) Hepatic lipid metabolism and oxidative stress responses of grass carp (*Ctenopharyngodon idella*) fed diets of two different lipid levels against *Aeromonas hydrophila* infection. Aquacult 509:149–158

Zhu XM, Li MY, Liu XY, Xia CG, Niu XT, Wang GQ, Zhang DM (2020) Effects of dietary astaxanthin on growth, blood biochemistry, antioxidant, immune and inflammatory response in lipopolysaccharide-challenged Channa argus. Aquac Res 51:1980–1991

Zmijewski JW, Zhao X, Xu Z, Abraham E (2007) Exposure to hydrogen peroxide diminishes NF-kappaB activation, IkappaB-alpha degradation, and proteasome activity in neutrophils. Am J Physiol Cell Physiol 293:C255–C266

# Chapter 3
# Oxidative Costs and Antioxidant Rewards of Nutrition

**Abstract** What you eat, when you eat, and where you eat have dramatic effects on many aspects of an individual life. Nutrition is a source of a number of substances that can increase the organismal resistance to oxidative stress. However, nutrition can also generate oxidative stress and trigger a cascade of molecular events that can disrupt cellular oxidative status. As a consequence, oxidative stress has also likely major implications for the evolution of foraging ecology and the strategies of maternal allocation of nutrients and antioxidants to their offspring. In this chapter, I have examined (i) the role that oxidative stress might have played in the evolution of foraging strategies and food selection, (ii) the links between diet and organismal resistance to oxidative stress, and (iii) the many ways through which mothers can shape offspring physiology through transfer of dietary antioxidants from them to their offspring.

## 3.1 Introduction

The environment supplies food to wildlife. Animals need this food because their body constituents become damaged with time, built, consumed, discarded, or replaced, hence there is continuous need of chemical building blocks (Hill et al. 2008). Body functions also require energy to work, and energy is derived from nutrients (Hill et al. 2008). However, this need for food also faces animals with a number of demands, such as time and energy budgets to allocate to foraging, nutritional requirements, or processes from food selection to ingestion, digestion, and absorption of nutrients. For example, feeding and digestion can cause a rise of metabolic rate (Specific Dynamic Action), as well as increased levels of oxidative damage markers (Butler et al. 2016; Schvezov et al. 2022). Costs of feeding may also increase when animals need to intensify their foraging activity, such as when covering long distances or increasing searching time because there is less food available or their nutritional requirements increase. Thus, nutrition can be considered undoubtedly a key driver of variation in many aspects of an animal life history,

© The Author(s), under exclusive license to Springer Nature Switzerland AG 2024

D. Costantini, *The Role of Organismal Oxidative Stress in the Ecology and Life-History Evolution of Animals*, Fascinating Life Sciences,
https://doi.org/10.1007/978-3-031-65183-0_3

behaviour, and physiology, such as growth rates and adult mass gain, body condition, probability of pregnancy, over-winter survival, timing of parturition, clutch size, neonatal birth mass, survival, development of sexual traits, immune function, antioxidant levels, and metabolic rate (e.g. Blount et al. 2003a, b; Byrne et al. 2008; Lee et al. 2008; Cohen et al. 2009; Eeva et al. 2009; Raubenheimer et al. 2009; Sorensen et al. 2009; Orledge et al. 2012; Deeming and Pike 2013; Stewart 2013; Wagner et al. 2013; García-Tarrasón et al. 2014; Schneeberger et al. 2014; Swanson et al. 2016; Birnie-Gauvin et al. 2017).

In this chapter, I have made the point that dietary antioxidants and propensity of certain nutrients to fuel oxidative stress are two relevant aspects of nutrition that have a significant impact on the life history of animals. In particular, I have examined (i) the role that oxidative stress might have played in the evolution of foraging strategies and food selection, (ii) the links between diet and organismal resistance to oxidative stress, and (iii) the many ways through which mothers can shape offspring physiology through transfer of dietary antioxidants from them to their offspring.

## 3.2 Oxidative Costs of Foraging Behaviour

The availability of food varies spatially and temporally. Needs for specific foods can also vary over time, depending on the nutritional requirements of the organism as it moves through different life stages, such as the transition from a non-reproductive to a reproductive state. Tracking food availability may sometimes require animals to perform long foraging trips; alternatively, animals might face periods of food shortage or even starvation, or switch slowly (or rapidly) to other trophic resources.

Spatio-temporal variation in foraging strategies is common in wildlife. We can infer that this variation in foraging strategies would be connected to the individual variation in oxidative status through different mechanisms, such as the metabolic demands of the physical effort to perform the foraging trip, and the nutritional and antioxidant compositions of food. Stable carbon and nitrogen isotope analyses in biological matrices are widely used to infer the foraging behaviour and the diet of an individual or a certain species. In nature, both carbon and nitrogen occur as two stable isotopes: 12C and 13C (ratio δ13C = 13C/12C), 14N and 15N (ratio δ15N = 15N/14N), respectively (DeNiro and Epstein 1978, 1981). Carbon isotope composition provides an indication of the foraging habitat; nitrogen isotope composition is used to draw inferences regarding both the protein source and trophic level of an individual. Studies linking oxidative status markers to stable isotope ratios in adults have offered insight on the (i) links between diet and oxidative status and (ii) physiological costs associated with a given feeding strategy (e.g. Woshner et al. 2008; Beaulieu et al. 2010, 2015; García-Tarrasón et al. 2014; Costantini et al. 2017; Tapia-Monsalve et al. 2018; Soldatini et al. 2020). For example, Beaulieu et al. (2010) found that the oxidative status of Adélie penguins *Pygoscelis adeliae* was related to their spatial distribution, rather than to the duration of their foraging trip, and therefore presumably to their diet. The plasma non-enzymatic antioxidant capacity was significantly higher in Scopoli's shearwaters *Calonectris diomedea* that had higher $\delta^{13}C$ values, suggesting that birds foraging closer to the mainland (where $\delta^{13}C$ values are higher) had a higher intake of or consumed endogenously less antioxidants than those foraging offshore (Fig. 3.1a; Costantini et al. 2017). The plasma non-enzymatic antioxidant capacity was significantly lower in wrinkle-lipped free-tailed bats *Chaerephon plicatus* that had lower δ13C values, suggesting that bats foraging over less heterogeneous landscapes (where δ13C values are lower) had a smaller intake of or consumed endogenously more antioxidants than those foraging in more heterogeneous landscapes (Fig. 3.1b; Costantini et al. 2019).

Diving animals are excellent models to study the interplay between oxidative stress and foraging behaviour because they have to cope with a reduced oxygen availability while diving, and rapid and strong reoxygenation of tissues at the end of an apnoea bout (e.g. Ramirez et al. 2007; Vázquez-Medina et al. 2012). It has been proposed that seals and diving birds maintain high constitutive activities and concentrations of endogenous antioxidants, and the hypoxia/reperfusion-induced oxidative stress would stimulate hormetic responses that protect tissues from damage (Vázquez-Medina et al. 2012).

Animals can change their foraging strategy as they mature and age. Thus, we can infer that animals have to adjust their physiological machinery to the novel circumstances to avoid or limit oxidative stress. For example, honey bee *Apis mellifera* workers clean the hive and feed the larvae during their first weeks of life. As they get older, honey bee workers begin to go out of the hive to collect nectar and pollen for the colony. Foraging may be highly demanding for forager bees as it can require them to fly over several kilometres to find food (Winston 1987). The metabolic rate of honey bees during flight increases dramatically, reaching one of the highest

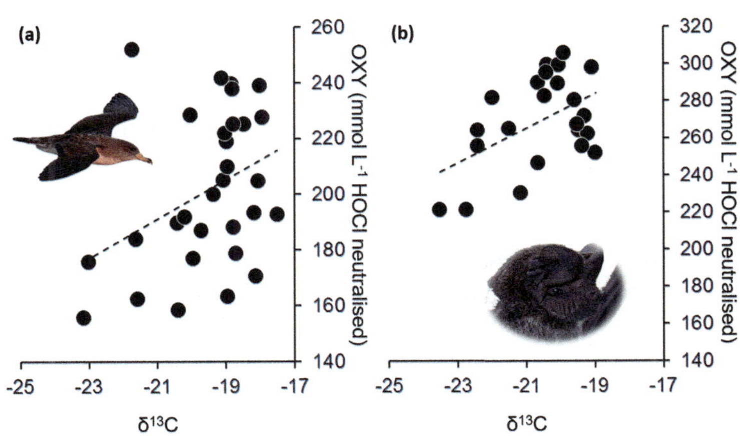


**Fig. 3.1** Both biplots show significant correlations between foraging habitat, estimated by $\delta^{13}C$ values, and plasma levels of non-enzymatic antioxidants. (**a**) Scopoli's shearwaters *Calonectris diomedea* that foraged closer to the mainland (higher $\delta^{13}C$ values) had higher plasma levels of non-enzymatic antioxidants (see Costantini et al. 2017); (**b**) wrinkle-lipped free-tailed bats *Chaerephon plicatus* that foraged over less heterogeneous landscapes (lower $\delta^{13}C$ values) had lower plasma levels of non-enzymatic antioxidants (see Costantini et al. 2019)

values ever recorded in any animal species (Suarez et al. 1996). Williams et al. (2008) showed that, compared to nurse stage, naturally occurring flight elicited an increase of the concentration of heat shock proteins in flight muscle (independently from induction of hyperthermia by muscle activity) in both young (had been foraging for 2–3 days) and old (had been foraging for at least 12 days) foragers. By contrast, they also found that young and old bees had similar levels of oxidative damage to proteins, while catalase and non-enzymatic antioxidant capacity increased only in young bees.

## 3.3 Searching for Antioxidant Rewards

Food is a source of a number of substances (tocopherols, carotenoids, polyphenols; for some species also ascorbic acid) with antioxidant properties. The roles that certain dietary antioxidants play in protection against oxidative stress was probably over-estimated in some taxa (birds: Costantini and Møller 2008; Simons et al. 2012; reptiles: Olsson et al. 2008, 2009; mammals: Selman et al. 2013). However, the so-called dietary antioxidants are relevant candidates as modulators of the expression of life-history traits because they have several physiological functions (e.g. immunostimulation, protection of brain or sperm cells) that make them relevant for the organism (von Schantz et al. 1999; Surai 2002; Catoni et al. 2008a; Cohen et al. 2012). Moreover, the protective effects of food components against oxidative stress may also depend on environmental circumstances, such as high environmental ammonia and salinity stress (Sinha et al. 2015) or ambient temperature (Andersson et al. 2018), which indicates that the antioxidant potential of dietary molecules in organisms may be circumstantial.

Lipophilic dietary antioxidants can be stored in certain tissues (e.g. liver, adipose tissue), while hydrophilic dietary antioxidants (e.g. polyphenols) are rapidly metabolised and excreted from the body. In this context, ascorbic acid is peculiar because many species can synthesise it (Chatterjee 1973; Martínez del Rio 1997). Lipophilic antioxidants can also be mobilised from tissues, where they are stored, and re-directed towards those sites that need them (Faivre et al. 2003; Costantini and Dell'Omo 2006). It is therefore plausible to infer that needs for specific antioxidants are contingent upon the life cycle stage or the condition of the individual. As a consequence, we might expect that mechanisms driving food selection are flexible to some degree in order to meet the organismal nutritional requirements. For example, some species can select those foods that provide substances for self-medication (e.g. Clayton and Wolfe 1993; Huffman 1997; Singer et al. 2009). Some studies suggested that self-medication might also be relevant to regulate the cellular oxidative status (Beaulieu and Schaefer 2013).

Irrespective of self-medication behaviour, some studies suggested that the concentration of antioxidants in food might have contributed to drive the evolution of food selection strategies because they are beneficial for many body functions (Carciofi et al. 2006; Catoni et al. 2008a; Schaefer et al. 2008; Beaulieu and Schaefer

2014). As for those dietary substances that do not act as pigments, some studies suggested that food choice would be driven by taste, odour, or physiological feedbacks (e.g. Elliott and Loudon 1987; Fromentin and Nicolaidis 1996; Pierce and McWilliams 2005; Senar et al. 2010). By contrast, both carotenoids and polyphenols act as pigments, so that the intensity of the food colouration would convey to the animal the extent to which that food is rich in antioxidants (Schaefer et al. 2008). For example, the frugivorous European blackcap *Sylvia atricapilla* preferentially feeds on food containing anthocyanins (a group of polyphenols). Schaefer et al. (2008) found that blackcaps were capable of discriminating anthocyanin concentrations in fruits based on colour because fruits rich in anthocyanins are black or reflect ultraviolet radiation that birds can perceive. Furthermore, in food choice experiments, Schaefer et al. (2008) found that blackcaps selected food containing anthocyanins over food without anthocyanins. Some experiments showed that blackcaps and other species of birds may benefit from a diet rich in polyphenols. For example, blackcaps supplemented for 4 weeks with polyphenols were more likely to mount a humoral immune response after being injected with sheep red blood cells compared to control blackcaps (Catoni et al. 2008b). Captive Gouldian finches *Erythrura gouldiae* fed more on seeds rich in polyphenols when they were housed at a room temperature below their thermoneutral zone than when they were housed at a room temperature that was within their thermoneutral zone (Beaulieu et al. 2014). Importantly, the larger intake of polyphenols at low ambient temperatures enabled birds to reduce oxidative damage. Carbeck et al. (2018) also found that a diet rich in anthocyanins stimulated the testosterone production in male starlings *Sturnus vulgaris*.

Selection of food rich in polyphenols might be particularly relevant for migrating animals given the high metabolic requirements that sustained flight imposes on them. For example, birds might mitigate oxidative stress accrued during the strenuous migratory flight by consuming prophylactically (before departure) or therapeutically (while en route at stop-over sites) food rich in polyphenols (Alan et al. 2013; Bolser et al. 2013; Cooper-Mullin and McWilliams 2016). In addition to fruit selection, birds might also select and exploit other sources of polyphenols, such as nectar of flowers. Cecere et al. (2011) recorded several opportunistic nectar-feeding species at a place that birds use to rest, refuel, and seek shelter during their migration. Further studies found that nectar can contain significant amounts of polyphenols. For example, the *Leucosceptrum canum* is a plant species that produces coloured nectar (Fig. 3.2) owing to the presence of a purple anthocyanidin, the 5-hydroxyflavylium (Zhang et al. 2012). In a food selection experiment, Zhang et al. (2012) found that individuals belonging to two species of avian pollinators, the blue-winged minla *Minla cyanouroptera* and the oriental white-eye *Zosterops palpebrosus*, preferentially fed on model flowers that contained coloured (richer in polyphenols) rather than uncoloured nectar. Whether polyphenols in nectar improved the antioxidant defences of these birds is unknown. Other experiments of supplementation of diet with polyphenols found contrasting evidence about their antioxidant relevance in both vertebrates and invertebrates (e.g. Beaulieu et al. 2016, 2017; Cecere et al. 2016; Sebastiano et al. 2018). Although antioxidants in

**Fig. 3.2** The *Leucosceptrum canum* is a plant species that produces coloured nectar owing to the presence of the pigment anthocyanidin. (**a**) dark purple nectar (arrow); (**b**) chemical structure of the anthocyanidin; (**c**) blue-winged minla *Minla cyanouroptera*, which is an important avian pollinator for the *Leucosceptrum canum*. It has been suggested that the dark colour of the nectar works as a foraging signal that is positively selected by birds because it increases the pollination efficiency through nectar visibility and palatability. Reprinted from Zhang et al. (2012) with permission of John Wiley and Sons

nectar occur in limited supply, nectarivorous might also use other nutrients to mitigate oxidative stress. Using respirometry coupled with $\delta^{13}C$ analyses, Levin et al. (2017) suggested that moths *Manduca sexta* might generate antioxidant potential in the form of nicotinamide adenine dinucleotide phosphate by shunting nectar glucose to the pentose phosphate pathway, resulting in a reduction of oxidative damage to the flight muscles as a result.

We have also relevant examples of connections between food selection and antioxidants in herbivore insects. For example, the abundance of herbivore insects is commonly higher on spring (or young) leaves than on mature tree leaves (Feeny 1970; Barbehenn et al. 2013). It has been hypothesised that this feeding strategy would increase the fitness of insects because spring leaves have a larger content of nutrients that would enable larvae to grow rapidly and to reach a large body size (e.g. Hough and Pimentel 1978). However, selection of leaves might also be driven by their content in antioxidant and pro-oxidant compounds. Barbehenn et al. (2013) used larvae of the moth *Lymantria dispar* to test the effects of feeding on leaves of two contrasting host plants in spring and summer: the red oak *Quercus rubra*, a high-quality host, and the sugar maple *Acer saccharum*, a low-quality host. On spring foliage, the combined effects of rapid consumption rate, efficient nutrient assimilation, and high nutritional quality enabled larvae to grow rapidly and attain a large body mass (Barbehenn et al. 2013). Ellagitannins, a major source of oxidative stress in the midgut of caterpillars (Barbehenn et al. 2006), occurred at higher concentrations in maple leaves in spring than in summer, but they were at negligible levels throughout the growing season in the oak. Generation of semiquinone free radicals and oxidation of glutathione in larvae feeding on maple remained at elevated levels during both spring and summer. By contrast, larvae that fed on oak had consistently low levels of oxidative stress (Barbehenn et al. 2013). Although feeding on leaves of different species resulted in caterpillars having different oxidative stress levels, feeding on spring leaves did not increase the capacity of larvae to cope with oxidative stress (Barbehenn et al. 2013).

## 3.4 Food Composition: Friend and Foe of Oxidative Status

Comparative studies on birds and mammals showed that the diet composition explains to some degree the variation among species in certain oxidative status markers. Cohen et al. (2009) measured circulating antioxidant levels in 95 bird species and examined their associations with diet. Bird species that consumed more invertebrates had higher plasma non-enzymatic antioxidant capacity and plasma concentration of uric acid, but lower plasma concentration of carotenoids. Bird species that had higher values of the seed-to-fruit ratio had also higher plasma non-enzymatic antioxidant capacity and plasma concentration of uric acid. By contrast, the diet did not explain the variation among bird species in the plasma concentration of vitamin E (Cohen et al. 2009). The authors suggested that their diet classification might not properly reflect the micronutrient levels in food. In addition, variation in the concentrations of carotenoids and vitamin E in plasma were partly explained by differences among species in certain metabolic traits (e.g. usage, storage, recycling; Cohen et al. 2009). It should not be forgotten that both carotenoids and vitamin E are lipophilic compounds that can be stored in tissues. Therefore, any differences among species in mechanisms of storage might also explain why the link between diet and lipophilic antioxidants in blood was not consistent. Another comparative study on birds found that insectivorous species had higher levels of glutathione and glutathione peroxidase than omnivorous species (Fig. 3.3; Messina et al. 2020). In bats, some authors found that frugivorous species had lower levels of oxidative damage markers than insectivorous, haematophagous, or nectivorous species (Schneeberger et al. 2014; Torquetti et al. 2023).

In addition to the amount and types of antioxidants that food contains, other substances (e.g. proteins, fatty acids) and the energetic content may also influence the organismal resistance against oxidative stress (e.g. Hulbert 2005; Pamplona and Barja 2007; Alan and McWilliams 2013; Krief et al. 2023). For example, protein intake might affect the synthesis of some antioxidants (e.g. thiols, enzymes), whose production depends on the availability of amino acids (Li et al. 2014). The fatty acid and protein contents of diet may also affect the oxidative status because both fatty acids and proteins differ in their susceptibility to oxidative damage, and in how they influence the susceptibility of cell membranes to oxidative damage (Hulbert 2005; Pamplona and Barja 2007; Yamato et al. 2007). For example, Alan and McWilliams (2013) found that white-throated sparrows *Zonotrichia albicollis* that fed on diets

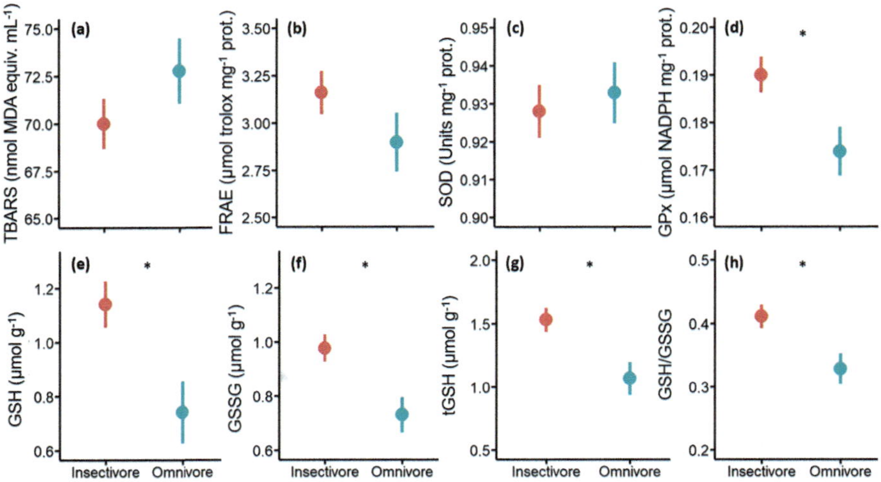


**Fig. 3.3** Comparisons of blood-based oxidative status markers between insectivore and omnivore bird species living in the tropical forests of Borneo. (**a**) TBARS, thiobarbituric acid reactive substances in plasma; (**b**) FRAE, non-enzymatic antioxidant capacity in erythrocytes; (**c**) SOD, superoxide dismutase in erythrocytes; (**d**) GPx, glutathione peroxidase in erythrocytes; (**e**) GSH, reduced glutathione in erythrocytes; (**f**) GSSG, oxidised glutathione in erythrocytes; (**g**) tGSH, total glutathione in erythrocytes; (**h**) ratio of reduced glutathione (GSH) to oxidised glutathione (GSSG). Asterisks indicate significant differences between trophic guilds: * = $P < 0.05$. Values are shown as least square mean and standard error. Reprinted with minor modifications from Messina et al. (2020) with permission of John Wiley and Sons

richer in polyunsaturated fatty acids had higher oxidative damage than birds that fed on diets richer in monounsaturated fatty acids, respectively. Krief et al. (2023) found that free-living chimpanzees *Pan troglodytes* with higher urinary levels of C-peptide (proxy of carbohydrate metabolism) had also higher urinary concentrations of 8-hydroxy-2′-deoxyguanosine (8-OHdG) and lipid peroxides, respectively. The direction and size of an effect of diet on oxidative status markers are not, however, easily predictable. For example, increased intake of dietary antioxidants may not affect levels of oxidative damage because any potential beneficial effects are offset by compensatory reductions in the expression of antioxidant genes and synthesis of enzymes (e.g. Selman et al. 2006; Wang et al. 2006). Finally, blood-feeding animals are exposed to a large intake of iron due to the haem groups that occur in red blood cells. Haem groups can increase the generation of reactive oxygen species (see Chap. 1), so that blood-feeding animals have evolved mechanisms to trap haem groups in order to limit the generation of reactive oxygen species (e.g. Toh et al. 2010; Whiten et al. 2017; Hernandez et al. 2022).

Other studies found significant effects of dietary antioxidants on oxidative damage, but also that these effects were influenced by the amount of antioxidants ingested. For example, guinea pigs *Cavia porcellus* cannot synthesise ascorbic acid, so that they rely on food as source of vitamin C. Barja et al. (1994) fed laboratory guinea pigs for 5 weeks with three different amounts of vitamin C [33 mg day$^{-1}$, which corresponded to an ingested amount of around 0.54–1.1 mg $day^{-1}$ (marginal deficiency); 660 mg $day^{-1}$ (standard optimum level for guinea pigs); 13,200 mg $day^{-1}$]. Compared to guinea pigs maintained on a diet low in vitamin C, animals that were fed larger quantities of vitamin C had lower levels of oxidative damage to both proteins and lipids (Barja et al. 1994).

de Ayala et al. (2006) reduced or increased the brood size of a certain number of couples of barn swallows *Hirundo rustica* in order to either reduce or increase food competition among siblings, respectively. Within each brood, they provided two different doses of tocopherols (comparable to those recorded in prey or daily ingested by nestlings) to some nestlings and compared them to their siblings that did not receive any doses of tocopherols. In so doing, de Ayala et al. (2006) found that the administration of tocopherols increased body mass, body condition, and growth of feathers compared to control nestlings. By contrast, Larcombe et al. (2010) found that the administration of either tocopherols or carotenoids to nestling blue tits *Cyanistes caeruleus* did not improve their growth, nor their body condition. Moreover, the administration of dietary antioxidants did not affect the concentration of malondialdehyde (end product of lipid peroxidation) in plasma. One explanation for this discrepancy between the two studies might lie with the availability of antioxidants in prey. The estimated average concentration of α-tocopherol in prey of barn swallows was 3.93 μg $g^{-1}$ (Ninni 2003; de Ayala et al. 2006), while that in prey of blue tits varied from less than 3.9 μg $g^{-1}$ to around 28 times higher depending on the season (Arnold et al. 2010). Thus, the effects of tocopherols might have been stronger in barn swallows because they occur in more limited supply in their prey than in those of blue tits.

Discrepancies among studies about the effects of tocopherols on organismal physiology or fitness are common in the literature. Deficiency of vitamin E may cause a number of detrimental effects on vertebrates, such as dysfunction of the neural pathways that coordinate movement or induction of myopathic lesions (e.g. Juan-Salles et al. 2003; Thomas et al. 1993). Vitamin E supplementation reduced the levels of oxidative damage markers in breast muscle in house finches *Haemorhous mexicanus*. Birds that received tocopherols or a cocktail of tocopherols and carotenoids had significantly lower levels of breast-muscle oxidative damage markers at the end of the experiment than birds who received carotenoids or were not given any dietary antioxidant (Giraudeau et al. 2013). Other studies on birds found that vitamin E increased the fledging probability in great tits *Parus major* (Maronde and Richner 2015) or protected telomeres from shortening in nestling pied flycatchers *Ficedula hypoleuca* (Pérez-Rodríguez et al. 2019). A higher intake of vitamin E also favoured compensatory growth in juvenile European seabass *Dicentrarchus labrax*, but this came at a cost in terms of increased oxidative damage. It also resulted in decreased activity of superoxide dismutase in erythrocytes and brain, increased activity of glutathione peroxidase in erythrocytes, and increased concentration of protein carbonyls in plasma (Costantini et al. 2018). The effects of vitamin E on the organism may also depend on its interaction with other antioxidants (see Chap. 1). For example, Lebold et al. (2013) fed the zebrafish *Danio rerio* a diet with or without vitamin E for 1 year. In the last 150 days of the experiment, they also decreased the amount of ascorbic acid in the food. They found that fish fed a diet deprived of vitamin E depleted ascorbic acid faster than fish fed a diet with vitamin E and, at the end of the experiment, they had also more oxidative damage, and showed sluggish behaviour and signs of degenerative myopathy of the skeletal muscle. The protective role of dietary antioxidants is therefore dependent on the availability of other antioxidants. This paradigm is well documented in the literature (see also Chap. 1). For example, animals fed with a diet poor in several antioxidants suffered more oxidative damage than animals fed with a diet poor in single types of antioxidants (Fig. 3.4; Hill et al. 2009), and supplementation of vitamin C enhanced body levels of vitamin E and reduced oxidative damage in sea bass larvae (Betancor et al. 2012). The synergism between dietary antioxidants is not necessarily related only to a general enhancement of antioxidant potential. For example, Selman et al. (2008) showed that the supplementation of vitamin E increased significantly both the median and maximum lifespan of mice compared to mice that did not receive any

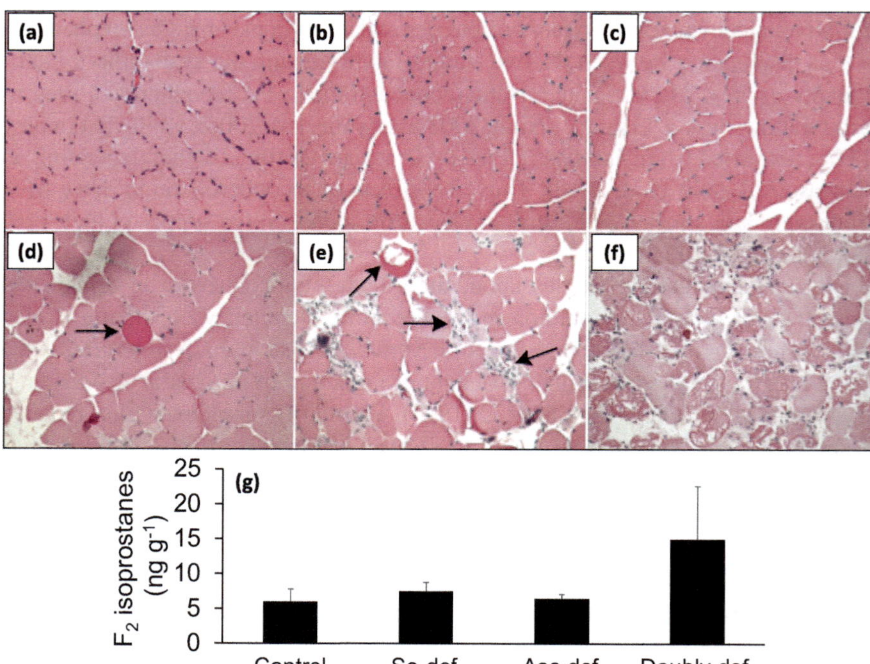


**Fig. 3.4** Photomicrographs of quadriceps muscle of guinea pigs *Cavia porcellus* that were fed four different experimental diets: control, selenium deficient (Se-def), ascorbate deficient (Asc-def), and doubly deficient (lacking both ascorbate and selenium, Doubly-def). Panels (**a**) to (**c**) show muscle of control, selenium-deficient, and ascorbate-deficient animals, respectively. Panels (**d**) to (**f**) show muscle of doubly deficient animals. The arrow in panel (**d**) points to an apoptotic cell, and the arrows in panel (**e**) point to areas of apoptotic and dying cells. Panel (**f**) shows extensive necrosis of the tissue section. Panel (**g**) shows the muscle concentration of $F_2$-isoprostanes in the four experimental groups. Data in panel (**g**) are shown as mean and standard deviation. Reprinted from Hill et al. (2009) with permission of Elsevier

surplus of vitamin E, but it did not influence levels of oxidative damage to DNA and to lipids. Selman et al. (2008) further showed that genes associated with xenobiotic metabolism were significantly upregulated in the livers of supplemented mice, suggesting that vitamin E might initially be metabolised as a xenobiotic. Thus, the effects of vitamin E on lifespan were probably not directly related to its antioxidant properties.

The antioxidant relevance of carotenoids is also controversial. Although they play a non-negligible role in the regulation of oxidative status during embryonic development and the first days after egg hatching in birds (Surai 2002; McGraw et al. 2005), a number of studies (involving three meta-analyses and a large number of species) showed that carotenoids are generally minor antioxidants in birds (Costantini and Møller 2008; Simons et al. 2012; Weaver et al. 2018; Mohr et al. 2019). It is important to point out that minor antioxidant effects of carotenoids were also recorded when their administration to experimental animals was carried out during demanding phases of life. For example, the period of growth and development is particularly demanding because of the high metabolic requirements coupled with immature endogenous antioxidant mechanisms. Thus, we might expect that an increased intake of carotenoids during the period of growth and development would carry some benefits in terms of protection against oxidative stress. Evidence in support of this hypothesis is, however, weak. Costantini et al. (2007) administered to nestling common kestrels *Falco tinnunculus* a dose of about 4 mg of lutein plus zeaxanthin (which are the only carotenoids identified in common kestrels, Casagrande et al. 2006). The administration increased the concentration of carotenoids in serum of about 1.5 times the concentrations recorded in control nestlings or those recorded in the same birds before the start of the experiment (Fig. 3.5). This increase in the levels of carotenoids was within the physiological range of circulating carotenoids previously recorded in both free-living nestling and adult kestrels (Casagrande et al. 2006, 2007; Costantini et al. 2006). Despite this increased cellular availability of carotenoids, the results of the experiment did not show any beneficial effects in terms of reduction of oxidative damage or increase of the serum non-enzymatic antioxidant capacity (Fig. 3.5). It might be that the availability of dietary antioxidants is relevant when intra-brood competition is particularly intense, as the case of strong asynchronous hatching that creates different castes of progeny (the so-called structured family; Hall et al. 2010). For example, Hall et al. (2010) found that there was no overall difference in the level of lipid peroxidation (malondialdehyde) in the plasma of antioxidant supplemented and control nestling red-winged blackbirds *Agelaius phoeniceus*; however, among marginal nestlings, those that received an antioxidant supplement had lower plasma levels of lipid peroxidation.

The relevance of dietary antioxidants is also contingent on the nutritional composition of the food. Plummer et al. (2013) examined the carry-over effects of winter food supplements (fat or fat and vitamin E) on the production of eggs in blue tits. Conversely to predictions, females that were supplemented with fat during winter produced eggs containing smaller yolks and lower amounts of carotenoids than control females. By contrast, females that were supplemented with fat and vitamin E

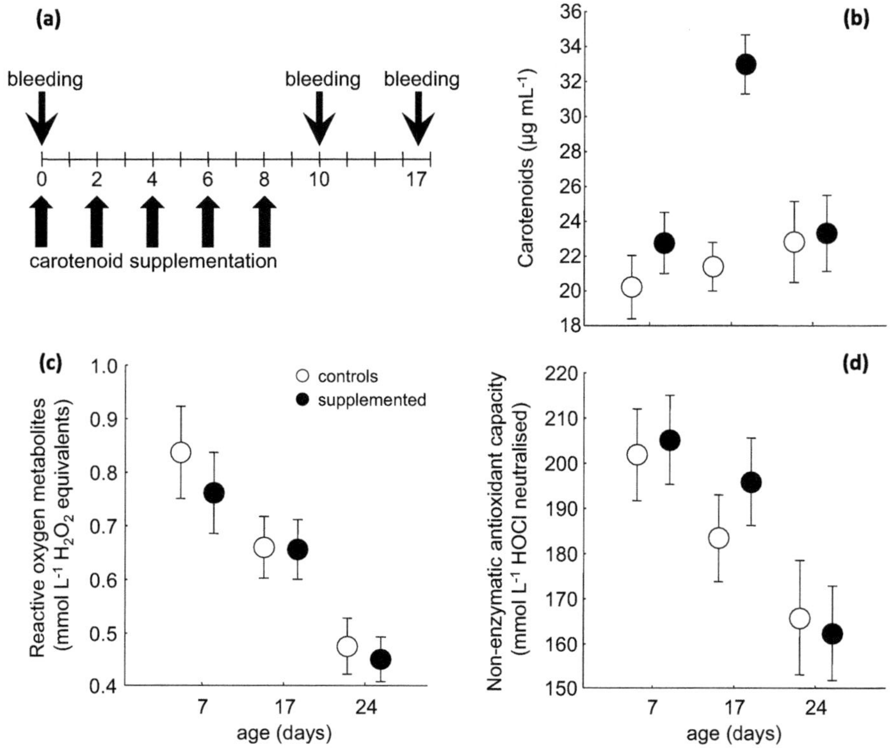


**Fig. 3.5** Supplementation of diet with carotenoids is a classic experimental approach that is used to test their physiological effects on the organismal oxidative status. (**a**) timeline of an experiment of carotenoid supplementation carried out on nestlings of the common kestrel *Falco tinnunculus* (Costantini et al. 2007). Numbers indicate the day; at day 0, nestlings were around 1 week old. (**b**) the concentrations of carotenoids in serum increased in supplemented nestlings, but they returned to pre-treatment values 1 week after the end of the supplementation period; (**c**) the supplementation of diet with carotenoids did not affect the serum levels of oxidative damage; (**d**) the supplementation of diet with carotenoids did not increase the serum non-enzymatic antioxidant capacity. Values are shown as mean and standard error. Reprinted with minor modifications from Costantini et al. (2007)

produced eggs similar to those of control females (Plummer et al. 2013). It might be that a diet rich only in fat increased lipid peroxidation of birds, but this effect was negligible when fat was supplemented together with vitamin E. We cannot exclude the possibility that vitamin E might have protected fat from oxidation (rancidity) catalysed by a continuous exposure to the atmospheric oxygen before the food was eaten by the birds. This mechanism could have enabled birds to ingest less lipid peroxides compared to those ingested by birds that fed only on fat.

In conclusion, these studies show that the roles of dietary antioxidants in the protection against oxidative stress depend on complex interactions between dietary antioxidants themselves, and between dietary antioxidants and nutrients that occur in the food. This is very relevant for our understanding of mechanisms underlying trade-offs in the transfer of substances from mothers to offspring or in the allocation of nutrients between life-history traits. It is also very timely in order to address the adaptability of species to environmental changes. It is increasingly recognised that human activities can impact on the physiology of free-living animals through changes in the quality of diet. For example, it has been found that birds feeding on invasive species or food wastes have altered levels of several oxidative status markers in eggs or in blood compared to conspecific individuals feeding on a more natural diet (García-Tarrasón et al. 2014; Tauler-Ametlller et al. 2019).

## 3.5 Food Availability: From Caloric Restriction to Fasting

As predicted by the food limitation hypothesis, animals that do not eat enough food during demanding phases of the life cycle may experience severe physiological and fitness costs. For example, female pigeons fed on a reduced quality diet had a decline in serum antioxidants compared to female pigeons fed on a standard quality diet during the nestling rearing period (Costantini 2010). Food supplementation reduced plasma levels of oxidative damage of female great tits during the nestling period (Fig. 3.6; Giordano et al. 2015a). Many animal species can even experience long periods of caloric restriction (a reduction in caloric intake) or even fasting (lack of food) while incubating, waning, migrating, or moulting (e.g. Mrosovsky and Sherry 1980; Cherel and Le Maho 1985; Cherel et al. 1988a, b; McCue 2010;

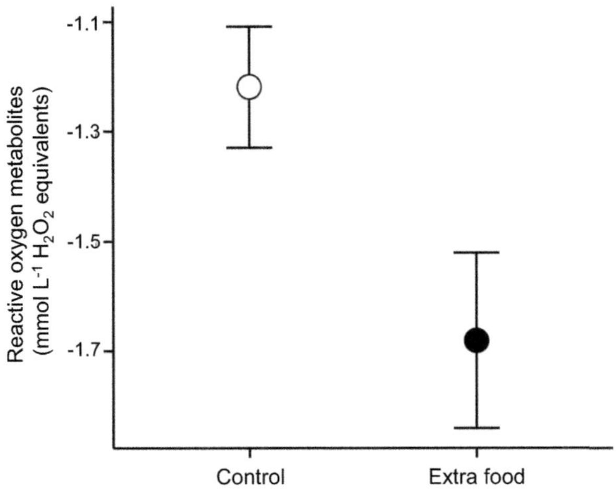


**Fig. 3.6** Female great tits that were food-supplemented (Extra food) during the egg-laying period (15 g of Diptera larvae every other day from before the deposition of the first egg until the clutch was completed) had lower levels of plasma reactive oxygen metabolites (log-transformed) during the nestling period than female great tits that were not food-supplemented (Control). Values are shown as mean and standard error. Reprinted with minor modifications from Giordano et al. (2015a) with permission of Springer Nature

McWilliams et al. 2004; Vázquez-Medina et al. 2011). Animals exhibit a number of biochemical and physiological responses to the lack of food. For example, during a fasting period, a metabolic switch from utilising glucose (phase I) to stored energy (glycogen and fat; phase II) for energy generation, and, eventually, to an enhanced protein catabolism (phase III) can be observed. Alongside, the condition of animals worsens (e.g. loss of body mass), which becomes critical in the phase III of fasting.

Seabirds have evolved a number of physiological adaptations to sustain long periods of caloric restriction or fasting (Cherel and Le Maho 1985; Cherel et al. 1988a). For example, many studies documented significant changes in circulating glucocorticoids, plasma concentration of uric acid, stress proteins, or prolactin (Cherel et al. 1988a; Spée et al. 2010; Thierry et al. 2013). Fasting can also increase the generation of superoxide free radical, hydrogen peroxide, and oxidative damage, or alter antioxidant levels in a variety of species (e.g. Richelle et al. 1999; Sorensen et al. 2006; Vázquez-Medina et al. 2011; Schull et al. 2016). However, other studies found contrasting results. For example, Geiger et al. (2012) showed that plasma oxidative damage decreased during fasting in mallards *Anas platyrhynchos*, and then increased to pre-fasting levels when the animals were re-fed. The non-enzymatic antioxidant capacity of plasma also tended to decrease during the period of fasting, and then increased after the end of the fasting period. These results indicated that fasting not only induces metabolic adjustments, but also modifies organismal levels of substrates (e.g. fatty acids), that could be oxidised by reactive oxygen species, and the levels of dietary antioxidants available, respectively.

In mammals, northern elephant seals *Mirounga angustirostris* experience prolonged periods (up to 3 months) of food and water deprivation while breeding, nursing, moulting, or weaning, with no apparent detrimental effects (Ortiz et al. 1978; Ortiz et al. 2006). Vázquez-Medina et al. (2010) found that elephant seals remained normothermic during their postweaning fast and maintained a high metabolic activity. They also found that fasting induced an increase in the expression of the reactive oxygen species-producing protein Nox4, along with NADPH oxidase activity. Although seals maintained a high metabolic rate and did not ingest any dietary antioxidants during the period of fasting, levels of oxidative damage to lipids and proteins in muscle did not increase at all. Vázquez-Medina et al. (2010) also found that levels of superoxide dismutase, catalase, and glutathione peroxidase increased in muscle and in erythrocytes of fasting seals. Overall, the results of this study suggested that upregulation of antioxidant enzymes was probably sufficient to protect against the generation of oxidative damage in seals that fasted for a prolonged period of time. However, upregulation of antioxidant enzymes during fasting does not always buffer against increased generation of oxidative damage in vertebrates (Figs. 3.7 and 3.8) (e.g. Pascual et al. 2003; Morales et al. 2004; Zhang et al. 2008).

These studies on mallards and seals suggest that certain animal species might have evolved mechanisms that enable them to save antioxidants for periods of need, such as fasting (*antioxidant-saving strategy hypothesis*). For example, burrowing petrels desert spontaneously their egg after periods of long fasting without any risk for the egg itself (Jouventin et al. 1985). The parental decision to desert the egg is to

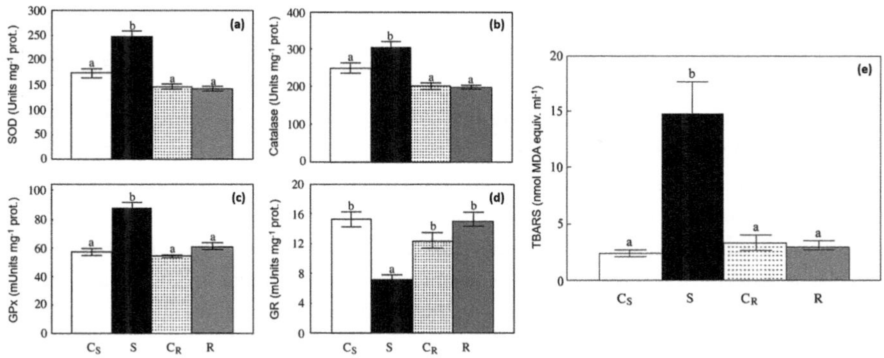


**Fig. 3.7** Effects of starvation (5 weeks) and refeeding (3 weeks) on levels of (**a**) superoxide dismutase, (**b**) catalase, (**c**) glutathione peroxidase, (**d**) glutathione reductase, and (**e**) thiobarbituric acid reactive substances in liver of common dentex *Dentex dentex*. $C_S$ = control-starvation; S = starvation; $C_R$ = control-refeeding; R = refeeding. Different letters indicate significant differences among groups. Values are shown as mean and standard error. Reprinted with minor modifications from Morales et al. (2004) with permission of Elsevier

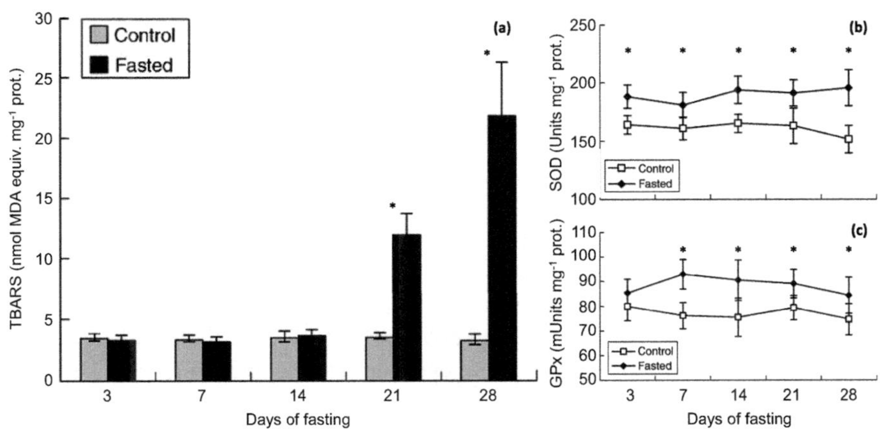


**Fig. 3.8** Effects of fasting on levels of (**a**) thiobarbituric acid reactive substances in muscle, (**b**) superoxide dismutase in liver, and (**c**) glutathione peroxidase in liver of yellow croakers *Pseudosciaena crocea*. Asterisks indicate significant differences between control and starvation groups within each sampling day. Values are shown as mean and standard deviation. Reprinted with minor modifications from Zhang et al. (2008) with permission of Elsevier

some extent related to the loss of body mass (Chaurand and Weimerskirch 1994a, b). Stress hormones could be important molecular modulators of the link between deserting behaviour and oxidative status. Glucocorticoids increase during the period of fasting (Cherel et al. 1988a), affect the regulation of oxidative status (Costantini et al. 2011), and modulate behavioural responses to stressful events (Wingfield et al. 1998). The observation that incubating blue petrels in worsen condition (proxy of fasting) had lower levels of circulating antioxidants suggests that the deserting behaviour might also provide benefits in terms of maintenance of the antioxidant capacity, a strategy that might also explain why petrels have a long lifespan (Costantini and Bonadonna 2010).

There is also growing evidence that in some species females and males may differ in their physiological responses to poor food conditions, possibly owing to their different reproductive strategies. Noiret et al. (2020) tested the occurrence of sexual differences in energy management and physiological consequences of fasting in the gray mouse lemur *Microcebus murinus* because females and males differ strongly in their reproductive strategies and seasonal body mass changes. The results of the experiment showed that females and males relied on different physiological strategies to cope with 2 weeks of caloric restriction at 60%. For example, while the concentration of 8-OHdG (marker of oxidative DNA damage) was similar between sexes, red blood cell levels of glutathione peroxidase and total thiols increased in females and decreased in males, respectively (Noiret et al. 2020). It might be that females maintained high constitutive levels of antioxidants because, conversely to males, they did not enter a hypometabolic state.

We have also evidence that some species are more vulnerable to the pro-oxidant effects of fasting than others. For example, larvae of the crested newts *Triturus macedonicus* and *Triturus ivanbureschi* that experienced food deprivation for 2 weeks had higher levels of thiobarbituric acid reactive substances and superoxide dismutase, and lower glutathione peroxidase, catalase, and glutathione compared to larvae that did not experience any food deprivation (Prokić et al. 2021).

Animals can also experience periods of reduced food (caloric) intake, for example when food occurs in short supply. A reduction of caloric intake for short periods of time during development seems to mould stress response mechanisms through hormetic priming, making the individual more resistant to stress in adulthood (Mangel 2008; Mattson 2008). Kaneko et al. (2011) found that rotifers *Brachionus plicatilis* under caloric restriction lived 50% longer, had enhanced oxidative stress resistance, and increased expression of genes for catalase and manganese superoxide dismutase compared to control rotifers. In addition, the daughters of the calorically restricted mothers lived 20% longer than those from the mothers fed ad libitum irrespective of nutritional conditions under which the daughters were maintained. Furthermore, the daughters of the caloric-restricted mothers were endowed at birth with a higher resistance against oxidative stress and increased expression of genes for catalase, but not for manganese superoxide dismutase. Finally, daughters of the caloric-restricted mothers laid eggs with higher contents of catalase and manganese superoxide dismutase. However, most of the work on the link between caloric restriction and hormesis is based on laboratory organisms maintained on ad libitum food, thus the effects of caloric restriction are probably exaggerated by using overweight controls. We have, however, some evidence that short-term stimulatory effects of caloric restriction might also occur in free-living animals (e.g. Noguera et al. 2011). In adults, food availability might not necessarily influence the organismal resistance to oxidative stress. For example, Stahlschmidt et al. (2020) found that sand field crickets *Gryllus firmus*, that had intermittent access to food, had a smaller body mass than control crickets, but they were resistant against an oxidative challenge as much as control crickets were.

Caloric restriction may also enhance the organismal capacity to withstand other types of stressors. For example, Kailasam et al. (2011) found that rotifers *Brachionus plicatilis* previously exposed to caloric restriction lived longer and had higher expression of the antioxidant enzyme genes, manganese superoxide dismutase and catalase, than controls when subsequently exposed to a chemical generator of reactive oxygen species. Thus, caloric restriction prepared rotifers for a future oxidative challenge through a prophylactic upregulation of the antioxidant machinery.

We do not have any definitive proof about the role of caloric restriction as a trigger of hormetic responses to oxidative stress. Most of the studies involved only the comparison between a control and a caloric-restricted group. Moreover, it is unclear if hormesis-like effects of caloric restriction occur only when it is the intake of certain nutrients to be limited (e.g. Piper et al. 2005). When nutritional quality is inadequate, studies showed that oxidative damage can be increased (malnutrition; Feoli et al. 2006). For example, it has been shown that zebra finches *Taeniopygia guttata* that were grown on a diet poor in carotenoids and vitamins had a strong reduction of plasma levels of carotenoids, vitamins A, and E in adulthood (Blount et al. 2003a), possibly owing to a reduced capacity in the assimilation or storage of these antioxidants.

Many food molecules can be both nutritionally relevant or toxic for organisms, depending on circumstances like amount ingested or needs of the organism (Ames et al. 1990; Berenbaum 1995; Raubenheimer and Simpson 2009). Reviews of the literature led several authors to conclude that this dichotomy between beneficial and toxic is owing to the hormetic potential that, at least, some dietary substances have (Hayes 2007; Mattson 2008; Calabrese and Mattson 2010). It has also been suggested that the dual pro-/anti-oxidant nature of some dietary substances underlies this hormetic potential. For example, some phytonutrients can cause a mild injury to the mitochondria; this injury would stimulate the activation of self-maintenance mechanisms through the signalling role of reactive oxygen species generated into the mitochondria (*mitohormesis hypothesis*; Tapia 2006).

There is also good evidence that non-essential dietary substances generate typical hormetic responses (e.g. Stipanovic et al. 1986; Calabrese and Blain 2005; Hayes 2007; Mattson 2008; de la Celorio-Mancera et al. 2011; Padró et al. 2018). Nutrients that generate hormetic responses (hormetins sensu Rattan 2008) can influence one or more pathways related to stress responses or cellular maintenance and repair (Ali and Rattan 2006). Dietary antioxidants provide an example of substances that can have beneficial effects at low doses, but that turn into pro-oxidants or toxic at high doses. For example, the antibody production in response to *Escherichia coli* or Newcastle disease was significantly higher in chicks that did not receive or received 10 mg $kg^{-1}$ added vitamin E, whereas it was significantly lower in birds that received 30 or 150 mg $kg^{-1}$ added vitamin E (Friedman et al. 1998).

It should not be forgotten that hormetic responses to food intake might also depend on the potential energy content of food. It has been suggested that low levels of energy intake activate hormetic responses in a variety of organisms, including yeast, insects, rats, mice, and humans (Mattson 2008). For example, it has been found that a low level of dietary energy restriction increased levels of the heat shock protein 70 (Heydari et al. 1993) and reduced oxidative stress (Ugochukwu and Figgers 2007) in the liver of laboratory rats. The potential hormetic effects induced by dietary energy restriction can also involve changes in regulatory systems of cellular energy metabolism (Rodgers et al. 2005; Hyun et al. 2006; Liu et al. 1997), in particular through changes of the mitochondrial activity (Tapia 2006).

## 3.6 Changing Eating Habits

In seasonal environments, food quality and availability vary across the year, so that physiological plasticity might be an important mechanism through which animals acclimate to the new trophic conditions. Moreover, colonisation of new environments may expose animals to new food resources, such as those that can be found in cities. Changes in food quality, quantity, and diversity may occur slowly or be abrupt as in the case of environments impacted by human encroachment (e.g. forest logging, agricultural expansion), which might rapidly and severely alter the diet of animals and be responsible for stress. For example, abrupt changes to the diet may disrupt physiological homeostasis with several metabolic consequences for the organism, such as alteration of microbiota observed in humans and other vertebrates (e.g. Walker et al. 2011; Garber et al. 2020). There are only a few studies in the literature that investigated the links among changes in diet, metabolic adaptation to food type, and oxidative stress. For example, Magnanou et al. (2009) compared the gene expression between a population of woodrats *Neotoma lepida* from the Great Basin, which feeds on the ancestral diet of juniper, and a population of woodrats from the Mojave Desert, which switched from feeding on juniper to feeding on creosote. Individual woodrats from each population were fed juniper and creosote diets, separately. Juniper and creosote have notable differences in secondary chemistry; consequently, there were large differences in expression of biotransformation genes when woodrats were fed juniper vs. creosote. In particular, Mojave woodrats had greater expression of biotransformation genes than Great Basin woodrats on a creosote diet (Magnanou et al. 2009). Moreover, superoxide dismutase was also expressed to a greater extent in Mojave woodrats, suggesting that upregulation of antioxidant enzymes might be important in protecting cells against reactive oxygen species generated either directly from plant secondary compounds in creosote resin or from biotransformation metabolites (Magnanou et al. 2009). Further work on woodrats also showed a significant impact of diet on the enzymes related to the glutathione pathway (Magnanou et al. 2013).

Changes in diet are not necessarily gradual; they can also be rapid. Costantini et al. (2021) carried out an individual-based longitudinal experiment under controlled conditions to test the hypothesis that an abrupt decrease in dietary antioxidants would increase oxidative damage generation in captive chimpanzees, while a rapid increase of antioxidant content in diet would enable chimpanzees to decrease oxidative damage accrued while on the low-antioxidant diet, respectively. This experimental design enabled to test how fast the regulation of oxidative status is in tackling rapid changes of antioxidant content in diet, which may occur in free-living primates. The results of the experiment showed that a sudden decrease in dietary antioxidants (estimated by the low oxygen radical absorption of food) increased urinary levels of lipid peroxides and of 8-OHdG, but not of 8-isoprostanes (Fig. 3.9;

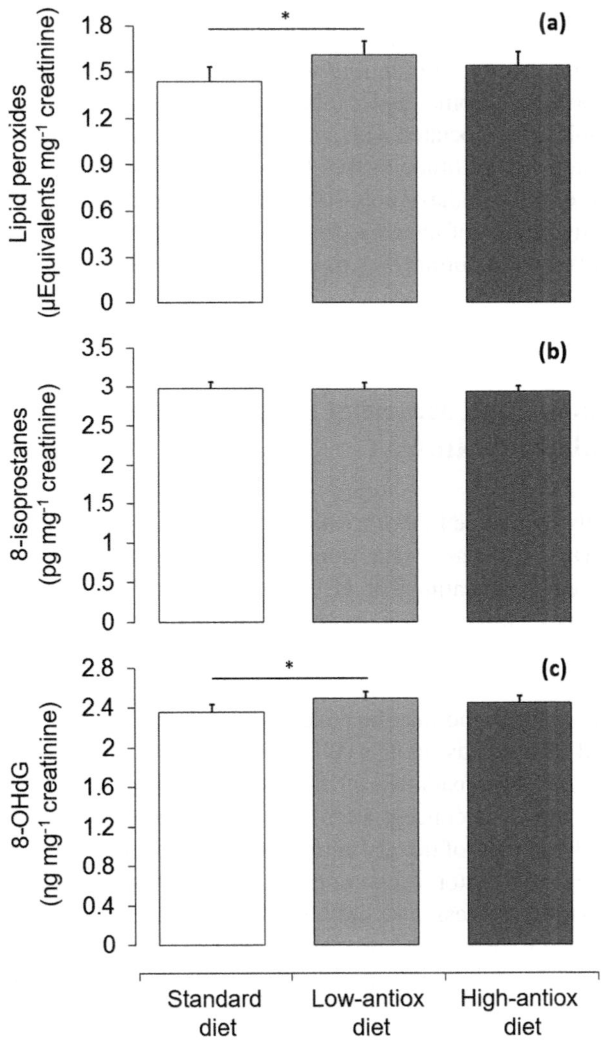


**Fig. 3.9** Comparison of urinary oxidative damage markers among chimpanzees that were on a control diet (Standard diet), a low-antioxidant diet (Low-antiox diet), and a high-antioxidant diet (High-antiox diet), respectively. (**a**) lipid peroxides, (**b**) 8-isoprostanes, and (**c**) 8-hydroxy-2′-deoxyguanosine (8-OHdG). Asterisks indicate significant differences between groups: * = $P < 0.05$. Values are log10-transformed and are shown as estimated marginal mean and standard error. Reprinted with minor modifications from Costantini et al. (2021). © 2021, The American Physiological Society

Costantini et al. 2021). Conversely, a short-term improvement of diet quality (i.e. more fruits and vegetables containing a higher concentration of antioxidants, determining a high oxygen radical absorption of food) could not sufficiently reduce urinary levels of oxidative damage markers to their prior values. The endogenous antioxidant mechanisms of chimpanzees did not appear to be sufficiently flexible to rapidly adjust to these sudden changes in diet because they did not prevent the increase of oxidative damage when chimpanzees were on the low-antioxidant diet. These results indicated that both intake of dietary antioxidants and rapidity/duration of dietary changes play non-negligible roles in the regulation of oxidative status.

As modern humans, we are familiar with the health effects of eating junk food. Obesity, metabolic syndrome, and cardiovascular diseases are some of the most renowned pathologies associated with bad eating habits. A growing number of studies are documenting the attitude of free-living animals to feed habitually on waste food of human origin. We have now some evidence that feeding on this food can alter the oxidative status of animals (Pineda-Pampliega et al. 2021; Bernat-Ponce et al. 2023). This is an important area of investigation to get insight about the impact of human activities on wildlife.

## 3.7 Nutritional and Maternal Programming of Oxidative Status

Mothers can rely on a wide toolkit to mould the oxidative status of their offspring. This toolkit involves (i) a certain amount of molecules that mothers passively deposit into the eggs, indicating that the composition of maternal diet influences the availability of these molecules (*resource limitation paradigm*) and (ii) another amount of molecules whose deposition into the egg is to some degree under female control (*active regulation paradigm*).

Royle et al. (1999) found that the concentration of the two dietary antioxidants, tocopherols and carotenoids, in egg yolk varied with the laying order of the eggs. Given the link between oxidative stress and testosterone metabolites, and that, within a same clutch, the concentration of testosterone in egg yolk may be negatively correlated with that of dietary antioxidants, Royle et al. (2001) hypothesised that these two maternal factors have complementary, but opposing effects: testosterone induces oxidative stress and dietary antioxidants mitigate oxidative stress, respectively. Royle et al. (2001) also suggested that the effect of testosterone on offspring hatched from last-laid eggs would be detrimental if food were scarce owing to the reduced availability of dietary antioxidants.

Effects of maternally derived antioxidants are not strictly limited to the embryonic period, but persist after hatching. For example, Karadas et al. (2005) found that the hepatic concentration of carotenoids soon after hatching was much higher in chicks *Gallus gallus domesticus* from mothers that were supplemented with carotenoids than in chicks from mothers that were not supplemented. Importantly, Karadas et al. (2005) also showed that the higher hepatic concentration of carotenoids in chicks from supplemented mothers persisted for the first week of the post-hatching life, even when chicks were not provided any dietary antioxidants. Although the long persistence of maternal carotenoids in the tissues of offspring might be advantageous in terms of protection against post-hatching oxidative stress (Karadas et al. 2005), a study on chicks of the yellow-legged gull *Larus michahellis* showed that an experimental increase of carotenoid concentration in egg yolk caused an increase in oxidative damage in nine-days old male chicks compared to control chicks (Saino et al. 2011). Moreover, (i) chicks from first-laid eggs enriched with carotenoids had higher oxidative damage than chicks from first-laid control eggs and (ii) chicks from last-laid control eggs had higher oxidative damage than chicks from last-laid eggs enriched with carotenoids. By contrast, carotenoid enrichment of eggs did not influence levels of oxidative damage recorded in chicks at hatching (Saino et al. 2011). Marri and Richner (2014) also showed that carotenoid enrichment of eggs did not influence oxidative damage levels of nestling great tits on day 12 post-hatch. Therefore, the expectation that carotenoids would reduce oxidative damage of chicks is oversimplistic. For example, carotenoids may boost the immune function, which may result in increased generation of reactive oxygen species and oxidative damage (Costantini 2022). Similarly, work on amphibians also found poor evidence in favour of beneficial effects of carotenoids in growing individuals. Increased carotenoid intake had no effect on survival to metamorphosis in gray treefrogs *Dryophytes versicolor* and caused lower survival to metamorphosis in wood frogs *Lithobates sylvaticus* (*Rana sylvatica*), respectively (Cothran et al. 2015). Increased intake of carotenoids also caused both frog species to experience slower development and growth. When exposed to the fungal pathogen *Batrachochytrium dendrobatidis* after metamorphosis, wood frogs experienced high mortality, and feeding on a diet rich in carotenoids did not have any mitigating effects on the infection (Cothran et al. 2015). The antioxidant relevance of carotenoids is probably mediated by complex interactions among multiple factors, such as laying order, maternal condition, concentration of other molecules deposited by mothers into the egg, sibling competition, or additional functions that carotenoids may perform in the organism.

Accurately predicting the effect of other maternally derived antioxidants is also challenging. For example, Lin et al. (2005) found that the protective effects of maternally derived vitamin E on chicks may last several days. In particular, Lin et al. (2005) found that (i) brain levels of thiobarbituric acid reactive substances of chicks decreased with the increase of the amount of vitamin E given to their mothers, (ii) the generation of reactive oxygen species decreased in chicks from supplemented mothers compared to controls, and (iii) the effects of vitamin E on antioxidant enzymes differed between tissues and doses of vitamin E administered to the mothers. However, other studies showed that the effects of vitamin E on nestlings vary among species (e.g. de Ayala et al. 2006; Larcombe et al. 2010).

The composition in fatty acids and amino acids of maternal diet has also a significant impact on the offspring oxidative status. For example, animals can actively control the molecular composition of their cell membranes and, in so doing, they can influence some properties of cell membranes, such as fluidity, but also resistance to oxidative damage (e.g. use of fatty acids, Hulbert et al. 2007; Buttemer et al. 2008). The nutrients that females pass over to their offspring can also influence the susceptibility of embryo tissues to oxidative stress because nutrients are potential targets of oxidation (Surai et al. 1996). For example, the arachidonic acid (a type of polyunsaturated omega-6 fatty acid) forms between 8 and 19% of the phospholipid fatty acids of egg yolk of the northern gannet *Morus bassanus*, the great skua *Catharacta skua*, the American white pelican *Pelecanus erythrorhynchos*, and the double-crested cormorant *Phalacrocorax auritus* (Surai et al. 2001). In pelicans and cormorants, the yolk composition of fatty acids is consistent with the consumption of freshwater fish in which arachidonic acid may be a significant acyl constituent. By contrast, it is difficult to explain why gannets and skuas have such an amount of arachidonic acid in yolk because they largely feed on marine fish, which have a low content of arachidonic acid. These results suggest that female birds may actively control the deposition of fatty acids in the egg (Surai et al. 1996). The composition and the quantity of fatty acids in diet have also relevant effects on the oxidative status of offspring. A meta-analysis involving studies on laboratory rodents showed that higher malondialdehyde levels occurred in offspring of females maintained on a high-fat diet (Moraes-Souza et al. 2021). Furthermore, offspring of mothers exposed to high-fat diet during pregnancy and/or lactation showed greater generation of reactive oxygen species and lower levels of both superoxide dismutase and catalase (Moraes-Souza et al. 2021).

Although mothers have some control on which nutrients are passed over to their offspring, the prolonged exposure of mothers to a poor diet may have detrimental consequences for the offspring (including increased oxidative stress). This is particularly relevant for mammals, where the physiological intimacy between the developing foetus and the mother occurs over a prolonged period of time. These detrimental effects of a poor maternal diet may involve malnutrition of offspring or the metabolic costs owing to a slowdown of growth followed by a post-natal catch-up growth (see Chap. 7). For example, offspring of female rats that were on a diet poor in proteins lived shorter than offspring of female rat controls (Jennings et al. 1999). This reduction in longevity might be explained by an accelerated rate of telomere shortening associated with growth retardation during the foetal life followed by a period of post-natal catch-up growth (Jennings et al. 1999) or by an increase of the oxidative damage (Langley-Evans and Sculley 2005), which were both observed in rats from protein-restricted pregnant mothers. The temporal window during which mothers experience poor nutrition is very important in terms of consequences for the offspring. For example, Tarry-Adkins et al. (2008) compared the oxidative status between 3- and 12-months old offspring of rat mothers from three experimental groups: controls in which mothers were always fed nutritionally adequate diets; protein restriction of mothers during gestation followed by adequate nutrition during lactation that enabled for rapid catch-up growth of offspring; and protein restriction of mothers only during lactation. Offspring from mothers that were on a diet poor in proteins during gestation, and then were cross-fostered to control diet-fed female rats, had higher levels of oxidative DNA damage and telomere shortening than both control offspring and offspring of mothers that were only

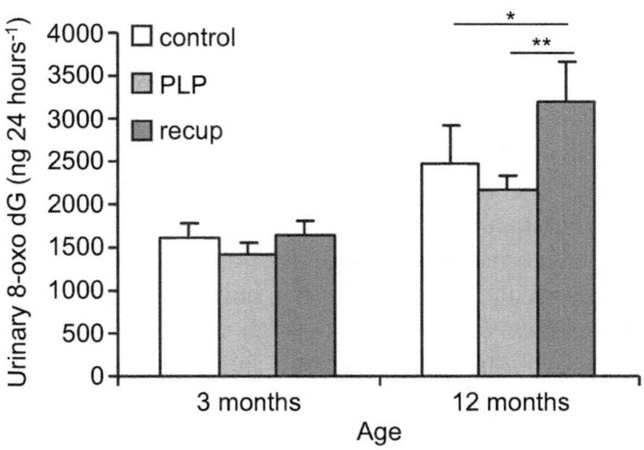


**Fig. 3.10** Comparison of urinary levels of oxidative DNA damage among three groups of rat offspring, whose mothers were maintained during pregnancy on a 20% protein diet (control) or an isocaloric low-protein (8%) diet, and that were then maintained on different post-natal nutritional conditions. Cross-fostering techniques were used to generate protein-restricted offspring during gestation or lactation. The post-natal low-protein (PLP) group were offspring of rat dams fed the control diet during pregnancy and then nursed by rat dams fed the low-protein diet. The recuperated group (recup) were offspring of mothers fed the low-protein diet during gestation and then cross-fostered to control diet-fed rat dams (followed by rapid catch-up growth). The control group were control diet-fed offspring that were suckled by control diet-fed rat dams. All animals were weaned onto a standard diet containing 20% proteins at 21 days of age, and remained on this diet until the end of the study. Asterisks indicate significant differences between groups: * $P < 0.05$, ** $P < 0.01$. Values are shown as mean and standard error. Reprinted with minor modifications from Tarry-Adkins et al. (2008) with permission of John Wiley and Sons

protein-restricted during the post-natal period (Fig. 3.10; Tarry-Adkins et al. 2008). These effects were partly driven by the pattern of growth: much faster growth rates were recorded in offspring from mothers that experienced a gestational protein restriction and then were cross-fostered to control diet-fed female rats (Tarry-Adkins et al. 2008). Offspring of mothers that were only protein-restricted during the post-natal period had also higher levels of superoxide dismutase and catalase in aorta homogenates than the other two experimental groups. Further work also showed that protein-restricted mother rats generated offspring that experienced accelerated telomere shortening, alterations of antioxidant defences, increased oxidative DNA damage, and reduced expression of molecules involved in the repair of DNA in the thymus (Tarry-Adkins et al. 2019).

We have also evidence that female and male offspring can be affected differently from maternal diet. Giordano et al. (2015b) found that experimentally modified pre- and early post-natal nutritional conditions affected the oxidative status of nestling great tits in a sex-specific way. Daughters of non-supplemented mothers, and daughters which did not receive extra food during the early post-natal period, had higher oxidative damage than sons, while no differences between sons and daughters were observed when extra food was provided pre- or post-natally. The results of this experiment showed that nutritional conditions experienced early in life, and, in particular, the resources transferred by the mother through the egg, may play an important sex-specific role in shaping the oxidative status of offspring.

Finally, we have evidence that early and late developmental nutritional conditions may interact in shaping multiple aspects of the antioxidant system. Zebra finches that had a low micronutrient diet during the growth phase initially had a lower plasma non-enzymatic antioxidant capacity than those reared under a high micronutrient diet, but then showed a compensatory response, so that by the end of the growth phase, the plasma non-enzymatic antioxidant capacity of the two groups was similar (Noguera et al. 2015). In addition, only those birds that continued with the same dietary treatment (low or high) throughout development showed a significant increase in their plasma non-enzymatic antioxidant capacity and in the red blood cell glutathione peroxidase during the period of sexual maturation (Noguera et al. 2015).

## 3.8 Conclusions and Future Directions

Access to food plays a key role in the life of animals because it provides nutrients to sustain a number of organismal functions, such as energy production, development, body maintenance, and reproduction. In this book chapter, I have illustrated how the need to regulate the organismal oxidative status has been one important proximate mechanism underlying the evolution of nutritional strategies. Animals are not simply passive absorbers of nutrients. They have evolved a number of strategies to look for those foods that provide the substances they need (e.g. self-medication) or to selectively absorb some nutrients at a cost of others because they are more important to meet the prevailing metabolic demands. Searching for food is, however, costly for the organism because it requires consumption of several resources, also including antioxidants that are needed to mitigate the action of reactive oxygen species generated by the physical activity (see Chap. 6).

Many studies found convincing support for a role of early in life diet composition in determining a number of long-lasting effects on the adult phenotype. This research has important implications in terms of the way mothers can use nutrients (including antioxidants) to mould offspring phenotype, including the mechanisms regulating the cellular oxidative status.

We have also evidence that both gradual and rapid fluctuations in the availability and quality of food may have a number of relevant consequences for the organismal resistance to oxidative stress. For example, mild restriction of caloric intake may stimulate self-maintenance mechanisms (hormesis) that can impact on the oxidative status of the individual, as well as of its offspring and nephews. In other circumstances, the endogenous antioxidant mechanisms might not be sufficiently flexible to rapidly adjust to any sudden changes in diet.

It remains elusive whether phytonutrients (e.g. the so-called dietary antioxidants) act as important antioxidants or whether some of their effects stem more from their pro-oxidant properties. Biomedical, nutritional, and zootechnical sciences shifted their focus from supplementations of single antioxidants to those of cocktails of antioxidants, in combination with different quantities and typologies of lipids, proteins, or carbohydrates, and under different circumstances (e.g. exposure to a stressor, pregnancy, weaning). This is absolutely necessary because it is increasingly recognised that the physiological actions of dietary antioxidants depend on complex molecular interactions and specific needs of the organism. Other disciplines, although aware of this theoretical background, still struggle in their experimental application. In addition, we need to consider that species differ in digestive function and flexibility, which may inform us on how and why the organism gives priority to the absorption of certain nutrients or antioxidants at a cost of others. For example, while flavonoids effectively decreased absorption of glucose in rats *Rattus norvegicus*, they did not influence the absorption of glucose in American robins *Turdus migratorius* (Skopec et al. 2010). One explanation for these results might lie in the coevolution of mechanisms that enable simultaneous absorption of glucose and flavonoids in American robins because, in contrast to rats, these birds need glucose to sustain their energy-demanding migratory flight and flavonoids to mitigate the impact of reactive oxygen species generated by the intense work of flight muscle, respectively. Thus, we could speculate that the absorption of flavonoids might constrain glucose metabolism more in sedentary than in migratory species.

Karasov (2011) argued that the matching between digestive function and diet composition might provide evidence of trade-offs that preclude doing well on all possible dietary substrates with a single digestive design. The flexibility of the digestive system varies largely among species, and having a reversible flexibility is very important because it enables species to increase exploitation of local resources to sustain a successful reproduction across different environments (Karasov 2011). This flexibility is likely to be also important when animals experience abrupt changes in diet quality and quantity as shown in the paragraph 3.6, and to explain why some species are better at exploiting food products made for human consumption or their remains, such as food waste (for an example about impact of foraging in landfills on oxidative status, see Pineda-Pampliega et al. 2021).

I have finally illustrated a number of scenarios under which mothers can rely on a wide nutritional toolkit to mould the oxidative status of their offspring. This toolkit involves both molecules that mothers passively deposit into the eggs, and molecules whose deposition into the egg is to some degree under female control. Evidence suggests that maternal antioxidants (i) play a relevant role in the protection of embryos and young against oxidative stress and (ii) can even influence the absorption and storage of dietary antioxidants in adulthood. However, we are still in need of a better understanding of the proximate mechanisms underlying the passive and active regulations of the deposition of maternal antioxidants into the egg.

## References

Alan RR, McWilliams SR (2013) Oxidative stress, circulating antioxidants, and dietary preferences in songbirds. Comp Biochem Physiol Part B 164:185–193

Alan RR, McWilliams SR, McGraw KJ (2013) The importance of antioxidants for avian fruit selection during autumn migration. Wilson J Ornithol 125:513–525

Ali RE, Rattan SIS (2006) Curcumin's biphasic hormetic response on proteasome activity and heat shock protein synthesis in human keratinocytes. Ann N Y Acad Sci 1067:394–399

Ames BN, Profet M, Gold LS (1990) Dietary pesticides (99.99% all natural). Proc Natl Acad Sci USA 87:7777–7781

Andersson MN, Nilsson J, Nilsson JÅ, Isaksson C (2018) Diet and ambient temperature interact to shape plasma fatty acid composition, basal metabolic rate and oxidative stress in great tits. J Exp Biol 221:jeb186759

Arnold KE, Ramsay SL, Henderson L, Larcombe SD (2010) Seasonal variation in diet quality: antioxidants, invertebrates and blue tits *Cyanistes caeruleus*. Biol J Linn Soc 99:708–717

Barbehenn RV, Jones CP, Hagerman AE, Karonen M, Salminen J-P (2006) Ellagitannins have greater oxidative activities than gallotannins and condensed tannins at high pH: potential impact on caterpillars. J Chem Ecol 32:2253–2267

Barbehenn RV, Niewiadomski J, Pecci C, Salminen J-P (2013) Physiological benefits of feeding in the spring by *Lymantria dispar* caterpillars on red oak and sugar maple leaves: nutrition versus oxidative stress. Chemoecol 23:59–70

Barja G, López-Torres M, Pérez-Campo R, Rojas C, Cadenas S, Prat J, Pamplona R (1994) Dietary vitamin C decreases endogenous protein oxidative damage, malondialdehyde, and lipid peroxidation and maintains fatty acid unsaturation in the Guinea pig liver. Free Radic Biol Med 17:105–115

Beaulieu M, Schaefer HM (2013) Rethinking the role of dietary antioxidants through the lens of self-medication. Anim Behav 86:17–24

Beaulieu M, Schaefer HM (2014) The proper time for antioxidant consumption. Physiol Behav 128:54–59

Beaulieu M, Ropert-Coudert Y, Le Maho Y, Ancel A, Criscuolo F (2010) Foraging in an oxidative environment: relationship between $\delta^{13}C$ values and oxidative status in Adélie penguins. Proc R Soc Lond B 277:1087–1092

Beaulieu M, Haas A, Schaefer HM (2014) Self-supplementation and effects of dietary antioxidants during acute thermal stress. J Exp Biol 217:370–375

Beaulieu M, González-Acuña D, Thierry AM, Polito MJ (2015) Relationships between isotopic values and oxidative status: insights from populations of gentoo penguins. Oecologia 177:1211–1220

Beaulieu M, Bischofberger I, Lorenz I, Scheelen L, Fischer K (2016) Reproducing butterflies do not increase intake of antioxidants when they could benefit from them. Biol Lett 12:20150941

Beaulieu M, Franke K, Fischer K (2017) Feeding on ripening and over-ripening fruit: interactions between sugar, ethanol and polyphenol contents in a tropical butterfly. J Exp Biol 220:3127–3134

Berenbaum MR (1995) Turnabout is fair play - secondary roles for primary compounds. J Chem Ecol 21:925–940

Bernat-Ponce E, Gil-Delgado JA, Guardiola JV, López-Iborra GM (2023) Eating in the city: experimental effect of anthropogenic food resources on the body condition, nutritional status, and oxidative stress of an urban bioindicator passerine. J Exp Zool A 339:803–815

Betancor MB, Caballero MJ, Terova G, Cora S, Saleh R, Benitez-Santana T, Bell JG, Hernandez-Cruz CM, Izquierdo M (2012) Vitamin C enhances vitamin E status and reduces oxidative stress indicators in sea bass larvae fed high DHA microdiets. Lipids 47:1193–1207

Birnie-Gauvin K, Peiman KS, Raubenheimer D, Cooke SJ (2017) Nutritional physiology and ecology of wildlife in a changing world. Conserv Physiol 5:cox030

Blount JD, Metcalfe NB, Arnold KE, Surai PF, Devevey GL, Monaghan P (2003a) Neonatal nutrition, adult antioxidant defences and sexual attractiveness in the zebra finch. Proc R Soc Lond B 270:1691–1696

Blount JD, Metcalfe NB, Birkhead TR, Surai PF (2003b) Carotenoid modulation of immune function and sexual attractiveness in zebra finches. Science 300:125–127

Bolser JA, Alan RR, Smith AD, Li L, Seeram NP, McWilliams SR (2013) Birds select fruits with more anthocyanins and phenolic compounds during autumn migration. Wilson J Ornithol 125:97–108

Butler MW, Lutz TJ, Fokidis HB, Stahlschmidt ZR (2016) Eating increases oxidative damage in a reptile. J Exp Biol 219:1969–1973

Buttemer W, Battam H, Hulbert AJ (2008) Fowl play and the price of petrel: long-living *Procellariiformes* have peroxidation resistant membrane composition compared with short-living *Galliformes*. Biol Lett 4:351–354

Byrne M, Sewell MA, Prowse TAA (2008) Nutritional ecology of sea urchin larvae: influence of endogenous and exogenous nutrition on echinopluteal growth and phenotypic plasticity in *Tripneustes gratilla*. Funct Ecol 22:643–648

Calabrese EJ, Blain R (2005) The occurrence of hormetic dose responses in the toxicological literature, the hormesis database: an overview. Toxicol Appl Pharmacol 202:289–301

Calabrese EJ, Mattson MP (2010) The hormetic pharmacy: the feature of natural products and mad-made drugs in disease prevention and treatment. In: Mattson MP, Calabrese EJ (eds) Hormesis: a revolution in biology, toxicology and medicine. Springer, New York, pp 177–198

Carbeck KM, DeMoranville KJ, D'Amelio PB, Goymann W, Trost L, Pierce B, Bryla A, Dzialo M, Bauchinger U, McWilliams SR (2018) Environmental cues and dietary antioxidants affect breeding behavior and testosterone of male European starlings (*Sturnus vulgaris*). Horm Behav 103:36–44

Carciofi AC, Duarte JMB, Mendes D, de Oliveira LD (2006) Food selection and digestibility in yellow-headed conure (*Aratinga jandaya*) and golden-caped conure (*Aratinga auricapilla*) in captivity. J Nutr 136:2014–2016

Casagrande S, Csermely D, Pini E, Bertacche V, Tagliavini J (2006) Skin carotenoid concentration correlates with male hunting skill and territory quality in the kestrel (*Falco tinnunculus*). J Avian Biol 37:190–196

Casagrande S, Costantini D, Fanfani A, Tagliavini J, Dell'Omo G (2007) Patterns of serum carotenoid accumulation and skin color variation in nestling kestrels in relation to breeding conditions and different terms of carotenoid supplementation. J Comp Physiol B 177:237–245

Catoni C, Peters A, Schaffer HM (2008a) Life history trade-offs are influenced by the diversity, availability and interactions of dietary antioxidants. Anim Behav 76:1107–1119

Catoni C, Schaffer HM, Peters A (2008b) Fruit for health: the effect of flavonoids on humoral immune response and food selection in a frugivorous bird. Funct Ecol 22:649–645

Cecere JG, Spina F, Jenni-Eiermann S, Boitani L (2011) Nectar: an energy drink used by European songbirds during spring migration. J Ornithol 152:923–931

Cecere JG, Caprioli M, Carnevali C, Colombo G, Dalle-Donne I, Mancuso E, Milzani A, Parolini M, Portanova A, Saino N, Serra L, Rubolini D (2016) Dietary flavonoids advance timing of moult but do not affect redox status of juvenile blackbirds (*Turdus merula*). J Exp Biol 219:3155–3162

Chatterjee IB (1973) Evolution and biosynthesis of ascorbic acid. Science 182:1271–1272

Chaurand T, Weimerskirch H (1994a) Incubation routine, body mass regulation and egg-neglect in the blue petrel *Halobaena caerulea*. Ibis 136:285–290

Chaurand T, Weimerskirch H (1994b) The regular alternation of short and long foraging trips in the blue petrel *Halobaena caerulea*: a previously undescribed strategy of food provisioning in a pelagic seabird. J Anim Ecol 62:275–282

Cherel Y, Le Maho Y (1985) Five months of fasting in king penguin chicks: body mass loss and fuel metabolism. Am J Phys 249:R387–R392

Cherel Y, Robin JP, Walch O, Karmann H, Netchitailo P, Le Maho Y (1988a) Fasting in king penguin. I. Hormonal and metabolic changes during breeding. Am J Phys 254:R170–R177

Cherel Y, Leloup J, Le Maho Y (1988b) Fasting in king penguin. II. Hormonal and metabolic changes during molt. Am J Phys 254:R178–R184

Clayton DH, Wolfe ND (1993) The adaptive significance of self-medication. Trends Ecol Evol 8:60–63

Cohen AA, McGraw KJ, Douglas Robinson W (2009) Serum antioxidant levels in wild birds vary in relation to diet, season, life history strategy, and species. Oecologia 161:673–683

Cohen AA, Martin LB, Wingfield JC, McWilliams SR, Dunne JA (2012) Physiological regulatory networks: ecological roles and evolutionary constraints. Trends Ecol Evol 27:428–435

Cooper-Mullin C, McWilliams SR (2016) The role of the antioxidant system during intense endurance exercise: lessons from migrating birds. J Exp Biol 219:3684–3695
Costantini D (2010) Effects of diet quality on serum oxidative status and body mass in male and female pigeons during reproduction. Comparative Biochemistry and Physiology Part A, 156:294–299
Costantini D (2022) A meta-analysis of impacts of immune response and infection on oxidative status in vertebrates. Conserv Physiol 10:coac018
Costantini D, Bonadonna F (2010) Patterns of variation of serum oxidative stress markers in two seabird species. Pol Res 29:30–35
Costantini D, Dell'Omo G (2006) Effects of T-cell-mediated immune response on avian oxidative stress. Comp Biochem Physiol A 145:137–142
Costantini D, Møller AP (2008) Carotenoids are minor antioxidants for birds. Funct Ecol 22:367–370
Costantini D, Casagrande S, De Filippis S, Brambilla G, Fanfani A, Tagliavini J, Dell'Omo G (2006) Correlates of oxidative stress in wild kestrel nestlings (*Falco tinnunculus*). J Comp Physiol B 176:329–337
Costantini D, Fanfani A, Dell'Omo G (2007) Carotenoid availability does not limit the capability of nestling kestrels (*Falco tinnunculus*) to cope with oxidative stress. J Exp Biol 210:1238–1244
Costantini D, Marasco V, Møller AP (2011) A meta-analysis of glucocorticoids as modulators of oxidative stress in vertebrates. J Comp Physiol B 181:447–456
Costantini D, Sebastiano M, Müller M, Eulaers I, Ambus P, Malarvannan G, Covaci A, Massa B, Dell'Omo G (2017) Individual variation of persistent organic pollutants in relation to stable isotope ratios, sex, reproductive phase and oxidative status in Scopoli's shearwaters (*Calonectris diomedea*) from the Southern Mediterranean. Sci Total Environ 598:179–187
Costantini D, Angeletti D, Strinati C, Trisolino P, Carlini A, Nascetti G, Carere C (2018) Dietary antioxidants, food deprivation and growth affect differently oxidative status of blood and brain in juvenile European seabass (*Dicentrarchus labrax*). Comp Biochem Physiol Part A 216:1–7
Costantini D, Czirják GÁ, Bustamante P, Bumrungsri S, Voigt CC (2019) Impacts of land use on an insectivorous tropical bat: the importance of mercury, physio-immunology and trophic position. Sci Total Environ 671:1077–1085
Costantini D, Masi S, Rachid L, Beltrame M, Rohmer M, Krief S (2021) Mind the food: rapid changes in antioxidant content of diet affect oxidative status of chimpanzees. Am J Phys 320:728–734
Cothran RD, Gervasi SS, Murray C, French BJ, Bradley PW, Urbina J, Blaustein AR, Relyea RA (2015) Carotenoids and amphibians: effects on life history and susceptibility to the infectious pathogen, *Batrachochytrium dendrobatidis*. Conserv Physiol 3:cov005
de Ayala RM, Martinelli R, Saino N (2006) Vitamin E supplementation enhances growth and condition of nestling barn swallows (*Hirundo rustica*). Behav Ecol Sociobiol 60:619–630
de la Celorio-Mancera MP, Ahn SJ, Vogel H, Heckel DG (2011) Transcriptional responses underlying the hormetic and detrimental effects of the plant secondary metabolite gossypol on the generalist herbivore *Helicoverpa armigera*. BMC Genomics 12:575
Deeming DC, Pike TW (2013) Embryonic growth and antioxidant provision in avian eggs. Biol Lett 9:20130757
DeNiro MJ, Epstein S (1978) Influence of diet on the distribution of carbon isotopes in animals. Geochim Cosmochim Acta 42:495–506
DeNiro MJ, Epstein S (1981) Influence of diet on the distribution of nitrogen isotopes in animals. Geochim Cosmochim Acta 45:341–351
Eeva T, Sillanpää S, Salminen J-P (2009) The effects of diet quality and quantity on plumage colour and growth of great tit *Parus major* nestlings: a food manipulation experiment along a pollution gradient. J Avian Biol 40:491–499
Elliott S, Loudon A (1987) Effects of monoterpene odors on food selection by red deer calves (*Cervus elaphus*). J Chem Ecol 13:1343–1349

Faivre B, Grégoire A, Préault M, Cézilly F, Sorci G (2003) Immune activation rapidly mirrored in a secondary sexual trait. Science 300:103

Feeny PP (1970) Seasonal changes in oak leaf tannins and nutrients as a cause of spring feeding by winter moth caterpillars. Ecology 51:565–581

Feoli AM, Siqueira IR, Almeida L, Tramontina AC, Vanzella C, Sbaraini S, Schweigert ID, Netto CA, Perry MLS, Gonçalves CA (2006) Effects of protein malnutrition on oxidative status in rat brain. Nutrition 22:160–165

Friedman A, Bartov I, Sklan D (1998) Humoral immune response impairment following excess vitamin E nutrition in the chick and Turkey. Poult Sci 77:956–962

Fromentin G, Nicolaidis S (1996) Rebalancing essential amino acids intake by self-selection in the rat. Br J Nutr 75:669–682

Garber A, Hastie P, McGuinness D, Malarange P, Murray JA (2020) Abrupt dietary changes between grass and hay alter faecal microbiota of ponies. PLoS One 15:e0237869

García-Tarrasón M, Sanpera C, Jover L, Costantini D (2014) Levels of antioxidants in breeding female Audouin's gulls and their deposition in eggs across different environments. J Exp Mar Biol Ecol 453:116–122

Geiger S, Kauffmann M, Le Maho Y, Robin J-P, Criscuolo F (2012) Of the importance of metabolic phases in the understanding of oxidative stress in prolonged fasting and refeeding. Physiol Biochem Zool 85:415–420

Giordano M, Costantini D, Tschirren B (2015a) Female oxidative status, egg antioxidant protection and eggshell pigmentation: a supplemental feeding experiment in great tits. Behav Ecol Sociobiol 69:777–785

Giordano M, Costantini D, Tschirren B (2015b) Sex-specific effects of prenatal and postnatal nutritional conditions on the oxidative status of great tit nestlings. Oecologia 177:123–131

Giraudeau M, Sweazea K, Butler MW, McGraw KJ (2013) Effects of carotenoid and vitamin E supplementation on oxidative stress and plumage coloration in house finches (*Haemorhous mexicanus*). Comp Biochem Physiol Part A 166:406–413

Hall ME, Blount JD, Forbes S, Royle NJ (2010) Does oxidative stress mediate the trade-off between growth and self-maintenance in structured families? Funct Ecol 32:365–373

Hayes DP (2007) Nutritional hormesis. Eur J Clin Nutr 61:147–159

Hernandez EP, Anisuzzaman AMA, Kawada H, Kwofie KD, Ladzekpo D, Koike Y, Inoue T, Sasaki S, Mikami F, Matsubayashi M, Tanaka T, Tsuji N, Hatta T (2022) Ambivalent roles of oxidative stress in triangular relationships among arthropod vectors, pathogens and hosts. Antioxidants 11:1254

Heydari AR, Wu B, Takahashi R, Strong R, Richardson A (1993) Expression of heat shock protein 70 is altered by age and diet at the level of transcription. Mol Cell Biol 13:2909–2918

Hill RW, Wyse GA, Anderson M (2008) Animal physiology, 2nd edn. Sinauer Associates, Sunderland

Hill KE, Motley AK, May JM, Burk RF (2009) Combined selenium and vitamin C deficiency causes cell death in Guinea pig skeletal muscle. Nutr Res 29:213–219

Hough JA, Pimentel D (1978) Influence of host foliage on development, survival and fecundity of the gypsy moth. Environ Entomol 7:97–102

Huffman MA (1997) Current evidence for self-medication in primates: a multidisciplinary perspective. Am J Physic Anthropol 140:171–200

Hulbert AJ (2005) On the importance of fatty acid composition of membranes for aging. J Theor Biol 234:277–288

Hulbert AJ, Pamplona R, Buffenstein R, Buttemer WA (2007) Life and death: metabolic rate, membrane composition, and life span of animals. Physiol Rev 87:1175–1213

Hyun DH, Emerson SS, Jo DG, Mattson MP, de Cabo R (2006) Calorie restriction up-regulates the plasma membrane redox system in brain cells and suppresses oxidative stress during aging. Proc Natl Acad Sci USA 103:19908–19912

Jennings BJ, Ozanne SE, Dorling MW, Hales CN (1999) Early growth determines longevity in male rats and may be related to telomere shortening in the kidney. FEBS Lett 448:4–8

Jouventin P, Mougin JL, Stahl JC, Weimerskirch H (1985) Comparative biology of the burrowing petrels of the Crozet Islands. Notornis 32:157–220

Juan-Salles C, Prats N, Resendes A, Domingo M, Hilton D, Ruiz JM, Garner MM, Valls X, Marco AJ (2003) Anemia, myopathy, and pansteatitis in vitamin E-deficient captive marmosets (*Callithrix* spp.). Vet Pathol 40:540–547

Kailasam M, Kaneko G, Oo AKS, Ozaki Y, Thirunavukkarasu AR, Watabe S (2011) Effects of calorie restriction on the expression of manganese superoxide dismutase and catalase under oxidative stress conditions in the rotifer *Brachionus plicatilis*. Fish Sci 77:403–409

Kaneko G, Yoshinaga T, Yanagawa Y, Ozaki Y, Tsukamoto K, Watabe S (2011) Calorie restriction-induced maternal longevity is transmitted to their daughters in a rotifer. Funct Ecol 25:209–216

Karadas F, Pappas AC, Surai PF, Speake BK (2005) Embryonic development within carotenoid-enriched eggs influences the post-hatch carotenoid status of the chicken. Comp Biochem Physiol Part B 141:244–251

Karasov WH (2011) Digestive physiology: a view from molecules to ecosystem. Am J Phys 301:R276–R284

Krief S, Spirhanzlova P, Masi S, Couturier C, Okwir E, Asalu E, Bustamante P, Costantini D (2023) High urinary oxidative DNA damage in wild chimpanzees ranging in proximity of agricultural fields in Sebitoli area, Uganda. Environ Sci Pollut Res 30:110600–110611

Langley-Evans SC, Sculley DV (2005) Programming of hepatic antioxidant capacity and oxidative injury in the ageing rat. Mech Ageing Develop 126:804–812

Larcombe SD, Mullen W, Alexander L, Arnold KE (2010) Dietary antioxidants, lipid peroxidation and plumage colouration in nestling blue tits *Cyanistes caeruleus*. Naturwissenschaften 97:903–913

Lebold KM, Löhr CV, Barton CL, Miller GW, Labut EM, Tanguay RL, Traber MG (2013) Chronic vitamin E deficiency promotes vitamin C deficiency in zebrafish leading to degenerative myopathy and impaired swimming behavior. Comp Biochem Physiol Part C 157:382–389

Lee KP, Simpson SJ, Wilson K (2008) Dietary protein-quality influences melanization and immune function in an insect. Funct Ecol 22:1052–1061

Levin E, Lopez-Martinez G, Fane B, Davidowitz G (2017) Hawkmoths use nectar sugar to reduce oxidative damage from flight. Science 355:733–735

Li C, Xu B, Wang Y, Yang Z, Yang W (2014) Protein content in larval diet affects adult longevity and antioxidant gene expression in honey bee workers. Entomol Exp Applicata 151:19–26

Lin Y-F, Tsai H-L, Lee Y-C, Chang S-J (2005) Maternal vitamin E supplementation affects the antioxidant capability and oxidative status of hatching chicks. J Nutr 135:2457–2461

Liu D, Diorio J, Tannenbaum B, Caldji C, Francis DD, Freedman A, Sharma S, Pearson D, Plotsky PM, Meaney MJ (1997) Maternal care, hippocampal glucocorticoid receptors, and hypothalamic-pituitary-adrenal responses to stress. Science 277:1659–1662

Magnanou E, Malenke JR, Dearing MD (2009) Expression of biotransformation genes in woodrat (Neotoma) herbivores on novel and ancestral diets: identification of candidate genes responsible for dietary shifts. Mol Ecol 18:2401–2414

Magnanou E, Malenke JR, Dearing MD (2013) Hepatic gene expression in herbivores on diets with natural and novel plant secondary compounds. Physiol Genomics 45:774–785

Mangel M (2008) Environment, damage and senescence: modelling the life-history consequences of variable stress and caloric intake. Funct Ecol 22:422–430

Maronde L, Richner H (2015) Effects of increased begging and vitamin E supplements on oxidative stress and fledging probability. Behav Ecol 26:465–471

Marri V, Richner H (2014) Yolk carotenoids increase fledging success in great tit nestlings. Oecologia 176:371–377

Martínez del Rio C (1997) Can Passerines synthesize vitamin C? Auk 114:513–516

Mattson MP (2008) Dietary factors, hormesis and health. Ageing Res Rev 7:43–48

McCue MD (2010) Starvation physiology: reviewing the different strategies animals use to survive a common challenge. Comp Biochem Physiol Part A 156:1–18

McGraw KJ, Adkins-Regan E, Parker RS (2005) Maternally derived carotenoid pigments affect offspring survival, sex-ratio, and sexual attractiveness in a colourful songbird. Naturwissenschaften 92:375–380

McWilliams SR, Guglielmo C, Pierce B, Klaassen M (2004) Flying, fasting, and feeding in birds during migration: a nutritional and physiological ecology perspective. J Avian Biol 35:377–393

Messina S, Edwards DP, AbdElgawad H, Beemster GTS, Cosset CCP, Tomassi S, Benedick S, Eens M, Costantini D (2020) Impacts of selective logging on the oxidative status of tropical understory birds. J Anim Ecol 89:2222–2234

Mohr AE, Girard M, Rowe M, McGraw KJ, Sweazea KL (2019) Varied effects of dietary carotenoid supplementation on oxidative damage in tissues of two waterfowl species. Comp Biochem Physiol Part B 231:67–74

Moraes-Souza RQ, Vesentini G, Paula VG, Sinzato YK, Soares TS, Gelaleti RB, Volpato GT, Damasceno DC (2021) Oxidative stress profile of mothers and their offspring after maternal consumption of high-fat diet in rodents: a systematic review and meta-analysis. Oxidative Med Cell Longev 2021:9073859

Morales AE, Pérez-Jiménez A, Hidalgo MC, Abellán E, Cardenete G (2004) Oxidative stress and antioxidant defenses after prolonged starvation in *Dentex dentex* liver. Comp Biochem Physiol Part C 139:153–161

Mrosovsky N, Sherry DF (1980) Animal anorexias. Science 207:837–842

Ninni P (2003) La fonction des carotenoids chez les hirondelles de cheminée (*Hirundo rustica*). PhD thesis. Université Pierre et Marie Curie, Paris

Noguera JC, Lores M, Alonso-Alvarez C, Velando A (2011) Thrifty development: early-life diet restriction reduces oxidative damage during later growth. Funct Ecol 25:1144–1153

Noguera JC, Monaghan P, Metcalfe NB (2015) Interactive effects of early and later nutritional conditions on the adult antioxidant defence system in zebra finches. J Exp Biol 218:2211–2217

Noiret A, Puch L, Riffaud C, Costantini D, Riou J-F, Aujard F, Terrien J (2020) Sex-specific response to caloric restriction after reproductive investment in Microcebus murinus: an integrative approach. Front Physiol 11:506

Olsson M, Wilson M, Isaksson C, Uller T, Mott B (2008) Carotenoid intake does not mediate a relationship between reactive oxygen species and bright coloration: an experimental test in a lizard. J Exp Biol 211:1257–1261

Olsson M, Wilson M, Isaksson C, Uller T (2009) Polymorphic ROS scavenging revealed by CCCP in a lizard. Naturwissenschaften 96:845–849

Orledge JM, Blount JD, Hoodless AN, Pike TW, Royle NJ (2012) Synergistic effects of supplementation of dietary antioxidants during growth on adult phenotype in ring-necked pheasants, *Phasianus colchicus*. Funct Ecol 26:254–264

Ortiz CL, Costa D, Le Boeuf BJ (1978) Water and energy flux in elephant seal pups fasting under natural conditions. Physiol Zool 51:166–178

Ortiz RM, Crocker DE, Houser DS, Webb PM (2006) Angiotensin II and aldosterone increase with fasting in breeding adult male northern elephant seals (*Mirounga angustirostris*). Physiol Biochem Zool 79:1106–1112

Padró J, De Panis DN, Vrdoljak J, Carmona PM, Colines B, Hasson E, Soto IM (2018) Experimental evolution of alkaloid tolerance in sibling Drosophila species with different degrees of specialization. Evol Biol 45:170–181

Pamplona R, Barja G (2007) Highly resistant macromolecular components and low rate of generation of endogenous damage: two key traits of longevity. Ageing Res Rev 6:189–210

Pascual P, Pedrajas JR, Toribio F, López-Barea J, Peinado J (2003) Effect of food deprivation on oxidative stress biomarkers in fish (*Sparus aurata*). Chem Biol Interact 145:191–199

Pérez-Rodríguez L, Redondo T, Ruiz-Mata R, Camacho C, Moreno-Rueda G, Potti J (2019) Vitamin E supplementation—but not induced oxidative stress—influences telomere dynamics during early development in wild passerines. Front Ecol Evol 7:173

Pierce BJ, McWilliams SR (2005) Seasonal changes in composition of lipid stores in migratory birds: causes and consequences. Condor 107:269–279

Pineda-Pampliega J, Ramiro Y, Herrera-Dueñas A, Martinez-Haro M, Hernández JM, Aguirre JI, Höfle U (2021) A multidisciplinary approach to the evaluation of the effects of foraging on landfills on white stork nestlings. Sci Total Environ 775:145197

Piper MDW, Mair W, Partridge L (2005) Counting the calories: the role of specific nutrients in extension of life span by food restriction. J Geront A 60:549–555

Plummer KE, Bearhop S, Leech DI, Chamberlain DE, Blount JD (2013) Fat provisioning in winter impairs egg production during the following spring: a landscape-scale study of blue tits. J Anim Ecol 82:673–682
Prokić MD, Petrović TG, Despotović SG, Vučić T, Gavrić JP, Radovanović TB, Gavrilović BR (2021) The effect of short-term fasting on the oxidative status of larvae of crested newt species and their hybrids. Comp Biochem Physiol Part A 251:110819
Ramirez JM, Folkow LP, Blix AS (2007) Hypoxia tolerance in mammals and birds: from the wilderness to the clinic. Annu Rev Physiol 69:113–143
Rattan SIS (2008) Hormesis in aging. Ageing Res Rev 7:63–78
Raubenheimer D, Simpson SJ (2009) Nutritional PharmEcology: doses, nutrients, toxins, and medicines. Integr Comp Biol 49:329–337
Raubenheimer D, Simpson SJ, Mayntz D (2009) Nutrition, ecology and nutritional ecology: toward an integrated framework. Funct Ecol 23:4–16
Richelle M, Turini ME, Guidoux R, Tavazzi I, Metairon S, Fay LB (1999) Urinary IP excretion is not confounded by the lipid content of the diet. FEBS Lett 459:259–262
Rodgers JT, Lerin C, Haas W, Gygi SP, Spiegelman BM, Puigserver P (2005) Nutrient control of glucose homeostasis through a complex of PGC-1alpha and SIRT1. Nature 434:113–118
Royle NJ, Surai PF, McCartney RJ, Speake BK (1999) Parental investment and egg yolk lipid composition in gulls. Funct Ecol 13:298–306
Royle NJ, Surai PF, Hartley IR (2001) Maternally derived androgens and antioxidants in bird eggs: complementary but opposing effects? Behav Ecol 12:381–385
Saino N, Romano M, Caprioli M, Rubolini D, Ambrosini R (2011) Yolk carotenoids have sex-dependent effects on redox status and influence the resolution of growth trade-offs in yellow-legged gull chicks. Behav Ecol 22:411–421
Schaefer HM, McGraw K, Catoni C (2008) Birds use fruit colour as honest signal of dietary antioxidant rewards. Funct Ecol 22:303–310
Schneeberger K, Czirják GA, Voigt CC (2014) Frugivory is associated with low measures of plasma oxidative stress and high antioxidant concentration in free-ranging bats. Naturwissenschaften 101:285–290
Schull Q, Viblanc VA, Stier A, Saadaoui H, Lefol E, Criscuolo F, Bize P, Robin JP (2016) The oxidative debt of fasting: evidence for short- to medium-term costs of advanced fasting in adult king penguins. J Exp Biol 219:3284–3293
Schvezov N, Wilson RW, Urbina MA (2022) Oxidative damages and antioxidant defences after feeding a single meal in rainbow trout. J Comp Physiol B 192:459–471
Sebastiano M, Eens M, Messina S, AbdElgawad H, Pineau K, Gerrit B, Chastel O, Costantini D (2018) Resveratrol supplementation reduces oxidative stress and modulates the immune response in free-living animals during a viral infection. Funct Ecol 32:2509–2519
Selman C, McLaren JS, Meyer C, Duncan JS, Redman P, Collins AR, Duthie GG, Speakman JR (2006) Life-long vitamin C supplementation in combination with cold exposure does not affect oxidative damage or lifespan in mice, but decreases expression of antioxidant protection genes. Mech Ageing Devel 127:897–904
Selman C, McLaren JS, Mayer C, Duncan JS, Collins AR, Duthie GG, Redman P, Speakman JR (2008) Lifelong alpha-tocopherol supplementation increases the median life span of C57BL/6 mice in the cold but has only minor effects on oxidative damage. Rejuven Res 11:83–96
Selman C, McLaren JS, Collins AR, Duthie GG, Speakman JR (2013) Deleterious consequences of antioxidant supplementation on lifespan in a wild-derived mammal. Biol Lett 9:20130432
Senar JC, Møller AP, Ruiz I, Negro JJ, Broggi J, Hohtola E (2010) Specific appetite for carotenoids in a colorful bird. PLoS One 5:e10716
Simons MJP, Cohen AA, Verhulst S (2012) What does carotenoid-dependent coloration tell? Plasma carotenoid level signals immunocompetence and oxidative stress state in birds–a meta-analysis. PLoS One 7:e43088
Singer MS, Mace KC, Bernays EA (2009) Self-medication as adaptive plasticity: increased ingestion of plant toxins by parasitized caterpillars. PLoS One 4:e4796

Sinha AK, AbdElgawad H, Zinta G, Dasan AF, Rasoloniriana R, Asard H, Blust R, De Boeck G (2015) Nutritional status as the key modulator of antioxidant responses induced by high environmental ammonia and salinity stress in European sea bass (*Dicentrarchus labrax*). PLoS One 10:e0135091

Skopec MM, Green AK, Karasov WH (2010) Flavonoids have differential effects on glucose absorption in rats (*Rattus norvegicus*) and American robins (*Turdus migratorius*). J Chem Ecol 36:236–243

Soldatini C, Sebastiano M, Albores-Barajas YV, AbdElgawad H, Bustamante P, Costantini D (2020) Mercury exposure in relation to foraging ecology and its impact on the oxidative status of an endangered seabird. Sci Total Environ 724:138131

Sorensen M, Sanz A, Gomez J, Pamplona R, Portero-Otin M, Gredilla R, Barja G (2006) Effects of fasting on oxidative stress in rat liver mitochondria. Free Radic Res 40:339–347

Sorensen MC, Hipfner JM, Kyser TK, Norris DR (2009) Carry-over effects in a Pacific seabird: stable isotope evidence that pre-breeding diet quality influences reproductive success. J Anim Ecol 78:460–467

Spée M, Beaulieu M, Dervaux A, Chastel O, Le Maho Y, Raclot T (2010) Should I stay or should I go? Hormonal control of nest abandonment in a long-lived bird, the Adélie penguin. Horm Behav 58:762–768

Stahlschmidt ZR, Jeong N, Johnson D, Meckfessel N (2020) From phenoloxidase to fecundity: food availability does not influence the costs of oxidative challenge in a wing-dimorphic cricket. J Comp Physiol B 190:17–26

Stewart L (2013) Dietary effects on adult performance and oxidative stress status in three-spined sticklebacks. Ph.D. Thesis. University of Glasgow, Glasgow

Stipanovic RD, Williams HJ, Smith LA (1986) Cotton terpenoid inhibition of *Heliothis virescens* development. Am Chem Soc Symp Ser 296:79–94

Suarez RK, Lighton JRB, Joos B, Roberts SP, Harrison JF (1996) Energy metabolism, enzymatic flux capacities and metabolic flux rates in flying honeybees. Proc Natl Acad Sci USA 93:12616–12620

Surai P (2002) Natural antioxidants in avian nutrition and reproduction. Nottingham University Press, Nottingham

Surai PF, Noble RC, Speake BK (1996) Tissue-specific differences in antioxidant distribution and susceptibility to lipid peroxidation during development of the chick embryo. Biochim Biophys Acta 1304:1–10

Surai PF, Bortolotti GR, Fidgett AL, Blount JD, Speake BK (2001) Effects of piscivory on the fatty acid profiles and antioxidants of avian yolk: studies on eggs of the gannet, skua, pelican and cormorant. J Zool 255:305–312

Swanson EM, Espeset A, Mikati I, Bolduc I, Kulhanek R, White WA, Kenzie S, Snell-Rood EC (2016) Nutrition shapes life-history evolution across species. Proc R Soc Lond B 283:20152764

Tapia PC (2006) Sublethal mitochondrial stress with an attendant stoichiometric augmentation of reactive oxygen species may precipitate many of the beneficial alterations in cellular physiology produced by caloric restriction, intermittent fasting, exercise and dietary phytonutrients: "Mitohormesis" for health and vitality. Medic Hypoth 66:832–843

Tapia-Monsalve R, Newsome SD, Sanchez-Hernandez JC, Bozinovic F, Nespolo R, Sabat P (2018) Terrestrial birds in coastal environments: metabolic rate and oxidative status varies with the use of marine resources. Oecologia 188:65–73

Tarry-Adkins JL, Martin-Gronert MS, Chen JH, Cripps RL, Ozanne SE (2008) Maternal diet influences DNA damage, aortic telomere length, oxidative stress, and antioxidant defense capacity in rats. FASEB J 22:2037–2044

Tarry-Adkins JL, Aiken CE, Ashmore TJ, Fernandez-Twinn DS, Chen JH, Ozanne SE (2019) A suboptimal maternal diet combined with accelerated postnatal growth results in an altered aging profile in the thymus of male rats. FASEB J 33:239–253

Tauler-Ametlller H, Pretus JL, Hernández-Matías A, Ortiz-Santaliestra ME, Mateo R, Real J (2019) Domestic waste disposal sites secure food availability but diminish plasma antioxidants in Egyptian vulture. Sci Total Environ 650:1382–1391

Thierry AM, Brajon S, Massemin S, Handrich Y, Chastel O, Raclot T (2013) Decreased prolactin levels reduce parental commitment, egg temperatures, and breeding success of incubating male Adélie penguins. Horm Behav 64:737–747

Thomas PK, Cooper JM, King RH, Workman JM, Schapira AH, Goss-Sampson MA, Muller DP (1993) Myopathy in vitamin E deficient rats: muscle fibre necrosis associated with disturbances of mitochondrial function. J Anat 183:451–461

Toh SQ, Glanfield A, Gobert GN, Jones MK (2010) Heme and blood-feeding parasites: friends or foes? Parasit Vectors 3:108

Torquetti CG, de Carvalho TP, de Freitas RMP, Freitas MB, Guimarães ATB, Soto-Blanco B (2023) Influence of landscape ecology and physiological implications in bats from different trophic guilds. Sci Total Environ 857:159631

Ugochukwu NH, Figgers CL (2007) Dietary caloric restriction improves the redox status at the onset of diabetes in hepatocytes of streptozotocin-induced diabetic rats. Chem Biol Inter 165:45–53

Vázquez-Medina JP, Crocker DE, Forman HJ, Ortiz RM (2010) Prolonged fasting does not increase oxidative damage or inflammation in postweaned northern elephant seal pups. J Exp Biol 213:2524–2530

Vázquez-Medina JP, Zenteno-Savín T, Forman HJ, Crocker DE, Ortiz RM (2011) Prolonged fasting increases glutathione biosynthesis in postweaned northern elephant seal pups. J Exp Biol 214:1294–1299

Vázquez-Medina JP, Zenteno-Savín T, Elsner R, Ortiz RM (2012) Coping with physiological oxidative stress: a review of antioxidant strategies in seals. J Comp Physiol B 182:741–750

von Schantz T, Bensch S, Grahn M, Hasselquist D, Wittzell H (1999) Good genes, oxidative stress and condition-dependent sexual signals. Proc R Soc Lond B 266:1–12

Wagner ND, Hillebrand H, Wacker A, Frost PC (2013) Nutritional indicators and their uses in ecology. Ecol Lett 16:535–544

Walker AW, Ince J, Duncan SH, Webster LM, Holtrop G, Ze X, Brown D, Stares MD, Scott P, Bergerat A, Louis P, McIntosh F, Johnstone AM, Lobley GE, Parkhill J, Flint HJ (2011) Dominant and diet-responsive groups of bacteria within the human colonic microbiota. ISME J 5:220–230

Wang Y-J, Chien Y-H, Pan C-H (2006) Effects of dietary supplementation of carotenoids on survival, growth, pigmentation, and antioxidant capacity of characins, *Hyphessobrycon callistus*. Aquacult 261:641–648

Weaver RJ, Santos ESA, Tucker AM, Wilson AE, Hill GE (2018) Carotenoid metabolism strengthens the link between feather coloration and individual quality. Nat Commun 2018:73

Whiten SR, Eggleston H, Adelman ZN (2017) Ironing out the details: exploring the role of iron and heme in blood-sucking arthropods. Front Physiol 8:1134

Williams JB, Roberts SP, Elekonich MM (2008) Age and natural metabolically-intensive behavior affect oxidative stress and antioxidant mechanisms. Exp Geront 43:538–549

Wingfield JC, Maney DL, Breuner CW, Jacobs JD, Lynn S, Ramenofsky M, Richardson RD (1998) Ecological bases of hormone-behavior interactions: the 'emergency life history stage'. Am Zool 38:191–206

Winston ML (1987) The biology of the honey bee. Harvard University Press, Cambridge

Woshner V, Knott K, Wells R, Willetto C, Swor R, O'Hara T (2008) Mercury and selenium in blood and epidermis of bottlenose dolphins (*Tursiops truncatus*) from Sarasota Bay, FL: interaction and relevance to life history and hematologic parameters. EcoHealth 5:360–370

Yamato M, Shiba T, Yoshida M, Ide T, Seri N, Kudou W, Kinugawa S, Tsutsui H (2007) Fatty acids increase the circulating levels of oxidative stress factors in mice with diet-induced obesity via redox changes of albumin. FEBS J 274:3855–3863

Zhang X-D, Zhu Y-F, Cai L-S, Wu T-X (2008) Effects of fasting on the meat quality and antioxidant defenses of market-size farmed large yellow croaker (*Pseudosciaena crocea*). Aquaculture 280:136–139

Zhang F-P, Cai X-H, Wang H, Ren Z-X, Larson-Rabin Z, Li D-Z (2012) Dark purple nectar as a foraging signal in a bird-pollinated Himalayan plant. New Phytol 193:188–195

# Chapter 4
# Diversity in Oxidative Status: Roles of Stress Response, Metabolism, Rhythmicity, and (Epi)Genetics

**Abstract** Many studies documented significant individual or species differences in oxidative status markers and resistance against oxidative stress over time and space. In this chapter, I have addressed the roles of several proximate mechanisms in the generation, maintenance, and evolution of diversity in oxidative status. In particular, I have examined the role of glucocorticoids as relevant mediators of this individual variation (coping style), because they drive the organismal response to stressors, such as the predation risk. I have also addressed other possible factors that generate differentiation in oxidative status, including the overlooked role of historical biogeographical events. Finally, I have focussed on the (i) endogenous and exogenous factors that determine circadian and circannual rhythmicity, (ii) relative contributions of parental effects and (epi)genetic background to variation in oxidative status, and (iii) comparative studies that assessed the links between oxidative status markers and pace of life of species.

## 4.1 Introduction

There is tremendous variation in oxidative status and resistance against oxidative stress between individuals, conspecific populations, or species. The bacterium *Deinococcus radiodurans* is one of the most resistant species against a number of stressors (e.g. radiation) in the world owing to its superior capacity to repair DNA damage or scavenge reactive oxygen species (Fig. 4.1). Life-history theory provides a valuable framework to understand variation in oxidative status between conspecific individuals or species. If we take a close look at this variation, we can identify a number of proximate factors that contribute to generate this variation, such as style of coping with stress, body size, biological rhythms, or genetic background. In addition, local selective pressures contribute in some way to build up variation among individuals, populations, or species (e.g. Hardeland et al. 2003; Pigliucci et al. 2006; Badyaev 2009, 2011; Cohen et al. 2009; Gilbert and Epel 2009; Wilson

© The Author(s), under exclusive license to Springer Nature Switzerland AG 2024
D. Costantini, *The Role of Organismal Oxidative Stress in the Ecology and Life-History Evolution of Animals*, Fascinating Life Sciences, https://doi.org/10.1007/978-3-031-65183-0_4

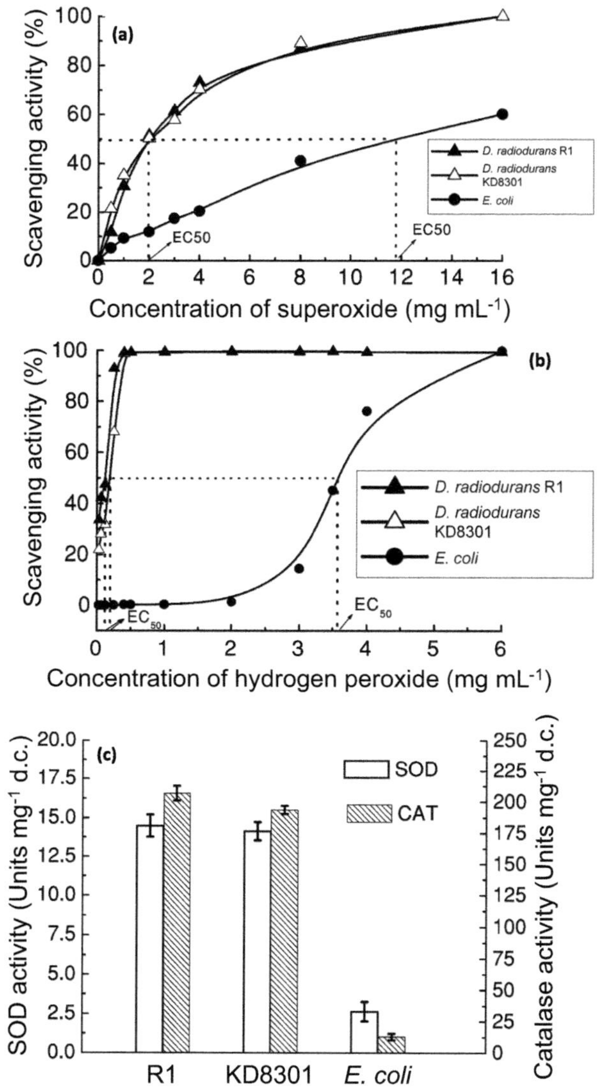


**Fig. 4.1** Scavenging effects of cellular protein preparations of *Deinococcus radiodurans* and *Escherichia coli* on (**a**) the superoxide free radical and (**b**) hydrogen peroxide. Dotted line shows the value of EC50 (50% effective concentration). Values are shown as average values of triplicate experiments. (**c**) Enzyme activity assay of superoxide dismutase (SOD) and catalase (CAT) in cellular protein preparations. Values are shown as mean and standard deviation. Reprinted with minor modifications from Tian et al. (2004) with permission of John Wiley and Sons

and Nussey 2010; Carere and Maestripieri 2013; Grether 2014; Konarzewski and Książek 2013; White and Kearney 2013).

In this chapter, I have examined (1) the link between oxidative status and stress responsiveness, with emphasis on the impact of stress induced by predation risk on the oxidative status, and (2) the possible factors that generate population differentiation in oxidative status, including the overlooked role of historical biogeographical events. I have also focussed on (3) the circadian and circannual rhythmicity of oxidative status, (4) the contributions of parental effects and (epi)genetic background to variation in oxidative status, and (v) the links between oxidative status markers and pace of life of species.

## 4.2 What Animals Tell Us About Coping with Stress

Conspecific individuals of the same sex or age often differ from each other in clusters of behavioural and physiological traits, even under standardised conditions. Most of this variation is non-random, and such individual differences are consistent across situations or contexts and across time (Carere and Eens 2005). Individual differences in behaviour and physiology in non-human animals have been often neglected as biologically meaningful variation. By contrast, the study of individual variation in humans has had a crucial role in biomedical research, because this variation is associated with individual disease vulnerability and response to medication (Carere and Eens 2005). Individual differences in non-human animals are now recognised as a real biological phenomenon (Wolf and Weissing 2012; Carere and Maestripieri 2013; Cabrera et al. 2021; Martin et al. 2023).

One of the fundamental features that characterises a behavioural type is the responsiveness to environmental stressors. Although some individuals tend to be highly responsive to such stimuli, others are unresponsive and show routine-like behaviours (Wolf et al. 2008). Variation in the way individuals behaviourally cope with stressful episodes can be modelled along an axis polarised at the two extremes by proactive and reactive responses (Koolhaas et al. 1999, 2010). Proactive individuals are generally bold, superficial explorers, active, and aggressive. Reactive individuals are lowly active and aggressive, and explore a novel environment thoroughly. It has been suggested that differences in coping styles reflect consistent behavioural differences among individuals that might underlie the expression of different personalities (Sih et al. 2004; Groothuis and Carere 2005; Réale et al. 2007; Carere and Maestripieri 2013; Cabrera et al. 2021; Martin et al. 2023).

There is increasing evidence that such individual variation in personalities has important ecological and evolutionary consequences (Carere and Eens 2005; Sih et al. 2004; Réale et al. 2007; Wolf and Weissing 2012; Cabrera et al. 2021; Martin et al. 2023). It is generally thought that variation among individuals is maintained because, while certain behavioural phenotypes do better than others under some environmental conditions, those same behavioural phenotypes may perform poorly under different conditions (e.g. Sih et al. 2004; Carere et al. 2010a). Furthermore, several studies have suggested that selection should have favoured integrated sets of behavioural, physiological, and morphological traits along with variation in life-history strategies (Sih et al. 2004; Wolf et al. 2008; Biro and Stamps 2008; Careau et al. 2008b; Réale et al. 2010). In other words, these integrated neuro-physiological and behavioural sets of traits may be part of more general life-history strategies with proactive individuals showing high growth rate, early sexual maturity, and short lifespan, and reactive individuals being characterised by a slower pace of life (Biro and Stamps 2008; Careau et al. 2008b; Wolf et al. 2008; Réale et al. 2010).

In this context, we should expect that natural selection has also led to some potential coevolution between coping styles and mechanisms involved in life-history trade-offs. Some of these mechanisms involve stress responsiveness modulated by glucocorticoids. When confronted to a challenging situation (e.g. predator attack or social conflict), proactive individuals tend to respond with a strong sympathetic activation and increase in noradrenergic stimulation, resulting in a general fight-or-flight behavioural response (e.g. Koolhaas et al. 2010; Carere et al. 2010a; Coppens et al. 2010; Cabrera et al. 2021; Martin et al. 2023). On the other hand, reactive individuals respond to a challenge with a strong hypothalamic–pituitary–adrenocortical (HPA) reactivity (Koolhaas et al. 2010; Carere et al. 2010a; Coppens et al. 2010), resulting in a freezing response to a stimulus and an increase in circulating glucocorticoids. However, the link between behavioural and physiological response is not so straightforward. In an updated version of the reactive–proactive model, Koolhaas et al. (2010); see also Steimer et al. (1997) have proposed the existence of two independent axes: coping style and stress reactivity. This new model seems to be supported by a study on coping styles in wild marmots *Marmota marmota* (Ferrari et al. 2013). Thus, the behavioural reaction to an environmental stimulus may reflect a certain coping style, but it does not need to be necessarily related to the glucocorticoid response to a stressor (stress reactivity, Koolhaas et al. 2010). Many studies on artificially selected or inbred lines in laboratory settings (Carere and Maestripieri 2013) and, more recently, in wild animals (Ferrari et al. 2013) suggest that an association between behavioural and glucocorticoid response may generally occur, but that its strength may be weaker in a natural context.

The link between coping style and reactivity of the HPA axis is very important in terms of consequences for the maintenance of oxidative status homeostasis and downstream effects related to any change in oxidative stress levels. Stress responsiveness may influence the generation of oxidative damage and the expression of several antioxidant mechanisms (Costantini et al. 2011). Importantly, the pro-oxidant effect of glucocorticoids appears to be consistent across multiple tissues (Fig. 4.2) (Costantini et al. 2011). In addition, reproductive and physical activities in socio-sexual contexts, which are known to differ among coping styles (e.g. mouse *Mus musculus*, Koolhaas et al. 1999; great tit *Parus major*, Groothuis and Carere 2005), could also contribute to significantly affect the oxidative status because they would expose individuals with differing coping styles to oxidative challenges of different magnitudes (e.g. Alessio 1993; Costantini et al. 2008). For example, changes in metabolic activity or energy expenditure, which are both correlated with some traits of personalities, have an impact on the generation of reactive oxygen species (Ros et al. 2006; Careau et al. 2008a; Biro and Stamps 2010).

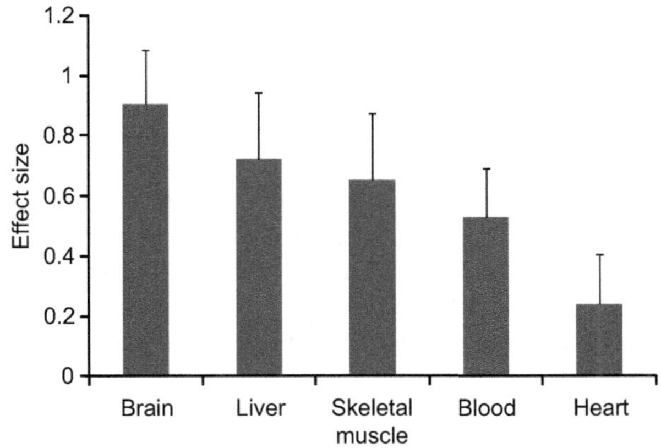


**Fig. 4.2** Effect of experimentally elevated glucocorticoids on oxidative stress across multiple tissues in vertebrates. The meta-analysis included 191 effect size estimates from 19 studies carried out on seven species (Costantini et al. 2011). Values are shown as mean and standard error. Reprinted from Costantini (2014) with permission of Springer Nature

Importantly, we have good evidence in laboratory models and humans that individual differences in behaviour and physiology may influence longevity and susceptibility to certain diseases and resilience to others (Cavigelli and McClintock 2003; Cavigelli 2005; Korte et al. 2005). For example, laboratory male rats identified as neophobic during infancy were more likely to die at any point in time than their less fearful brothers, although they died of similar pathologies (Cavigelli and McClintock 2003). It seems therefore reasonable to expect that, if a link between personality and oxidative stress occurs, oxidative stress might be one cellular mechanism that underlies the link between personality and longevity or health span.

Research on a number of vertebrate species showed that personality types could indeed differ in oxidative status, although results do not appear to be consistent across species, different personality traits, or markers of oxidative status. Mice selected for aggression and coping (long attack latency = reactive coping strategy; short attack latency = proactive coping strategy) are a useful model for studying the physiological background of animal personalities. These mice also show differential stress responsiveness, especially in terms of HPA axis reactivity, to various challenges. Costantini et al. (2008) also found that reactive mice had higher serum non-enzymatic antioxidant capacity than proactive mice, while they had similar levels of serum reactive oxygen metabolites. Moreover, the two mouse lines showed inverse relationships between serum reactive oxygen metabolites or non-enzymatic antioxidant capacity and body mass corrected for age, respectively. Levels of reactive oxygen metabolites decreased with body mass in proactive mice, while the antioxidant capacity increased with body mass in reactive mice, respectively. The results of this study indicated that (1) individual variation in oxidative status may be linked to that in coping styles and (2) oxidative status may have a (epi)genetic basis because all subjects were adolescent and experimentally naive concerning social challenges. Moreover, the results of this study suggested that reactive mice might invest more than proactive mice in the long-term physiological homeostasis by maintaining higher constitutive antioxidant levels in blood because (1) they have to cope with a higher generation of reactive oxygen species, or (2) they prepare the organism (anticipatory response) for the repeated stressful events that they will experience throughout adulthood because of their higher stress responsiveness.

Another study on laboratory mice also found a correlation between aggressiveness, one of the most reliable proxies of personality type, and a marker of oxidative status. Mice with a higher basal generation of reactive oxygen species in blood granulocytes were those with a shorter latency time and a higher number of attacks, respectively (Rammal et al. 2010). This study also suggested that the more aggressive mice had a higher basal activity of immune cells. This result might be very important if aggressive mice are more exposed to injuries and inflammation consequently to fights among males. Other studies on mice suggested that the aggressiveness level can also be influenced by the oxidative status, indicating a possible continuous cross-talk between organismal oxidative status and behaviour. Garratt and Brooks (2015) found that laboratory male mice with experimentally elevated sensitivity to oxidative stress (through the inhibition of copper-zinc superoxide dismutase (SOD1)) were more aggressive than both wild-type males and males that expressed 50% of this antioxidant enzyme. Finally, in less exploratory mound-building mice *Mus spicilegus*, oxidative DNA damage and the activity of both catalase and superoxide dismutase in the muscle were increased, but only in individuals with higher relative starting mass measured nine days after birth (Fig. 4.3) (Rödel et al. 2022). This interaction between the within-litter body mass ranking and the exploration tendency of mice may be explained by the typically higher adrenocortical activity in less exploratory phenotypes and by the higher growth in relatively heavier pups, because both factors may increase oxidative stress.

Other studies also found evidence for a link between aggressiveness and oxidative status in other vertebrates. For example, in a study on the White's skink *Egernia whitii*, Isaksson et al. (2011) found that more aggressive males had higher plasma non-enzymatic antioxidant capacity, but not higher oxidative damage, than less aggressive males. They also showed that the link between aggressiveness and antioxidants was independent of testosterone, suggesting that other behavioural patterns associated with an aggressive phenotype (e.g. territory defence, mate acquisition) might influence the individual oxidative status. Similarly, after 20 min of continuous conspecific aggressive behaviour, both females and males of the rufous horneros *Furnarius rufus* had lower concentrations of non-enzymatic antioxidants than control individuals, independently of testosterone (Mentesana and Adreani 2021). Short-term exposure to aggressive behaviour also caused changes in some types of antioxidants in round goby *Neogobius melanostomus* individuals, without causing any increase in oxidative damage (Błońska et al. 2023). However, Butler and Ligon (2015) found that the visual exposure of male veiled chameleons *Chamaeleo calyptratus* to a potential combatant resulted in decreased body mass during the subsequent 2 days, but levels of oxidative damage in plasma, liver, and spleen were unaffected. Similarly, Quque et al. (2022) did not find any evidence of increased oxidative stress in male zebra finches *Taeniopygia guttata* emitting aggression. Rather, they found that received aggression caused increased oxidative damage generation and reduced telomere length in male zebra finches.

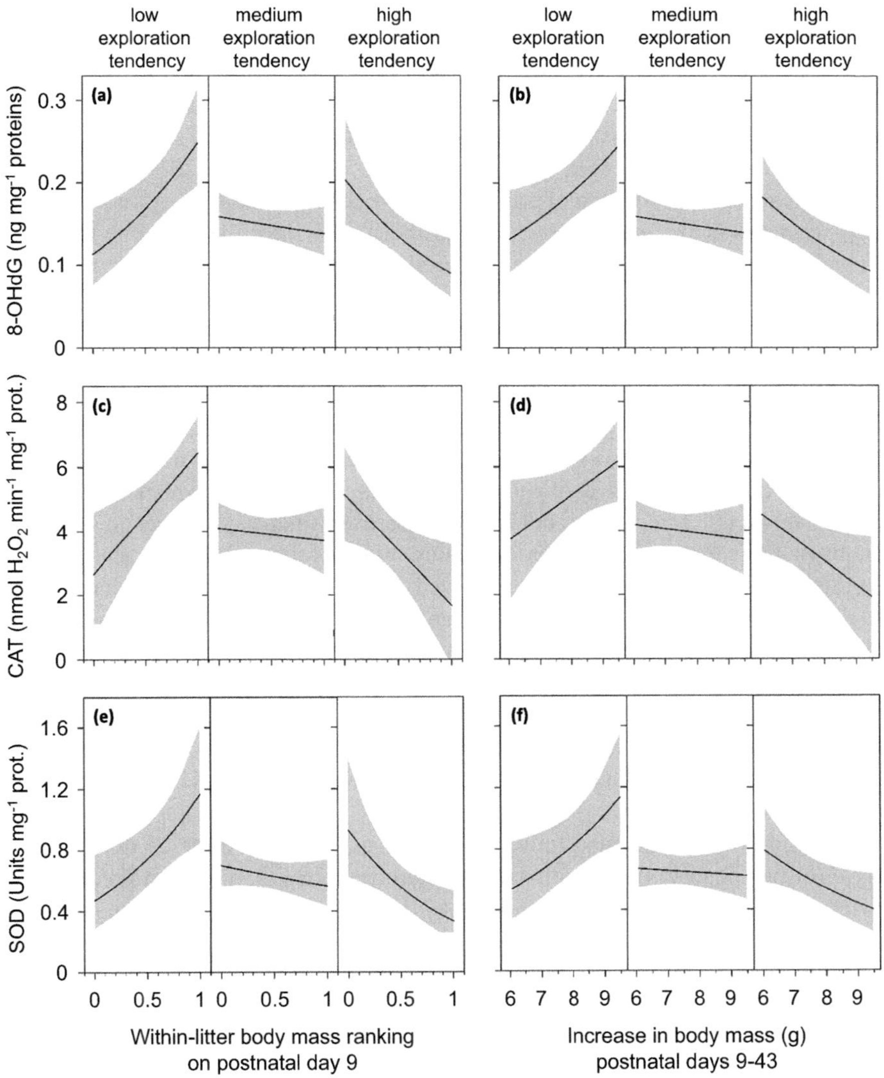


**Fig. 4.3** In female mound-building mice *Mus spicilegus* with low exploration tendency, (**a**–**b**) oxidative DNA damage (8-hydroxy-2′-deoxyguanosine, 8-OHdG), (**c**–**d**) catalase (CAT), and (**e**–**f**) superoxide dismutase (SOD) in the muscle were notably higher in individuals with increasing growth, whereas in females with higher exploration tendency, these associations between markers and growth were inverse. Within-litter pup body mass: 0 = lightest, 1 = heaviest pup per litter. The three conditions plotted for each significant interaction are exemplary categorisations (low: 10% percentile, medium: 50% percentile, high: 90% percentile) of the females' within-litter body mass ranking prior to weaning. The regression lines with 95% confidence intervals (grey shading) are based on parameter estimates of linear mixed-effects models. Reprinted with minor modifications from Rödel et al. (2022). © 2022 The Authors; distributed under the terms of the Creative Commons Attribution License (CC BY)

We can identify some limitations in current research on the link between coping style and oxidative stress. For example, all studies on mice were carried out using laboratory lines, whilst studies on other vertebrates were only carried out on wild-caught animals maintained in captivity. Under wild conditions, however, the social experience of an individual may affect its behavioural profile and stress responsiveness to subsequent challenges (Jansen et al. 2010; Sachser et al. 2010), and there is no the potential confounding of artificial genetic selection. In a study on free-living Alpine marmots *Marmota marmota*, it has been shown for the first time that individuals with different coping styles differed in both pre-restraint and acute stress-induced levels of oxidative status markers under natural conditions (Costantini et al. 2012). A first model (confirmatory path analysis) on baseline values of markers showed that marmots with a more proactive coping style had higher levels of plasma non-enzymatic antioxidant capacity, possibly through a positive action of plasma cortisol that contributed to the increased antioxidant levels (Fig. 4.4). The coping style of marmots was, however, not associated with plasma levels of two markers of oxidative damage (reactive oxygen metabolites and 4-hydroxynonenal-histidine protein adducts). A second model including restraint-induced changes in oxidative status markers (i.e. a restraint period lasting 30 min followed by an open-field test lasting 3 min, respectively) further showed that proactive marmots experienced a higher increase in plasma reactive oxygen metabolites and plasma non-enzymatic antioxidant capacity than reactive marmots (Fig. 4.4). In contrast to pre-restraint values, changes in plasma levels of cortisol induced by the restraint were not associated with the coping style of the marmot nor with any of the markers of oxidative status analysed. Further analyses showed that the plasma levels of 4-hydroxynonenal-histidine protein adducts were also not associated with the coping style of the marmot.

A number of studies also found interesting associations between other coping style (or personality) traits and oxidative status markers. A report on greenfinches *Carduelis chloris* showed that (1) neophobic individuals had higher plasma levels of reactive oxygen metabolites and lower levels of plasma non-enzymatic antioxidant capacity than neophilic greenfinches, (2) fast exploring individuals had higher plasma non-enzymatic antioxidant capacity and lower plasma malondialdehyde than slow exploring greenfinches, and (3) greenfinches with extremely high or low neophobia had lower malondialdehyde than intermediate responders, respectively (Herborn et al. 2011). Wintering blue tits *Cyanistes caeruleus* that were both highly neophobic and exploratory had low plasma non-enzymatic antioxidant capacity, and individuals that showed low neophobia and low exploration had high plasma non-enzymatic antioxidant capacity (Arnold et al. 2015). Similarly, more exploratory rufous-collared sparrows *Zonotrichia capensis* had lower circulating antioxidants, but only in those populations that were exposed to extreme environmental temperatures and/or dehydration (Poblete et al. 2023). A study on free-living male rhesus macaques *Macaca mulatta* found that males who attacked others at higher rates had higher plasma levels of oxidative DNA damage (8-OHdG), indicating a cost of male competition (Georgiev et al. 2015). Mallards *Anas platyrhynchos* and northern pintails *Anas acuta* with higher concentrations of circulating carotenoids entered the

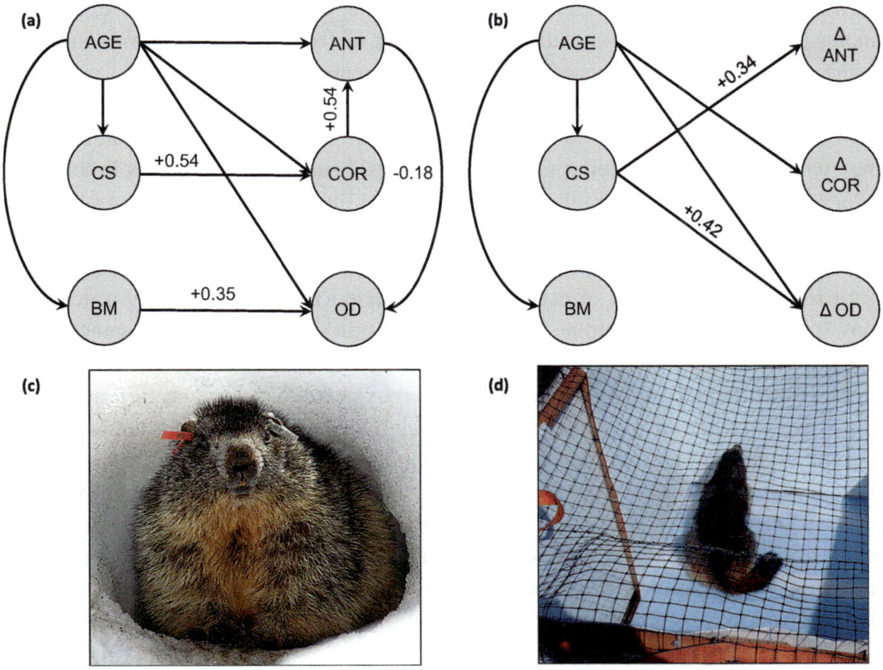


**Fig. 4.4** Best fitting path models of relationships among (**a**) coping style (CS), pre-restraint values of cortisol (COR), plasma oxidative damage (OD), and plasma non-enzymatic antioxidant capacity (ANT), and (**b**) coping style, restraint-induced changes in cortisol (ΔCOR), restraint-induced changes in plasma oxidative damage (ΔOD), and restraint-induced changes in plasma non-enzymatic antioxidant capacity (ΔANT) in Alpine marmots *Marmota marmota*, while controlling for individual age and body mass (BM); (**c**) the picture (Photograph courtesy of Caterina Ferrari) shows a marmot from Orvieille (Valsavarenche, Aosta, Gran Paradiso National Park, North-Western Italian Alps, 45°34'N/7°11'E) that was tagged for individual recognition; (**d**) the picture (Photograph by David Costantini) shows a marmot within an open-field arena, where the individual behavioural response to a novel environment was tested. Panels (**a**) and (**b**) were reprinted from Costantini et al. (2012). Panels (**c**) and (**d**) were reprinted from Costantini (2014) with permission of Springer Nature

test arena sooner and approached and entered predator-scented bedding material more frequently than birds with low carotenoid concentrations (Rowe et al. 2015). These results suggested that dietary carotenoids may promote aspects of exploratory behaviour either by providing antioxidant protection to tissues in the body that are relevant to navigation and locomotion (e.g. muscle, brain) or by boosting systematic health and immune function that indirectly affect the exploratory behaviour (Rowe et al. 2015).

In conclusion, these studies showed both similarities and discrepancies. Studies on free-living animals showed that, generally, more aggressive individuals tend to have higher basal levels of circulating non-enzymatic antioxidants, probably because they mobilise antioxidants from internal stores when being challenged (e.g. Mentesana and Adreani 2021). By contrast, there was no consistent association between coping style and levels of oxidative damage across studies. These discrepancies among studies may result from a number of factors, such as strains included in the experiment or markers of oxidative status analysed. For example, reactive and proactive mice come from artificial selection lines, which represent the two most extreme phenotypes/genotypes of the range existing in the original population, and they were maintained under optimal and standard husbandry conditions in the laboratory. By contrast, Costantini et al. (2012) relied on a natural population of Alpine marmots. For example, reactive and proactive mice were naïve concerning resident-intruder tasks, whereas Alpine marmots had totally different social experiences prior to the analyses. Any effects of social experiences and life-history trajectories may make difficult to compare results across studies and identify any general patterns in free-living animals that would be consistent with those identified in animals under controlled laboratory experiments. Moreover, frequencies of different personalities within a social group might be relevant to determine the individual physiological effects. For example, levels of oxidative damage to lipids increased in high- and low-exploratory groups, respectively; however, they decreased in variable and random groups of house sparrows *Passer domesticus* (Vágási et al. 2021). Thus, group- or collective personalities, as determined by the personality composition of the group (i.e. frequency of occurrence of the different personalities), might play a non-negligible role in explaining variation of oxidative status between individuals.

## 4.3 Landscape of Fear

The landscape of fear is a model based on the spatial distribution of perceived predation risk. The risk of predation has a strong impact on life histories of prey because it can induce significant increases in the secretion of stress hormones and allocation of resources to support organismal responses to life-threatening events, which can come at a cost for reproduction (e.g. Sheriff et al. 2009; Hawlena and Schmitz 2010; Clinchy et al. 2013). A number of studies found significant associations between oxidative status markers and predation risk. Thus, oxidative stress might be a relevant sublethal consequence of predator exposure that would impact on behaviour and the expression of life-history traits. For example, if organisms slow down their growth in environments where predators abound, this strategy might carry benefits in terms of reduced generation of molecular oxidative damage. On the other hand, if growing up quickly is a best solution because the individual attains sexual maturity earlier, this might carry costs in terms of accelerated rate of cellular senescence. Slos and Stoks (2008) found that, under predation risk, larvae of the common blue damselfly *Enallagma cyathigerum* reduced their growth rate, but increased their consumption of oxygen compared to larvae that were not exposed to predators. In presence of predators, larvae had also higher levels of heat shock proteins 70 and lower levels of catalase; by contrast, heat shock proteins 60 and superoxide dismutase were similar between control and predator-exposed larvae (Slos and Stoks 2008). Other experiments found that damselfly larvae exposed to predation risk had

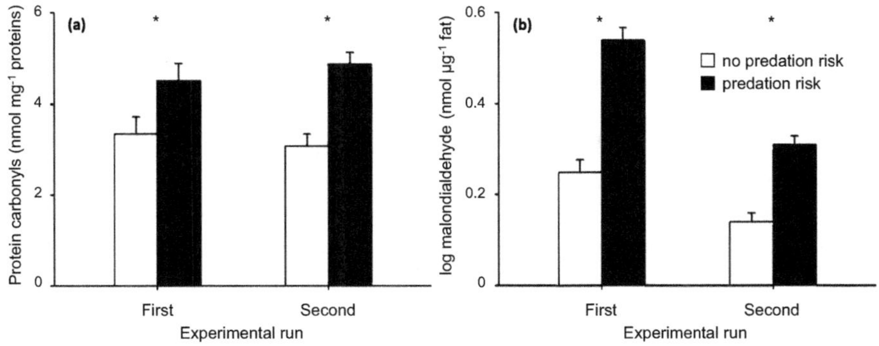


**Fig. 4.5** Effect of predation risk on (**a**) protein carbonyls and (**b**) malondialdehyde (log-transformed data) in *Coenagrion puella* damselfly larvae. Asterisks indicate significant differences between groups. Values are shown as mean and standard error. Reprinted with minor modifications from Janssens and Stoks (2014); distributed under the terms of the Creative Commons Attribution License (CC BY)

(1) lower activity of superoxide dismutase, higher generation of the superoxide free radical, and higher levels of lipid oxidative damage (Janssens and Stoks 2013a) and (2) higher levels of protein carbonyls and lipid peroxidation (Fig. 4.5) (Janssens and Stoks 2014) than larvae that were not exposed to predation risk. Moreover, we have evidence that co-exposure to other stressors can either exacerbate or mitigate the pro-oxidant effect of predation risk as shown in larvae of the common blue damselfly (Janssens and Stoks 2013b, 2017) or in tadpoles of the western spadefoot *Pelobates cultripes* (Florencio et al. 2020). These findings indicated that predation stress might sum up with other stressors. However, the emergence of these hidden physiological synergisms between stressors did not occur for the growth rate and food intake (Janssens and Stoks 2013b). Similarly, exposure to predation risk affected antioxidant levels but not growth rate in Nile tilapia *Oreochromis niloticus* (Cruz et al. 2021). By contrast, while exposure to chemical cues from predators decreased the levels of superoxide dismutase and of glutathione reductase in tadpoles of the western spadefoot *Pelobates cultripes*, levels of both enzymes did not change in tadpoles that were exposed simultaneously to a combination of predator clues and low concentration of the herbicide glyphosate (Burraco et al. 2013). When melon aphids *Aphis gossypii* were exposed to the odours of their predator, the ladybird *Menochilus sexmaculatus*, (1) their reproductive period and offspring production were significantly decreased, (2) their longevity was significantly shortened, and (3) their activities of superoxide dismutase and catalase, and levels of lipid peroxidation significantly increased (Lin et al. 2024). By contrast, the predation risk did not affect levels of catalase, glutathione peroxidase, and thiobarbituric acid reactive substances (Burraco et al. 2013). Other studies also did not find evidence for significant oxidative costs associated with predation risk (e.g. Jermacz et al. 2020; Roux et al. 2021).

Another route through which predation risk might influence the oxidative status of prey lies in the energetic demands of escaping behaviour. Philipp et al. (2008)

simulated an escape from a predator by exposing queen scallops *Aequipecten opercularis* to one of their main predators, the common sea star *Asterias rubens*. Predation exposure increased the swimming activity of scallops, resulting in an increased generation of reactive oxygen species in muscle. Guerra et al. (2013) found that Catarina scallops *Argopecten ventricosus* experimentally exposed to the predator the blue crab *Callinectes sapidus* had reduced levels of superoxide dismutase and catalase, but also of protein carbonyls and thiobarbituric acid reactive substances in both gills and mantle compared to individuals that were not exposed to predation risk. By contrast, catalase, protein carbonyls, and thiobarbituric acid reactive substances were higher in the swimming muscle of predator-exposed scallops than in predator-free scallops. Another study on the escape response through tail-flipping until fatigue in whiteleg shrimps *Litopenaeus vannamei* found a significant increase in glutathione peroxidase in abdominal muscle, without any changes in protein carbonyls nor in thiobarbituric acid reactive substances (Robles-Romo et al. 2016).

Some studies suggested that predation-induced oxidative stress might constrain to some degree the habitat selection decisions. For example, pied flycatchers *Ficedula hypoleuca* breeding in closer proximity to nests of sparrowhawks *Accipiter nisus* had higher levels of heat shock proteins 60 and 70 (Thomson et al. 2010). However, another study on female pied flycatchers did not find any effect of increased perceived predation risk on the oxidative status (Ruuskanen et al. 2017). Pinya et al. (2016) found that the presence of exudates of the invasive viperine water snake *Natrix maura* induced a significant increase in catalase, glutathione reductase, and glutathione in tadpoles of the Majorcan midwife toad *Alytes muletensis*, whereas it did not affect any oxidative status marker in tadpoles of the Balearic green toad *Bufotes balearicus*. Finally, Morosinotto et al. (2018) found that levels of glutathione-S-transferase and superoxide dismutase in willow tits *Poecile montanus* were higher in high predation risk areas than in low predation risk areas.

One may wonder whether the increased production of glucocorticoids links predation risk to changes in the oxidative status. A number of studies documented increased levels of cortisol or corticosterone in animals exposed to predation risk (e.g. Canoine et al. 2002; Sheriff et al. 2009; Dulude-de Broin et al. 2020). Many studies also found that (1) glucocorticoids can induce a number of changes in cellular oxidative status, (2) some species are particularly resistant to pro-oxidant effects of glucocorticoids, and (3) oxidative damage tends to increase with duration of the physiological stress (Costantini et al. 2011; Birnie-Gauvin et al. 2017; Vágási et al. 2018, 2020; Majer et al. 2019; Stier et al. 2019; Josserand et al. 2020; Ensminger et al. 2021). Thus, increased production of glucocorticoids might be responsible for the changes in antioxidant and oxidative damage markers recorded in individuals exposed to predation risk.

## 4.4 Population Differentiation in Oxidative Status and the Legacy of Historical Biogeography

Individual variation in oxidative status has been implicated in processes of ecological and evolutionary importance in a wide range of organisms. However, while the temporal component of this variation has been the subject of much research, we do not know yet whether a spatial component exists within species in the dynamics of cellular oxidative status. This spatial component can include variation owing to differences in local conditions, where conspecific populations live, and the legacy of events that occurred in the past (historical biogeography) that shaped the current distribution of a certain species. Our understanding of the population differentiation in oxidative status is still limited, particularly in relation to past biogeographical events.

Galápagos land iguanas *Conolophus subcristatus* are large lizards that live on different islands of the Galápagos Archipelago. Costantini et al. (2005, 2009) found that populations from the different islands differed in plasma levels of carotenoids, non-enzymatic antioxidant capacity, and reactive oxygen metabolites. These differences may result from a combination of factors that differed among islands, such as the peak of the mating season, the availability of food, or the local pressures due to human activities (Snell et al. 1984). In particular, the comparison between the populations from Santa Cruz and Venecia provided interesting indications because these two populations originated from a small stock of iguanas from Santa Cruz that survived an abrupt population size reduction due to wild dogs (Snell et al. 1984). Both populations have been recovering over the last decades, however, contrary to Santa Cruz, the limited area of Venecia caused high-density conditions. Additionally, suitably large areas for nesting were not available at Venecia, where nests can be excavated only on artificially created areas. Although Costantini et al. (2009) did not find any significant differences in plasma reactive oxygen metabolites between males, they found significantly higher levels of plasma reactive oxygen metabolites in females from Santa Cruz than in females from Venecia during the reproductive season. Levels of plasma non-enzymatic antioxidant capacity were also higher in males and females from Santa Cruz compared to those from Venecia. These results might reflect the high metabolic costs incurred for egg production and, possibly, for nest excavation in females from Santa Cruz. By contrast, males from Venecia seemed to experience higher oxidative stress than those from Santa Cruz because, whereas reactive oxygen metabolites were similar between islands, levels of plasma non-enzymatic antioxidant capacity were much lower in July compared to December. This result could mirror a possible combined effect of scarce availability of food and high population density, which would be consistent with the smaller values of the body condition index recorded in iguanas from Venecia (Costantini et al. 2009).

Population differentiation in oxidative status can also be the result of coevolution with life history. Robert and Bronikowski (2010) addressed this point in populations of western terrestrial garter snakes *Thamnophis elegans*, which are differentiated in a lakeshore fast (fast growth, earlier sexual maturation, intense reproduction, no trematode infection, lower longevity) and a meadow slow ecotype (slow growth, later sexual maturation, moderate reproduction, high trematode infection, higher longevity). Ecotype differentiation has been probably driven by different predation pressures (higher for the fast ecotype) and food availability (fluctuating for the slow ecotype) between areas. Robert and Bronikowski (2010) found that life-history differences between ecotypes are also translated at physiological level. Compared to the short-lived phenotype, offspring snakes from the long-lived ecotype consumed equal amounts of oxygen, produced more energy for a given amount of consumed oxygen, had lower mitochondrial generation of reactive oxygen species, and had DNA in red blood cells that was damaged more readily but was also repaired more efficiently. Differences between ecotypes were enhanced in offspring from mothers that were given corticosterone before parturition, which supported the role of non-genetic maternal effects in shaping the oxidative stress resistance of offspring (Robert and Bronikowski 2010). The results of this study also supported the idea that there may be selection on physiological mechanisms that enable individuals to respond to the demands of life-history strategies in the more advantageous way under the specific circumstances. In another study, Schwartz and Bronikowski (2013) sought to identify which nodes of the molecular stress network were more divergent between the two populations of western terrestrial garter snakes under both non-stressful and stressful conditions, respectively. They found that two common markers of stress (plasma concentration of corticosterone and liver gene expression of heat shock proteins) increased under stressful conditions in both ecotypes (Schwartz and Bronikowski 2013). By contrast, the two ecotypes diverged at four nodes under both non-stressful and stressful conditions, such as the levels of superoxide and hydrogen peroxide or the level of DNA damage in erythrocytes. These results supported the hypothesis that life-history phenotypes of western terrestrial garter snakes have diverged at the molecular level in how they respond to stressors, particularly in those nodes that regulate the oxidative status (Schwartz and Bronikowski 2013). Hence, oxidative stress has probably represented an important agent of microevolutionary differentiation in garter snakes. Another finding of the study was that differences appeared to be stronger in females than in males. It has been suggested that this difference between sexes might be due to females that invested more into reproduction than do males due to their viviparity. Thus, the fitness costs of mismatched regulation of their stress responses might be predicted to be higher if stress networks have pleiotropic effects on longevity and reproduction (Schwartz and Bronikowski 2013). Finally, the results of these studies on garter snakes supported the idea that the long-lived ecotype invested more in protective mechanisms to promote longevity rather than in keeping low the basal generation of reactive oxygen species (Robert and Bronikowski 2010; Schwartz and Bronikowski 2013). This possible coevolution between life history (or pace of life) and mechanisms of protection against oxidative stress has also been observed in other organisms. For example, using within-species contrasts of eight natural populations of *Nothobranchius* fishes in common garden experiments, Blažek et al. (2017) found that populations originating from dry regions (with short life expectancy) had shorter lifespans and a greater increase in mortality with age, more pronounced cellular and physiological deterioration (oxidative damage, tumour load), and a faster decline in fertility than populations from wetter regions.

Studies on other organisms also found similar associations between ecotype and regulation of oxidative status. Philipp et al. (2012) found that changes in the expression of regulatory genes that govern the response to pro-oxidant agents differed between short- and long-lived populations (living less than 50 and until 410 years, respectively, as determined by shell ring counts) of the bivalve *Arctica islandica*. Compared to the long-lived populations, the short-lived populations are exposed to frequent and prolonged episodes of hypoxia and of anoxia. Most of the hypoxia/anoxia related genes (including antioxidant genes) investigated in the long-lived population were downregulated under experimentally forced anoxia and hypoxia, and in animals with anoxia self-induced by shell closure, whereas they were upregulated in individuals from the short-lived populations. Such changes in the gene expression were transient and disappeared within 6 h of reoxygenation, suggesting a high flexibility of the transcriptional regulation (Philipp et al. 2012). It has also been observed that an increase in the expression of a certain gene was not always associated with a proportional increase in the level of the synthesised protein, especially for bivalves from the short-lived populations (Philipp et al. 2012). This result might indicate that (1) enzymes lose function with time and are replaced by new synthesised molecules, (2) the availability of amino acids or energy might have constrained the synthesis of proteins, because bivalves also needed to use them to sustain other physiological functions, (3) short-lived populations are being exposed to stronger endogenous or environmental constraints, which might explain their minor capability to cope with stressors, or (4) post-translation mechanisms are influenced by environmental conditions so that they are differentially expressed between populations.

Population differentiation in the generation rate of oxidative damage, levels of antioxidant defences, or generation of reactive oxygen species suggests that those processes that regulate the cellular oxidative status might be important targets of selection leading to adaptive divergence. For example, Nikinmaa et al. (2013) found that there was divergence among three populations of three-spined sticklebacks *Gasterosteus aculeatus* in both genetic and protein markers linked to the antioxidant response. Physiological mechanisms of oxidative status regulation might be under disruptive selection due to adaptations to the local environmental conditions. In turn, these mechanisms might influence selection on other fitness-related traits and vice versa, possibly contributing to evolutionary diversification and speciation. For example, outcrossing between populations with divergent mitochondrial DNA of the copepod *Tigriopus californicus* can exacerbate cellular oxidative stress in hybrid offspring, resulting in low fecundity (Barreto and Burton 2013).

Cichlid fish living in the East African great lakes offer another notable example of divergence in physiological performance among closely related species. For example, when body colourations and aggression are experimentally maximised, red males of *Pundamilia nyererei* experienced more oxidative stress than blue males of *Pundamilia pundamilia* (Dijkstra et al. 2011). It has been hypothesised that a higher oxidative stress in red males has negative fitness consequences and, as a result, this fitness cost may counterbalance the advantages that red males have in aggressive interactions, stabilising the coexistence of red and blue fish in the population (Dijkstra et al. 2011). By contrast, levels of two plasma markers of oxidative status did not differ among three sympatrically occurring female colour morphs of the cichlid *Neochromis omnicaeruleus*, which are known to differ in agonistic behaviour (Dijkstra et al. 2016). Similarly, Border et al. (2019) found that dominant males of the cichlid fish *Astatotilapia burtoni* had higher concentrations of circulating reactive oxygen metabolites without notably increasing their antioxidant defences. In particular, yellow but not blue males showed signs of increased oxidative damage with increasing agonistic effort.

It is often forgotten that past historical events could have moulded the current patterns of variation in oxidative status between conspecific populations. For example, past biogeographic events, such as appearance of geographic barriers, range expansion, or island colonisation, could have played non-negligible roles in shaping the current spatial structure of the cellular mechanisms that regulate the organismal oxidative status. Relying on a longitudinal common garden experiment lasting 1 year, Liparoto et al. (2020) analysed the spatial structure of blood oxidative status in the Tyrrhenian tree frog *Hyla sarda*, a species that underwent a well-documented major northward range expansion during the Late Pleistocene from Sardinia (core population) to Corsica (edge population). The two populations from Sardinia and Corsica have been isolated from each other from the last glacial epoch (Bisconti et al. 2011a). At the beginning of the experiment, Sardinian and Corsican frogs had similar concentrations of all oxidative status markers analysed (see Fig. 10.8). One year later, frogs from Corsica had significantly higher serum reactive oxygen metabolites, lower serum non-enzymatic antioxidant capacity, and lower red blood cell glutathione peroxidase than frogs from Sardinia. By contrast, levels of red blood cell superoxide dismutase increased similarly in frogs from both islands during the course of the experiment. Corsican frogs had also higher values of the SOD/GPx ratio than Sardinian frogs (Liparoto et al. 2020). These results were robust for any selective mortality of frogs having either very high or very low values of markers compared with the overall distribution, because oxidative status markers were not associated with mortality rate. Frogs were also maintained under the same common garden conditions, thus limiting the confounding effects of phenotypic plasticity due to environmental variation. The results of the study indicated that, while age-related changes in red blood cell superoxide dismutase might be a shared (or conserved) mechanism (sensu Partridge and Gems 2002) of physiological ageing, age-related changes in other oxidative status markers might be private (i.e. peculiar to particular evolutionary lineages) because these changes were expressed only in Corsican frogs (Liparoto et al. 2020). The possible conserved level in the expression of the superoxide dismutase across the two islands is also supported by the lack of genetic variation between Sardinian and Corsican frogs in the gene SOD-1 (Bisconti et al. 2011b). Overall, these results suggested a differentiation in the pace of physiological ageing between conspecific populations that have been geographically and genetically isolated since the end of the Pleistocene epoch (Bisconti et al. 2011a).

A number of factors can potentially confound our inferences about the causes and consequences of population differentiation in oxidative status. For example, we have a growing number of studies documenting the role of urbanisation in determining differences between conspecific populations in many traits, including behaviour, reproduction, and oxidative status (e.g. Partecke et al. 2006; Møller et al. 2010; Lucas and French 2012; Miranda et al. 2013; Sol et al. 2013; Costantini et al. 2014; Herrera-Dueñas et al. 2014; see paragraph 5.9). In addition, local exposure to certain pollutants may also hide biological patterns or bias our conclusions (Carere et al. 2010b; Isaksson 2010). For example, Labrada-Martagón et al. (2011) found that antioxidant levels differed between populations of East Pacific green turtles *Chelonia mydas*, and these differences were more likely driven by the different levels of contamination that occur across the coastal lagoons, where turtles live (Labrada-Martagón et al. 2011). Similarly, variation in human presence can also generate variation across populations. For example, Semeniuk et al. (2009) found that southern stingrays *Dasyatis americana* suffered higher levels of oxidative stress in tourist than in non-tourist sites. Thus, in order to infer about any ongoing selective pressure that generates differences in oxidative status between populations, it is very important to identify any potential confounding factors.

## 4.5 Circadian–Circannual Changes and Rhythmicity

The circadian clock is a highly conserved biological timing mechanism shared by most organisms from cyanobacteria to humans. Reactive oxygen species play an important role as regulators of clock gene expression mediated by D-box enhancer elements (Pagano et al. 2018). The many components that shape the organismal oxidative status also show some degree of circadian and circannual rhythmicity (Fig. 4.6). Diurnal fluctuations in the generation of reactive oxygen species, oxidative damage, activity of antioxidant enzymes, and levels of low-molecular weight antioxidants including both endogenous and dietary antioxidants have been described in a range of taxa, establishing in some cases a tight connection between the clock and the cellular redox signalling (e.g. van den Driessche et al. 2000; Hardeland et al. 2003; Costantini and Bonadonna 2010; van de Crommenacker et al. 2011; Beaver et al. 2012; Giannetto et al. 2012; Xu et al. 2012; Peek et al. 2013; Patel et al. 2014; Hidalgo et al. 2017; Tartarotti et al. 2019; Mezhnina et al. 2022). This connection has also been described in species that express a polymorphism of winter phenotypes, i.e. responding and non-responding individuals that change their physiological status or maintain their summer phenotype, respectively (Jefimow et al. 2020). For example, rhythms of nitric oxide synthase activity were recorded in various tissues of mice (Tunçtan et al. 2002). Rhythmicity in the generation of protein carbonyls was recorded in many taxa, such as the dinoflagellate *Lingulodinium polyedra* (Burkhardt et al. 1999), *Drosophila melanogaster* (Coto-Montes and Hardeland 1997), and Syrian hamsters *Mesocricetus auratus* (Coto-Montes et al. 2001). Extracellular ascorbate and glutathione peroxidase were

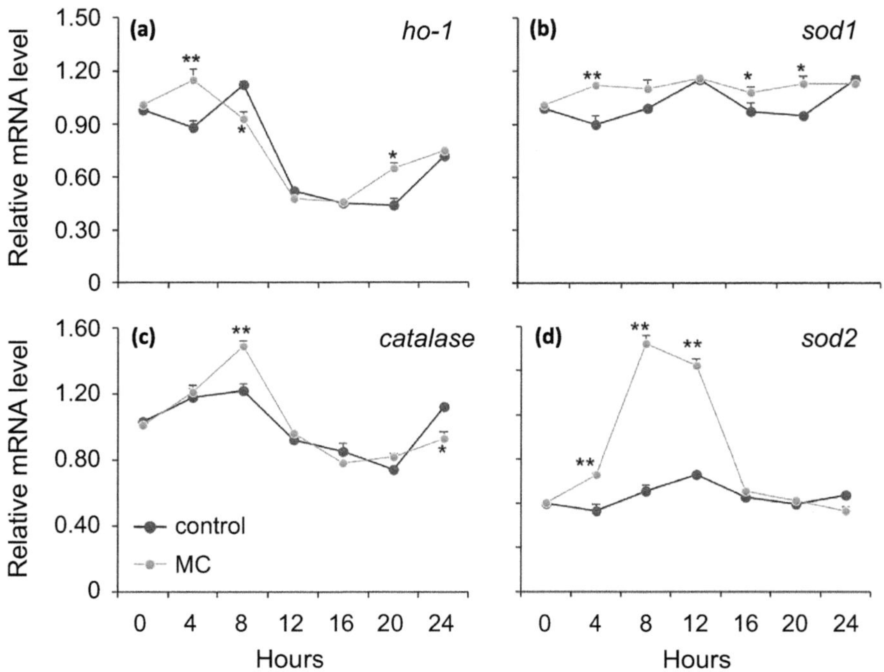


**Fig. 4.6** Temporal expression of the genes (**a**) haem oxygenase-1, (**b**) superoxide dismutase 1, (**c**) catalase, and (**d**) superoxide dismutase 2, after the addition of microcystin-LR (microcystins can enhance the generation of reactive oxygen species) in cell cultures of H9C2 cardiomyocytes. Asterisks indicate significant differences between groups: * $P < 0.05$, ** $P < 0.01$. Values are shown as mean and standard error. Reprinted with minor modifications from Xu et al. (2018); distributed under the terms of the Creative Commons Attribution License (CC BY)

observed to peak at night in the brain of rats (O'Neill et al. 1982; Barlow-Walden et al. 1995). The results of many studies also show that the rhythmicity of some antioxidants can differ between tissues, sexes, or species (Hardeland et al. 2003; Xu et al. 2012). Daily changes in retinol, α-tocopherol, and β-carotene appeared to be of low amplitude in human blood (Nierenberg and Stukel 1987). By contrast, β-carotene strongly oscillated over the day in the dinoflagellate *Lingulodinium polyedra*, a species capable of synthesising this particular carotenoid (Di Mascio et al. 1995; Hollnagel et al. 1996).

Edgar et al. (2012) suggested that sensing and responding to oxidative cycles in cellular environments could have driven the evolution of circadian rhythms and maintained the intrinsic link between clocks and metabolism. As a consequence, organisms that lack detoxification systems for reactive oxygen species will not have circadian rhythms. Interestingly, Wilking et al. (2013) suggested that oxidative status regulation and circadian rhythms are together intricately involved in the healthy body, as well as in disease. However, we have limited understanding of the reasons that explain circadian variation of oxidative status markers. Many other traits vary across the daily cycle (e.g. physical activity, food intake; Coughlan et al. 2023), so that it is not always clear the extent to which diurnal changes in oxidative status is driven by internal clocks.

A number of studies also documented circannual changes in markers of oxidative status (e.g. Costantini et al. 2009, 2010; Serra et al. 2012; Lemieux et al. 2019). For example, Aras et al. (2009) found that hepatic levels of superoxide dismutase, catalase, glutathione peroxidase, and glutathione *S*-transferase were higher in female trout *Salmo trutta* in autumn than during other seasons. This autumnal peak of antioxidant levels coincided with an increase in the gonado-somatic index, which indicates the proportion of gonad mass of the total body mass of the animal (Aras et al. 2009), which suggests a possible direct or indirect regulatory role of sexual hormones in trout (Hoogenboom et al. 2012). Female great tits and female starlings *Sturnus vulgaris* that bred later in the season had higher plasma non-enzymatic antioxidant capacity than those that bred earlier in the season (Costantini et al. 2010). Female starlings that bred later in the season had also lower plasma levels of reactive oxygen metabolites than female starlings that bred earlier in the season (Costantini et al. 2010). Other studies found that levels of glutathione peroxidase were lower in late broods of great tits (Norte et al. 2009), and the in vitro resistance of red blood cells to haemolysis induced by a free radical was lower in late broods of yellow-legged gulls *Larus cachinnans* (Kim et al. 2010). It has also been found that second-brood nestling starlings had lower plasma non-enzymatic antioxidant capacity than first-brood nestling starlings within a same breeding season (Serra et al. 2012), or had higher oxidative damage in low quality seasons (Cornell and Williams 2017). We have also evidence that the concentration of antioxidants in eggs can vary across the season. Yellow-legged gulls that bred later in the season laid eggs with a lower content of antioxidants (Rubolini et al. 2006). The interpretation of these seasonal changes in the levels of oxidative damage markers or antioxidants is, however, not straightforward. Although we cannot exclude a possible control mediated by an internal clock, we can identify a number of possible reasons for these seasonal effects, such as changes in ambient temperature and humidity, food quality and quantity, differential reproductive investment of females between first and second broods, or different qualities of early and late breeders. Moreover, many animal species go through a moulting phase, so that some seasonal variation of levels of oxidative damage and antioxidants might also be induced by moult, which is a high energy-demanding process. Correlative evidence in barn swallows *Hirundo rustica* showed that individuals with nearly completed moult had lower lipid hydroperoxides and activity of glutathione-S-transferase than those in the initial stages of moult (Raja-aho et al. 2012).

An important question then regards the proximate mechanisms that modulate the daily or seasonal changes in oxidative status, independently of ultimate factors. Several studies showed that melatonin and glucocorticoids are relevant hormonal regulators of these changes in oxidative status (Goncharova et al. 2006; Ashkenazi and Haim 2012, 2013; Kharwar and Haldar 2012). Rhythmicity of antioxidants or in the generation of reactive oxygen species is also probably affected by several other molecular mechanisms, including a direct genetic control through differential expression of genes, a post-translational regulation by means of other mechanisms (e.g. daily fluctuations in the secretion of stress hormones or melatonin, mitochondrial activity), or temporal variation in food intake. Experimental induction of oxidative stress can also disrupt the circadian pattern of the expression of antioxidant genes (Fig. 4.6) (Xu et al. 2018). Experimental quantification of the contributions of these factors is an important area for future research.

Regardless of mechanisms, the occurrence of circadian and circannual changes in antioxidant levels suggests that there might be time-dependent selection on behavioural schedules, i.e. a certain activity would be better done at some times of the day or of the year than in other times in order to buffer or reduce the potential oxidative costs associated with the expression of a certain behaviour. Circadian and circannual changes in some antioxidants might also reflect anticipatory responses in order to prepare the organism for regular tasks. Biological clocks exert a feed-forward control over effectors: they initiate changes in physiological systems, rather than correcting for changes after they happen (Hill et al. 2008). The extent to which a feed-forward control over the redox system is important would depend on whether an individual or a species is strictly dependent on external cues. Selective advantages of antioxidant rhythmicity are still far from being understood, and so are the perturbations of such rhythmicity induced by oxidative stress itself. Research on mutants suggested that the antioxidant rhythmicity can be important for evolutionary fitness. For example, it has been found that the loss of rhythmicity reduced lifespan (Klarsfeld and Rouyer 1998; Krishnan et al. 2012) and reproductive performance (number of eggs laid, quantity of sperm, success of egg fertilisation; Beaver et al. 2002) in *Drosophila melanogaster*.

## 4.6 Genetic Background, Heritability, and Selection of Oxidative Status

It is well established that some components of the organismal oxidative status are under strong genetic control; antioxidant enzymes being a renowned example (e.g. Larsen 1993; Ito et al. 2004; Wang et al. 2001; Miyazaki et al. 2005; Martin et al. 1996; Weber et al. 2012; Borel et al. 2015). Genetic variants associated with susceptibility to oxidative stress have been identified in human populations, as many alleles associated with increased risk for common diseases and ageing have been identified in genes regulating oxidative stress response pathways (e.g. Finkel 2005; Dani and Poggi 2014; Hernández-Guerrero et al. 2018). The hypothesis that genetic polymorphisms of antioxidant enzymes or reactive oxygen species generators occur in natural animal populations is supported to some degree by some studies, such as those that recorded allelic variants in cytochrome P450 enzymes, which influence generation rates of reactive oxygen species and oxidative damage (Daly et al. 1993; Roy and Wirgin 1997). However, very limited information is available on how resistance to oxidative stress in free-living animals is determined by genetic, epigenetic, maternal, or environmental factors and by interactions between them. This paucity of studies limits our capacity to estimate the potential for evolutionary change in organismal capability to resist or tolerate oxidative stress. To do so, we need to assess the magnitude of additive genetic variance and genetic heritability (proportion of phenotypic variance explained by additive genetic variance; Falconer and Mackay 1996). It is also very important to assess the interaction between environment and genetics in order to estimate the degree to which heritability and reaction norms vary across environments (Lynch and Walsh 1998).

Establishing the extent to which the concentration of a certain antioxidant or of a marker of oxidative damage is a stable trait across time within an individual can, at least partially, provide some clue about its endogenous control, which might be a target of selection. Despite being labile traits that change over time and space, markers of oxidative status are often repeatable within individuals. In particular, some studies showed that, while certain markers of oxidative status changed significantly even on a short-time scale, other markers were significantly repeatable within individuals across time, ranging from a few weeks to years (e.g. Costantini et al. 2007, 2021; Galván and Alonso-Alvarez 2009; Saino et al. 2011; Sepp et al. 2012; Récapet et al. 2019). The biological meaning of these intrinsic individual differences in oxidative status has not been elucidated yet. This type of information is relevant to the end of determining whether metrics of oxidative status integrate individual history, are heritable, underlie individual reaction norms, and are plastic or subject to a strict endogenous control. This is an important research area that requires further investigation.

Studies on reptiles and birds indicated that individual differences in certain oxidative status markers have an inherited component, but also that very often the environment is the main driver of this variation. However, the data were collected in such a manner that these heritability estimates are sometimes based on a combination of additive genetic effects and parental effects (i.e. indirect genetic or environmental effects on the phenotype of the offspring, caused by the mother or father). In reptiles, Olsson et al. (2008) found significant heritability in the generation of reactive oxygen species in painted dragon lizards *Ctenophorus pictus*. In birds, relying on a partial cross-fostering experiment, Costantini and Dell'Omo (2006) found that the nest of origin (reflecting the genetic but also the early maternal effects) explained 23.5% of the variance in serum reactive oxygen metabolites in nestling common kestrels *Falco tinnunculus*. By contrast, the nest of rearing (reflecting the environmental contribution) explained 52.8% of the variance in serum non-enzymatic antioxidant capacity. Rubolini et al. (2006) showed that the non-enzymatic antioxidant capacity of yolk in yellow-legged gulls had a significant genetic-maternal contribution (nest of origin explained more than 48% of the variance). In gull chicks, variation in plasma non-enzymatic antioxidant capacity was mostly explained by the rearing environment when they were four (18.7%) or eight (56.5%) days of age. These results also suggested that chicks become more sensitive to environmental variation with age, possibly because the passive antioxidant defences inherited by mothers through the yolk decline with time. Norte et al. (2009) found that the rearing environment explained 17.7% of the variance of glutathione peroxidase levels in great tits. Using a cross-fostering design that limited more the influence of maternal effects compared to previous studies, Kim et al. (2010) found that heritability and common environmental effects were low and non-significant for the in vitro resistance of red blood cells to haemolysis induced by a free radical exposure in one-day old yellow-legged gulls. However, the resistance of red blood cells to a free radical exposure showed significant heritability in eight-days old chicks. Hence, the contribution of genetic background to the generation of the variation in the resistance of red blood cells to free radicals increased with age in gull chicks, which is not in line with previous results from Rubolini et al. (2006).

The expression of genetic variance for resistance to oxidative stress can also be influenced by the environmental conditions (genotype–environment interaction). For example, in yellow-legged gulls, the availability of exogenous antioxidants in the environment influenced the expression of genetic variance for the plasma levels of antioxidants (Kim et al. 2013). As we saw in Chap. 1, the supplementation of dietary antioxidants can induce the downregulation of antioxidant enzymes; thus, some of the chicks might have reduced the expression of endogenous antioxidants in response to an increased intake of dietary antioxidants. On the other hand, the change in the genetic variance for plasma non-enzymatic antioxidant capacity suggested that individuals might have differed in their ability to capture, absorb, and transport dietary antioxidants into their body (Kim et al. 2013). Although genetic variance influenced to some extent the plasma levels of non-enzymatic antioxidants, two markers of oxidative damage (reactive oxygen metabolites and malondialdehyde) in plasma did not show any genetic variance (Kim et al. 2013). In another study on wild great tits, it has been found that the common environment played a major role in determining variation of two blood-based markers of oxidative status (in vitro resistance of red blood cells to haemolyse and plasma concentration of malondialdehyde, respectively) early in life, suggesting a low potential for evolution of individual capability to resist against oxidative stress (Losdat et al. 2014). By contrast, Kahar et al. (2016) found evidence for the heritability of the two antioxidant enzymes superoxide dismutase and glutathione S-transferase in the Atlantic salmon *Salmo salar*, suggesting that these two enzymes have the potential to evolve by selection across generations.

Studies on invertebrates provided stronger evidence for a significant heritability of resistance against oxidative stress. Reynolds and Phillips (2013) quantified the genetic variation for lifespan and stress response using the *Caenorhabditis remanei* as a study species. Heritability (and the coefficient of additive genetic variation) was high for resistance against oxidative stress (tested as the effects of the exposure to hydrogen peroxide and heat stress), low but significant for longevity, and essentially zero for stress response to ultraviolet radiation. Although the resistance against oxidative stress is linked to some degree to longevity, heritability estimates differed between the two traits. One explanation for this result might be that the suite of genes that determined longevity differed from those that determined the resistance against oxidative stress because longevity was measured under non-stressful conditions (Reynolds and Phillips 2013).

Research on inbreeding (mating of individuals or organisms that are genetically closely related, resulting in increased homozygosity) also suggested that there is significant heritability of the resistance against oxidative stress. Inbreeding depression results in reduced evolutionary fitness in a given population (e.g. Lynch and Walsh 1998). Inbreeding makes the organism less resistant to environmental stressors, including oxidative stress (e.g. Pedersen et al. 2008; Okada et al. 2011). For example, Okada et al. (2011) found that the levels of testicular oxidative stress were significantly elevated in inbred fruit fly males, although sperm viability did not differ between inbred and outbred males. De Boer et al. (2018) found small significant long-term effects of inbreeding–environment interactions on adult oxidative status in female canaries *Serinus canaria*. Postnatal components of early life conditions were most persuasively reflected on adult oxidative status, with inbred females that experienced disadvantageous postnatal conditions upregulating enzymatic antioxidants in adulthood. The effects of inbreeding were, however, rather small and limited to females; thus, oxidative stress might have a limited role as a mechanism underlying inbreeding–environment interactions.

If we look at the role of genetic background in moulding the organismal oxidative status from a different angle, we find evidence in support of its relevance. As we saw in other chapters of this book, hormesis is a biological organismal response strictly related to cellular oxidative status. Thus, if hormesis has a genetic basis, it could increase the relevance of additive genetic variance for oxidative stress resistance. Rodriguez et al. (2012) investigated the genetic variation of heat-shock-dependent hormetic effects on lifespan and the associated quantitative trait loci in various strains of *Caenorhabditis elegans*. Wild type CB4856 worms exposed to heat stress survived 18% longer than controls of the same strain, while there was not any hormetic effect on lifespan in wild-type Bristol N2 worms. Using recombinant inbred lines (RILs) derived from a cross between wild types N2 and CB4856, Rodriguez et al. (2012) also found natural variation in stress-response hormesis in lifespan. Some 28% of the RILs displayed a hormetic effect on lifespan. Importantly, the ability to recover from heat shock mapped to a significant quantitative trait locus on chromosome II. The QTL was confirmed by introgressing relatively small CB4856 regions into chromosome II of N2.

Hormetic responses might also be under selection in case they were influenced by non-genetic inheritance, such as epigenetic modifications. Many hormetic responses induced by a variable range of agents have been characterised by large-scale changes in gene expression (Vaiserman 2008, 2010, 2011; Sarup et al. 2013). For example, hormetic responses to heat stress have been routinely found in various animal species, and, in 1998, it has been suggested to be a plausible molecular mechanism underlying the emergence of epigenetic effects in natural animal populations (Rutherford and Lindquist 1998).

In conclusion, we have contrasting evidence about the contribution of genetics in shaping the organismal oxidative status. A low heritability of a trait is usually related to its positive contribution to fitness, since strong directional and constant selection is supposed to erode genetic variation driving the alleles that confer the highest fitness outcomes to fixation by natural selection (e.g. Fisher 1930; Kimura 1958; Gustafsson 1986; Mousseau and Roff 1987; Kruuk et al. 2000). However, many examples in the literature showed that fitness-related traits can have high additive genetic variation, whose expression is influenced by environmental conditions (e.g. Hoffman and Parsons 1991; Price and Schluter 1991; Merilä and Sheldon 1999; Charmantier and Garant 2005). Environmental conditions can hide the influence of genetic variation or drive its expression (e.g. Pigliucci 2001; West-Eberhard 2003; Badyaev 2009). Finally, there is growing awareness that even in the case of a low additive genetic variance, epigenetic heritability can still generate potential for the evolution of oxidative stress resistance (e.g. parental effects, epigenetics: Bonduriansky and Day 2009; Danchin et al. 2011). This is a very important area of research that has been poorly explored in natural animal populations.

## 4.7 Oxidative Status, Metabolism, and Pace of Life

Many comparative studies documented significant associations between the pace of life (or life history with which coevolves) and concentrations of certain antioxidants or markers of oxidative damage. Cohen et al. (2008) analysed various serum non-enzymatic antioxidants (carotenoids, vitamin E, uric acid, non-enzymatic antioxidant capacity) in 95 bird species (mostly passerines). They found that those species characterised by a more rapid development, a lower survival, a smaller body size, a larger clutch size, and a higher mass-adjusted metabolic rate had higher serum concentrations of antioxidants. However, the strength of the link between pace of life and antioxidants differed between types of antioxidants; vitamin E had the smaller number of significant connections with life-history traits. The strength of the relationship between pace of life and antioxidants was also quite variable between taxa. Phylogenetic subsampling and division into tropical and temperate subsets produced changes in the correlations between life-history traits and antioxidants (Cohen et al. 2008). Moreover, the results of the study suggested that, while larger species may have low levels of circulating antioxidants because they would increase them only when needed, smaller species might have evolved high constitutive levels in order to anticipate any stressors they will encounter in life. Hence, smaller and short-lived species would invest in long-term physiological homeostasis because they experience repeated stressors on a daily basis if compared to larger ones (Cohen et al. 2008). The question then is why a species with low perspectives of survival would invest resources in physiological homeostasis, diverting them away from reproduction. It might also be that short-lived species maintain high constitutive levels of antioxidants because they need to avoid excessive oxidative stress due to an intense reproductive investment, which might have a negative impact on fertility or offspring.

A subsequent comparative study found different results from those of Cohen et al. (2008). Galván et al. (2012) analysed the connections between life-history traits and the concentrations of carotenoids or vitamin E in the liver of 125 bird species. They found that high carotenoid concentrations in liver were typical of those species with large bodies, brains, and eggs, high absolute metabolic rates, and resident lifestyles. By contrast, high concentrations of vitamin E were typical of those species with large brain sizes, and long lifespans and incubation periods. Thus, high liver antioxidant concentrations were associated with life-history strategies characterised by a "live slow, die old" pattern. Why are these two comparative studies in clear disagreement with each other? One reason for this contrast might lie in the different metabolic regulation of vitamin E in blood and in liver. Plasma vitamin E is immediately available in case the organism needs to quench any reactive oxygen species and can be embedded into cell membranes. By contrast, variation in hepatic vitamin E levels probably better reflects the evolution of storing strategies that might not be directly connected to the amount of vitamin E that is made available to contrast reactive oxygen species.

In another comparative study, Tella et al. (2004) also found that the diet composition significantly explained variation in plasma concentration of carotenoids among bird species. Phylogeny was even more relevant than diet; it explained 65% of the variation in plasma carotenoids. Although there might be some collinearity between diet and phylogeny, the results of this comparative study suggested that species have evolved different mechanisms to regulate the absorption and metabolism of carotenoids (e.g. Trams 1969; Yang and Tume 1993; Rock et al. 1996; Surai et al. 1998; Tella et al. 2004).

We have also a number of studies that compared the oxidative status between two or a few more related species. Although two-species comparative studies raise several logical and statistical problems (e.g. the appropriate null hypothesis for comparing two species is something closer to a difference rather than no difference; Garland and Adolph 1994) with respect to inferring adaptation, they provided some interesting indications about the link between pace of life and oxidative status. For example, López-Cruz et al. (2010) found that the two shark species *Isurus oxyrinchus* and *Carcharhinus falciformis* with intense swimming habits and a diet mainly based on highly active prey had higher generation of the superoxide free radical and levels of glutathione peroxidase and glutathione-S-transferase in skeletal muscle than the shark species *Sphyrna zygaena* with less active swimming habits. However, the concentration of thiobarbituric acid reactive substances was lower in *I. oxyrinchus* than in the other two species. Hence, sharks with high demanding lifestyles may need to maintain high constitutive antioxidant protection in order to mitigate oxidative damage. By contrast, the comparison of five blood-based markers of oxidative status between sympatric cheetahs *Acinonyx jubatus* and leopards *Panthera pardus* (Fig. 4.7) living on Namibian farmlands did not corroborate the assumption that two species with a different pace of life would consistently differ in key physiological traits (Costantini et al. 2019).

Parasite helminths express a comprehensive repertoire of both cytoplasmic and secreted antioxidant enzymes, but there are significant differences between species and life-cycle stages. For example, adult individuals of *Brugia malayi*, *Dirofilaria immitis*, and *Onchocerca volvulus* secrete copper-zinc superoxide dismutase; *Brugia malayi* and *Dirofilaria immitis* also secrete a selenocysteine-independent glutathione peroxidase (Selkirk et al. 1998). Adult *Onchocerca volvulus* may compensate for an apparent deficiency in the expression of glutathione peroxidase by means of a variant of glutathione S-transferase (Liebau et al. 1994; Selkirk et al.

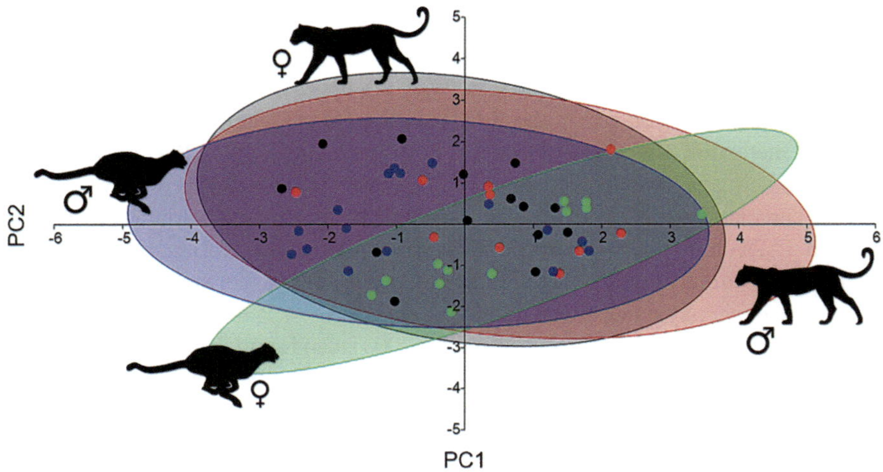


**Fig. 4.7** Comparison of the multivariate space based on blood-based markers of oxidative status between sympatric cheetahs *Acinonyx jubatus* and leopards *Panthera pardus*. The plot of the first two principal components shows a strong overlap of the multivariate spaces among male cheetahs and both female and male leopards; by contrast, female cheetahs have a different multivariate space characterised by lower variability of the oxidative status markers. Black dot = female leopard, red dot = male leopard, green dot = female cheetah, blue dot = male cheetah. Reprinted from Costantini et al. (2019) with permission of Elsevier

1998). It has also been found that helminth phylogenetic lineages also differed in genes coding for glutathione peroxidases and peroxiredoxins, leading to the selection of different isoforms (Kim et al. 2009; Bae et al. 2011).

Closely related algal species can rely on very different strategies to withstand oxidative stress. For example, conversely to the green alga *Ulva procera*, the sympatric green alga *Cladophora glomerata* responded to an increase in photosynthetic active radiation by upregulating the synthesis of carotenoids and activating the violaxanthin xanthophyll cycle, respectively (Choo et al. 2005). Moreover, *Cladophora glomerata* had higher baseline activities of catalase and ascorbate peroxidase than *Ulva procera* (Choo et al. 2004). These differences in antioxidant mechanisms would make *Ulva procera* more sensitive to oxidative stress than *Cladophora glomerata* (Choo et al. 2005). *Ulva procera* algae produce high quantities of hydrogen peroxide when being exposed to light. Given their low levels of hydrogen peroxide removing antioxidant enzymes, *Ulva procera* algae have evolved the capacity to remove hydrogen peroxide through diffusion out of the cells into the seawater (Choo et al. 2005). These differences in antioxidant strategies reflect ecological differentiation of the two algal species. Although both species have opportunistic habits, *Cladophora glomerata* is a persistent annual alga, while *Ulva procera* has ephemeral habits, appearing or disappearing in a same site depending on the current local conditions (Choo et al. 2005). We might argue that *Cladophora glomerata* invests more than *Ulva procera* into self-maintenance mechanisms to support their higher persistency in the environment. By contrast, *Ulva procera* algae invest more in growth and reproduction over their short periods of activity. *Ulva procera* algae have in fact very high photosynthetic activity compared to other algal species, including *Cladophora glomerata* (Johansson and Snoeijs 2002).

Variation in oxidative stress resistance was also documented for closely related plant species. Hu et al. (2013) showed that foliar antioxidant and free amino acid levels were significantly affected by drought and ambient temperature. However, this effect differed between species, with the drought-tolerant species *Quercus pubescens* that showed a higher scavenging capacity of reactive oxygen species than the less drought-tolerant species *Q. robur* and *Q. petraea.* Li et al. (2013) investigated the effects of ozone (a pro-oxidant agent) on winter wheat *Triticum aestivum* and on its natural competitor, the flixweed *Descurainia sophia.* The generation of hydrogen peroxide and oxidative damage to lipids was increased in both plant species, but the effect was stronger when plants were grown in mixture than singly. Importantly, the winter wheat was less resistant to the pro-oxidant agent than the flixweed. The higher susceptibility to oxidative stress of winter wheat resulted in lower growth and grain yield than those of the flixweed (Li et al. 2013). Competition between plants has probably exacerbated the pro-oxidant effect of an environmental stressor, but the different resistance of the two plants against oxidative stress determined their success in the competition.

Finally, a largely neglected question is whether hybridisation between genetic lineages of a given species or between different species results in changes in the regulation of oxidative status that influence the fitness of hybrids. An oxidative cost can be expected because genomic incompatibilities can influence different biochemical and metabolic processes in hybrids. For example, Meng et al. (2021) found that levels of oxidative damage in hybrid strains of Pacific oysters *Crassostrea gigas* were similar to those of parental strains. Zheng et al. (2022) found that intraspecific hybridisation of noble scallops *Chlamys nobilis* from different populations resulted in slightly higher carotenoid levels (from ca. 4–18%) in inbred than cross scallops under both ambient and acidified seawater conditions (Fig. 4.8).

Studies on interspecific hybridisation provided stronger results than those on intraspecific hybridisation. Prokić et al. (2018) found that hybrid individuals of newt species, *Triturus macedonicus* and *T. ivanbureschi*, displayed higher levels of superoxide dismutase, catalase, glutathione peroxidase, and thiol groups in comparison to both parental species, suggesting that they may differ in how they regulate their oxidative status. Fasting caused greater oxidative damage in hybrids than in parental species, indicating that they were more susceptible to fasting-induced oxidative stress (Prokić et al. 2021). Similarly, when being exposed to thermal stress, parental species had enhanced antioxidant levels, which enabled them to alleviate temperature-induced stress (revealed by the absence of oxidative damage), while hybrids suffered higher lipid peroxidation (Petrović et al. 2023). These findings pointed to a greater disruption of oxidative status homeostasis in hybrid newts, probably reflecting a direct cost of hybridisation.

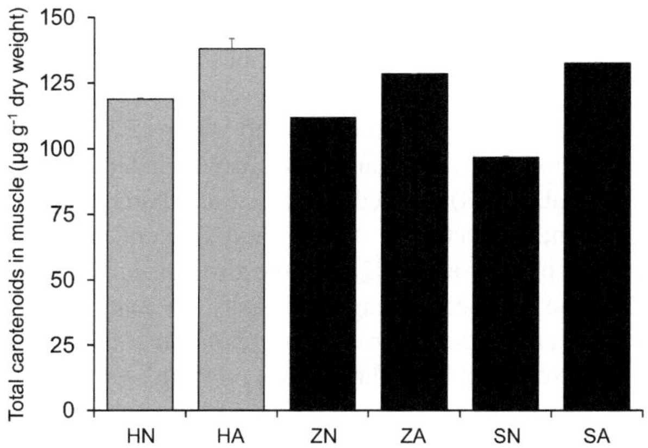


**Fig. 4.8** Effects of ocean acidification on the concentration of total carotenoids in the adductor muscle of inbred (black histograms) and cross (grey histograms) marine bivalve noble scallops *Chlamys nobilis* juveniles. HN = cross in ambient seawater; HA = cross in acidified seawater; ZN = Zhanjiang inbred in ambient seawater; ZA = Zhanjiang inbred in acidified seawater; SN = Shantou inbred in ambient seawater; SA = Shantou inbred in acidified seawater. Values are shown as mean and standard deviation. Data were collected from Table 6 in Zheng et al. (2022)

## 4.8 Conclusions and Future Directions

We have now good evidence to say that endogenous factors linked to stress response pathways and molecular clocks are relevant proximate causes of individual variation in the generation of reactive oxygen species, oxidative damage, and expression of antioxidant mechanisms. For example, individuals largely differ in the way they physiologically respond to stressful situations (coping style or personality). Living in a landscape of fear also exposes individuals to a predation risk, which can impact on the oxidative status through the release of glucocorticoids (modulators of the stress response). Additionally, a number of studies documented circadian and circannual changes in oxidative status. Although many of these documented changes are due to the influence of exogenous factors (e.g. food availability, abiotic conditions), some endogenous regulators have also been identified.

Conspecific populations can strongly differ in levels of reactive oxygen species generation or expression of antioxidants, as well as in the regulation of oxidative status. Although variation of environmental conditions (e.g. phenotypic plasticity) is one important generator of population differentiation in the oxidative status, we now start to appreciate that current physiological differences between conspecific populations can also be the legacy of past biogeographic events.

One of the unsolved questions that scientists face today lies in the role of (epi) genetics in determining individual, population, and species differences in oxidative status. This is mainly due to a shortage of clear and robust evidence. Most of the data were collected in such a manner that heritability estimates are sometimes based on a combination of additive genetic effects and maternal/paternal effects. Moreover, evolutionary theory predicts that persistent selection should deplete genetic and phenotypic variance. Thus, if we assume that oxidative status markers possess heritable variation that is under selection, how can we explain the paradox about how variation in oxidative status phenotypes is maintained? One explanation might lie in the role of context-dependent selection, whereby the fitness benefits of a given phenotype change across space and time. However, we have very limited understanding of whether oxidative status is a target of selection. Phenotypic responses to selection depend on the genetic basis of traits, but also on the genetic relationship among different traits. Oxidative status markers are often correlated with other phenotypic traits, so that selection on one trait can indirectly lead to selection and phenotypic change in the correlated marker of oxidative status, even if it is not under direct selection. This could be particularly relevant in case of genetic correlations between oxidative status components and other traits that rise due to pleiotropy or linkage disequilibrium. Pleiotropy occurs when a single gene is involved in the regulation of multiple different traits; linkage disequilibrium refers to the co-inheritance of two genes that affect two different traits, either by chance if the two genes are physically close in the genome or when previous selection favoured a specific combination of phenotypic traits (Lande 1975, 1980, 1984). Thus, genetic correlations can prevent traits from evolving independently and influence transmission to the next generation, a prerequisite for the evolution to occur. We definitely need more work on this topic if we aim at estimating the potential for evolution of oxidative status regulation and oxidative stress resistance.

Finally, the pace of life is strictly connected to the metabolic rate, so that we could expect that values of oxidative status markers of individuals or species with faster metabolisms should differ from those with slower metabolisms. However, comparative studies provided mixed support for a link between oxidative status and pace of life or even phylogenetic history. Implementation of comparative studies and experimental testing of the link between pace of life and oxidative status are important areas for future research.

## References

Alessio HM (1993) Exercise-induced oxidative stress. Med Sci Sports Exerc 25:218–224

Aras NM, Bayir A, Sirkecioglu AN, Bayir M, Aksakal E, Haliloglu HI (2009) Seasonal changes in antioxidant defence system of liver and gills of *Salmo trutta caspius*, *Salmo trutta labrax* and *Salmo trutta macrostigma*. J Fish Biol 74:842–856

Arnold KE, Herborn KA, Adam A, Alexander L (2015) Individual variation in the oxidative costs of personality traits. Funct Ecol 29:522–530

Ashkenazi L, Haim A (2012) Light interference as a possible stressor altering HSP70 and its gene expression levels in brain and hepatic tissues of golden spiny mice. J Exp Biol 215:4034–4040

Ashkenazi L, Haim A (2013) Effect of light at night on oxidative stress markers in Golden spiny mice (*Acomys russatus*) liver. Comp Biochem Physiol A 165:353–357

Badyaev AV (2009) Evolutionary significance of phenotypic accommodation in novel environments: an empirical test of the Baldwin effect. Phil Trans R Soc Lond B 364:1125–1141

Badyaev AV (2011) Origin of the fittest: link between emergent variation and evolutionary change as a critical question in evolutionary biology. Proc R Soc Lond B 278:1921–1929

Bae Y-A, Kim S-H, Lee E-G, Sohn W-M, Kong Y (2011) Identification and biochemical characterization of two novel peroxiredoxins in a liver fluke, *Clonorchis sinensis*. Parasitology 138:1143–1153

Barlow-Walden LR, Reiter RJ, Abe M, Pablos M, Menendez-Pelaez A, Chen L-D, Poeggeler B (1995) Melatonin stimulates brain glutathione peroxidase activity. Neurochem Int 26:497–502

Barreto FS, Burton RS (2013) Elevated oxidative damage is correlated with reduced fitness in interpopulation hybrids of a marine copepod. Proc R Soc Lond B 280:20131521

Beaver LM, Gvakharia BO, Vollintine TS, Hege DM, Stanewsky R, Giebultowicz JM (2002) Loss of circadian clock function decreases reproductive fitness in males of *Drosophila melanogaster*. Proc Natl Acad Sci USA 99:2134–2139

Beaver LM, Klichko VI, Chow ES, Kotwica-Rolinska J, Williamson M, Orr WC, Radyuk SN, Giebultowicz JM (2012) Circadian regulation of glutathione levels and biosynthesis in *Drosophila melanogaster*. PLoS One 7:e50454

Birnie-Gauvin K, Peiman KS, Larsen MH, Baktoft H, Aarestrup K, Willmore WG, Cooke SJ (2017) Oxidative stress and partial migration in brown trout (*Salmo trutta*). Can J Zool 95:829–835

Biro PA, Stamps JA (2008) Are animal personality traits linked to life-history productivity? Trends Ecol Evol 23:361–368

Biro PA, Stamps JA (2010) Do consistent individual differences in metabolic rate promote consistent individual differences in behavior? Trends Ecol Evol 25:653–659

Bisconti R, Canestrelli D, Colangelo P, Nascetti G (2011a) Multiple lines of evidence for demographic and range expansion of a temperate species (*Hyla sarda*) during the last glaciation. Mol Ecol 20:5313–5327

Bisconti R, Canestrelli D, Nascetti G (2011b) Genetic diversity and evolutionary history of the Tyrrhenian treefrog Hyla sarda (Anura: Hylidae): adding pieces to the puzzle of Corsica–Sardinia biota. Biol J Linn Soc 103:159–167

Blažek R, Polačik M, Kačer P, Cellerino A, Řežucha R, Methling C, Tomášek O, Syslová K, Terzibasi Tozzini E, Albrecht T, Vrtílek M, Reichard M (2017) Repeated intraspecific divergence in life span and aging of African annual fishes along an aridity gradient. Evolution 71:386–402

Błońska D, Tarkan AS, Janic B, Tszydel M, Bukowska B (2023) Does intraspecific competition cause oxidative stress? Influence of biotic and abiotic factors on antioxidant system of an invasive round goby. Ecol Evol 13:e10795

Bonduriansky R, Day T (2009) Nongenetic inheritance and its evolutionary implications. Annu Rev Ecol Evol System 40:103–125

Border SE, DeOliveira GM, Janeski HM, Piefke TJ, Brown TJ, Dijkstra PD (2019) Social rank, color morph, and social network metrics predict oxidative stress in a cichlid fish. Behav Ecol 30:490–499

Borel P, Desmarchelier C, Nowicki M, Bott R, Tourniaire F (2015) Can genetic variability in α-tocopherol bioavailability explain the heterogeneous response to α-tocopherol supplements? Antioxid Redox Signal 22:669–678

Burkhardt S, Coto-Montes A, Hardeland R (1999) Diurnal rhythm of protein carbonyl in *Gonyaulax polyedra*. In: Hardeland R (ed) Studies on antioxidants and their metabolites. Cuvillier, Göttingen, pp 10–13

Burraco P, Duarte LJ, Gomez-Mestre I (2013) Predator-induced physiological responses in tadpoles challenged with herbicide pollution. Curr Zool 59:475–484

Butler MW, Ligon RA (2015) Interactions between biliverdin, oxidative damage, and spleen morphology after simulated aggressive encounters in veiled chameleons. PLoS One 10:e0138007

Cabrera D, Nilsson JR, Griffen BD (2021) The development of animal personality across ontogeny: a cross-species review. Anim Behav 173:137–144

Canoine V, Hayden TJ, Rowe K, Goymann W (2002) The stress response of European stonechats depends on the type of stressor. Behaviour 139:1303–1311

Careau V, Thomas D, Humphries M, Réale D (2008a) Energy metabolism and animal personality. Oikos 117:641–653
Careau V, Bininda-Emonds ORP, Thomas DW, Réale D, Humphries MM (2008b) Exploration strategies map along fast–slow metabolic and life-history continua in muroid rodents. Funct Ecol 23:150–156
Carere C, Eens M (2005) Unravelling animal personalities: how and why individuals consistently differ. Behaviour 142:1149–1157
Carere C, Maestripieri D (2013) Animal personalities: behavior, physiology, and evolution. The University of Chicago Press, Chicago
Carere C, Caramaschi D, Fawcett T (2010a) Covariation between personalities and individual differences in coping with stress: converging evidence and hypotheses. Curr Zool 56:728–740
Carere C, Costantini D, Sorace A, Santucci D, Alleva E (2010b) Bird populations as sentinels of endocrine disrupting chemicals. Ann Ist Super Sanità 46:81–88
Cavigelli SA (2005) Animal personality and health. Behaviour 142:1223–1244
Cavigelli SA, McClintock MK (2003) Fear of novelty in infant rats predicts adult corticosterone dynamics and an early death. Proc Natl Acad Sci USA 100:16131–16136
Charmantier A, Garant D (2005) Environmental quality and evolutionary potential: lessons from wild populations. Proc R Soc Lond B 272:1415–1425
Choo KS, Snoeijs P, Pedersén M (2004) Oxidative stress tolerance in the filamentous green algae *Cladophora glomerata* and *Enteromorpha ahlneriana*. J Exp Mar Biol Ecol 298:111–123
Choo KS, Nilsson J, Pedersén M, Snoeijs P (2005) Photosynthesis, carbon uptake and antioxidant defence in two coexisting filamentous green algae under different stress conditions. Mar Ecol Progr Ser 292:127–138
Clinchy M, Sheriff MJ, Zanette LY (2013) Predator-induced stress and the ecology of fear. Funct Ecol 27:56–65
Cohen AA, McGraw KJ, Wiersma P, Williams JB, Robinson WD, Robinson TR, Brawn JD, Ricklefs RE (2008) Interspecific associations between circulating antioxidant levels and life-history variation in birds. Am Nat 172:178–193
Cohen AA, McGraw KJ, Robinson W (2009) Serum antioxidant levels in wild birds vary in relation to diet, season, life history strategy, and species. Oecologia 161:673–683
Coppens CM, De Boer SF, Koolhaas JM (2010) Coping styles and behavioural flexibility: towards underlying mechanisms. Phil Trans R Soc Lond B 365:4021–4028
Cornell A, Williams TD (2017) Variation in developmental trajectories of physiological and somatic traits in a common songbird approaching fledging. J Exp Biol 220:4060–4067
Costantini D (2014) Oxidative stress and hormesis in evolutionary ecology and physiology: a marriage between mechanistic and evolutionary approaches. Springer-Verlag, Berlin Heidelberg, p 348
Costantini D, Bonadonna F (2010) Patterns of variation of serum oxidative stress markers in two seabird species. Polar Res 29:30–35
Costantini D, Dell'Omo G (2006) Environmental and genetic components of oxidative stress in wild kestrel nestlings (*Falco tinnunculus*). J Comp Physiol B 176:575–579
Costantini D, Dell'Omo G, Casagrande S, Fabiani A, Carosi M, Bertacche V, Marquez C, Snell H, Snell H, Tapia W, Gentile G (2005) Inter-population variation of carotenoids in Galápagos land iguanas (*Conolophus subcristatus*). Comp Biochem Physiol B 142:239–244
Costantini D, Coluzza C, Fanfani A, Dell'Omo G (2007) Effects of carotenoid supplementation on colour expression, oxidative stress and body mass in rehabilitated captive adult kestrels (*Falco tinnunculus*). J Comp Physiol B 177:723–731
Costantini D, Carere C, Caramaschi D, Koolhaas JM (2008) Aggressive and non-aggressive personalities differ in oxidative status in selected lines of mice (*Mus musculus*). Biol Lett 4:119–122
Costantini D, Dell'Omo G, De Filippis PS, Marquez C, Snell H, Snell H, Tapia W, Brambilla G, Gentile G (2009) Temporal and spatial covariation of gender and oxidative stress in the Galápagos land iguana *Conolophus subcristatus*. Physiol Biochem Zool 82:430–437

Costantini D, Carello L, Fanfani A (2010) Relationships among oxidative status, breeding conditions and life-history traits in free-living great tits *Parus major* and common starlings *Sturnus vulgaris*. Ibis 152:793–802
Costantini D, Marasco V, Møller AP (2011) A meta-analysis of glucocorticoids as modulators of oxidative stress in vertebrates. J Comp Physiol B 181:447–456
Costantini D, Ferrari C, Pasquaretta C, Cavallone E, Carere C, von Hardenberg A, Réale D (2012) Interplay between plasma oxidative status, cortisol and coping styles in wild alpine marmots, *Marmota marmota*. J Exp Biol 215:374–383
Costantini D, Greives TJ, Hau M, Partecke J (2014) Does urban life change blood oxidative status in birds? J Exp Biol 217:2994–2997
Costantini D, Czirják GÁ, Melzheimer J, Menges V, Wachter B (2019) Sex and species differences of stress markers in sympatric cheetahs and leopards in Namibia. Comp Biochem Physiol A 227:8–13
Costantini D, Masi S, Rachid L, Beltrame M, Rohmer M, Krief S (2021) Mind the food: rapid changes in antioxidant content of diet affect oxidative status of chimpanzees. Am J Phys 320:728–734
Coto-Montes A, Hardeland R (1997) Diurnal time patterns of protein carbonyl in *Drosophila melanogaster*: comparison of wild-type flies and clock mutants. In: Hardeland R (ed) Biological rhythms and antioxidative protection. Cuvillier, Göttingen, pp 119–126
Coto-Montes A, Boga JA, Tomás-Zapico C, Rodríguez-Colunga MJ, Martínez-Fraga J, Tolivia-Cadrecha J, Menéndez G, Hardeland R, Tolivia D (2001) Porphyric enzymes in hamster Harderian gland, a model of damage by porphyrins and their precursors. A chronobiological study on the role of sex differences. Chem Biol Interact 134:135–149
Coughlan K, Sadowska ET, Bauchinger U (2023) Repeat sampling of female passerines during reproduction reveals surprising higher plasma oxidative damage during resting compared to active state. Integr Comp Biol 63:1197–1208
Cruz FM, Carneiro CLS, Oliveira JM, Valbon WR, Lima GDA, Freitas MB, Oliveira EE, Salaro AL (2021) "Fearing the enemy": growth and stress biomarker responses of sexually reversed *Oreochromis niloticus* in the presence of aquatic predatory insects. Physiol Behav 228:113202
Daly AK, Cholerton S, Gregory W, Idle JR (1993) Metabolic polymorphisms. Pharmacol Ther 57:129–160
Danchin E, Charmantier A, Champagne FA, Mesoudi A, Pujol B, Blanchet S (2011) Beyond DNA: integrating inclusive inheritance into an extended theory of evolution. Nat Rev Genet 12:475–486
Dani C, Poggi C (2014) The role of genetic polymorphisms in antioxidant enzymes and potential antioxidant therapies in neonatal lung disease. Antioxid Redox Signal 21:1863–1880
De Boer RA, Costantini D, Casasole G, AbdElgawad H, Asard H, Eens M, Müller W (2018) Sex-specific effects of inbreeding and early life conditions on the adult oxidative balance. Curr Zool 64:631–639
Di Mascio P, Hollnagel HC, Sperança M, Colepicolo P (1995) Diurnal rhythm of β-carotene in photosynthetic alga *Gonyaulax polyedra*. Biol Chem Hoppe Seyler 376:297–301
Dijkstra PD, Wiegertjes GF, Forlenza M, van der Sluijs I, Hofmann HA, Metcalfe NB, Groothuis TGG (2011) The role of physiology in the divergence of two incipient cichlid species. J Evol Biol 24:2639–2652
Dijkstra PD, Pierotti ME, Seehausen O, Metcalfe NB (2016) Metabolism, oxidative stress and territorial behaviour in a female colour polymorphic cichlid fish. Behav Ecol Sociobiol 70:99–109
Dulude-de Broin F, Hamel S, Mastromonaco GF, Côté SD (2020) Predation risk and mountain goat reproduction: evidence for stress-induced breeding suppression in a wild ungulate. Funct Ecol 34:1003–1014
Edgar RS, Green EW, Zhao Y, van Ooijen G, Olmedo M, Qin X, Xu Y, Pan M, Valekunja UK, Feeney KA, Maywood ES, Hastings MH, Baliga NS, Merrow M, Millar AJ, Johnson CH, Kyriacou CP, O'Neill JS, Reddy AB (2012) Peroxiredoxins are conserved markers of circadian rhythms. Nature 485:459–464

Ensminger DC, Crocker DE, Lam EK, Allen KN, Vázquez-Medina JP (2021) Repeated stimulation of the HPA axis alters white blood cell count without increasing oxidative stress or inflammatory cytokines in fasting elephant seal pups. J Exp Biol 224:jeb243198

Falconer DS, Mackay TFC (1996) Introduction to quantitative genetics. Longman, Essex

Ferrari C, Pasquaretta C, Carere C, Cavallone E, von Hardenberg A, Réale D (2013) Testing for the presence of coping styles in a wild mammal. Anim Behav 85:1385–1396

Finkel T (2005) Radical medicine: treatment ageing to cure disease. Nat Rev Mol Cell Biol 6:971–976

Fisher RA (1930) The genetical theory of natural selection. Dover, New York

Florencio M, Burraco P, Rendón MÁ, Díaz-Paniagua C, Gomez-Mestre I (2020) Opposite and synergistic physiological responses to water acidity and predator cues in spadefoot toad tadpoles. Comp Biochem Physiol A 242:110654

Galván I, Alonso-Alvarez C (2009) The expression of melanin-based plumage is separately modulated by exogenous oxidative stress and a melanocortin. Proc R Soc Lond B 276:3089–3097

Galván I, Erritzøe J, Karadas F, Møller AP (2012) High levels of liver antioxidants are associated with life-history strategies characteristics of slow growth and high survival rates in birds. J Comp Physiol 182:947–959

Garland T Jr, Adolph SC (1994) Why not to do two-species comparative studies: limitations on inferring adaptation. Physiol Zool 67:797–828

Garratt M, Brooks RC (2015) A genetic reduction in antioxidant function causes elevated aggression in mice. J Exp Biol 218:223–227

Georgiev AV, Thompson ME, Mandalaywala TM, Maestripieri D (2015) Oxidative stress as an indicator of the costs of reproduction among free-ranging rhesus macaques. J Exp Biol 218:1981–1985

Giannetto C, Fazio F, Casella S, Assenza A, Marafioti S, Piccione G (2012) Comparison of daily rhythms of oxygen metabolites and serum barrier to oxidation in domestic animals. Centr Eur J Biol 6:91–98

Gilbert SF, Epel D (2009) Ecological developmental biology: integrating epigenetics, medicine, and evolution. Sinauer Associates

Goncharova ND, Shmaliy AV, Bogatyrenko TN, Koltorev VK (2006) Correlation between activity of antioxidant enzymes and circadian rhythms of corticosteroids in *Macaca mulatta* monkeys of different age. Exp Gerontol 41:778–783

Grether GF (2014) Redesigning the genetic architecture of phenotypically plastic traits in a changing environment. Biol J Linn Soc 112:276–286

Groothuis TGG, Carere C (2005) Avian personalities: characterization and epigenesis. Neurosci Biobehav Rev 29:137–150

Guerra C, Zenteno-Savín T, Maeda-Martínez AN, Abele D, Philipp EER (2013) The effect of predator exposure and reproduction on oxidative stress parameters in the Catarina scallop *Argopecten ventricosus*. Comp Biochem Physiol A 165:89–96

Gustafsson L (1986) Lifetime reproductive success and heritability: empirical support for Fisher's fundamental theorem. Am Nat 128:761–764

Hardeland R, Coto-Montes A, Poeggeler B (2003) Circadian rhythms, oxidative stress, and antioxidative defense mechanisms. Chronobiol Intern 20:921–962

Hawlena D, Schmitz OJ (2010) Physiological stress as a fundamental mechanism linking predation to ecosystem functioning. Am Nat 176:537–556

Herborn KA, Coffey J, Larcombe SD, Alexander L, Arnold KE (2011) Oxidative profile varies with personality in European greenfinches. J Exp Biol 214:1732–1739

Hernández-Guerrero C, Parra-Carriedo A, Ruiz-de-Santiago D, Galicia-Castillo O, Buenrostro-Jáuregui M, Díaz-Gutiérrez C (2018) Genetic polymorphisms of antioxidant enzymes CAT and SOD affect the outcome of clinical, biochemical, and anthropometric variables in people with obesity under a dietary intervention. Genes Nutr 13:1

Herrera-Dueñas A, Pineda J, Antonio MT, Aguirre JI (2014) Oxidative stress of house sparrow as bioindicator of urban pollution. Ecol Indic 42:6–9

Hidalgo MC, Trenzado CE, Furné M, Beltrán A, Manzaneda C, García-Gallego M, Domezain A, Sanz A (2017) Tissue-specific daily variation in the oxidative status of sturgeon (*Acipenser naccarii*) and rainbow trout (*Oncorhynchus mykiss*): a comparative study. Fish Physiol Biochem 43:1105–1115

Hill RW, Wyse GA, Anderson M (2008) Animal physiology, 2nd edn. Sinauer Associates, Sunderland

Hoffman AA, Parsons PA (1991) Evolutionary genetics and environmental stress. Oxford University Press, Oxford

Hollnagel HC, Di Mascio P, Asano CS, Okamoto OK, Stringher CG, Oliveira MC, Colepicolo P (1996) The effect of light on the biosynthesis of β-carotene and superoxide dismutase activity in the photosynthetic alga *Gonyaulax polyedra*. Braz J Med Biol Res 29:105–110

Hoogenboom MO, Metcalfe NB, Groothuis TGG, de Vries B, Costantini D (2012) Relationship between oxidative stress and circulating testosterone and cortisol in pre-spawning female brown trout. Comp Biochem Physiol A 163:379–387

Hu B, Simon J, Rennenberg H (2013) Drought and air warming affect the species-specific levels of stress-related foliar metabolites of three oak species on acidic and calcareous soil. Tree Physiol 33:489–504

Isaksson C (2010) Pollution and its impact on wild animals: a meta-analysis on oxidative stress. EcoHealth 7:342–350

Isaksson C, While GM, McEvoy J, van de Crommenacker J, Olsson M, Groothuis TGG, Komdeur J, Wapstra E (2011) Aggression, but not testosterone, is associated to oxidative status in a free-living vertebrate. Behaviour 148:713–731

Ito K, Hirao A, Arai F, Matsuoka S, Takubo K, Hamaguchi I, Nomiyama K, Hosokawa K, Sakurada K, Nakagata N, Ikeda Y, Mak TW, Suda T (2004) Regulation of oxidative stress by ATM is required for self-renewal of haematopoietic stem cells. Nature 431:997–1002

Jansen F, Heiming R, Lewejohann L, Touma C, Palme R, Schmitt A, Lesch K, Sachser N (2010) Modulation of behavioural profile and stress response by 5-HTT genotype and social experience in adulthood. Behav Brain Res 207:21–29

Janssens L, Stoks R (2013a) Predation risk causes oxidative damage in prey. Biol Lett 9:20130350

Janssens L, Stoks R (2013b) Synergistic effects between pesticide stress and predator cues: conflicting results from life history and physiology in the damselfly *Enallagma cyathigerum*. Aquat Toxicol 132-133:92–99

Janssens L, Stoks R (2014) Chronic predation risk reduces escape speed by increasing oxidative damage: a deadly cost of an adaptive antipredator response. PLoS One 9:e101273

Janssens L, Stoks R (2017) Chlorpyrifos-induced oxidative damage is reduced under warming and predation risk: explaining antagonistic interactions with a pesticide. Environ Pollut 226:79–88

Jefimow M, Przybylska-Piech AS, Wojciechowski MS (2020) Predictive and reactive changes in antioxidant defence system in a heterothermic rodent. J Comp Physiol B 190:479–492

Jermacz Ł, Kletkiewicz H, Nowakowska A, Dzierżyńska-Białończyk A, Klimiuk M, Kobak J (2020) Continuity of chronic predation risk determines changes in prey physiology. Sci Rep 10:6972

Johansson G, Snoeijs P (2002) Macroalgal photosynthetic responses to light in relation to thallus morphology and depth zonation. Mar Ecol Progr Ser 244:63–72

Josserand R, Haussy C, Agostini S, Decencière B, Le Galliard JF, Meylan S (2020) Chronic elevation of glucorticoids late in life generates long lasting changes in physiological state without a life history switch. Gen Comp Endocrinol 285:113288

Kahar S, Debes PV, Vuori KAM, Vähä J-P, Vasemägi A (2016) Heritability, environmental effects, and genetic and phenotypic correlations of oxidative stress resistance-related enzyme activities during early life stages in Atlantic salmon. Evol Biol 43:215–226

Kharwar RK, Haldar C (2012) Daily variation in antioxidant enzymes and lipid peroxidation in lungs of a tropical bird *Perdicula asiatica*: role of melatonin and nuclear receptor RORα. Comp Biochem Physiol A 162:296–302

Kim S-H, Cai G-B, Bae Y-A, Lee E-G, Lee Y-S, Kong Y (2009) Two novel phospholipid hydroperoxide glutathione peroxidase genes of *Paragonimus westermani* induced by oxidative stress. Parasitology 136:553–565

Kim S-Y, Noguera JC, Morales J, Velando A (2010) Heritability of resistance to oxidative stress in early life. J Evol Biol 23:769–775

Kim S-Y, Noguera JC, Tato A, Velando A (2013) Vitamins, stress and growth: the availability of antioxidants in early life influences the expression of cryptic genetic variation. J Evol Biol 26:1341–1352

Kimura M (1958) On the change of population fitness by natural selection. Heredity 12:145–167

Klarsfeld A, Rouyer F (1998) Effects of circadian mutations and LD periodicity on the life span of *Drosophila melanogaster*. J Biol Rhythm 13:471–478

Konarzewski M, Książek A (2013) Determinants of intra-specific variation in basal metabolic rate. J Comp Physiol B 183:27–41

Koolhaas JM, Korte SM, de Boer SF, van der Vegt BJ, van Reenen CG, Hopster H, de Jong IC, Ruis MAW, Blokhuis HJ (1999) Coping styles in animals: current status in behavior and stress-physiology. Neurosci Biobehav Rev 23:925–935

Koolhaas JM, De Boer SF, Coppens CM, Buwalda B (2010) Neuroendocrinology of coping styles: towards understanding the biology of individual variation. Front Neuroendocrinol 31:307–321

Korte SM, Koolhaas JM, Wingfield JC, McEwen BS (2005) The Darwinian concept of stress: benefits of allostasis and costs of allostatic load and the trade-offs in health and disease. Neurosci Biobehav Rev 29:3–38

Krishnan N, Rakshit K, Chow ES, Wentzell JS, Kretzschmar D, Giebultowicz JM (2012) Loss of circadian clock accelerates aging in neurodegeneration-prone mutants. Neurobiol Dis 45:1129–1135

Kruuk LEB, Clutton-Brock TH, Slate J, Pemberton JM, Brotherstone S, Guinness FE (2000) Heritability of fitness in a wild mammal population. Proc Natl Acad Sci USA 97:698–703

Labrada-Martagón V, Rodríguez PA, Méndez-Rodríguez LC, Zenteno-Savín T (2011) Oxidative stress indicators and chemical contaminants in East Pacific green turtles (*Chelonia mydas*) inhabiting two foraging coastal lagoons in the Baja California peninsula. Comp Biochem Physiol C 154:65–75

Lande R (1975) The maintenance of genetic variability by mutation in a polygenic character with linked loci. Genet Res 26:221–235

Lande R (1980) The genetic covariance between characters maintained by pleiotropic mutations. Genetics 94:203–215

Lande R (1984) The genetic correlation between characters maintained by selection, linkage and inbreeding. Genet Res 44:309–320

Larsen PL (1993) Aging and resistance to oxidative damage in *Caenorhabditis elegans*. Proc Natl Acad Sci USA 90:8905–8909

Lemieux V, Garant D, Reale D, Bergeron P (2019) Spatio-temporal variation in oxidative status regulation in a small mammal. PeerJ 7:e7801

Li C-H, Wang T-Z, Li Y, Zheng Y-H, Jiang G-M (2013) Flixweed is more competitive than winter wheat under ozone pollution: evidences from membrane lipid peroxidation, antioxidant enzymes and biomass. PLoS One 8:e60109

Liebau E, Wildenburg G, Walter RD, Henkle-Dührsen K (1994) A novel type of glutathione *S*-transferase in *Onchocerca volvulus*. Infect Immunol 92:4762–4767

Lin X, Cui X, Tang J, Zhu J, Li J (2024) Predation risk effects of lady beetle *Menochilus sexmaculatus* (Fabricius) on the melon aphid, *Aphis gossypii* glover. Insects 15:13

Liparoto A, Canestrelli D, Bisconti R, Carere C, Costantini D (2020) Biogeographic history moulds population differentiation in ageing of oxidative status in an amphibian. J Exp Biol 223:jeb235002

López-Cruz RI, Zenteno-Savín T, Galván-Magaña F (2010) Superoxide production, oxidative damage and enzymatic antioxidant defenses in shark skeletal muscle. Comp Biochem Physiol A 156:50–56

Losdat S, Helfenstein F, Blount JD, Richner H (2014) Resistance to oxidative stress shows low heritability and high common environmental variance in a wild bird. J Evol Biol 27:1990–2000
Lucas LD, French SS (2012) Stress-induced tradeoffs in a free-living lizard across a variable landscape: consequences for individuals and populations. PLoS One 7:e49895
Lynch M, Walsh B (1998) Genetics and analysis of quantitative traits. Sinauer, Sunderland
Majer AD, Fasanello VJ, Tindle K, Frenz BJ, Ziur AD, Fischer CP, Fletcher KL, Seecof OM, Gronsky S, Vassallo BG, Reed WL, Paitz RT, Stier A, Haussmann MF (2019) Is there an oxidative cost of acute stress? Characterization, implication of glucocorticoids and modulation by prior stress experience. Proc R Soc Lond B 286:20191698
Martin GM, Austad SN, Johnson TE (1996) Genetic analysis of ageing: role of oxidative damage and environmental stresses. Nat Genet 13:25–34
Martin JS, Jaeggi AV, Koski SE (2023) The social evolution of individual differences: future directions for a comparative science of personality in social behavior. Neurosci Biobehav Rev 144:104980
Meng L, Li Q, Xu C, Liu S, Kong L, Yu H (2021) Hybridization improved stress resistance in the pacific oyster: evidence from physiological and immune responses. Aquacult 545:737227
Mentesana L, Adreani NM (2021) Acute aggressive behavior perturbates the oxidative status of a wild bird independently of testosterone and progesterone. Horm Behav 128:104913
Merilä J, Sheldon BC (1999) Genetic architecture of fitness and nonfitness traits: empirical patterns and development of ideas. Heredity 83:103–109
Mezhnina V, Ebeigbe OP, Poe A, Kondratov RV (2022) Circadian control of mitochondria in reactive oxygen species homeostasis. Antioxid Redox Signal 37:647–663
Miranda AC, Schielzeth H, Sonntag T, Partecke J (2013) Urbanization and its effects on personality traits: a result of microevolution or phenotypic plasticity? Glob Chang Biol 19:2634–2644
Miyazaki S, Nevo E, Bohnert HJ (2005) Adaptive oxidative stress in yeast *Saccharomyces cerevisiae*: interslope genetic divergence in 'evolution canyon'. Biol J Linn Soc 84:103–117
Møller AP, Erritzøe J, Karadas F (2010) Levels of antioxidants in rural and urban birds and their consequences. Oecologia 163:35–45
Morosinotto C, Rainio M, Ruuskanen S, Korpimäki E (2018) Antioxidant enzyme activities vary with predation risk and environmental conditions in free-living passerine birds. Physiol Biochem Zool 91:837–848
Mousseau TA, Roff DA (1987) Natural selection and the heritability of fitness components. Heredity 59:181–197
Nierenberg DW, Stukel T (1987) Diurnal variation in plasma levels of retinol, tocopherol, and b-carotene. Am J Med Sci 294:187–190
Nikinmaa M, McCairns RJS, Nikinmaa MW, Vuori KA, Kanerva M, Leinonen T, Primmer CR, Merilä J, Leder EH (2013) Transcription and redox enzyme activities: comparison of equilibrium and disequilibrium levels in the three-spined stickleback. Proc R Soc Lond B 280:20122974
Norte AC, Sheldon BC, Sousa JP, Ramos JA (2009) Environmental and genetic variation in body condition and blood profile of great tit *Parus major* nestlings. J Avian Biol 40:157–165
O'Neill RD, Fillenz M, Albery WJ (1982) Circadian changes in homovanillic acid and ascorbate levels in the rat striatum using microprocessor-controlled voltammetry. Neurosci Lett 34:189–193
Okada K, Blount JD, Sharma MD, Snook RR, Hosken DJ (2011) Male attractiveness, fertility and susceptibility to oxidative stress are influenced by inbreeding in *Drosophila simulans*. J Evol Biol 24:363–371
Olsson M, Wilson M, Uller T, Mott B, Isaksson C, Healey M, Wanger T (2008) Free radicals run in lizard families. Biol Lett 4:186–188
Pagano C, Siauciunaite R, Idda ML, Ruggiero G, Ceinos RM, Pagano M, Frigato E, Bertolucci C, Foulkes NS, Vallone D (2018) Evolution shapes the responsiveness of the D-box enhancer element to light and reactive oxygen species in vertebrates. Sci Rep 8:13180

Partecke J, Schwabl I, Gwinner E (2006) Stress and the city: urbanization and its effects on the stress physiology in European blackbirds. Ecology 87:1945–1952

Partridge L, Gems D (2002) Mechanisms of ageing: public or private? Nat Rev Genet 3:165–175

Patel SA, Velingkaar NS, Kondratov RV (2014) Transcriptional control of antioxidant defense by the circadian clock. Antioxid Redox Signal 20:2997–3006

Pedersen KS, Kristensen TN, Loeschcke V, Petersen BO, Duus JØ, Nielsen NC, Malmendal A (2008) Metabolomic signatures of inbreeding at benign and stressful temperatures in Drosophila melanogaster. Genetics 180:1233–1243

Peek CB, Affinati AH, Ramsey KM, Kuo HY, Yu W, Sena LA, Ilkayeva O, Marcheva B, Kobayashi Y, Omura C, Levine DC, Bacsik DJ, Gius D, Newgard CB, Goetzman E, Chandel NS, Denu JM, Mrksich M, Bass J (2013) Circadian clock $NAD^+$ cycle drives mitochondrial oxidative metabolism in mice. Science 342:1243417

Petrović TG, Vučić T, Burraco P, Gavrilović BR, Despotović SG, Gavrić JP, Radovanović TB, Šajkunić S, Ivanović A, Prokić MD (2023) Higher temperature induces oxidative stress in hybrids but not in parental species: a case study of crested newts. J Therm Biol 112:103474

Philipp E, Schmidt M, Gsottbauer C, Saenger AM, Abele D (2008) Size-and-age-dependent changes in adductor muscle swimming physiology of the scallop *Aequipecten opercularis*. J Exp Biol 211:2492–2501

Philipp EER, Wessels W, Gruber H, Strahl J, Wagner AE, Ernst IMA, Rimbach G, Kraemer L, Schreiber S, Abele D, Rosenstiel P (2012) Gene expression and physiological changes of different populations of the long-lived bivalve *Arctica islandica* under low oxygen conditions. PLoS One 7:e44621

Pigliucci M (2001) Phenotypic plasticity: beyond nature and nurture. Johns Hopkins University Press, Baltimore

Pigliucci M, Murren CJ, Schlichting CD (2006) Phenotypic plasticity and evolution by genetic assimilation. J Exp Biol 209:2362–2367

Pinya S, Tejada S, Capó X, Sureda A (2016) Invasive predator snake induces oxidative stress responses in insular amphibian species. The Science of the total environment, 566–567, 57–62

Poblete Y, Contreras C, Fernández C, Flores CR, Vega P, Ávila M, Sabat P (2023) Geographic variation in the altitudinal migration patterns, body size, oxidative status and exploratory behavior in a neotropical bird. Ecol Evol 13:e9941

Price T, Schluter D (1991) On the low heritability of life history traits. Evolution 45:853–861

Prokić MD, Despotović SG, Vučić TZ, Petrović TG, Gavrić JP, Gavrilović BR, Radovanović TB, Saičić ZS (2018) Oxidative cost of interspecific hybridization: a case study of two Triturus species and their hybrids. J Exp Biol 221:jeb182055

Prokić MD, Petrović TG, Despotović SG, Vučić T, Gavrić JP, Radovanović TB, Gavrilović BR (2021) The effect of short-term fasting on the oxidative status of larvae of crested newt species and their hybrids. Comp Biochem Physiol A 251:110819

Quque M, Ferreira C, Sosa S, Schull Q, Zahn S, Criscuolo F, Bleu J, Viblanc VA (2022) Cascading effects of conspecific aggression on oxidative status and telomere length in zebra finches. Physiol Biochem Zool 95:416–429

Raja-aho S, Kanerva M, Eeva T, Lehikoinen E, Suorsa P, Gao K, Vosloo D, Nikinmaa M (2012) Seasonal variation in the regulation of redox state and some biotransformation enzyme activities in the barn swallow (*Hirundo rustica* L.). Physiol Biochem Zool 85:148–158

Rammal H, Bouayed J, Soulimani R (2010) A direct relationship between aggressive behavior in the resident/intruder test and cell oxidative status in adult male mice. Eur J Pharmacol 627:173–176

Réale D, Reader SM, Sol D, McDougall P, Dingemanse NJ (2007) Integrating animal temperament within ecology and evolutionary biology. Biol Rev 82:291–318

Réale D, Garant D, Humphries M, Careau V, Bergeron P, Montiglio PO (2010) Personality and the emergence of a pace-of-life syndrome at the level of the population. Philos Trans R Soc Lond B 365:4051–4063

Récapet C, Arrivé M, Doligez B, Bize P (2019) Antioxidant capacity is repeatable across years but does not consistently correlate with a marker of peroxidation in a free-living passerine bird. J Comp Physiol B 189:283–298

Reynolds RM, Phillips PC (2013) Natural variation for lifespan and stress response in the nematode *Caenorhabditis remanei*. PLoS One 8:e58212

Robert KA, Bronikowski AM (2010) Evolution of senescence in nature: physiological evolution in populations of garter snake with divergent life histories. Am Nat 175:147–159

Robles-Romo A, Zenteno-Savín T, Racotta IS (2016) Bioenergetic status and oxidative stress during escape response until exhaustion in whiteleg shrimp *Litopenaeus vannamei*. J Exp Mar Biol Ecol 478:16–23

Rock CL, Fada RD, Jacob RA, Bowen PE (1996) Update on the biological characteristics of the antioxidant micronutrients: vitamin C, vitamin E, and the carotenoids. J Am Diet Assoc 96:693–702

Rödel HG, Jardim V, Rangassamy M, Jaravel L, Jacquet D, Monclús R, Féron C, Costantini D (2022) Early life parameters and personality affect oxidative status during adulthood in an altricial rodent. Physiol Rep 10:e15427

Rodriguez M, Basten Snoek L, Riksen JAG, Bevers RP, Kammenga JE (2012) Genetic variation for stress-response hormesis in *C. elegans* lifespan. Exp Geront 47:581–587

Ros AFH, Becker K, Oliveira RF (2006) Aggressive behaviour and energy metabolism in a cichlid fish, *Oreochromis mossambicus*. Physiol Behav 89:164–170

Roux O, Renault D, Mouline K, Diabaté A, Simard F (2021) Living with predators at the larval stage has differential long-lasting effects on adult life history and physiological traits in two anopheline mosquito species. J Insect Physiol 131:104234

Rowe M, Pierson KL, McGraw KJ (2015) Exploratory behavior is associated with plasma carotenoid accumulation in two congeneric species of waterfowl. Behav Process 115:181–190

Roy NK, Wirgin I (1997) Characterization of the aromatic hydrocarbon receptor gene and its expression in Atlantic tomcod. Arch Biochem Biophys 344:373–386

Rubolini D, Romano M, Bonisoli Alquati A, Saino N (2006) Early maternal, genetic and environmental components of antioxidant protection, morphology and immunity of yellow-legged gull (*Larus michahellis*) chicks. J Evol Biol 19:1571–1584

Rutherford SL, Lindquist S (1998) Hsp90 as a capacitor for morphological evolution. Nature 396:336–342

Ruuskanen S, Morosinotto C, Thomson RL, Ratnayake CP, Korpimäki E (2017) Food supplementation, but not predation risk, alters female antioxidant status during breeding. Behav Ecol Sociobiol 71:69

Sachser N, Hennessy MB, Kaiser S (2010) Adaptive modulation of behavioural profiles by social stress during early phases of life and adolescence. Neurosci Biobehav Rev 35:1518–1533

Saino N, Caprioli M, Romano M, Boncoraglio G, Rubolini D, Ambrosini R, Bonisoli-Alquati A, Romano A (2011) Antioxidant defenses predict long-term survival in a passerine bird. PLoS One 6:e19593

Sarup P, Sørensen P, Loeschcke V (2013) The long-term effects of a life-prolonging heat treatment on the *Drosophila melanogaster* transcriptome suggest that heat shock proteins extend lifespan. Exp Gerontol 50:34–39

Schwartz TS, Bronikowski AM (2013) Dissecting molecular stress networks: identifying nodes of divergence between life-history phenotypes. Mol Ecol 22:739–756

Selkirk ME, Smith VP, Thomas GR, Gounaris K (1998) Resistance of filarial nematode parasites to oxidative stress. Int J Parasitol 28:1315–1332

Semeniuk CAD, Bourgeon S, Smith SL, Rothley KD (2009) Hematological differences between stingrays at tourist and non-visited sites suggest physiological costs of wildlife tourism. Biol Conserv 142:1818–1829

Sepp T, Sild E, Blount JD, Männiste M, Karu U, Hõrak P (2012) Individual consistency and covariation of measures of oxidative status in greenfinches. Physiol Biochem Zool 85:299–307

Serra L, Pirrello S, Caprioli M, Griggio M, Andreotti A, Romano A, Pilastro A, Saino N, Sacchi R, Galeotti P, Fasola M, Spina F, Rubolini D (2012) Seasonal decline of offspring quality in the European starling *Sturnus vulgaris*: an immune challenge experiment. Behav Ecol Sociobiol 66:697–709

Sheriff MJ, Krebs CJ, Boonstra R (2009) The sensitive hare: sublethal effects of predator stress on reproduction in snowshoe hares. J Anim Ecol 78:1249–1258

Sih A, Bell A, Johnson JC (2004) Behavioral syndromes: an ecological and evolutionary overview. Trends Ecol Evol 19:372–378

Slos S, Stoks R (2008) Predation risk induces stress proteins and reduces antioxidant defense. Funct Ecol 22:637–642

Snell HL, Snell HM, Richard Tracy C (1984) Variation among populations of Galapagos land iguanas (*Conolophus*): contrasts of phylogeny and ecology. Biol J Linn Soc 21:185–207

Sol D, Lapiedra O, González-Lagos C (2013) Behavioural adjustments for a life in the city. Anim Behav 85:1101–1112

Steimer T, La Fleur FS, Schulz PE (1997) Neuroendocrine correlates of emotional reactivity and coping in male rats from the Roman high (RHA/Verh)- and low (RLA/Verh)-avoidance lines. Behav Genet 27:503–512

Stier A, Schull Q, Bize P, Lefol E, Haussmann M, Roussel D, Robin JP, Viblanc VA (2019) Oxidative stress and mitochondrial responses to stress exposure suggest that king penguins are naturally equipped to resist stress. Sci Rep 9:8545

Surai PF, Ionov I, Kuchmistova E, Noble RC, Speake BK (1998) The relationship between the levels of tocopherol and carotenoids in the maternal feed, yolk and neonatal tissues: comparison between the chicken, Turkey, duck and goose. J Sci Food Agric 76:593–598

Tartarotti B, Sommaruga R, Saul N (2019) Contrasting diurnal patterns in antioxidant capacities, but not in expression of stress protein genes among copepod populations from clear versus glacially fed alpine and subalpine lakes. J Plankton Res 41:897–908

Tella JL, Figuerola J, Negro JJ, Blanco G, Rodríguez-Estrella R, Forero MG, Blázquez MC, Green AJ, Hiraldo F (2004) Ecological, morphological and phylogenetic correlates of interspecific variation in plasma carotenoid concentration in birds. J Evol Biol 17:156–164

Thomson RL, Tomas G, Forsman JT, Broggi J, Monkkonen M (2010) Predator proximity as a stressor in breeding flycatchers: mass loss, stress protein induction, and elevated provisioning. Ecology 91:1832–1840

Tian B, Wu Y, Sheng D, Zheng Z, Gao G, Hua Y (2004) Chemiluminescence assay for reactive oxygen species scavenging activities and inhibition on oxidative damage of DNA in *Deinococcus radiodurans*. Luminescen 19:78–84

Trams EG (1969) Carotenoid transport in the plasma of the scarlet ibis (*Eudocimus ruber*). Comp Biochem Physiol 8:1177–1184

Tunçtan B, Weigl Y, Dotan A, Peleg L, Zengil H, Ashkenazi I, Abacioğlu N (2002) Circadian variation of nitric oxide synthase activity in mouse tissue. Chronobiol Int 19:393–404

Vágási CI, Pătraş L, Pap PL, Vincze O, Mureşan C, Németh J, Lendvai ÁZ (2018) Experimental increase in baseline corticosterone level reduces oxidative damage and enhances innate immune response. PLoS One 13:e0192701

Vágási CI, Tóth Z, Pénzes J, Pap PL, Ouyang JQ, Lendvai ÁZ (2020) The relationship between hormones, glucose, and oxidative damage is condition and stress dependent in a free-living passerine bird. Physiol Biochem Zool 93:466–476

Vágási CI, Fülöp A, Osváth G, Pap PL, Pénzes J, Benkő Z, Lendvai ÁZ, Barta Z (2021) Social groups with diverse personalities mitigate physiological stress in a songbird. Proc R Soc Lond B 288:20203092

Vaiserman AM (2008) Epigenetic engineering and its possible role in anti-aging intervention. Rejuv Res 11:39–42

Vaiserman AM (2010) Hormesis, adaptive epigenetic reorganization, and implications for human health and longevity. Dose Resp 8:16–21

Vaiserman AM (2011) Hormesis and epigenetics: is there a link? Ageing Res Rev 10:413–421 van de Crommenacker J, Komdeur J, Burke T, Richardson DS (2011) Spatio-temporal variation in territory quality and oxidative status: a natural experiment in the Seychelles warbler (*Acrocephalus sechellensis*). J Anim Ecol 80:668–680

van den Driessche T, Guisset J-L, Petiau-de Vries GM (2000) The redox state and circadian rhythms. Kluwer Academic Publishers, Dordrecht

Wang XL, Rainwater DL, VandeBerg JF, Mitchell BD, Mahaney MC (2001) Genetic contributions to plasma total antioxidant activity. Arterioscler Thromb Vas Biol 21:1190–1195

Weber AL, Khan GF, Magwire MM, Tabor CL, Mackay TF, Anholt RR (2012) Genome-wide association analysis of oxidative stress resistance in Drosophila melanogaster. PLoS One 7:e34745

West-Eberhard MJ (2003) Developmental plasticity and evolution. Oxford University Press, New York

White CR, Kearney MR (2013) Determinants of inter-specific variation in basal metabolic rate. J Comp Physiol B 183:1–26

Wilking M, Ndiaye M, Mukhtar H, Ahmad N (2013) Circadian rhythm connections to oxidative stress: implications for human health. Antioxid Redox Signal 19:192–208

Wilson AJ, Nussey DH (2010) What is individual quality? An evolutionary perspective. Trends Ecol Evol 25:207–214

Wolf M, Weissing FJ (2012) Animal personalities: consequences for ecology and evolution. Trends Ecol Evol 27:452–461

Wolf M, van Doorn GS, Weissing FJ (2008) Evolutionary emergence of responsive and unresponsive personalities. Proc Natl Acad Sci USA 105:15825–15830

Xu Y-Q, Zhang D, Jin T, Cai D-J, Wu Q, Lu Y, Liu J, Klaassen CD (2012) Diurnal variation of hepatic antioxidant gene expression in mice. PLoS One 7:e44237

Xu Y, Wang X, Jiang S, Men C, Xu D, Guo Y, Wu J (2018) Microcystin-LR regulates circadian clock and antioxidant gene expression in cultured rat cardiomyocytes. Cell Mol Biol Lett 23:50

Yang A, Tume RK (1993) A comparison of β-carotene splitting activity isolated from intestinal mucosa of pasture-grazed sheep, goats and cattle. Biochem Mol Biol Int 30:209–217

Zheng H, Tan K, Zhang H, Ma H, Li S, Zheng H (2022) Intraspecific hybridization as a mitigation strategy of ocean acidification in marine bivalve noble scallop *Chlamys nobilis*. Sci Total Environ 832:154736

# Chapter 5
# Exposome and Regulation of Oxidative Status Across Different Environments

**Abstract** Abiotic factors are elements of the environment that influence the way organisms work, including the regulation of organismal oxidative status. These factors vary largely within and among habitats, and across time, generating environmentally driven variation in exposure of organisms to different challenges for the homeostasis of oxidative status. Elucidating the physiological mechanisms underlying the responses of organisms to environmental stressors is key to understand their adaptability to changing environments. In this chapter, I have synthesised the literature that describes the responses of organisms to ambient temperature, ultraviolet radiation, partial pressure of oxygen, acidification, salinity stress, and availability of water. I have also illustrated the responses of organisms to exposure to multiple abiotic stressors (cocktail effect), and to the ongoing environmental changes, such as deforestation and urbanisation. Finally, I have provided examples of physiological responses that are peculiar to particular evolutionary lineages, and the role of hormesis in priming these responses against abiotic stressors.

## 5.1 Introduction: Life and Death in Oxidising Environments

In Chap. 4, I have examined the role that proximate factors, like style of coping with stress or genetic background, have in generating variation in oxidative status between conspecific individuals or species. Abiotic factors are another important source of variation. Abiotic factors are elements of the environment, such as temperature, salinity, partial pressure of oxygen, or solar radiation, that influence the way organism work. They can vary largely within and among environments, and across time, subjecting organisms to different oxidation threats that require similar or circumstantial solutions to maintain homeostasis of the oxidative status (e.g. Cheviron et al. 2008; Freire et al. 2012; González et al. 2012; Lesser 2012; Regoli et al. 2012; Tattersall et al. 2012; Welker et al. 2013; Reguera et al. 2015; Zhang et al. 2015; Burraco and Gomez-Mestre 2016). Further environmentally driven variation in exposure of organisms to different challenges for the homeostasis of

© The Author(s), under exclusive license to Springer Nature Switzerland AG 2024

D. Costantini, *The Role of Organismal Oxidative Stress in the Ecology and Life-History Evolution of Animals*, Fascinating Life Sciences,
https://doi.org/10.1007/978-3-031-65183-0_5

oxidative status can be generated by seasonality, altitude, latitude, daily variations in sunlight exposure, drops in oxygen availability, or osmotic stress. Environmental changes are often gradual, but can sometimes be abrupt (e.g. heat waves), which requires organisms to quickly mount a physiological response. For example, after a tropical storm, aged trees may come crashing to the ground leaving a gap in the canopy above. This enables sunlight to reach the forest floor, so that plants or animals become suddenly exposed to it. Sunlight represents an important risk factor for low-light plants or animals because ultraviolet radiation can cause oxidative stress.

Species have evolved a number of behavioural and physiological adaptations to resist or tolerate fluctuations in abiotic conditions in order to maintain a stable oxidative status. For example, many invertebrate and fish marine species live near volcanic outflows, where warm waters poor in oxygen and rich in hydrogen sulphide merge with colder, oxygen-rich oceanic waters, creating very variable and extreme environmental conditions (Abele et al. 2012). Other species live in environments, like deserts, that can become really hot during the day. Species like the desert moss *Syntrichia caninervis* may tolerate an experimental exposure to a temperature of 120 °C for 30 min (Stark et al. 2009).

In this chapter, I have examined a wide variety of environmental contexts in which the interplay between oxidative stress and abiotic conditions has contributed to generate variation in physiological phenotypes and life-history strategies. I have also illustrated the responses of organisms to exposure to multiple abiotic stressors (cocktail effect), and to the ongoing environmental changes, such as deforestation and urbanisation. Finally, I have provided examples of physiological responses that are peculiar to particular evolutionary lineages, and discussed the role of hormesis in priming the organisms to resist or tolerate environmentally induced oxidative stress.

## 5.2 Thermal Relations: The Good, the Bad, and the Ugly

Research on thermal relations of organisms with their environment has provided relevant examples of how phenotypes can be adjusted to the prevailing environmental conditions, and why certain strategies are favoured over others. Ambient temperature has a strong effect on several traits, including morphology, physiology, behaviour, survival, and seasonal timing of life-history stages (e.g. Somero 1995, 2004; Hill et al. 2008; Angilletta 2009; Tattersall et al. 2012; Storey and Tanino 2012; Caro et al. 2013; Costantini et al. 2014b; Uehling et al. 2020; Takatsu et al. 2022). Ambient temperature influences a number of physiological traits, such as the metabolic rate, the cellular biophysical or biochemical processes, and the physical state of cell membranes (Somero 1995, 2004). Synthesis of enzymes demands energy and building blocks (amino acids), and the expression of enzymatic isoforms also requires multiple copies of a gene (Tattersall et al. 2012). Having more routes through which to respond to a thermal stressor may give more flexibility to the organism. However, this is likely to vary across species because not all species can, for example, express various enzymatic isoforms, and changes in isozyme patterns may occur for some enzymes but not for others (Shaklee et al. 1977; Tattersall et al. 2012). This wide variation among species raises many questions related to selective pressures that acted on the evolution of enzymatic antioxidant mechanisms activated in response to changes in ambient temperature, and the consequences for the regulation of organismal oxidative status.

Fluctuations in ambient temperature expose organisms to a number of trade-offs. Let us take a flying insect as an example. Flight muscles of an insect have to be warm in order to work properly so the insect can fly. The insect consumes energy to generate heat that is needed to warm up the muscles. At high ambient temperature, the insect can exploit environmental heat to keep muscles warm, so that it has much more of its own energy available to be invested into other body functions. Animals may also respond to hot conditions increasing intake of dietary antioxidants, thereby avoiding an imbalance between generation of reactive oxygen species and antioxidant defences (Beaulieu et al. 2015).

The study of the diversity in the response of animals to thermal challenges offers a unique opportunity to determine the links between oxidative stress and other organismal functions in comparative physiology. The thermal physiology largely varies across animal species, from those that are incapable of generating heat in order to regulate body temperature (ectotherms) to those that warm their body up using metabolically produced heat (endotherms). Homeothermic species have a thermoneutral zone, which is a range of ambient temperatures within which the organism does not need to modify its metabolic rate to maintain body temperature. Within this thermoneutral zone, normal metabolism provides enough heat to maintain an essentially constant body temperature. When the ambient temperature increases beyond the upper limit of the thermoneutral zone, organisms need to increase their metabolic rate to meet the energetic demands that are needed to increase the rate of water evaporation from the body surface (active evaporative cooling) and, at the same time, to counterbalance the body cooling in order to avoid an excessive decrease in body temperature (Hill et al. 2008). The capacity to do so varies across species because larger species have higher thermal inertia (higher capacity to tolerate progressive hyperthermia) than smaller species. The extent to which protection against hyperthermia-induced oxidative stress is critical for species may therefore be expected to vary in relation to thermal inertia (Al-Otaiba et al. 2010). Thermal kinetic energy and collisions among molecules tend to increase with ambient temperature (Tattersall et al. 2012), enhancing reactions between enzymes and their substrates as a consequence. Activity of enzymes, therefore, increases with temperature. Conformational changes of enzymes, however, occur at critically high temperatures, which impair their catalytic capacity, for example, through loss in structural integrity (Somero 1995, 2004; Abele et al. 2002; Tattersall et al. 2012). There are, however, many enzymes that are catalytically more effective at low temperatures (e.g. psychrophilic enzymes), such as those of organisms that live in cold environments (Gianese et al. 2001; Feller 2013).

Generally, animals become heat-stressed when they are unable to balance production and loss of heat. In homeotherms, the level of heat stress can be considered mild when the ambient temperature is slightly higher than the upper critical temperature of the thermoneutral zone, and high when the ambient temperature is well above the upper critical temperature, respectively. Although ectotherm species do not have a thermoneutral region, they can also become heat-stressed and may die when they are exposed to strong increases in ambient temperature (Hill et al. 2008; Abele 2012). In ectotherms, temperature can accelerate the rate of generation of reactive oxygen species simply by increasing the mitochondrial respiration. For example, it has been found that the in vitro generation of reactive oxygen species in mitochondria of the bivalve *Mya arenaria* doubled when the ambient temperature was increased from 15 to 25 °C (Abele et al. 2002). A recent meta-analysis including studies on aquatic ectotherms found that augmented water temperature increased significantly levels of oxidative damage and of antioxidant enzymes (Fig. 5.1a) (Messina et al. 2023). Exposure to acute increases in temperature of ectotherms adapted to stable environments generally produced significant effects on their oxidative status, whilst such effects were smaller in ectotherms adapted to

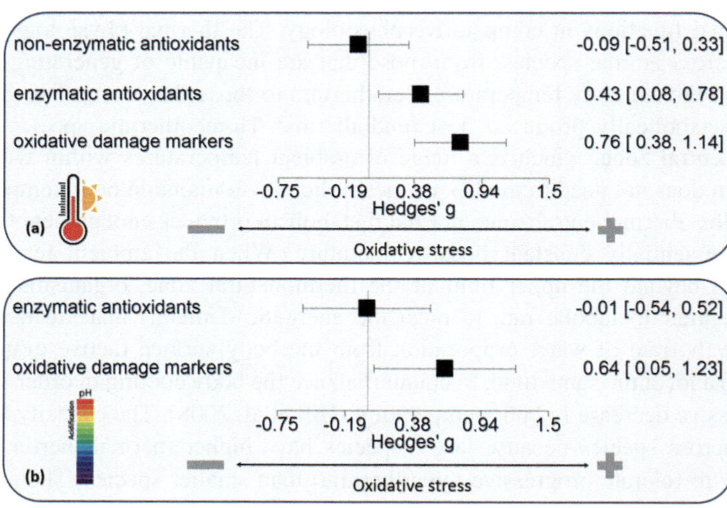


**Fig. 5.1** Forest plots show predictions of models for the impacts of increased (**a**) temperatures and (**b**) water acidification on the oxidative status of aquatic ectotherms. Effect size estimates (Hedges'g) are statistically significant when their own 95% credible intervals do not include zero. The meta-analyses included the following studies: Vinagre et al. (2012), Akhtar et al. (2013, 2020), Enzor and Place (2014), Mueller et al. (2014), Carney Almroth et al. (2015, 2019), Simčič et al. (2015), Burraco and Gomez-Mestre (2016), Freitas and Almeida (2016), Rosa et al. (2016), Silva et al. (2016), Freitas et al. (2017a, b), Klein et al. (2017), Pereira et al. (2017), Rossi et al. (2017), Faria et al. (2018), Jesus et al. (2018), Sampaio et al. (2018), Defo et al. (2019), Kim et al. (2019a, b), Liu et al. (2019), Pinto et al. (2019), Antonopoulou et al. (2020), Dettleff et al. (2020), Florencio et al. (2020), Madeira et al. (2020), Chang et al. (2021), Giuliani et al. (2021), Kim et al. (2021), Li et al. (2021), Lopes et al. (2021), Pereira et al. (2021), Roychowdhury et al. (2021), Singh et al. (2021), Vieira et al. (2021), Yang et al. (2021), Guillen et al. (2022), Lee et al. (2022a, b), O'Brien et al. (2022). Reprinted with minor modifications from Messina et al. (2023) with permission of Elsevier

variable conditions (Ritchie and Friesen 2015). Moreover, acclimation studies on ectotherms tended to find that oxidative stress was stronger at the lower acclimation temperatures (Ritchie and Friesen 2015). Importantly, although oxidative stress can be a cost of thermal stress, it can also constrain the capacity for acclimation (Loughland and Seebacher 2020), a topic that has so far received little experimental attention.

In addition to the metabolic adjustments, a state of heat stress is also characterised by a number of other physiological changes, such as increased synthesis of heat shock proteins, increased generation of reactive oxygen species or oxidative damage, and changes in the expression of antioxidant enzymes (e.g. Flanagan et al. 1998; Feder and Hofmann 1999; Lord-Fontaine and Averill-Bates 2002; Mujahid et al. 2007; Vinagre et al. 2012; Ding et al. 2023). For example, Japanese black porgy *Acanthopagrus schlegelii* had higher expression of genes and activity of the antioxidant enzymes copper-zinc superoxide dismutase, catalase, and glutathione peroxidase when they were housed at 30 °C than when they were housed at 20 °C (An et al. 2010). The upregulation of antioxidant enzymes was probably stimulated by the increased generation of reactive oxygen species, such as hydrogen peroxide. However, this upregulated antioxidant response did not prevent the increase in oxidative damage generation (An et al. 2010). Closely related species with similar environmental requirements may also significantly differ in their antioxidant response to heat stress. For example, Vinagre et al. (2014) found that the activity of antioxidant enzymes and levels of thiobarbituric acid reactive substances showed stronger changes over the temperature gradient 20–33 °C in the rockpool shrimp *Palaemon elegans* than in the related common prawn *Palaemon serratus* (Fig. 5.2). Finally, many studies recorded that the concentration or activity of antioxidants can show non-linear relationships with the intensity of thermal stress (Abele et al. 2002), and tissues may differ in their sensitivity to thermal stress (Bagnyukova et al. 2007).

This research on the impact of heat stress on organismal oxidative status is particularly relevant in the light of global warming. It may only take an increase of a few degrees Celsius to induce pervasive effects on the oxidative status of an organism. Rosa et al. (2012) showed that the projected near-future ocean warming of 2 °C may be already above the optimal thermal tolerance boundaries of European squid *Loligo vulgaris*. In this species, embryonic development was dramatically shortened with a concomitant negative effect on growth and survival, an increase in oxidative damage levels, and a consequent upregulation of antioxidant defences. Squids will probably be faced with higher demands in terms of homeostatic maintenance during development in order to increase their thermal tolerance windows. The results of this study suggested that hormetic priming (i.e. exposure to a mild stressful stimulus improves the future response to a subsequent stressful stimulus) during the early stages of development might increase the width of the thermal tolerance window. This is suggested from the fact that summer embryos of squids had stronger heat shock and antioxidant responses than autumn embryos, indicating that these phenotypic changes were stable over time and not transient as those induced by acclimatisation (Wilson and Franklin 2002; Costantini et al. 2010; Costantini 2014). Other factors (e.g. differential seasonal investment of mothers in their eggs,

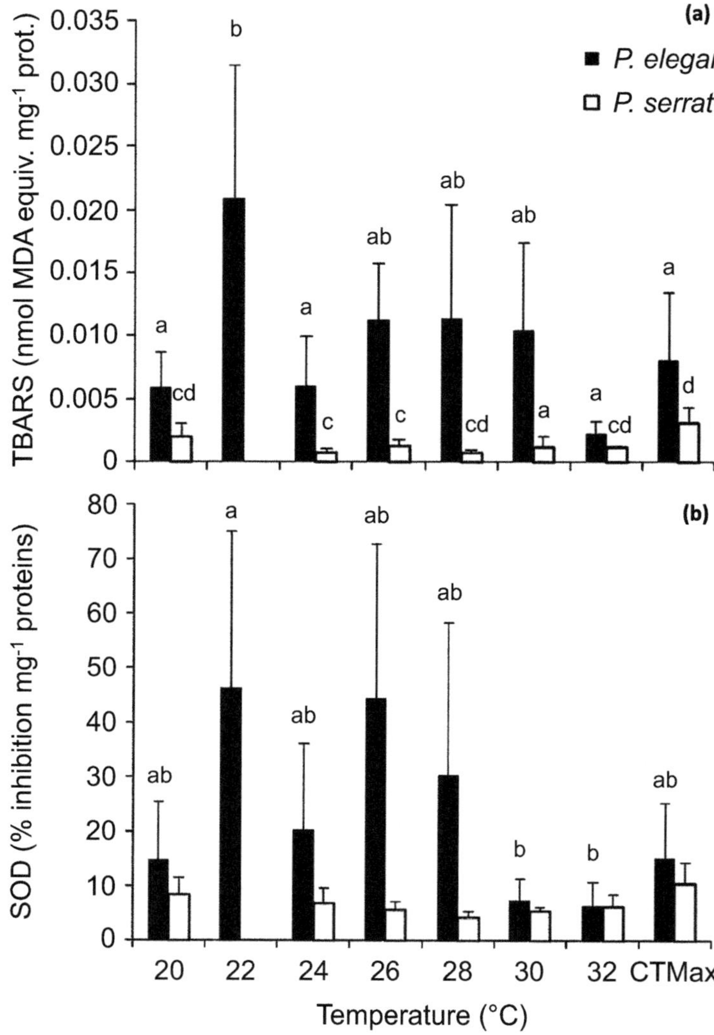


**Fig. 5.2** Effects of different water temperatures on (**a**) thiobarbituric acid reactive substances and (**b**) superoxide dismutase in dorsal muscle of the two closely related shrimp species *Palaemon elegans* and *Palaemon serratus*. Different letters indicate significantly different comparisons in the post-hoc tests. Values are shown as mean and standard deviation. Reprinted with minor modifications from Vinagre et al. (2014) with permission of Elsevier

larger nutritional supply of molecular dietary antioxidants in summer) may also have contributed to drive the organismal physiological adjustments to heat stress.

Many studies also documented transient or long-term changes in oxidative status markers in animals exposed to cold stress (e.g. Grubor-Lajsic et al. 1997; Selman et al. 2000; Abele and Puntarulo 2004; Koštál and Tollarová-Borovanská 2009;

Kammer et al. 2011). Cold stress can stimulate a number of physiological responses, such as ice formation in tissues, a process that may jeopardise the oxidative status by changing the partial pressure of oxygen (e.g. by decreasing the volume of liquid available for dissolving the gas; Hermes-Lima and Storey 1993), decreased mitochondrial membrane fluidity (Hazel 1995), and increased metabolic rate (Selman et al. 2000). For example, compared to gilthead sea bream *Sparus aurata* acclimated to 20 °C, fish maintained for 10 days at 8 °C had higher levels of liver thiobarbituric acid reactive substances and nitric oxide (Ibarz et al. 2010). Analyses of hepatic protein profiles also showed that the expression of many proteins, including antioxidant enzymes, was downregulated following cold exposure (Ibarz et al. 2010). Short-term effects of cold stress on the oxidative status were also documented. Selman et al. (2002) tested whether exposure to acute cold (7 ± 3 °C) affected protein oxidation and proteasome activity, compared to warm controls (22 ± 3 °C), across different tissues of a small mammal model, the short-tailed field vole *Microtus agrestis*, whose thermoneutral zone is in the range of 25–30 °C. The resting metabolic rate of voles increased significantly with duration of cold exposure. In skeletal muscle and liver, levels of protein carbonyls also increased, while proteasome activity decreased, with duration of cold exposure. By contrast, protein carbonyls did not change in brown adipose tissue, possibly because of upregulation of proteasome activity. It might be that voles prioritised the protection of the brown adipose tissue against oxidative damage to proteins because this tissue has a crucial thermogenic role during cold exposure in small mammals (Selman et al. 2002). Regulation of mitochondrial mechanisms might also be relevant in the defence of certain tissues. We have evidence that mitochondrial uncoupling may prevent increased generation of reactive oxygen species (Rey et al. 2010) or of oxidative damage (Stier et al. 2014) in animals exposed to short-term cold conditions. Finally, Chen et al. (2015, 2019) found that the regulation of antioxidant defences in response to acute cold exposure in juvenile Chinese soft-shelled turtle *Pelodiscus sinensis sinensis* may not be achieved at the transcriptional level, but may rely mainly on high constitutive antioxidant enzyme activities, high level of tissue-specific ascorbic acid storage, and the efficient plasma ascorbic acid transport potential.

When ambient temperatures drop below critical values, certain animal species can enter a state of adaptive or pathological hypothermia, which are distinguished by the capability and incapability of the organism to rewarm to normothermia, respectively (Tattersall et al. 2012). Having the capability of spontaneous rewarming is therefore important, although it is probably not cost free. Many species have also evolved behavioural mechanisms to help rewarming, such as solar basking in reptiles (Tattersall et al. 2012), which may, however, result in increased oxidative damage (Mészáros et al. 2017). Behavioural strategies may be constrained by habitat thermal heterogeneity (e.g. availability of shelters or water bodies) and prioritisation of other activities (e.g. reproduction). Hence, oxidative costs of spontaneous rewarming and strategies of mitigating them are likely to vary across species. Exposure to cold stress can also involve only certain body parts. For example, peripheral tissues can be allowed to cool down noticeably, to near freezing temperatures. We might expect that, under these circumstances, activation of protective mechanisms is only prioritised in certain tissues.

Much of what we know on the effects of low or high temperatures on oxidative stress is based on experiments carried out under laboratory conditions. Thus, we need to know more about heat- or cold-induced oxidative stress in free-living animals, since they can experience significant thermal variations. The phenotype is programmed to deal with a wide range of temperatures, yet its flexibility may vary among individuals, populations, or species. The capability to cope with unpredictable very high or low temperatures is likely to have significant evolutionary fitness consequences (examples on ectotherms in Abele 2012). Furthermore, the heat- or cold-induced oxidative stress is a possible key mediator of the fitness of organisms in the wild, given that the resultant tissue degradation may influence reproductive performance, growth patterns, cellular senescence, and survival (Costantini 2008; Abele 2012). This is particularly important nowadays because one of the most dramatic manifestations of global climatic changes is the increased occurrence of life-threating heatwaves (or cold waves), i.e. extended periods of hot (cold) weather relative to the expected conditions of the area at that time of the year. For example, Beaulieu et al. (2020) found that the plasma non-enzymatic antioxidant capacity of greater mouse-eared bats *Myotis myotis* decreased, while plasma reactive oxygen metabolites increased with increasing temperature anomalies occurring in the 2 weeks before blood sampling. Moreover, the antioxidant defences of these bats reflected vegetation indices, which themselves reflected the thermal conditions experienced by bats in their foraging habitat (Beaulieu et al. 2020).

The timescale over which the organism is being exposed to these temperature extremes is probably relevant. For example, levels of heat stress encountered in early life will vary among individuals, depending on the time of the season in which they are born and the prevailing weather conditions. Those individuals that experience heat stress early in life could actually be better placed to cope with subsequent temperature extremes if the initial exposure induced hormetic priming (or conditioning). There is some empirical support for hormetic priming of the individual to withstand high temperatures across life. Many studies on *Drosophila* fruit flies, *Caenorhabditis elegans*, and *Saccharomyces cerevisiae* found that exposure to mild heat stress early in life can increase longevity (e.g. Shama et al. 1998; Le Bourg et al. 2001; Cypser and Johnson 2002; Olsen et al. 2006). A meta-analysis on the life extension promoted by thermal hormesis showed that the response to heat stress was highly conserved across 12 species of invertebrates (Lagisz et al. 2013). The results of the meta-analysis showed that (1) heat shock temperature had the largest influence on experimental outcomes, with higher temperatures, longer duration of heat shock, or more heat shock repeats all decreasing the likelihood of any life extension, (2) longer periods of recovery between repeated heat shocks increased the likelihood of life extension, and (3) hormetic effects on longevity were stronger in young than adult individuals. The results of the meta-analysis also confirmed the importance of individual age at which the mild stress is experienced, and of the strength/duration of heat stress as determinants of the life-extension promotion typical of hormesis (Lagisz et al. 2013). Studies on birds also showed that early life hormetic priming to mild heat stress prepared individuals to withstand high heat stress later in life, resulting in lower mortality (Yahav and McMurtry 2001; Costantini et al. 2014a) and higher resistance to oxidative stress (Costantini et al. 2012, see also Chap. 7) compared to individuals that were not exposed to heat stress or experienced high heat stress early in life. Hormetic effects are not limited to mild heat stress. Exposure to mild cold stress can also enhance subsequent cold tolerance and survival (Lee et al. 1987; Czajka and Lee 1990) and preserve courtship behaviours (Shreve et al. 2004). For example, Yi et al. (2007) found that cold-conditioned young *Drosophila melanogaster* fruit flies (exposed to 5 °C for 2 h prior to −5 °C or −7 °C for 2 h) had higher survival probability than cold-shocked fruit flies (directly exposed to −5 °C or −7 °C for 2 h). They also found that the higher tolerance of acute cold stress of cold-conditioned fruit flies was associated with a block of apoptosis through the upregulation of the anti-apoptotic protein Bcl-2 (and possibly of other factors, such as heat shock proteins; Yi et al. 2007). In another study on fruit flies, Shreve et al. (2004) found that mild cold stress (16 °C for 2 h) increased courting and mating success compared to controls (23 °C), suggesting that hormetic stimulation of survival may not necessarily come at a cost for reproduction. For species that have short lives, like fruit flies, having the capability to adjust to changes in environmental temperatures may be critical for successful reproduction.

Thermal stress experiences may also shape the resistance of symbiotic marine species to oxidative challenges. Coral bleaching is a well-known phenomenon characterised by a loss of colouration caused by the loss of zooxanthellae, which are the algal endosymbionts of reef corals. Both field and laboratory studies showed that the increase in water temperature causes accumulation of oxidative damage, which would be a mechanism implicated in the process of coral bleaching. Bleaching reflects a poor condition of corals that results in reduced growth rates, suppression of sexual reproduction, impaired healing after mechanical damage, increased susceptibility to diseases, or mass mortality (e.g. Szmant and Gassman 1990; Meesters and Bak 1993; Brown et al. 2002a; Dunn et al. 2002; Lesser 2006). Although for certain coral species, the susceptibility to bleaching may be influenced by how much the algal genotype is resistant to thermal stress (Rowan et al. 1997), variation in bleaching susceptibility within colonies bearing genetically homogenous algal populations may also be induced by small-scale variation in environmental conditions. In colonies of the massive coral *Goniastrea aspera* occurring at Phuket (Thailand), it has been observed that during periods of maximal sea temperatures bleaching occurred in the east but not in the west side of the colonies (Brown et al. 2002b). The genotype of algae extracted from the corals did not appear to differ between east and west corals. However, when exposed to elevated water temperatures and low irradiance in the laboratory, east but not west corals bleached in response to treatment (Brown et al. 2002b). These findings indicated that the bleaching responses could not be attributed to genetic differentiation, but were probably shaped by the different local conditions that corals experienced (Brown et al. 2002b). Hawkins and Warner (2017) demonstrated that experimental conditioning to warm conditions protected the symbiotic sea anemone *Exaiptasia pallida* against acute heat-induced respiratory dysfunction and delayed bleaching. Similarly,

Martell (2023) found experimental support for a potential thermal hormetic priming in corals that resulted in greater symbiont retention when corals were exposed to severe thermal stress. The results of these studies suggest that hormetic priming may explain why corals containing genetically similar algae differ in their resistance to bleaching. Hormesis-induced host transcriptomic changes (e.g. gene expression) might represent one mechanism underlying the increase of thermal tolerance in corals (Bellantuono et al. 2012).

There is some evidence that thermal stress can also influence the expression of body colourations because the induced oxidative stress would cause depletion of pigments, such as carotenoids. Eraud et al. (2007) exposed male zebra finches *Taeniopygia guttata* to two thermal regimes (6 and 26 °C) over a period of 4 weeks. Simultaneously, half of the males from each of the two experimental groups were supplemented with carotenoids, whereas the other half of the males were not. Those males that received carotenoids developed a redder bill; however, this effect was smaller when males were exposed to cold conditions (Eraud et al. 2007). Although these males developed duller bills compared to carotenoid-supplemented birds, at the end of the experiment they had plasma levels of carotenoids higher than those recorded before the start of the experiment (Eraud et al. 2007). Thus, birds might have allocated carotenoids to sustain self-maintenance mechanisms at the expense of the ornament production.

## 5.3 Ultraviolet Radiation: The Dark Side of Light

The organic matter present in seawater may reduce the incidence of ultraviolet radiation on organisms living in shallow waters, such as corals (Zepp et al. 2008). However, the absorption of this ultraviolet radiation can generate reactive oxygen species that are free to move into the seawater (Mopper and Kieber 2000). These reactive oxygen species can represent a threat for microorganisms because they are able of oxidising their cell membranes. For example, degradation of organic matter generates hydrogen peroxide, a reactive oxygen species with a long life in seawater and a low propensity to react with other molecules that enables the hydrogen peroxide to easily move through cell membranes and enter into microorganisms (Mopper and Kieber 2000; Halliwell and Gutteridge 2015).

Ultraviolet radiation can also have more direct effects on the physiology and ecology of a species (e.g. Bancroft et al. 2007, 2008). Laboratory experiments carried out on the green sea urchin *Strongylocentrotus droebachiensis* found that exposure to ultraviolet radiation can reduce fertilisation success and influence the development time of embryos, probably because ultraviolet radiation caused strong oxidative stress (Adams and Shick 1996, 2001; Lesser et al. 2006; Meng et al. 2009). The results of these experiments received support from field studies on green sea urchin embryos, which showed that oxidative stress could be induced by radiation at depths deeper than 5 m (Lesser 2010). Ultraviolet radiation represents indeed an environmental hazard for many aquatic organisms, and increased generation of reactive oxygen species is one physiological mechanism through which ultraviolet radiation causes damage to organisms (Fig. 5.3) (e.g. Kim et al. 2015). However, resistance against damage induced by light also depends on the evolutionary history of the population or species. For example, lake and sea populations of the opossum shrimp *Mysis relicta* are genetically very similar, but their eyes differ markedly in their resistance against light-induced damage. Specifically, their eyes contain ommochromes (or visual pigments), which have a very pronounced antioxidant activity, and the eyes of lake shrimps have lower concentrations of these pigments and accumulate higher levels of lipid peroxidation than the eyes of sea shrimps, respectively (Dontsov et al. 1999).

Ultraviolet radiation can also affect the oxidative status of vertebrates. Exposure of zebra finches to simulated sunlight ultraviolet radiation was associated with low levels of carotenoids in liver, and elevated levels of carotenoids in blood plasma, suggesting that carotenoids stored in the liver were probably mobilised into the bloodstream (Blount and Pike 2012). However, exposure to sunlight did not cause any oxidative damage, nor did it reduce the intensity of bill colouration (Blount and Pike 2012). These results might indicate that carotenoids have a protective role against sunlight-induced damage (Alaluf et al. 2002; Sies and Stahl 2007). However, there is little evidence that supports a protective role of skin carotenoids against reactive oxygen species (Ogilvy and Preziosi 2012).

Many studies on vertebrates also documented increased levels of antioxidant enzymes in response to ultraviolet radiation. For example, Charron et al. (2000) found increased levels of superoxide dismutase and catalase in adult zebrafish *Brachydanio rerio* exposed to ultraviolet B radiation. Likewise, embryos of the

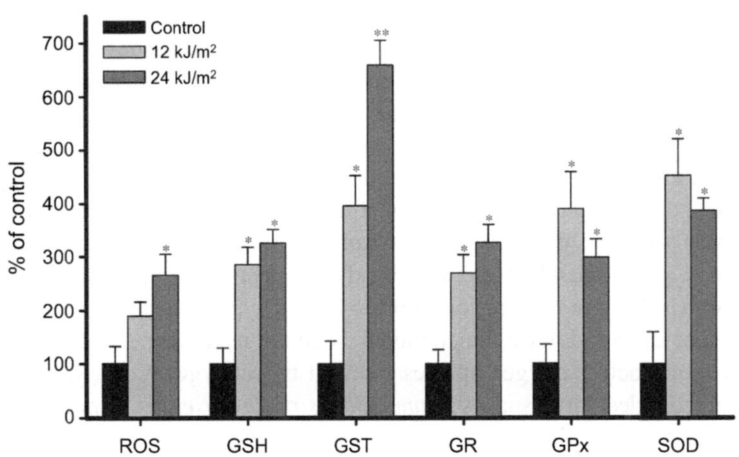


**Fig. 5.3** Effects of ultraviolet B irradiation exposure for 96 h at two intensities on intracellular generation of reactive oxygen species (ROS), glutathione (GSH), glutathione S-transferase (GST), glutathione reductase (GR), glutathione peroxidase (GPx), and superoxide dismutase (SOD) of the benthic copepod *Tigriopus japonicus*. Asterisks indicate significant differences between controls and UV-B groups: * $P < 0.05$, ** $P < 0.01$. Values are shown as mean and standard deviation. Reprinted with minor modifications from Kim et al. (2015) with permission of Elsevier

salamander *Ambystoma maculatum* and *Bufo arenarum* tadpoles had enhanced levels of superoxide dismutase when exposed to low doses of ultraviolet radiation (Lesser et al. 2001; Herkovits et al. 2006). These patterns of response of endogenous antioxidants to ultraviolet radiation are similar to hormetic patterns or to biochemical responses reported for many animal species challenged by extreme environments (e.g. Geihs et al. 2020; see also the *preparation for oxidative stress hypothesis* in the following paragraph 5.4).

## 5.4 Ways of Living With(out) Oxygen

### *5.4.1 Hypoxia–Anoxia and Reoxygenation*

Many animal species can survive long periods of reduced oxygen availability (mild hypoxia) or complete deprivation (anoxia). The partial pressure of oxygen can actually dramatically change daily or seasonally in certain environments, exposing organisms to potential oxidative stress. For example, cycles of dissolved oxygen in estuaries can range from anoxia (complete lack of oxygen), hypoxia (oxygen concentrations lower than normoxia) to various levels of supersaturation (200–300% higher than normoxia) over short periods of time (Ross et al. 2001). Many other species experience changes in oxygen concentration in their tissues because they have evolved certain behavioural strategies, such as freezing, dehydration, and aestivation, to survive harsh seasons that require small amounts of oxygen to function, but that then expose the organism to very high levels of reactive oxygen species upon awakening (for more details on this topic, see Chap. 6).

How can individuals with low levels of oxygen in their tissues suffer oxidative stress? We know that oxidative stress occurs when cells are being reoxygenated, i.e. when hibernation or aestivation end, and the breathing rate of animals return to normal. Upon reoxygenation, the generation of reactive oxygen species in cells increases dramatically, which means that the organism needs antioxidants to protect cells against their potentially detrimental effects. Many animal species have evolved mechanisms to resist the damage that can potentially arise from conditions of anoxia–reoxygenation, which are notoriously detrimental for many species, including humans (ischaemia-reperfusion). Following the hypothesis that anoxia/hypoxia-tolerant species could have biochemical adaptations to cope with this strong generation of reactive oxygen species induced by reoxygenation (Reischl 1986), studies on red-sided garter snakes *Thamnophis sirtalis parietalis*, exposed to 10 h of anoxia at 5 °C, showed that these animals had higher activity of superoxide dismutase and higher concentration of glutathione in both muscle and liver than controls (Hermes-Lima and Storey 1993). At that time, researchers thought that the generation of reactive oxygen species could not occur under anoxic conditions. A few years later, it has been discovered that reactive oxygen species can be even generated under low oxygen conditions (e.g. Chandel et al. 1998; Clanton 2007; Hermes-Lima et al. 2015; Giraud-Billoud et al. 2019). Thus, it has been suggested that the upregulation of antioxidant defences during the hypoxia/anoxia period might be relevant to control this minimal generation of reactive oxygen species, and mainly to mount an anticipatory protective response to the upcoming burst of reactive oxygen species caused by reoxygenation, which would be followed by the upregulation of other antioxidant mechanisms (Fig. 5.4). This anticipatory response would be stimulated by reactive oxygen species through the activation of specific transcription factors (*preparation for oxidative stress hypothesis*; Hermes-Lima and Storey 1993; Hermes-Lima et al. 1998, 2015; Lushchak et al. 2001, 2005a, b; Lushchak and Bagnyukova 2006a, b, c, 2007; Moreira et al. 2016, 2017; Yin et al.

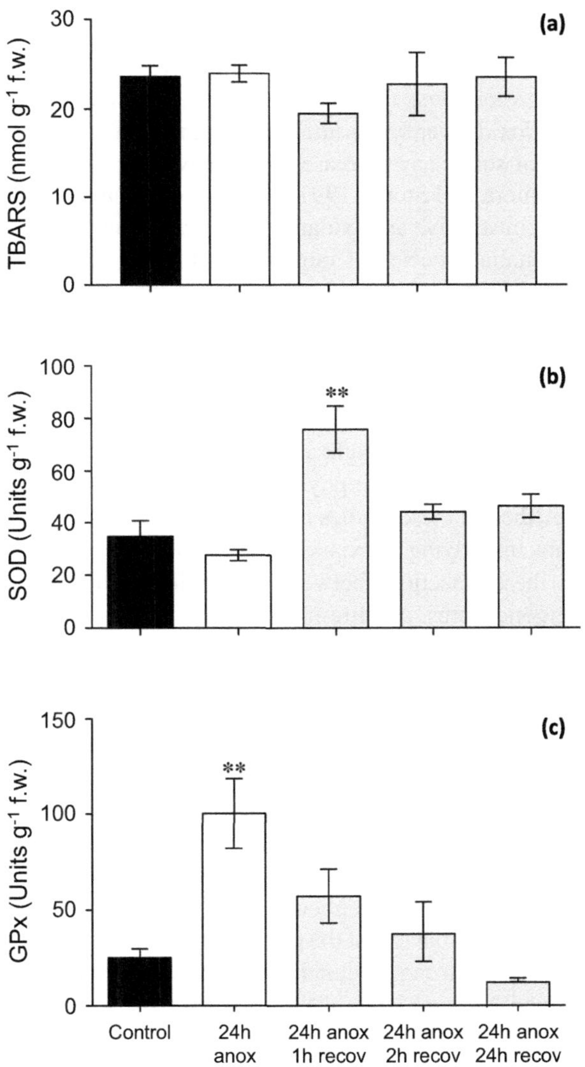


**Fig. 5.4** Effects of short- and long-term drastic variations in oxygen level on (**a**) the concentration of thiobarbituric acid reactive substances (TBARS), (**b**) activity of superoxide dismutase (SOD), and (**c**) activity of glutathione peroxidase (GPx) in the subterranean crustacean *Niphargus rhenorhodanensis*. Anox = anoxia; recov = recovery. Asterisks indicate significant differences between controls and experimental groups: ** $P < 0.01$. Values are shown as mean and standard error. Reprinted with minor modifications from Lawniczak et al. (2013) with permission of Elsevier

2017; González-Ruiz et al. 2023; Yang et al. 2023). For example, over-expression of endogenous antioxidants prevented post-reoxygenation injury in laboratory rodents (Wang et al. 1998; Tanaka et al. 2004), while inhibition of catalase during hypoxia caused an increase in the ratio of oxidised glutathione to reduced glutathione in Nile tilapia *Oreochromis niloticus* (Welker et al. 2012). However, experimental inhibition of glutathione synthesis did not significantly influence the impact of anoxia exposure and reoxygenation on levels of protein carbonyls, thiobarbituric acid reactive substances, or antioxidant enzymes in land snail *Helix aspersa* (Ferreira-Cravo et al. 2023).

This anticipatory response has been documented in a wide range of organisms. However, we have also many studies that did not find any evidence in support of this anticipatory response to reoxygenation following anoxia nor an increase in oxidative damage upon reoxygenation, such as in red-eared slider turtles *Trachemys scripta elegans* (Willmore and Storey 1997a, b) or in cave-dwelling salamanders *Proteus anguinus* (Issartel et al. 2009). Red-eared slider turtles had reduced levels of antioxidant enzymes after exposure to 20 h of anoxic conditions. However, turtles did not suffer any increase in oxidative damage during anoxia nor reoxygenation (Willmore and Storey 1997a, b). One explanation for this result might lie with the high constitutive antioxidant levels that turtles have in comparison with other non-mammalian species. It might be that the diving habits of red-eared slider turtles expose them to repeated episodes of anoxia–reoxygenation that could have positively selected individuals with high constitutive antioxidants (Hermes-Lima et al. 2001; Hermes-Lima and Zenteno-Savín 2002). Therefore, turtles might mainly rely on constitutive antioxidant levels to buffer the pro-oxidant effects of reoxygenation, so that they do not need to upregulate the expression of antioxidant genes. Different shell morphologies might also generate variation in how different turtle species strategically regulate their physiological homeostasis (Zhang et al. 2019).

Although these studies have provided relevant insight into the molecular mechanisms underlying anoxia–reoxygenation episodes, we need more studies that analyse the connections between expression of the anticipatory response, individual energetic status, and life-history stage. We also need to analyse in more detail the generation of reactive oxygen species in order to provide robust evidence about how levels of reactive oxygen species change across different environmental conditions (e.g. Rivera-Ingraham et al. 2013a, b). For example, using live imaging techniques, Rivera-Ingraham et al. (2013a) found that marine platyhelminths *Macrostomum lignano* exposed to hyperoxic, normoxic, or anoxic conditions generated similar levels of the superoxide free radical. However, the total mitochondrial generation of reactive oxygen species was higher in platyhelminths after hyperoxic exposure and lower under near-anoxia conditions compared with normoxic controls, respectively.

Organisms that live in coral reefs, or in intertidal and estuarine habitats, are particularly suitable to test the preparation for oxidative stress hypothesis, because they experience strong fluctuations in environmental levels of oxygen (Nilsson and Östlund-Nilsson 2004). Marine invertebrates and fish that inhabit these environments are exposed daily to episodes of anoxia and reoxygenation because they become exposed to air during low tides, and their tissues are reoxygenated when the water level increases during high tides. These animals have evolved a number of metabolic adjustments to tolerate short-term anoxia, such as the use of fermentable fuels to produce energy or the depression of metabolic rate (Storey and Storey 1990; Brooks and Storey 1997). For example, Pannunzio and Storey (1998) found that adaptation to post-anoxic oxidative stress is one important physiological mechanism that has been evolved by the gastropod *Littorina littorea*. However, surprisingly, Pannunzio and Storey (1998) found that several components of the antioxidant machinery were suppressed during the anoxic period. It might be that the prolonged anoxic period experienced by the gastropods was rather inhibitory and not stimulatory of the antioxidant response (Hermes-Lima and Zenteno-Savín 2002). These findings also suggest that the response to anoxia may follow an hormetic pattern, with short episodes of anoxia being stimulatory of the antioxidant response, while long episodes of anoxia being inhibitory of protective mechanisms, respectively.

Similarly to anoxia, the duration of the reoxygenation phase is also important in terms of consequences for the oxidative status. For example, de Oliveira et al. (2005) found that the duration of the anoxia recovery affected levels of both thiobarbituric acid reactive substances and of antioxidants in estuarine crabs *Chasmagnathus granulata*. These biphasic responses might indicate that the generation of reactive oxygen species differs between the anoxic and normoxic phases. For example, Dirmeier et al. (2002) found in yeast that certain proteins were carbonylated during a shift to anoxia, and some of these proteins also were carbonylated during exposure to hydrogen peroxide. Nitric oxide can also activate hypoxic genes because it may combine with the superoxide free radical, generating peroxynitrite, which in turn may nitrosylate proteins and induce the expression of hypoxic genes (Poyton et al. 2009).

Other studies showed that exposure to high levels of oxygen (hyperoxia) can also jeopardise the homeostasis of the oxidative status and, depending on intensity and duration, induce hormetic responses of antioxidants (e.g. Ritola et al. 2002a, b, c; Dabrowski et al. 2004; Olsvik et al. 2006). For example, the activity of the superoxide dismutase in gills and digestive glands was higher in blue mussels *Mytilus edulis* living in high-shore areas than in low-shore areas, probably because animals in high-shore areas experienced longer and more frequent periods of emersion through the tidal cycle (Letendre et al. 2008). Similarly, epaulette sharks *Hemiscyllium ocellatum*, which inhabit the reef platforms, go through periodic cycles of hypoxia–reoxygenation because platforms repeatedly become cut off from the ocean during periods of low tides forming large tide pools (Nilsson and Renshaw 2004). During nocturnal low tides, the concentration of oxygen in water can fall by 80% due to respiration of the corals and associated organisms (Nilsson and Renshaw 2004). The intensity of the hypoxic period is, however, quite variable across the tidal cycle. This might be relevant for young developing sharks: the stage of development at which they experience hypoxia might determine the future resistance against or tolerance of oxidative stress in adulthood. If this were the case, adjusting reproductive decisions to tides might confer an advantage to sharks in terms of maintenance of oxidative status homeostasis. For example, these sharks reproduce during periods of lower water temperatures (Heupel et al. 1999), which might limit the magnitude of hypoxia because water oxygen concentration is lower at higher water temperatures.

Intertidal marine species may also be exposed to additional oxidative stressors, such as those caused by the photosynthetic oxygen production of benthic algae. Oxygen supersaturation favours the generation of hydrogen peroxide in sediments (Zika et al. 1985; Szymczak and Waite 1988). The polychaete *Nereis diversicolor* is an invertebrate species that lives into burrows of intertidal sediments. During low tides, worms respond behaviourally by dragging air bubbles inside their burrows and exchanging for fresh bubbles when needed (Lindroth 1941). Moreover, worms may also switch to anaerobiosis and reduce their activity when levels of exogenous peroxides become very high (Schöttler 1979). However, additional physiological mechanisms are needed to resist oxidative stressors. This is particularly important during the spring spawning time because the worms have to cope with an increase in energy and oxygen demands and, therefore, take higher amounts of externally formed peroxides. Beyond keeping enzymatic antioxidant defences high, spawning worms may also rely on non-enzymatic antioxidants. When worms are incubated with peroxides under aerobic conditions during spawning, there is oxidative decomposition of haemoglobin to biliverdin, which has non-negligible antioxidant properties and may offer additional protection against reactive oxygen species (Abele-Oeschger et al. 1994). However, under anoxic conditions, the biochemical decomposition of haem is inhibited, as it requires high quantities of oxygen to decompose haemoglobin to biliverdin (Abele-Oeschger et al. 1994). There is also autoxidation of haemoglobin and consequent generation of reactive oxygen species (Abele-Oeschger and Oeschger 1995). Therefore, oxidative stress may represent a relevant constraint on reproductive investment of worms when the availability of oxygen in the environment is limited.

Many organisms have also evolved other strategies to limit oxidative stress. For example, many marine invertebrates with open circulatory systems establish low and constant oxygen partial pressure around their tissues (Abele et al. 2010). Adjustment of mantle cavity partial pressure of oxygen to lower than ambient levels through controlled pumping should prevent the formation of high oxygen gradients between bivalve tissues and the surrounding fluid, limiting oxygen flux across the body surface and, possibly, any increases of oxidative stress (Abele et al. 2010).

### *5.4.2 The Curious Case of Symbiotic Species*

Given the pleiotropic role of reactive oxygen species in cell–cell communication, development, immunity, and reproduction, but also their associated physiological costs (e.g. generation of molecular damage), their regulation can influence the establishment, the maintenance, and the breakdown of various symbiotic associations (e.g. Moné et al. 2014; Monnin et al. 2016, 2017; Capobianco et al. 2018). Maintenance of oxygen equilibrium in tissues is not only challenged by changes in partial pressure of oxygen that occur in the environment external to the body, but also by those that occur within it. This is particularly relevant for symbiotic species. Invertebrates that live in symbiosis with unicellular algae experience higher partial pressures of oxygen in their tissues than aposymbiotic species, because the endosymbiotic algae produce oxygen through photosynthesis, and the rate of oxygen consumption is low (e.g. D'Aoust et al. 1976; Shick and Dykens 1985; Shick 1990; Shick et al. 1996; Richier et al. 2005). Thus, symbiotic species are being exposed to high concentrations of oxygen. One consequence of this exposure to high concentrations of oxygen might be the increase in the rate of generation of oxidative damage, which might impact on cellular senescence or reproduction. However, symbiotic species do not suffer higher oxidative stress than aposymbiotic species because they are equipped with efficient antioxidant defences (e.g. Dykens and Shick 1982, 1984; Shick 1990; Shick et al. 1996; Richier et al. 2005; Moné et al. 2014).

Pro-oxidant effects of hyperoxic conditions in tissues may be exacerbated in symbiotic species that live in environments with high solar irradiance because oxygen can undergo univalent reduction, giving rise to many types of reactive oxygen species. However, the toxicity of sunlight for symbiotic species seems to differ among environments (Regoli et al. 2000) and might also be influenced by hormetic priming. For example, the Mediterranean demosponge *Petrosia ficiformis* lives symbiotically with the cyanobacterium *Aphanocapsa feldmanni*, but individuals living in submarine caves can be aposymbiotic. Aposymbiotic specimens of this species die within a week when transplanted from caves to cliffs exposed to sunlight (Regoli et al. 2000). By contrast, the mortality does not increase in symbiotic specimens transplanted to areas where solar irradiance is higher, even when they are deprived of most endosymbionts through the grazing activity of nudibranchs (Regoli et al. 2000). Therefore, the pro-oxidant effect of cyanobacteria could have prepared the host's antioxidant defences to better withstand the pro-oxidant effects of ultraviolet radiation. Many other studies found that symbiotic forms have higher antioxidant levels than conspecific aposymbiotic forms (Fig. 5.5) (e.g. aggregating anemone *Anthopleura elegantissima*, Dykens and Shick 1982, Mitchelmore et al. 2003; maxima clam *Tridacna maxima*, Shick and Dykens 1985; Mediterranean sea anemone *Anemonia viridis*, Richier et al. 2003, Plantivaux et al. 2004). We might hypothesise that the evolution of symbiotic forms has been favoured in those environments where the abiotic conditions challenge the oxidative status homeostasis.

### 5.4.3 Into the Deep Sea

Most oceans are characterised by higher oxygen concentrations in the euphotic zone (uppermost layer of a body of water that receives sunlight) and lower oxygen levels in the deep sea, respectively. As a consequence, it is expected that deep-sea organisms would face a reduced threat from oxidative stress. Relying on the analysis of antioxidant enzymes in the tissues of 16 species of meso- and bathypelagic fish that live between the surface and a depth of 1300 m, Janssens et al. (2000) found that the reduced metabolic requirements of bathypelagic species compared to those of

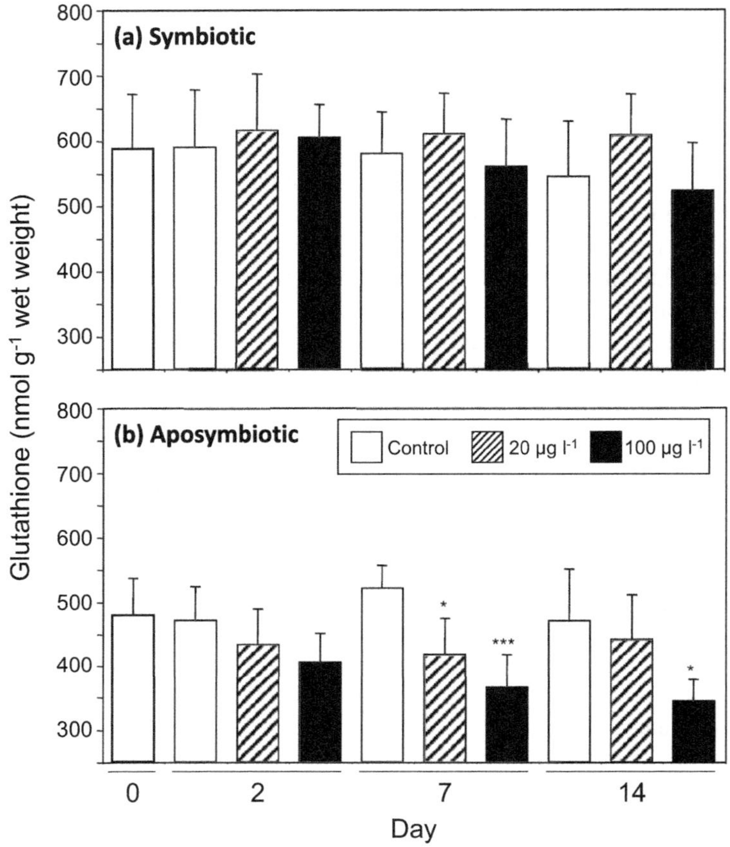


**Fig. 5.5** Levels of glutathione in (**a**) symbiotic and (**b**) aposymbiotic sea anemones *Anthopleura elegantissima* following an exposure to 20 or 100 μg $L^{-1}$ of cadmium for a period of 2 weeks, respectively. Asterisks indicate significant differences between controls and experimental groups: * $P < 0.05$, *** $P < 0.001$. Values are shown as mean and standard deviation. Reprinted with minor modifications from Mitchelmore et al. (2003) with permission of Elsevier

mesopelagic species were associated with lower activities of superoxide dismutase and glutathione peroxidase, but not of catalase. However, there are organisms that live in the deep sea that do vertical migrations and, consequently, become exposed to a transition between hypoxia and reoxygenation states. Huge numbers of marine organisms daily move from deeper to surface waters and back. This vertical migration faces organisms with metabolic requirements induced by the physical effort, and with the need of dynamically adjusting their physiological machinery in order to tolerate the changes in water parameters, such as the partial pressure of oxygen and the temperature. This is particularly relevant because it would require a flexible antioxidant machinery. Euphausiacea crustaceans (the so-called krill) are renowned for performing the deepest migrations; interestingly, they have also very high levels of antioxidant enzymes (Tremblay et al. 2012). There are relevant differences among species in the expression of antioxidant enzymes with depth. For example, the tropical krill *Euphausia eximia* showed higher activity of superoxide dismutase at the surface than in the deeper waters, while in the subtropical *Nematoscelis simplex* and the temperate *Nematoscelis difficilis* the activity of superoxide dismutase was higher in deeper than in shallower waters (Tremblay et al. 2012).

Myctophids (or lanternfishes) are important members of mesopelagic communities that undergo extensive diurnal vertical migrations from epipelagic layers to mesopelagic depths at night. Lopes et al. (2013) found that activities of catalase and glutathione-S-transferase were lower in fish caught at the surface compared to those caught in deep waters. By contrast, the activity of superoxide dismutase was higher in fish caught at the surface compared to those caught in deep waters for *Benthosema panamense*, and it was similar between *Triphoturus mexicanus* fish caught at the surface and in deep waters, respectively. Levels of thiobarbituric acid reactive substances were higher in fish caught at the surface compared to those caught in deep waters for both fish species, but the difference between groups was statistically significant only for *Triphoturus mexicanus* (Fig. 5.6) (Lopes et al. 2013). Thus, both fish species appeared to differ in how they regulate the expression of antioxidant levels in response to the changes in oxygen exposure during vertical migrations.

These study systems offer interesting opportunities to test the effects of environmental hormetic priming and oxidative stress on life-history strategies. It might be relevant to examine those species that reproduce at the sea surface and then larvae migrate into the deep waters, because larvae would be exposed to different environmental conditions. Developing in these two contrasting environments might have a strong impact on the selective role of oxidative stress for these animals. For example, palolo worms of the *Eunicidae* family are sea annelids that have a unique spawning behaviour. In preparation for spawning, they generate a tail containing eggs and sperm (epitoke). At spawning, all epitokes are released simultaneously and make their way to the surface, where fertilisation of eggs occurs. The larvae hatch quickly and then migrate to the ocean floor. They have to withstand costs of migration straightaway after hatching, an important temporal window for selection to operate, when being prepared to tolerate oxidative stress might be vital.

## 5.5 Partial Pressure of Carbon Dioxide

Exposure to a high partial pressure of carbon dioxide ($CO_2$) in freshwater or seawater causes an increase in the concentration of $CO_2$ in the blood, which increases acidification (decrease of pH) of tissues (Brauner and Baker 2010). Animals have evolved several mechanisms to regulate the pH of tissues, but pH regulation is energy demanding (e.g. Pörtner et al. 2004; Ishimatsu et al. 2008). Moreover, exposure to high levels of $CO_2$ in water may also jeopardise the homeostasis of oxidative status because the decrease of pH in tissues may trigger the Fenton reaction (see Chap. 1), and the interaction between $CO_2$ and reactive oxygen species can generate new pro-oxidant molecules (Veselá and Wilhelm 2002). For example,

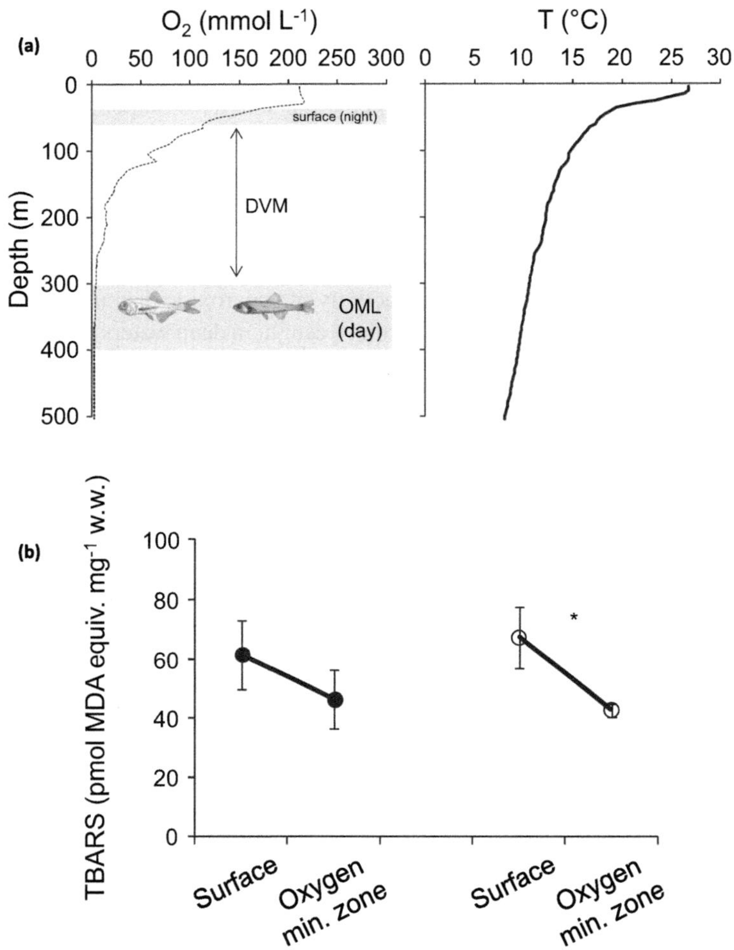


**Fig. 5.6** Impact of diel vertical migration on the oxidative status of two fish species. (**a**) oxygen levels and temperature conditions along a depth gradient in the study area. Grey bars indicate the capture depth of fish (at surface waters between 40 and 50 m at night-time, and in the upper limit of the mesopelagic oxygen minimum zone, OMZ, between 300 and 400 m during the daytime). DVM = diel vertical migration. (**b**) levels of thiobarbituric acid reactive substances in two species of myctophid fishes (black closed circles, *Benthosema panamense*; open circles, *Triphoturus mexicanus*) caught at the surface and in the oxygen minimum zone. Asterisks indicate significant differences between groups. Values are shown as mean and standard deviation. Reprinted with minor modifications from Lopes et al. (2013) with permission of Elsevier

Eastern oysters *Crassostrea virginica* live in estuarine ecosystems that are characterised by large fluctuations in $CO_2$ levels (from ca. 40–4700 Pa) that can occur even over short-time scales (Tomanek et al. 2011). Experimental exposure to a high partial pressure of $CO_2$ (~357 Pa) for 2 weeks caused a significant change of the proteome in mantle tissues of oysters, compared with control conditions (Tomanek et al. 2011). In particular, the results of this study showed that the main functional categories of proteins that were upregulated in response to hypercapnia included those associated with the cytoskeleton (e.g. several actin isoforms) and those associated with oxidative stress (e.g. superoxide dismutase, peroxiredoxins, and thioredoxin-related nucleoredoxin). However, other studies found contrasting results, indicating that the oxidative status of some species is particularly resilient to high $CO_2$ conditions (e.g. 9 months at pCO2 ~ 900 μatm, small-spotted catshark *Scyliorhinus canicula* in Pegado et al. 2020). A recent meta-analysis including studies on aquatic ectotherms found that water acidification increased significantly markers of oxidative damage, but it did not influence levels of antioxidants (Fig. 5.1b) (Messina et al. 2023). It might be that (1) the performance of antioxidant enzymes was reduced at acidic conditions, (2) the upregulation of the antioxidant response was too demanding for the organism, or (3) animals could tolerate the increased levels of oxidative damage.

Exposure to high concentrations of $CO_2$ may also trigger hormetic mechanisms, and any effects may persist over generations in both invertebrates and vertebrates (e.g. Miller et al. 2012; Parker et al. 2015). Experimental exposure to ocean conditions projected for the end of the century (approximately 1000 μatm $CO_2$) caused an increase in metabolic rate and a decrease in length, weight, condition, and survival of juvenile cinnamon clownfish *Amphiprion melanopus* (Miller et al. 2012). However, these effects were absent or reversed when parents experienced similar high concentrations of $CO_2$. It might be that these $CO_2$ levels were stimulatory of the protective molecular mechanisms in adults, and somehow adults passed this information to their offspring, making them more resistant against water acidification.

In a related experiment, breeding pairs of cinnamon clownfish were held under three treatments of $CO_2$ [current-day control (430 μatm), moderate (584 μatm), and high (1032 μatm)] over a period of 9 months that included the summer breeding season (Miller et al. 2013). Surprisingly, increased $CO_2$ significantly stimulated the breeding activity and the production of eggs of clownfish compared to controls. This study provided the first evidence of a potential hormetic effect caused by exposure to an increased partial pressure of $CO_2$ (Miller et al. 2013).

## 5.6 Salinity Stress

Salinity stress owing to increased or decreased salt exposure requires activation of several physiological pathways underlying osmoregulation. This is well exemplified by marine organisms living in coastal lagoons, mangrove swamps, deltas, estuaries, or intertidal areas that are frequently exposed to large fluctuations in salinity (Rivera-Ingraham and Lignot 2017; Bal et al. 2021). When being exposed to salinity stress, osmoregulatory organisms increase their energy expenditure in order to osmoregulate (Rivera-Ingraham and Lignot 2017). Aerobic ATP production results in increased generation of reactive oxygen species, implying a certain degree of cross-talk among osmoregulation, energy balance, and oxidative status. For example, Rivera-Ingraham et al. (2016a) found an increase in reactive oxygen species generation in Mediterranean green crab *Carcinus aestuarii* in diluted marine water. Rivera-Ingraham et al. (2016b) also found that flatworm *Macrostomum lignano* increased both their respiration rates and the generation of the superoxide free radical when being exposed to hypersalinity. Importantly, reactive oxygen species may reduce osmoregulatory capacity through the inactivation of membrane-bound transporters or of enzymes involved in osmolyte synthesis (Rivera-Ingraham and Lignot 2017).

Terrestrial organisms also need physiological adjustments in response to salinity stress (e.g. intake from water or food). Studies on laboratory rodents found that salt intake may indeed cause oxidative stress (e.g. Lenda and Boegehold 2002; Kitiyakara et al. 2003). In free-living terrestrial vertebrates, however, we have very limited work on this topic. van de Crommenacker et al. (2011) found that Seychelles warbler *Acrocephalus sechellensis* living in territories with high salinity had also higher levels of a marker of oxidative damage. Sabat et al. (2017) documented changes in the oxidative status in rufous-collared sparrow *Zonotrichia capensis* that persistently drank saltwater, and showed that these physiological adjustments ranged in scale from biochemical endpoints (e.g. non-enzymatic antioxidants and oxidative damage to lipids (Fig. 5.7) to whole organismal traits (e.g. metabolic rate)).

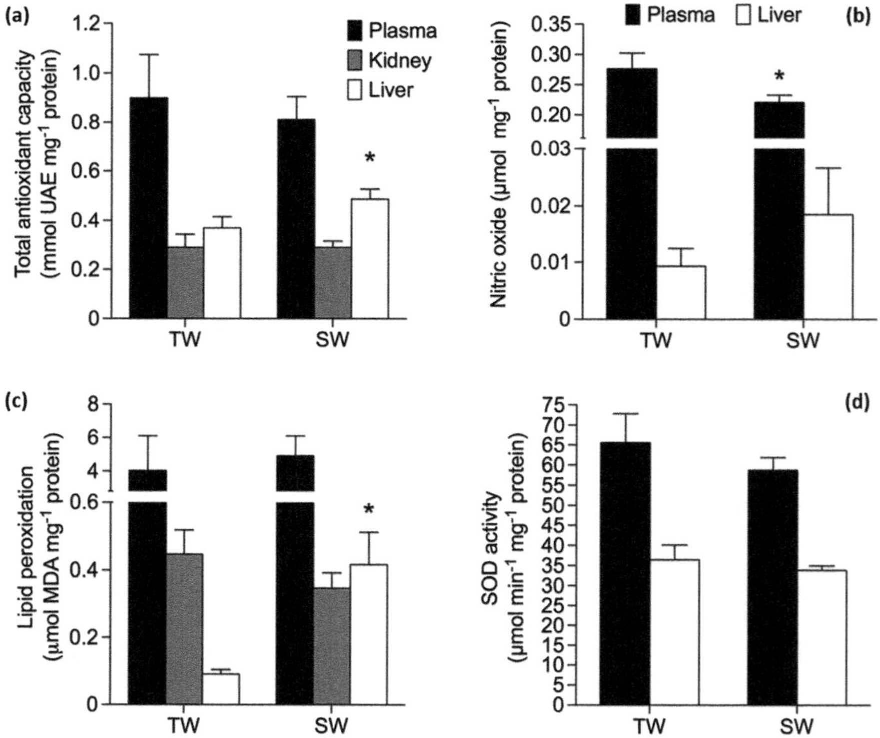


**Fig. 5.7** Impact of salinity stress on (**a**) non-enzymatic antioxidant capacity, (**b**) nitric oxide, (**c**) thiobarbituric acid reactive substances, and (**d**) superoxide dismutase in rufous-collared sparrows *Zonotrichia capensis*. Birds were given tap (TW) or 200 mM NaCl (SW) water for 30 days. Asterisks indicate significant differences between groups per tissue, separately. Values are shown as mean and standard error. Reprinted with minor modifications from Sabat et al. (2017). © 2017 Sabat, Narváez, Peña-Villalobos, Contreras, Maldonado, Sanchez-Hernandez, Newsome, Nespolo and Bozinovic; distributed under the terms of the Creative Commons Attribution License (CC BY)

## 5.7 Water Balance and Anhydrobiosis

Maintenance of water balance is one component of osmoregulation so that it is also strictly connected to salt balance. Petrović et al. (2021) examined how the oxidative status of yellow-bellied toad *Bombina variegata* larvae changed during metamorphosis (Gosner stages of development: 42, 44, and 46) in relation to constant or decreasing water availability. At stage 42, water availability did not affect the oxidative status of tadpoles. By contrast, at stage 44, tadpoles exposed to desiccation had a lower concentration of glutathione and a higher concentration of thiobarbituric acid reactive substances than control tadpoles, respectively. At stage 46, larvae exposed to decreasing water had lower glutathione and higher levels of thiobarbituric acid reactive substances, as well as lower values for catalase, glutathione peroxidase, glutathione-S-transferase, and total thiols compared to controls.

Effects of water shortage on oxidative status are not necessarily direct. For example, tadpoles of the western spadefoot toad *Pelobates cultripes* exposed to low water levels had higher catalase, superoxide dismutase, and glutathione peroxidase, and an acceleration of development (32.3%) compared to control tadpoles (Gomez-Mestre et al. 2013). Thus, the accelerated development, likely needed to evade drying ponds, might have contributed to influence the oxidative status of tadpoles further, irrespective of the costs associated with osmoregulation.

Short periods of water restriction can also cause long-lasting effects. Dupoué et al. (2020) exposed yearling female and male common lizards *Zootoca vivipara* either to a period of 51 days of water restriction or to water ad libitum, followed by 45 days in common garden outdoor conditions. They found that water-restricted males had enhanced plasma non-enzymatic antioxidant capacity and decreased levels of plasma reactive oxygen metabolites at day 36, whereas females did not immediately respond. A month and a half after water restriction, both females and males experienced a decline in antioxidant capacity, but only males exhibited significant telomere shortening. Water shortage is also an issue for adults during demanding phases of their life cycle. For example, Dezetter et al. (2021) found that a short-term water deprivation during pregnancy increased levels of reactive oxygen metabolites in captive common European adder *Vipera berus*.

Finally, some animals have evolved the capacity to enter a particular state of dormancy called anhydrobiosis, where organisms lose all of their body water and turn their metabolic machinery off. Several studies showed that the management of the oxidative status plays a relevant role in the regulation of cell homeostasis during desiccation and rehydration (e.g. França et al. 2007; Rebecchi 2013; Giovannini et al. 2022).

## 5.8 Exposome: How Organisms Respond to a Cocktail of Environmental Stressors

One of the major limitations of the studies illustrated in this chapter lies in the focus on single environmental stressors. In a natural context, a change in a certain environmental parameter, say water temperature, is associated with changes in other parameters, such as oxygen concentration in water. The overall effect on the organismal oxidative status can therefore be exacerbated or mitigated, depending on whether the action of the different stressors is additive, synergistic, or antagonistic. It is therefore important to focus on the exposome, i.e. the exposure of the organisms to a cocktail of environmental stressors and the associated consequences for the health and reproductive fitness.

Deschaseaux et al. (2010, 2011) examined the cellular responses of embryos and hatched larvae of three rocky shore intertidal gastropod species (*Bembicium nanum*, *Dolabrifera brazieri*, and *Siphonaria denticulata*) to concomitant changes of both temperature and salinity. Encapsulated embryos of each species were exposed for 72 h to six combinations of ecologically realistic values of temperature (22 °C and 30 °C) and salinity (25, 35, and 45 ppt). Both extremes of salinity and high temperature negatively affected the embryos and hatched larvae, causing oxidative stress and increasing their mortality. Ghanizadeh-Kazerouni et al. (2017) found that mosquitofish *Gambusia holbrooki* exposed to ultraviolet B radiation and kept in flowing water had increased sustained swimming performance, increased levels of catalase and glutathione, and reduced concentrations of thiobarbituric acid reactive substances and protein carbonyls compared with fish exposed to ultraviolet B radiation and kept in still water. Other studies found limited support for an interaction between salinity stress and ambient temperature (Sabat et al. 2019).

Hypoxia may also prepare the organism to better cope with different environmental stressors through hormetic mechanisms (e.g. Wu et al. 2002; Visser et al. 2018). For example, Wu et al. (2002) found that an acute exposure of young male locust *Locusta migratoria* to anoxia had protective effects during a subsequent exposure to high ambient temperatures, resulting in higher survival compared to locusts that did not previously experience any anoxia. Under normoxia, exposure of silver catfish *Rhamdia quelen* to manganese increased levels of thiobarbituric acid reactive substances in brain and kidney, increased the concentration of glutathione in brain, and decreased levels of catalase in both brain and kidney (Dolci et al. 2013). By contrast, fish exposed to manganese at moderate hypoxia did not suffer any lipid peroxidation in brain, had attenuated increased levels of thiobarbituric acid reactive substances in kidney, had higher concentrations of glutathione in brain, but had lower levels of catalase in both brain and kidney. Similarly, Zambonino-Infante et al. (2013) showed that thermal experiences early in life can shape a phenotype more resistant to hypoxia. A warmer temperature experienced during the larval stage had a positive effect on body mass and resistance to hypoxia at the juvenile stage in common soles *Solea solea*, suggesting a potential hormetic priming of metabolic pathways.

Early in life exposure to stressors can also induce an increase of the thermal tolerance. For example, Brown et al. (2002a) found that at experimentally elevated water temperatures, reef corals *Goniastrea aspera* from the surfaces of colonies collected from shallow waters exposed in the wild to solar radiation lost fewer symbiotic algae, had lower levels of lipid peroxidation, had higher levels of superoxide dismutase, and heat shock proteins 60 and 70 than corals from surfaces of colonies that were less exposed to solar radiation in the wild. In addition, corals from highly exposed surfaces showed less chronic photoinhibition and greater photosystem II recovery potential than corals from less exposed surfaces when both corals were experimentally exposed to high irradiance at ambient water temperature in the laboratory. The results of this experiment suggested that the priming of host tissues might be important for maintaining symbiosis under thermal stress (Brown et al. 2002a). A number of protective physiological mechanisms are activated when corals are exposed to air. The activation of these protective mechanisms might reflect an anticipatory response to the oxidative stress induced by reoxygenation that corals would experience during re-immersion, which would enable them to thrive in intertidal habitats and withstand air exposure during low tides (Teixeira et al. 2013).

## 5.9 Habitat Quality and Land-Cover Changes: From Forests to Cities

The quality of the habitat in which an animal lives can have major consequences for its physiology through different routes. For example, the availability of food is very important because it may influence the individual oxidative status through the intake of nutrients and antioxidants, as well as through the effort needed for foraging activity (see Chap. 3). Seychelles warblers inhabit a fixed feeding territory year-round and experience profound and local fluctuations in food availability. van de Crommenacker et al. (2011) found that the spatio-temporal variation in territory quality can be associated with different levels of oxidative damage: it was higher when the quality of territory was low. This study has been carried out in such a way that it is complicate to disentangle the effect of territory quality per se from those of low-quality individuals in case they are forced to occupy lower quality territories by high quality competitors. Nevertheless, this study provided the first evidence that the level of organismal oxidative damage can actually reflect to some degree the quality of the territory. Thus, the level of oxidative damage is not a fixed individual trait, rather it may reflect individual responses to environmental changes. Consequences for reproductive fitness might differ if costs of sustaining high levels of oxidative damage differ among individuals of different quality.

Breeding in a high-quality habitat does not necessarily pay in terms of better oxidative status whether, for example, individuals are pushed to over-invest in reproduction or to make other costly decisions. For example, great tits *Parus major* breeding in deciduous dominated habitats (i.e. higher abundance of caterpillars) had higher baseline levels of oxidative damage than birds that bred in evergreen habitats (Isaksson 2013). Nestling blue tits *Cyanistes caeruleus* had lower levels of glutathione peroxidase in mixed forests, higher in Scots pine forests, and intermediate in holm oak and Pyrenean oak forests (Garrido-Bautista et al. 2021). As compared to native oak, eucalyptus leaf litter impaired growth and development of *Rana temporaria* larvae, inhibited their antipredator responses, and altered their oxidative status (Burraco et al. 2018).

Environmental quality may also impinge on cellular senescence through mechanisms sensitive to oxidative stress, such as the telomere dynamics (Haussmann and Marchetto 2010). Angelier et al. (2013) reported that telomeres of male American redstarts *Setophaga ruticilla* wintering in a low-quality non-breeding habitat had shorted faster than those of individuals wintering in a high-quality non-breeding habitat. As with the study on Seychelles warblers, this study was correlative, hence it does not enable to disentangle the effect of non-breeding habitat from a potential effect of individual quality on the rate of telomere shortening. This would be the case if low-quality American redstarts winter in low-quality habitats. However, prior experimental work showed that the quality of the non-breeding habitat may actually determine the individual performance of American redstarts: wintering birds that were upgraded from a low-quality habitat to a high-quality habitat had higher breeding success and return rates in the next year than control birds in the low-quality habitat (Studds and Marra 2005, 2007, 2011; Reudink et al. 2009). Similarly, Mizutani et al. (2013) found that telomere dynamics in black-tailed gull *Larus crassirostris* were influenced by dramatic changes in environmental conditions, highlighting the importance of environmental fluctuations in determining stress levels and lifespan. The local availability of food is one important factor that can generate differentiation in oxidative status between conspecific populations. For example, Wilcoxen et al. (2015) found that birds at forested sites with feeders had significantly greater circulating non-enzymatic antioxidant capacity than birds at forested sites without feeders, and that these positive effects on antioxidant status disappeared 10 months after feeders were removed, suggesting that the different levels of antioxidants between birds living in the two forest types were determined by food availability.

Research on plants provided strong evidence for an important role of habitat in determining resistance to oxidative stress. Studies on three closely related oak species (*Quercus petraea*, *Q. pubescens*, and *Q. robur*) transplanted randomly to different soil typologies showed that changes in the generation of reactive oxygen species and levels of foliar antioxidants in response to stressful events differed between calcareous and acidic soil: reactive oxygen species and antioxidants were lower and higher in plants living on calcareous than on acidic soil, respectively (Contran et al. 2013; Hu et al. 2013). These results indicated a better capacity of plants to avoid oxidative stress on calcareous compared with acidic soil; however, this effect varied between metabolites and species. Hu et al. (2013) suggested that one of the reasons for such soil-dependent differences in resistance against oxidative stress was the lower water and nutrient availability that plants experienced on calcareous compared with acidic soil (Marion et al. 1993; Xie et al. 1998). Regardless of mechanisms, these studies clearly showed a strong link between habitat characteristics and organismal oxidative status.

Research on the impact of land-cover changes on organismal oxidative stress offers an excellent opportunity to assess the role of habitat quality in moulding individual physiology. Changes in land-cover have serious implications for ecological systems and the distribution of species. Although certain species have been rather successful in exploiting the new ecological opportunities offered by human-modified environments, other species have suffered strong population declines and even local extinctions. Recent work has focussed on the link between oxidative status and land-cover changes because, in these altered environments, animals are being exposed to a number of novel conditions that may cause cellular oxidative stress (e.g. food, parasites, pollutants, abiotic conditions) (e.g. Isaksson 2015). Common scale-backed antbirds *Willisornis poecilinotus* sampled at forest gaps had higher levels of thiobarbituric acid reactive substances in plasma than individuals sampled in undisturbed forests (Gomes et al. 2014). Relying on a correlational cross-sectional approach involving 15 understorey bird species, Messina et al. (2020) found that birds living in logged forests of Borneo (last logging rotation occurred ~15 years prior to the study) had higher levels of the antioxidant enzyme superoxide dismutase and higher values of the GSH/GSSG ratio compared to conspecific birds living in unlogged forests (Fig. 5.8). However, neither oxidative damage nor oxidised glutathione differed between birds living in the two forest types, indicating that the

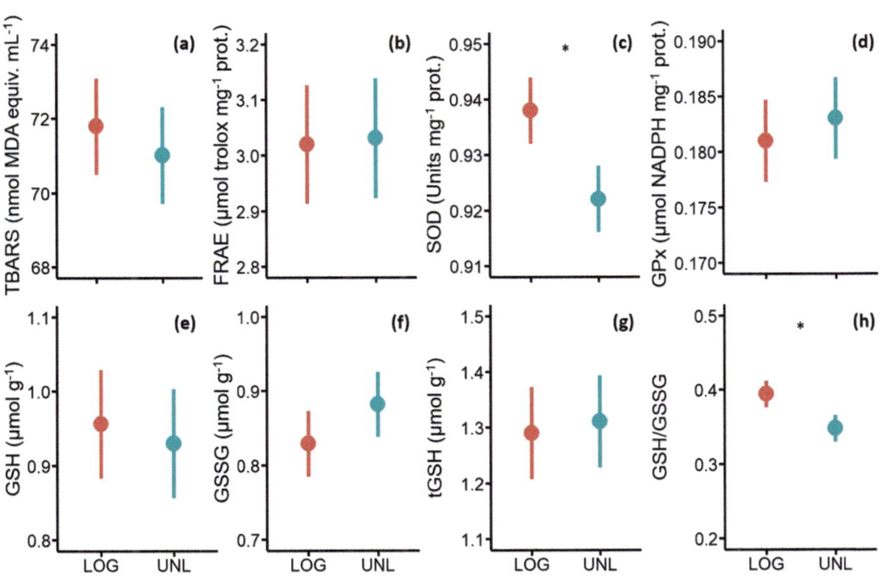


**Fig. 5.8** Differences in blood-based oxidative status markers between birds living in unlogged (UNL) and logged (LOG) tropical forests of Borneo. (**a**) TBARS, thiobarbituric acid reactive substances in plasma; (**b**) FRAE, non-enzymatic antioxidant capacity in erythrocytes; (**c**) SOD, superoxide dismutase in erythrocytes; (**d**) GPx, glutathione peroxidase in erythrocytes; (**e**) GSH, reduced glutathione in erythrocytes; (**f**) GSSG, oxidised glutathione in erythrocytes; (**g**) tGSH, total glutathione in erythrocytes; (**g**) ratio of reduced glutathione (GSH) to oxidised glutathione (GSSG). Asterisks indicate significant differences between forest types. Values are shown as least square mean and standard error. Reprinted with minor modifications from Messina et al. (2020) with permission of John Wiley and Sons

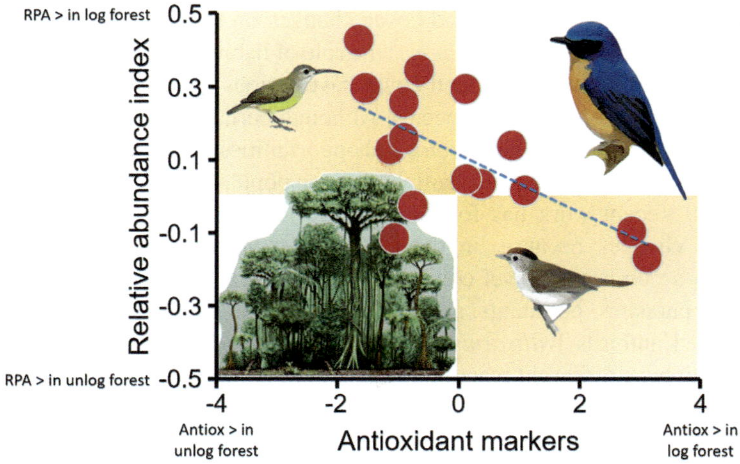


**Fig. 5.9** Relationship between changes in the relative population abundance (RPA) of 15 understorey species (positive values indicate higher abundance of a given species in logged than in unlogged forest) and markers of antioxidant capacity represented by the principal component axis [positive values mean higher antioxidants (glutathione, glutathione peroxidase, non-enzymatic antioxidant capacity in erythrocytes) of a given species in logged than in unlogged forest]. Reprinted with minor modifications from Messina et al. (2020) with permission of John Wiley and Sons

impact of forest logging on oxidative status might have been mild. Importantly, those bird species that had higher levels of certain antioxidant markers (glutathione peroxidase, glutathione, and non-enzymatic antioxidant capacity) in a given type of forest compared to the other forest type were also less abundant in that forest type compared to the other forest type (Fig. 5.9). In conclusion, these correlative results pointed to a potential negative association between some antioxidants and population outcomes.

Differences in behaviour or stress physiology were also found between urban and rural populations, suggesting the importance of human activities as new selective agents for wildlife (e.g. Partecke et al. 2006; Herrera-Dueñas et al. 2014; Miranda et al. 2013; Sol et al. 2013; Costantini et al. 2014a; Watson et al. 2017; Murray et al. 2019). Compared to rural populations, urban populations of common side-blotched lizards *Uta stansburiana* showed higher levels of stress hormones, which were associated with higher oxidative damage levels and suppressed immunity. Moreover, urban lizards had also a higher reproductive output and a lower survival than rural lizards, suggesting a trade-off between reproduction and life maintenance processes (Lucas and French 2012). A comparative study on birds found that rural avian populations had higher concentrations of vitamin E and carotenoids in liver than conspecific urban populations, probably owing to the different diets of rural and urban birds (Møller et al. 2010).

In addition to changes in trophic ecology or reproduction, there are also likely endogenous reasons that underlie differences in oxidative status between rural and urban conspecific populations. Costantini et al. (2014a) found that, under common garden conditions, city blackbirds *Turdus merula* had lower levels of reactive oxygen metabolites during stressful conditions compared with rural blackbirds, but also tended to generally maintain lower levels of non-enzymatic and enzymatic antioxidants in blood than rural blackbirds. Interestingly, previous work found that rural blackbirds exposed to a standardised stressor responded with a greater secretion of stress hormones than urban conspecifics (Partecke et al. 2006). These findings suggest that blackbirds from urban and rural areas differed intrinsically in the regulation of blood oxidative status, possibly as an adaptation to city life. Herrera-Dueñas et al. (2017) found that urban sparrows *Passer domesticus* had higher levels of oxidative damage and higher activity of antioxidant enzymes, but lower antioxidant capacity in comparison with the rural sparrows, particularly during the breeding season. Analysis of differentially expressed genes (transcriptome) showed that urban great tits faced greater oxidative stress than rural great tits (Watson et al. 2017). Using a cross-fostering experiment, Salmón et al. (2018) found that being reared in the urban environment triggered an increase in red blood cell levels of superoxide dismutase in nestling great tits, independently from the natal habitat. In addition, plasma levels of malondialdehyde increased with hatching date in urban-reared nestlings, but there was little seasonal change in rural-reared nestlings. Finally, a common garden study involving 62 genotypes of *Daphnia magna* from replicated urban and rural populations revealed that rural *Daphnia* had significantly higher baseline levels of the antioxidant enzymes superoxide dismutase and glutathione-S-transferase compared with city populations, yet urban animals were equally well protected against lipid peroxidation (Brans et al. 2018). In conclusion, the regulation of oxidative status homeostasis plays a relevant role in the adaptability of animals to urban environments.

## 5.10 Special Adaptations to Local Environments

We are now familiar with a number of studies that documented significant differences in organismal oxidative status between conspecific populations or species living in different environments. But do we have evidence that it was the local environmental pressures to mould the organismal resistance against or tolerance of oxidative stress? Or is this population differentiation owing to variation in occupant quality among habitats?

Ultraviolet radiation largely varies across environments, with low levels normally occurring at polar circles. Hence, polar species have evolved under conditions of low radiation, so that they are expected to be more vulnerable to ultraviolet radiation than species living elsewhere. Lamare et al. (2007) compared the effects of ultraviolet radiation on the rate of DNA damage (cyclobutane pyrimidine dimer production) and abnormal development between embryos of four sea urchin species: *Sterechinus neumayeri* from polar regions, *Evechinus chloroticus* from temperate regions, *Diadema savignyi* and *Tripneustes gratilla* from tropical regions. Levels of DNA damage were significantly higher in embryos of the polar sea urchin compared to the other species (Fig. 5.10). Moreover, embryos of the polar sea urchin suffered higher rates of abnormal development compared to the other species. Further work showed that polar sea urchins are equipped with low concentrations of sunscreens and slow mechanisms of DNA repair compared with temperate and tropical species (Lamare et al. 2006, 2007; Lesser et al. 2006). The degree of resistance against ultraviolet radiation can also be influenced by prior experiences of radiation stress. For example, resistance to UV radiation of *Hydra oligactis* was increased by a prior short and sublethal irradiation (Tökölyi et al. 2017). There is, however, variation in how *Hydra* species respond to environmental stressors, and this variation appears to be linked to reproductive strategies, because *Hydra* species with a higher rate of asexual reproduction showed lower resistance to experimentally induced oxidative stress (Tökölyi et al. 2016).

Ambient temperature is another abiotic factor that largely varies across environments. Synthesis of heat shock proteins is a highly conserved response to heat stress. However, we have evidence that some species lack or have an attenuated heat shock response (e.g. Bosch et al. 1988; Pöhlmann et al. 2011). For example, Bosch et al. (1988) found that the freshwater cnidarian brown hydra *Hydra oligactis* did not synthesise any heat shock proteins in response to thermal stress. One reason for

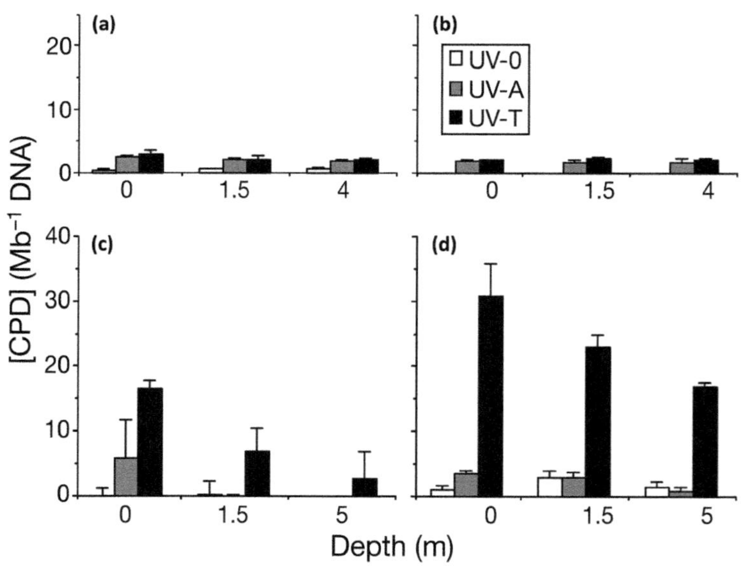


**Fig. 5.10** Concentrations of cyclobutane pyrimidine dimer (CPD, marker of DNA damage) in embryos of sea urchins (**a**) *Diadema savignyi* (Cook Islands), (**b**) *Tripneustes gratilla* (Cook Islands), (**c**) *Evechinus chloroticus* (from New Zealand), and (**d**) *Sterechinus neumayeri* (from Antarctica) as a function of UV-treatment at three experimental depths (0.5, 1.5, and 4.0 or 5.0 m). UV-0 treatment: photosynthetically active radiation (PAR, 400–700 nm), but not ultraviolet radiation (UV); UV-A treatment: PAR + UV-A (320–400 nm) but no UV-B (300–320 nm); UV-T treatment: PAR + UV-A + UV-B. Values are shown as mean and standard error. Reprinted with minor modifications from Lamare et al. (2007) with permission of Inter-Research (copyright holder)

this result might lie with the cold and thermally stable conditions that characterise the environments where *Hydra oligactis* lives.

Similarly, *Notothenioidei* polar fish are extreme stenotherms that have evolved in the cold and thermally stable waters of coastal Antarctica (e.g. Eastman 1993; Regoli et al. 2012). Hofmann et al. (2000) found that emerald rockcods *Trematomus bernacchii* are unable of upregulating the synthesis of heat shock proteins in response to increased water temperature. Polar white-blooded icefishes (those that lost haemoglobin) are more vulnerable to temperature-induced oxidative stress than temperate red-blooded notothenioids. Antarctic icefishes have higher densities of phospholipids in their mitochondria compared with red-blooded species (O'Brien and Mueller 2010). This high density of phospholipids in oxidative muscles enhances the capacity of icefishes to deliver oxygen, even in the absence of haemoglobin and myoglobin (O'Brien 2011). However, these phospholipids are rich in polyunsaturated fatty acids, which make tissues more vulnerable to lipid peroxidation (Mueller et al. 2011). In addition, icefishes have lower levels of antioxidants compared to related temperate species (Witas et al. 1984; Cassini et al. 1993; Abele et al. 2012). Mueller et al. (2012) found that levels of protein carbonyls and thiobarbituric acid reactive substances in the heart ventricle increased in icefishes compared to red-blooded species when exposed to increased water temperature. Despite an increase in oxidative damage, icefishes did not upregulate their antioxidant response. Similarly, exposure of notothenioid fishes to both high temperature and partial pressure of $CO_2$ increased their levels of protein carbonyls, but it did not influence levels of antioxidant enzymes (Enzor and Place 2014). Although the notothenioid *Pagothenia borchgrevinki* has the ability to increase levels of endogenous antioxidants in response to an increase in water temperature, this response was transient and insufficient to protect against oxidative damage (Fig. 5.11) (Carney Almroth et al. 2015). Overall, the results of these studies indicated that notothenioid fishes have limited capacity to cope with oxidative stress.

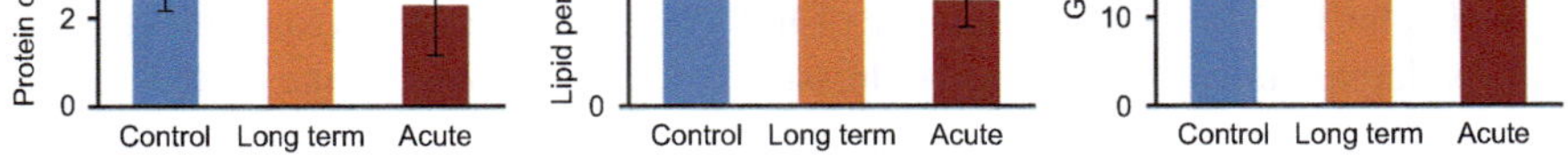


**Fig. 5.11** Levels of (**a**) protein oxidation measured as protein carbonyls, (**b**) lipid peroxidation measured as a mixture of different compounds, and (**c**) oxidised glutathione (GSSG) in liver of bald notothen *Pagothenia borchgrevinki* exposed to ambient temperature (−1.6 °C) for 3 weeks (Control), 4 °C for 3 weeks (Long term), and 4 °C for 12 h (acute), respectively. Letters indicate statistically significant differences. Values are shown as mean and standard error. Reprinted with minor modifications from Carney Almroth et al. (2015) with permission of Elsevier

Temperate and tropical birds also show many differences in their physiology that are likely linked to their life histories. Tropical birds have a slower pace of life compared to temperate species, which is characterised by smaller clutches, slower grow rates, and longer lives (Ricklefs 1976; Cardillo 2002). Wiersma et al. (2007a, b, 2012) found that tropical birds have lower basal, exercise-induced, and cold-induced metabolic rates, and smaller organs than temperate species. Comparative studies on conspecific populations that live in both tropical and temperate regions provided similar results. Relying on a common garden experiment, Wikelski et al. (2003) found that stonechats *Saxicola torquata* from tropical regions had lower resisting metabolic rates compared to stonechats from other regions. Further work also found physiological differences at cellular level between tropical and temperate birds. Finally, Jimenez et al. (2013) found that cultured dermal fibroblasts of tropical birds were more resistant against a number of pro-oxidant agents than cultured dermal fibroblasts of temperate birds.

Comparative studies of conspecific populations also provided a wealth of information about the role of local environmental pressures in moulding the regulation of oxidative status. For example, Nikinmaa et al. (2013) found divergence in both constitutive and induced levels of antioxidant mechanisms among three populations of three-spined stickleback *Gasterosteus aculeatus*. Amphibians are particularly suitable to assess adaptive divergence because conspecific populations of species with a large distribution may experience highly contrasted environments in terms of abiotic conditions. A relevant example comes from a study of Burraco et al. (2017), who found that tadpoles of the common frog *Rana temporaria* from islands with temporary pools had accelerated development compared to tadpoles from permanent pools, but they had lower levels of catalase and glutathione reductase, and also shorter telomeres as a consequence. Frogs also show remarkable physiological adaptations to life at high altitude. For example, high-altitude adult Xizang plateau frog *Nanorana parkeri* showed lower levels of lipid peroxidation and protein carbonyls, and higher levels of superoxide dismutase, catalase, glutathione peroxidase, glutathione-S-transferase, and non-enzymatic antioxidants in the liver compared to low-altitude adults (Niu et al. 2022).

## 5.11 Conclusions and Future Directions

This chapter has illustrated a number of links between abiotic factors and organismal oxidative status. It has also addressed how a dramatic change in the habitat structure (e.g. degradation, transformation) generates a number of downstream challenges for the organismal homeostasis. We have seen that many species express shared (or conserved) mechanisms in response to changes in abiotic conditions. Many comparative studies involving different species or conspecific populations also showed that the expression of certain physiological responses is peculiar or private to particular evolutionary lineages. Understanding how these physiological mechanisms are expressed in response to abiotic stressors and why they vary across taxa is very important to infer about the organismal capability to adjust to the ongoing environmental changes. For example, the rate at which environments change may be too rapid for genetic adaptation to kick in, particularly in those species with long generation times. Hence, plasticity of oxidative status regulation may represent one crucial short-term compensatory response (e.g. effect of a simulated heat wave on oxidative status in Stahlschmidt et al. 2017) that enables organisms to survive.

This chapter has also identified a number of limitations for our understanding of how and why changes in abiotic conditions jeopardise the homeostasis of organismal oxidative status. One of the main limitations is the shortage of experimental studies that tested the additive, synergistic, or antagonistic effects of multiple abiotic factors (cocktail effect) on organismal physiology. Another of these main limitations is that most studies exposed animals to constant conditions. This experimental setting complicates the extrapolation of experimental results to natural habitats, where conditions are much more variable (e.g. Niehaus et al. 2012; Sørensen et al. 2013; Jimenez et al. 2015).

It is also very important to assess both short- and long-term physiological responses to environmental stressors because we know little about the pace and timing of acclimatisation. For example, short-term acclimatisation to decreased seawater pH and increased temperature resulted in significant increases in oxidative damage to proteins in three notothenioid species (Enzor and Place 2014). By contrast, levels of oxidative damage returned to near basal levels or, even decreased, after long-term acclimatisation (Enzor and Place 2014).

Interactions occurring between the insect and its host plant are another interesting, but poorly explored topic, for understanding the processes underlying physiological responses of organisms to their environments. For example, Durak et al. (2018) found that switching a host plant by aphids activates the upregulation of the antioxidant enzymes superoxide dismutase, catalase, and glutathione S-transferase in the tissues of *Cinara tujafilina*. Finally, a range of amphibian species are able of modifying their body colourations in order to match background colour (chromatic adaptation). Radovanović et al. (2023) found that (1) prolonged exposure of larvae of the European tree frog *Hyla arborea* to a constantly dark habitat led to significant changes in antioxidant levels compared to larvae kept continuously on a white background, (2) darker-pigmented larvae exhibited greater oxidative damage to lipids, and (3) a change in body colouration induced by experimental change of the background colouration caused increased oxidative damage to lipids. Thus, in addition to macro-scale investigations of the relationships between abiotic conditions and organismal oxidative status, we need more investigations of the significance of micro-scale environmental variation in generating diversity in physiological functions.

## References

Abele D (2012) Temperature adaptation in changing climate: marine fish and invertebrates. In: Storey KB, Tanino KK (eds) Temperature adaptation in a changing climate: nature at risk. Springer, Berlin, pp 67–79

Abele D, Puntarulo S (2004) Formation of reactive species and induction of antioxidant defence systems in polar and temperate marine invertebrates and fish. Comp Biochem Physiol A 138:405–415

Abele D, Heise K, Pörtner HO, Puntarulo S (2002) Temperature-dependence of mitochondrial function and production of reactive oxygen species in the intertidal mud clam *Mya arenaria*. J Exp Biol 205:1831–1841

Abele D, Kruppe M, Brey T, Philipp EER (2010) Mantle cavity water oxygen partial pressure (PO2) in marine molluscs aligns with lifestyle. Can J Fisher Aquat Sci 67:977–986

Abele D, Vázquez-Medina JP, Zenteno-Savín T (2012) Oxidative stress in aquatic ecosystems. Wiley-Blackwell

Abele-Oeschger D, Oeschger R (1995) Hypoxia-induced autoxidation of haemoglobin in the benthic invertebrates *Arenicola marina* (Polychaeta) and *Astarte borealis* (Bivalvia) and the possible effects of sulphide. J Exp Mar Biol Ecol 187:63–80

Abele-Oeschger D, Oeschger R, Theede H (1994) Biochemical adaptations of *Nereis diversicolor* (Polychaeta) to temporarily increased hydrogen peroxide levels in intertidal sandflats. Mar Ecol Prog Ser 106:101–110

Adams N, Shick JM (1996) Mycosporine-like amino acids provide protection against ultraviolet radiation in eggs of the green sea urchin *Strongylocentrotus droebachiensis*. Photochem Photobiol 64:149–158

Adams N, Shick JM (2001) Mycosporine-like amino acids prevent UVB induced abnormalities during early development of the green sea urchin *Strongylocentrotus droebachiensis*. Mar Biol 138:267–280

Akhtar MS, Pal AK, Sahu NP, Ciji A, Mahanta PC (2013) Thermal tolerance, oxygen consumption and haemato-biochemical variables of Tor putitora juveniles acclimated to five temperatures. Fish Physiol Biochem 39:1387–1398

Akhtar MS, Rajesh M, Kamalam BS, Ciji A (2020) Effect of photoperiod and temperature on indicators of immunity and wellbeing of endangered golden mahseer (Tor putitora) broodstock. J Therm Biol 93:102694

Alaluf S, Heinrich U, Stahl W, Tronnier H, Wiseman S (2002) Dietary carotenoids contribute to normal human skin colour and UV photosensitivity. J Nutr 132:399–403

Al-Otaiba A, John A, Al-Belooshi T, Raza H (2010) Redox homeostasis and respiratory metabolism in camels (*Camelus dromedaries*): comparisons with domestic goats and laboratory rats and mice. J Comp Physiol B 180:1121–1132

An KW, Kim NN, Shin HS, Kil GS, Choi CY (2010) Profiles of antioxidant gene expression and physiological changes by thermal and hypoosmotic stresses in black porgy (*Acanthopagrus schlegelii*). Comp Biochem Physiol A 156:262–268

Angelier F, Vleck CM, Holberton RL, Marra PP (2013) Telomere length, non-breeding habitat and return rate in male American redstarts. Funct Ecol 27:342–350

Angilletta MJ (2009) Thermal adaptation: a theoretical and empirical synthesis. Oxford University Press, Oxford

Antonopoulou E, Chatzigiannidou I, Feidantsis K, Kounna C, Chatzifotis S (2020) Effect of water temperature on cellular stress responses in meagre (*Argyrosomus regius*). Fish Physiol Biochem 46:1075–1091

Bagnyukova TV, Danyliv SI, Zin'ko OS, Lushchak VI (2007) Heat shock induces oxidative stress in rotan *Perccottus glenii* tissues. J Therm Biol 32:255–260

Bal A, Panda F, Pati SG, Das K, Agrawal PK, Paital B (2021) Modulation of physiological oxidative stress and antioxidant status by abiotic factors especially salinity in aquatic organisms. Comp Biochem Physiol C 241:108971

Bancroft BA, Baker NJ, Blaustein AR (2007) Effects of UVB radiation on marine and freshwater organisms: a synthesis through meta-analysis. Ecol Lett 10:332–345

Bancroft BA, Baker NJ, Blaustein AR (2008) A meta-analysis of the effects of ultraviolet B radiation and its synergistic interactions with pH, contaminants, and disease on amphibian survival. Cons Biol 22:987–996

Beaulieu M, Gillen E, Hahn S, Pape JM, Fischer K (2015) Behavioural antioxidant strategies to cope with high temperatures: a study in a tropical butterfly. Anim Behav 109:89–99

Beaulieu M, Touzalin F, Dool SE, Teeling EC, Puechmaille SJ (2020) Timescale and colony-dependent relationships between environmental conditions and plasma oxidative markers in a long-lived bat species. Conserv Physiol 8:coaa083

Bellantuono AJ, Granados-Cifuentes C, Miller DJ, Hoegh-Guldberg O, Rodriguez-Lanetty M (2012) Coral thermal tolerance: tuning gene expression to resist thermal stress. PLoS One 7:e50685

Blount JD, Pike TW (2012) Deleterious effects of light exposure on immunity and sexual coloration in birds. Funct Ecol 26:37–45

Bosch TCG, Krylow SM, Bode HR, Steele RE (1988) Thermotolerance and synthesis of heat shock proteins: these responses are present in *Hydra attenuata* but absent in *Hydra oligactis*. Proc Natl Acad Sci USA 85:7927–7931

Brans KI, Stoks R, De Meester L (2018) Urbanization drives genetic differentiation in physiology and structures the evolution of pace-of-life syndromes in the water flea Daphnia magna. Proc R Soc Lond B 285:20180169

Brauner CJ, Baker DW (2010) Patterns of acid-base regulation during exposure to hypercarbia in fishes. In: Glass ML, Woods SC (eds) Cardi-respiratory control in vertebrates. Springer, Berlin, pp 43–63

Brooks SP, Storey KB (1997) Glycolytic controls in estivation and anoxia: a comparison of metabolic arrest in land and marine molluscs. Comp Biochem Physiol A 118:1103–1114

Brown BE, Downs CA, Dunne RP, Gibb SW (2002a) Exploring the basis of thermotolerance in the reef coral *Goniastrea aspera*. Mar Ecol Prog Ser 242:119–129

Brown BE, Dunne RP, Goodson MS, Douglas AE (2002b) Experience shapes the susceptibility of a reef coral to bleaching. Coral Reefs 21:119–126

Burraco P, Gomez-Mestre I (2016) Physiological stress responses in amphibian larvae to multiple stressors reveal marked anthropogenic effects even below lethal levels. Physiol Biochem Zool 89:462–472

Burraco P, Valdés AE, Johansson F, Gomez-Mestre I (2017) Physiological mechanisms of adaptive developmental plasticity in Rana temporaria Island populations. BMC Evol Biol 17:164

Burraco P, Iglesias-Carrasco M, Cabido C, Gomez-Mestre I (2018) Eucalypt leaf litter impairs growth and development of amphibian larvae, inhibits their antipredator responses and alters their physiology. Conserv Physiol 6:coy066

Capobianco F III, Nandkumar S, Parker JD (2018) *Wolbachia* affects survival to different oxidative stressors dependent upon the genetic background in *Drosophila* melanogaster. Physiol Entomol 43:239–244

Cardillo M (2002) The life-history basis of latitudinal diversity gradients: how do species traits vary from the poles to the equator? J Anim Ecol 71:79–87

Carney Almroth B, Asker N, Wassmur B, Rosengren M, Jutfelt F, Gräns A, Sundell K, Axelsson M, Sturve J (2015) Warmer water temperature results in oxidative damage in an Antarctic fish, the bald notothen. J Exp Mar Biol Ecol 468:130–137

Carney Almroth B, de Souza KB, Jonsson E, Sturve J (2019) Oxidative stress and biomarker responses in the Atlantic halibut after long term exposure to elevated CO2 and a range of temperatures. Comp Biochem Physiol B 238:110321

Caro SP, Schaper SV, Hut RA, Ball GF, Visser ME (2013) The case of the missing mechanism: how does temperature influence seasonal timing in endotherms? PLoS Biol 11:e1001517

Cassini A, Favero M, Albergoni V (1993) Comparative studies of antioxidant enzymes in red-blooded and white-blooded Antarctic teleost fish, *Pagothenia bernacchii* and *Chionodraco hamatus*. Comp Biochem Physiol C 106:333–336

Chandel NS, Maltepe E, Goldwasser E, Mathieu CE, Simon MC, Schumacker PT (1998) Mitochondrial reactive oxygen species trigger hypoxia-induced transcription. Proc Natl Acad Sci USA 95:11715–11720

Chang CH, Mayer M, Rivera-Ingraham G, Blondeau-Bidet E, Wu WY, Lorin-Nebel C, Lee TH (2021) Effects of temperature and salinity on antioxidant responses in livers of temperate

(*Dicentrarchus labrax*) and tropical (*Chanos Chanos*) marine euryhaline fish. J Therm Biol 99:103016

Charron RA, Fenwick JC, Lean DR, Moon TW (2000) Ultraviolet-B radiation effects on antioxidant status and survival in the zebrafish, Brachydanio rerio. Photochem Photobiol 72:327–333

Chen BJ, Niu CJ, Yuan L (2015) Ascorbic acid regulation in stress responses during acute cold exposure and following recovery in juvenile Chinese soft-shelled turtle (*Pelodiscus sinensis*). Comp Biochem Physiol A 184:20–26

Chen BJ, Zhang WY, Niu CJ, Li WJ, Jia H, Storey KB (2019) Antioxidant response to acute cold exposure and during recovery in juvenile Chinese soft-shelled turtles (*Pelodiscus sinensis*). J Exp Biol 222:jeb197863

Cheviron ZA, Whitehead A, Brumfield RT (2008) Transcriptomic variation and plasticity in rufous-collared sparrows (Zonotrichia capensis) along an altitudinal gradient. Mol Ecol 17:4556–4669

Clanton TL (2007) Hypoxia-induced reactive oxygen species formation in skeletal muscle. J Appl Physiol 102:2379–2388

Contran N, Guenthardt-Georg MS, Kuster TM, Cerana R, Crosti P, Paoletti E (2013) Physiological and biochemical responses of *Quercus pubescens* to air warming and drought on acidic and calcareous soils. Plant Biol 15:157–168

Costantini D (2008) Oxidative stress in ecology and evolution: lessons from avian studies. Ecol Lett 11:1238–1251

Costantini D (2014) Oxidative stress and hormesis in evolutionary ecology and physiology: a marriage between mechanistic and evolutionary approaches. Springer-Verlag, Berlin, p 348

Costantini D, Metcalfe NB, Monaghan P (2010) Ecological processes in a hormetic framework. Ecol Lett 13:1435–1447

Costantini D, Monaghan P, Metcalfe N (2012) Early life experience primes resistance to oxidative stress. J Exp Biol 215:2820–2826

Costantini D, Greives TJ, Hau M, Partecke J (2014a) Does urban life change blood oxidative status in birds? J Exp Biol 217:2994–2997

Costantini D, Monaghan P, Metcalfe N (2014b) Prior hormetic priming is costly under environmental mismatch. Biol Lett 10:20131010

Cypser JR, Johnson TE (2002) Multiple stressors in *Caenorhabditis elegans* induce stress hormesis and extended longevity. J Gerontol A 57:B109–B114

Czajka M, Lee RE (1990) A rapid cold-hardening protection against cold shock injury in *Drosophila melanogaster*. J Exp Biol 148:245–254

D'Aoust BG, White R, Wells JM, Olsen DA (1976) Coral-algal associations: capacity for producing and sustaining elevated oxygen tensions in situ. Undersea Biomed Res 3:35–40

Dabrowski K, Lee KJ, Guz L, Verlhac V, Gabaudan J (2004) Effects of dietary ascorbic acid on oxygen stress (hypoxia or hyperoxia), growth and tissue vitamin concentrations in juvenile rainbow trout (*Oncorhynchus mykiss*). Aquaculture 233:383–392

de Oliveira UO, da Rosa Araújo AS, Belló-Klein A, da Silva RSM, Kucharskia LC (2005) Effects of environmental anoxia and different periods of reoxygenation on oxidative balance in gills of the estuarine crab *Chasmagnathus granulata*. Comp Biochem Physiol B 140:51–57

Defo MA, Gendron AD, Head J, Pilote M, Turcotte P, Marcogliese DJ, Houde M (2019) Cumulative effects of cadmium and natural stressors (temperature and parasite infection) on molecular and biochemical responses of juvenile rainbow trout. Aquat Toxicol 217:105347

Deschaseaux ESM, Taylor AM, Maher WA, Davis AR (2010) Cellular responses of encapsulated gastropod embryos to multiple stressors associated with climate change. J Exp Mar Biol Ecol 383:130–136

Deschaseaux E, Taylor A, Maher W (2011) Measure of stress response induced by temperature and salinity changes on hatched larvae of three marine gastropod species. J Exp Mar Biol Ecol 397:121–128

Dettleff P, Zuloaga R, Fuentes M, Gonzalez P, Aedo J, Estrada JM, Molina A, Valdes JA (2020) Physiological and molecular responses to thermal stress in red cusk-eel (*Genypterus chilensis*) juveniles reveals atrophy and oxidative damage in skeletal muscle. J Therm Biol 94:102750

Dezetter M, Le Galliard JF, Guiller G, Guillon M, Leroux-Coyau M, Meylan S, Brischoux F, Angelier F, Lourdais O (2021) Water deprivation compromises maternal physiology and reproductive success in a cold and wet adapted snake *Vipera berus*. Conserv Physiol 9:coab071

Ding Z, Wang X, Zou T, Hao X, Zhang Q, Sun B, Du W (2023) Climate warming has divergent physiological impacts on sympatric lizards. Sci Total Environ 912:168992

Dirmeier R, O'Brien KM, Engle M, Dodd A, Spears E, Poyton RO (2002) Exposure of yeast cells to anoxia induces transient oxidative stress implications for the induction of hypoxic genes. J Biol Chem 277:34773–34784

Dolci GS, Dias VT, Roversi K, Roversi K, Pase CS, Segat HJ, Teixeira AM, Benvegnú DM, Trevizol F, Barcelos RC, Riffel AP, Nunes MA, Dressler VL, Flores EM, Baldisserotto B, Bürger ME (2013) Moderate hypoxia is able to minimize the manganese-induced toxicity in tissues of silver catfish (*Rhamdia quelen*). Ecotoxicol Environ Saf 91:103–109

Dontsov A, Fedorovich I, Lindström M, Ostrovsky MA (1999) Comparative study of spectral and antioxidant properties of pigments from the eyes of two *Mysis relicta* (Crustacea, Mysidacea) populations, with different light damage resistance. J Comp Physiol B 169:157–164

Dunn SR, Bythell JC, Le Tessier DA, Burnett WJ, Thomason JC (2002) Programmed cell death and necrosis activity during hyperthermic stress-induced bleaching of the symbiotic sea anemone *Aiptasia* sp. J Exp Mar Biol Ecol 272:29–53

Dupoué A, Angelier F, Ribout C, Meylan S, Rozen-Rechels D, Decencière B, Agostini S, Le Galliard J-F (2020) Chronic water restriction triggers sex-specific oxidative stress and telomere shortening in lizards. Biol Lett 16:20190889

Durak R, Molon M, Durak T, Chrzanowski G (2018) The enzymatic markers of the adaptation of *Cinara tujafilina* to changing the host plant. Ethol Ecol Evol 30:416–429

Dykens JA, Shick JM (1982) Oxygen production by endosymbiotic algae controls superoxide dismutase activity in their animal host. Nature 297:579–580

Dykens JA, Shick JM (1984) Photobiology of the symbiotic sea anemone *Anthopleura elegantissima*: defenses against photodynamic effects and seasonal photoacclimatization. Biol Bull 167:683–697

Eastman JT (1993) Antarctic fish biology: evolution in a unique environment. Academic Press, New York

Enzor LA, Place SP (2014) Is warmer better? Decreased oxidative damage in notothenioid fish after long-term acclimation to multiple stressors. J Exp Biol 217:3301–3310

Eraud C, Devevey G, Gaillard M, Prost J, Sorci G, Faivre B (2007) Environmental stress affects the expression of a carotenoid-based sexual trait in male zebra finches. J Exp Biol 210:3571–3578

Faria AM, Lopes AF, Silva CSE, Novais SC, Lemos MFL, Gonçalves EJ (2018) Reproductive trade-offs in a temperate reef fish under high pCO2 levels. Mar Environ Res 137:8–15

Feder ME, Hofmann GE (1999) Heat-shock proteins, molecular chaperones, and the stress response: evolutionary and ecological physiology. Annu Rev Physiol 61:243–282

Feller G (2013) Psychrophilic enzymes: from folding to function and biotechnology. Scientifica 2013:512840

Ferreira-Cravo M, Moreira DC, Hermes-Lima M (2023) Glutathione depletion disrupts redox homeostasis in an anoxia-tolerant invertebrate. Antioxidants 12:1197

Flanagan SW, Moseley PL, Buettner GR (1998) Increased flux of free radicals in cells subjected to hyperthermia: detection by electron paramagnetic resonance spin trapping. FEBS Lett 431:285–286

Florencio M, Burraco P, Rendon MA, Diaz-Paniagua C, Gomez-Mestre I (2020) Opposite and synergistic physiological responses to water acidity and predator cues in spadefoot toad tadpoles. Comp Biochem Physiol A 242:110654

França MB, Panek AD, Eleutherio ECA (2007) Oxidative stress and its effects during dehydration. Comp Biochem Physiol A 146:621–631

Freire CA, Welker AF, Storey JM, Storey KB, Hermes-Lima M (2012) Oxidative stress in estuarine and intertidal environments (temperate and tropical). In: Abele D, Vázquez-Medina JP, Zenteno-Savín T (eds) Oxidative stress in aquatic ecosystems. Wiley-Blackwell, pp 41–57

Freitas JS, Almeida EA (2016) Antioxidant defense system of tadpoles (*Eupemphix nattereri*) exposed to changes in temperature and pH. Zool Sci 33:186–194

Freitas JS, Felício AA, Teresa FB, Alves de Almeida E (2017a) Combined effects of temperature and clomazone (Gamit®) on oxidative stress responses and B-esterase activity of *Physalaemus nattereri* (Leiuperidae) and *Rhinella schneideri* (Bufonidae) tadpoles. Chemosphere 185:548–562

Freitas JS, Teresa FB, de Almeida EA (2017b) Influence of temperature on the antioxidant responses and lipid peroxidation of two species of tadpoles (*Rhinella schneideri* and *Physalaemus nattereri*) exposed to the herbicide sulfentrazone (Boral 500SC I). Comp Biochem Physiol C 197:32–44

Garrido-Bautista J, Soria A, Trenzado CE, Pérez-Jiménez A, Ros-Santaella JL, Pintus E, Bernardo N, Comas M, Moreno-Rueda G (2021) Oxidative status of blue tit nestlings varies with habitat and nestling size. Comp Biochem Physiol A 258:110986

Geihs MA, Moreira DC, López-Martínez G, Minari M, Ferreira-Cravo M, Carvajalino-Fernández JM, Hermes-Lima M (2020) Commentary: ultraviolet radiation triggers "preparation for oxidative stress" antioxidant response in animals: similarities and interplay with other stressors. Comp Biochem Physiol A 239:11058

Ghanizadeh-Kazerouni E, Franklin CE, Seebacher F (2017) Living in flowing water increases resistance to ultraviolet B radiation. J Exp Biol 220:582–587

Gianese G, Argos P, Pascarella S (2001) Structural adaptation of enzymes to low temperatures. Prot Eng 14:141–148

Giovannini I, Boothby TC, Cesari M, Goldstein B, Guidetti R, Rebecchi L (2022) Production of reactive oxygen species and involvement of bioprotectants during anhydrobiosis in the tardigrade *Paramacrobiotus spatialis*. Sci Rep 12:1938

Giraud-Billoud M, Rivera-Ingraham GA, Moreira DC, Burmester T, Castro-Vazquez A, Carvajalino-Fernández JM, Dafre A, Niu C, Tremblay N, Paital B, Rosa R, Storey JM, Vega IA, Zhang W, Yepiz-Plascencia G, Zenteno-Savin T, Storey KB, Hermes-Lima M (2019) Twenty years of the 'preparation for oxidative stress' (POS) theory: ecophysiological advantages and molecular strategies. Comp Biochem Physiol A 234:36–49

Giuliani ME, Nardi A, Di Carlo M, Benedetti M, Regoli F (2021) Transcriptional and catalytic responsiveness of the Antarctic fish *Trematomus bernacchii* antioxidant system toward multiple stressors. Antioxidants 10:410

Gomes ADLS, Gonçalves AFG, Vieira JLF, Marceliano MLV, Silva JMCD (2014) Natural gaps associated with oxidative stress in *Willisornis poecilinotus* (Aves: Thamnophilidae) in a tropical forest. Acta Amaz 44:207–212

Gomez-Mestre I, Kulkarni S, Buchholz DR (2013) Mechanisms and consequences of developmental acceleration in tadpoles responding to pond drying. PLoS One 8:e84266

González PM, Wilhelms-Dick D, Abele D, Puntarulo S (2012) Iron in coastal marine ecosystems: role in oxidative stress. In: Abele D, Vázquez-Medina JP, Zenteno-Savín T (eds) Oxidative stress in aquatic ecosystems. Wiley-Blackwell, pp 115–126

González-Ruiz R, Leyva-Carrillo L, Molina DC, López JH, Yepiz-Plascencia G (2023) Antioxidant stress response to fluctuations of dissolved oxygen and temperature in a semi-intensive aquaculture shrimp farm during high summer temperature. Invert Surv J 20:65–78

Grubor-Lajsic G, Block W, Telesmanic M, Jovanovic A, Stevanovic D, Baca F (1997) Effect of cold acclimation on the antioxidant defense system of two larval lepidoptera (noctuidae). Arch Insect Biochem Physiol 36:1–10

Guillen AC, Borges ME, Herrerias T, Kandalski PK, de Souza M, Donatti L (2022) Gradual increase of temperature trigger metabolic and oxidative responses in plasma and body tissues in the Antarctic fish *Notothenia rossii*. Fish Physiol Biochem 48:337–354

Halliwell BH, Gutteridge JMC (2015) Free radicals in biology and medicine, 5th edn. Oxford University Press, Oxford

Haussmann MF, Marchetto NM (2010) Telomeres: linking stress and survival, ecology and evolution. Curr Zool 56:703–713

Hawkins TD, Warner ME (2017) Warm preconditioning protects against acute heat-induced respiratory dysfunction and delays bleaching in a symbiotic sea anemone. J Exp Biol 220:969–983
Hazel JR (1995) Thermal adaptation in biological membranes: is homeoviscous adaptation the explanation? Annu Rev Physiol 57:19–42
Herkovits J, D'Eramo JL, Fridman O (2006) The effect of UV-B radiation on *Bufo arenarum* embryos survival and superoxide dismutase activity. Int J Environ Res Public Health 3:43–47
Hermes-Lima M, Storey KB (1993) Antioxidant defenses in the tolerance of freezing and anoxia by garter snakes. Am J Phys 265:R646–R652
Hermes-Lima M, Zenteno-Savín T (2002) Animal response to drastic changes in oxygen availability and physiological oxidative stress. Comp Biochem Physiol C 133:537–556
Hermes-Lima M, Storey JM, Storey KB (1998) Antioxidant defenses and metabolic depression. The hypothesis of preparation for oxidative stress in land snails. Comp Biochem Physiol B 120:437–448
Hermes-Lima M, Storey JM, Storey KB (2001) Antioxidant defenses and animal adaptation to oxygen availability during environmental stress. In: Storey KB, Storey JM (eds) Cell and molecular responses to stress. Elsevier, Amsterdam, pp 263–287
Hermes-Lima M, Moreira DC, Rivera-Ingraham GA, Giraud-Billoud M, Genaro-Mattos TC, Campos ÉG (2015) Preparation for oxidative stress under hypoxia and metabolic depression: revisiting the proposal two decades later. Free Radic Biol Med 89:1122–1143
Herrera-Dueñas A, Pineda J, Antonio MT, Aguirre JI (2014) Oxidative stress of house sparrow as bioindicator of urban pollution. Ecol Indic 42:6–9
Herrera-Dueñas A, Pineda-Pampliega J, Antonio-García MT, Aguirre JI (2017) The influence of urban environments on oxidative stress balance: a case study on the house sparrow in the Iberian Peninsula. Front Ecol Evol 5:106
Heupel MR, Whittier JM, Bennett MB (1999) Plasma steroid hormone profiles and reproductive biology of the epaulette shark, *Hemiscyllium ocellatum*. J Exp Zool 284:586–594
Hill RW, Wyse GA, Anderson M (2008) Animal physiology, 2nd edn. Sinauer Associates, Sunderland
Hofmann GE, Buckley BA, Airaksinen S, Keen JE, Somero GN (2000) Heat-shock protein expression is absent in the Antarctic fish *Trematomus bernacchii* (family Nototheniidae). J Exp Biol 203:2331–2339
Hu B, Simon J, Rennenberg H (2013) Drought and air warming affect the species-specific levels of stress-related foliar metabolites of three oak species on acidic and calcareous soil. Tree Physiol 33:489–504
Ibarz A, Martín-Pérez M, Blasco J, Bellido D, de Oliveira E, Fernández-Borràs J (2010) Gilthead Sea bream liver proteome altered at low temperatures by oxidative stress. Proteomics 10:963–975
Isaksson C (2013) Opposing effects on glutathione and reactive oxygen metabolites of sex, habitat, and spring date, but no effect of increased breeding density in great tits (Parus major). Ecol Evol 3:2730–2738
Isaksson C (2015) Urbanization, oxidative stress and inflammation: a question of evolving, acclimatizing or coping with urban environmental stress. Funct Ecol 29:913–923
Ishimatsu A, Hayashi M, Kikkawa T (2008) Fishes in high-$CO_2$, acidified oceans. Mar Ecol Progr Ser 373:295–302
Issartel J, Hervant F, de Fraipont M, Clobert J, Voituron Y (2009) High anoxia tolerance in the subterranean salamander *Proteus anguinus* without oxidative stress nor activation of antioxidant defenses during reoxygenation. J Comp Physiol B 179:543–551
Janssens BJ, Childress JJ, Baguet F, Rees JF (2000) Reduced enzymatic antioxidative defense in deep-sea fish. J Exp Biol 203:3717–3725
Jesus TF, Rosa IC, Repolho T, Lopes AR, Pimentel MS, Almeida-Val VMF, Coelho MM, Rosa R (2018) Different ecophysiological responses of freshwater fish to warming and acidification. Comp Biochem Physiol A 216:34–41

Jimenez AG, Harper JM, Queenborough SA, Williams JB (2013) Linkages between the life-history evolution of tropical and temperate birds and the resistance of cultured skin fibroblasts to oxidative and non-oxidative chemical injury. J Exp Biol 216:1373–1380

Jimenez AG, Jayawardene S, Alves S, Dallmer J, Dowd WW (2015) Micro-scale environmental variation amplifies physiological variation among individual mussels. Proc R Soc Lond B 282:20152273

Kammer AR, Orczewska JI, O'Brien KM (2011) Oxidative stress is transient and tissue specific during cold acclimation of threespine stickleback. J Exp Biol 214:1248–1256

Kim BM, Rhee JS, Lee KW, Kim MJ, Shin KH, Lee SJ, Lee YM, Lee JS (2015) UV-B radiation-induced oxidative stress and p38 signaling pathway involvement in the benthic copepod *Tigriopus japonicus*. Comp Biochem Physiol C 167:15–23

Kim JH, Kim SK, Hur YB (2019a) Hematological parameters and antioxidant responses in olive flounder *Paralichthys olivaceus* in biofloc depend on water temperature. J Therm Biol 82:206–212

Kim JH, Kim SK, Hur YB (2019b) Temperature-mediated changes in stress responses, acetylcholinesterase, and immune responses of juvenile olive flounder *Paralichthys olivaceus* in a bio-floc environment. Aquacult 506:453–458

Kim JH, Kim SR, Kim SK, Kang HW (2021) Effects of pH changes on blood physiology, antioxidant responses and Ig M of juvenile olive flounder, Paralichthys olivaceus. Aquacult Rep 21:100790

Kitiyakara C, Chabrashvili T, Chen Y, Blau J, Karber A, Aslam S, Welch WJ, Wilcox CS (2003) Salt intake, oxidative stress, and renal expression of NADPH oxidase and superoxide dismutase. J Am Soc Nephrol 14:2775–2782

Klein RD, Borges VD, Rosa CE, Colares EP, Robaldo RB, Martinez PE, Bianchini A (2017) Effects of increasing temperature on antioxidant defense system and oxidative stress parameters in the Antarctic fish *Notothenia coriiceps* and *Notothenia rossii*. J Therm Biol 68:110–118

Koštál V, Tollarová-Borovanská M (2009) The 70 kDa heat shock protein assists during the repair of chilling injury in the insect, *Pyrrhocoris apterus*. PLoS One 4:e4546

Lagisz M, Hector KL, Nakagawa S (2013) Life extension after heat shock exposure: assessing meta-analytic evidence for hormesis. Ageing Res Rev 12:653–660

Lamare MD, Baker MF, Lesser MP, Marshall C (2006) DNA photorepair in echinoid embryos: effects of temperature on repair rate in Antarctic and non-Antarctic species. J Exp Biol 209:5017–5028

Lamare MD, Baker MF, Lesser MP (2007) In situ rates of DNA damage and abnormal development in Antarctic and non-Antarctic Sea urchin embryos. Aquat Biol 1:21–32

Lawniczak M, Romestaing C, Roussel D, Maazouzi C, Renault D, Hervant F (2013) Preventive antioxidant responses to extreme oxygen level fluctuation in a subterranean crustacean. Comp Biochem Physiol A 165:299–303

Le Bourg E, Valenti P, Lucchetta P, Payre F (2001) Effects of mild heat shocks at young age on aging and longevity in *Drosophila melanogaster*. Biogerontology 2:155–164

Lee RE, Chen C-P, Denlinger DL (1987) A rapid cold-hardening process in insects. Science 238:1415–1417

Lee DW, Choi YU, Park HS, Park YS, Choi CY (2022a) Effect of low pH and salinity conditions on the antioxidant response and hepatocyte damage in juvenile olive flounder Paralichthys olivaceus. Mar Environ Res 175:105562

Lee DW, Song JA, Park HS, Choi CY (2022b) The effects of low pH and high water temperature on oxidative stress and cell damage in juvenile olive flounder Paralichthys olivaceus: comparison of single and combined environmental conditions. Fish Physiol Biochem 48:1251–1264

Lenda DM, Boegehold MA (2002) Effect of a high salt diet on microvascular antioxidant enzymes. J Vasc Res 39:41–50

Lesser MP (2006) Oxidative stress in marine environments: biochemistry and physiological ecology. Annu Rev Physiol 68:253–278

Lesser MP (2010) Survivorship, oxidative stress, and DNA damage of sea urchin (*Strongylocentrotus droebachiensis*) embryos and larvae exposed to ultraviolet radiation (290-400 nm) in the Gulf of Maine. Photochem Photobiol 86:382–388

Lesser MP (2012) Oxidative stress in tropical marine ecosystems. In: Abele D, Vázquez-Medina JP, Zenteno-Savín T (eds) Oxidative stress in aquatic ecosystems. Wiley-Blackwell, pp 9–19

Lesser MP, Turtle SL, Farrell JH, Walker CW (2001) Exposure to ultraviolet radiation (290-400 nm) causes oxidative stress, DNA damage, and expression of p53/p73 in laboratory experiments on embryos of the spotted salamander, *Ambystoma maculatum*. Physiol Biochem Zool 74:733–741

Lesser MP, Barry TM, Lamare MD, Barker MF (2006) Biological weighting functions for DNA damage in sea urchin embryos exposed to ultraviolet radiation. J Exp Mar Biol Ecol 328:10–21

Letendre J, Chouquet B, Rocher B, Manduzio H, Leboulenger F, Durand F (2008) Differential pattern of Cu/Zn superoxide dismutase isoforms in relation to tidal spatio-temporal changes in the blue mussel *Mytilus edulis*. Comp Biochem Physiol C 148:211–216

Li SR, Li JH, Chen W, Xu ZW, Xie L, Zhang YP (2021) Effects of simulated heat wave on oxidative physiology and immunity in Asian yellow pond turtle (*Mauremys mutica*). Front Ecol Evol 9:704105

Lindroth A (1941) Atrnungsventllation der polychaeten. Z vgl Physiol 28:485–532

Liu F, Wang MJ, Chu TQ, Zhan W, Loul B (2019) The effect of water temperature on antioxidant and digestive flexibility in miiuy croaker, *Miichthys miiuy*. Israeli J Aquacult 71:9

Lopes AR, Trübenbach K, Teixeira T, Lopes VM, Pires V, Baptista M, Repolho T, Calado R, Diniz M, Rosa R (2013) Oxidative stress in deep scattering layers: heat shock response and antioxidant enzymes activities of myctophid fishes thriving in oxygen minimum zones. Deep-Sea Res I Oceanogr Res Pap 82:10–16

Lopes AR, Figueiredo C, Sampaio E, Diniz M, Rosa R, Grilo TF (2021) Impaired antioxidant defenses and DNA damage in the European glass eel (Anguilla Anguilla) exposed to ocean warming and acidification. Sci Total Environ 774:145499

Lord-Fontaine S, Averill-Bates DA (2002) Heat shock inactivates cellular antioxidant defenses against hydrogen peroxide: protection by glucose. Free Radic Biol Med 32:752–765

Loughland I, Seebacher F (2020) Differences in oxidative status explain variation in thermal acclimation capacity between individual mosquitofish (*Gambusia holbrooki*). Funct Ecol 34:1380–1390

Lucas LD, French SS (2012) Stress-induced tradeoffs in a free-living lizard across a variable landscape: consequences for individuals and populations. PLoS One 7:e49895

Lushchak VI, Bagnyukova TV (2006a) Effects of different environmental oxygen levels on free radical processes in fish. Comp Biochem Physiol B 144:283–289

Lushchak VI, Bagnyukova TV (2006b) Temperature increase results in oxidative stress in goldfish tissues. 1. Indices of oxidative stress. Comp Biochem Physiol C 143:30–35

Lushchak VI, Bagnyukova TV (2006c) Temperature increase results in oxidative stress in goldfish tissues. 2. Antioxidant and associated enzymes. Comp Biochem Physiol C 143:36–41

Lushchak VI, Bagnyukova TV (2007) Hypoxia induces oxidative stress in tissues of a goby, the rotan Perccottus glenii. Comp Biochem Physiology B 148:390–397

Lushchak VI, Lushchak LP, Mota AA, Hermes-Lima M (2001) Oxidative stress and antioxidant defenses in goldfish *Carassius auratus* during anoxia and reoxygenation. Am J Phys 280:R100–R107

Lushchak VI, Bagnyukova TV, Husak VV, Luzhna LI, Lushchak OV, Storey KB (2005a) Hyperoxia results in transient oxidative stress and an adaptive response by antioxidant enzymes in goldfish tissues. Int J Biochem Cell Biol 37:1670–1680

Lushchak VI, Bagnyukova TV, Lushchak OV, Storey JM, Storey KB (2005b) Hypoxia and recovery perturb free radical processes and antioxidant potential in common carp (*Cyprinus carpio*) tissues. Int J Biochem Cell Biol 37:1319–1330

Madeira D, Madeira C, Costa PM, Vinagre C, Pörtner H-O, Diniz MS (2020) Different sensitivity to heatwaves across the life cycle of fish reflects phenotypic adaptation to environmental niche. Mar Environ Res 162:105192

Marion GM, van Cleve K, Dyrness CT, Black CH (1993) The soil chemical environment along a forest primary successional sequence on the Tanana River floodplain, interior Alaska. Can J For Res 23:914–922

Martell HA (2023) Thermal priming and bleaching hormesis in the staghorn coral, *Acropora cervicornis* (Lamarck 1812). J Exp Mar Biol Ecol 560:151820
Meesters EH, Bak RPM (1993) Effects of coral bleaching on tissue regeneration potential and colony survival. Mar Ecol Prog Ser 96:189–198
Meng JY, Zhang CY, Zhu F, Wang XP, Lei CL (2009) Ultraviolet light-induced oxidative stress: effects on antioxidant response of *Helicoverpa armigera* adults. J Insect Physiol 55:588–592
Messina S, Edwards DP, AbdElgawad H, Beemster GTS, Cosset CCP, Tomassi S, Benedick S, Eens M, Costantini D (2020) Impacts of selective logging on the oxidative status of tropical understory birds. J Anim Ecol 89:2222–2234
Messina S, Costantini D, Eens M (2023) Impacts of rising temperatures and water acidification on the oxidative status and immune system of aquatic ectothermic vertebrates: a meta-analysis. Sci Total Environ 868:161580
Mészáros B, Herczeg G, Bajer K, Török J, Molnár O (2017) Effects of energy and thermoregulation time on physiological state and sexual signal in a lizard. J Exp Zool A 327:570–578
Miller GM, Watson SA, Donelson JM, McCormick MI, Munday PL (2012) Parental environment mediates impacts of increased carbon dioxide on a coral reef fish. Nat Clim Change 2:858–861
Miller GM, Watson SA, McCormick MI, Munday PL (2013) Increased $CO_2$ stimulates reproduction in a coral reef fish. Glob Chang Biol 19:3037–3045
Miranda AC, Schielzeth H, Sonntag T, Partecke J (2013) Urbanization and its effects on personality traits: a result of microevolution or phenotypic plasticity? Glob Chang Biol 19:2634–2644
Mitchelmore CL, Ringwood AH, Weis VM (2003) Differential accumulation of cadmium and changes in glutathione levels as a function of symbiotic state in the sea anemone *Anthopleura elegantissima*. J Exp Mar Biol Ecol 284:71–85
Mizutani Y, Tomita N, Niizuma Y, Yoda K (2013) Environmental perturbations influence telomere dynamics in long-lived birds in their natural habitat. Biol Lett 9:20130511
Møller AP, Erritzøe J, Karadas F (2010) Levels of antioxidants in rural and urban birds and their consequences. Oecologia 163:35–45
Moné Y, Monnin D, Kremer N (2014) The oxidative environment: a mediator of interspecies communication that drives symbiosis evolution. Proc R Soc Lond B 281:20133112
Monnin D, Kremer N, Berny C, Henri H, Dumet A, Voituron Y, Desouhant E, Vavre F (2016) Influence of oxidative homeostasis on bacterial density and cost of infection in *Drosophila–Wolbachia* symbioses. J Evol Biol 29:1211–1222
Monnin D, Kremer N, Desouhant E, Vavre F (2017) Impact of Wolbachia on oxidative stress sensitivity in the parasitic wasp *Asobara japonica*. PLoS One 12:e0175974
Mopper K, Kieber DJ (2000) Marine photochemistry and its impact on carbon cycling. In: De Mora S, Demers S, Vernet M (eds) The effects of UV radiation in the marine environment. Cambridge University Press, Cambridge, pp 101–130
Moreira DC, Venancio LPR, Sabino MACT, Hermes-Lima M (2016) How widespread is preparation for oxidative stress in the animal kingdom? Comp Biochem Physiol A 200:64–78
Moreira DC, Oliveira MF, Liz-Guimarães L, Diniz-Rojas N, Campos ÉG, Hermes-Lima M (2017) Current trends and research challenges regarding “preparation for oxidative stress”. Front Physiol 8:702
Mueller IA, Grim JM, Beers JM, Crockett EL, O’Brien KM (2011) Inter-relationship between mitochondrial function and susceptibility to oxidative stress in red- and white-blooded Antarctic notothenioid fishes. J Exp Biol 214:3732–3741
Mueller IA, Devor DP, Grim JM, Beers JM, Crockett EL, O’Brien KM (2012) Exposure to critical thermal maxima increases oxidative stress in hearts of white- but not red-blooded Antarctic notothenioid fishes. J Exp Biol 215:3655–3664
Mueller IA, Hoffman M, Dullen K, O’Brien K (2014) Moderate elevations in temperature do not increase oxidative stress in oxidative muscles of Antarctic notothenioid fishes. Polar Biol 37:311–320
Mujahid A, Akiba Y, Toyomizu M (2007) Acute heat stress induces oxidative stress and decreases adaptation in young white Leghorn cockerels by downregulation of avian uncoupling protein. Poult Sci 86:364–371

Murray MH, Sánchez CA, Becker DJ, Byers KA, Worsley-Tonks KEL, Craft ME (2019) City sicker? A meta- analysis of wildlife health and urbanization. Front Ecol Environ 17:575–583

Niehaus AC, Angilletta MJ Jr, Sears MW, Franklin CE, Wilson RS (2012) Predicting the physiological performance of ectotherms in fluctuating thermal environments. J Exp Biol 215:694–701

Nikinmaa M, McCairns RJS, Nikinmaa MW, Vuori KA, Kanerva M, Leinonen T, Primmer CR, Merilä J, Leder EH (2013) Transcription and redox enzyme activities: comparison of equilibrium and disequilibrium levels in the three-spined stickleback. Proc R Soc Lond B 280:20122974

Nilsson GE, Östlund-Nilsson S (2004) Hypoxia in paradise: widespread hypoxia tolerance in coral reef fishes. Proc R Soc Lond B 271:S30–S33

Nilsson GE, Renshaw GM (2004) Hypoxic survival strategies in two fishes: extreme anoxia tolerance in the North European crucian carp and natural hypoxic preconditioning in a coral-reef shark. J Exp Biol 207:3131–3139

Niu Y, Zhang X, Xu T, Li X, Zhang H, Wu A, Storey KB, Chen Q (2022) Physiological and biochemical adaptations to high altitude in Tibetan frogs, *Nanorana parkeri*. Front Physiol 13:942037

O'Brien KM, Oldham CA, Sarrimanolis J, Fish A, Castellini L, Vance J, Lekanof H, Crockett EL (2022) Warm acclimation alters antioxidant defences but not metabolic capacities in the Antarctic fish, Notothenia coriiceps. Conserv Physiol 10:coac054

O'Brien KM (2011) Mitochondrial biogenesis in cold-bodied fishes. J Exp Biol 214:275–285

O'Brien KM, Mueller IA (2010) The unique mitochondrial form and function of Antarctic channichthyid icefishes. Integr Comp Biol 50:993–1008

Ogilvy V, Preziosi RF (2012) Can carotenoids mediate the potentially harmful effects of ultraviolet light in Silurana (Xenopus) tropicalis larvae? J Anim Physiol Anim Nutr 96:693–699

Olsen A, Vantipalli MC, Lighgow GJ (2006) Lifespan extension of *Caenorhabditis elegans* following repeated mild hormetic heat treatments. Biogerontology 7:221–230

Olsvik PA, Kristensen T, Waagbo R, Tollefsen KE, Rosseland BO, Toften H (2006) Effects of hypo- and hyperoxia on transcription levels of five stress genes and the glutathione system in liver of Atlantic cod *Gadus morhua*. J Exp Biol 209:2893–2901

Pannunzio TM, Storey KB (1998) Antioxidant defenses and lipid peroxidation during anoxia stress and aerobic recovery in the marine gastropod *Littorina littorea*. J Exp Mar Biol Ecol 221:277–292

Parker LM, O'Connor WA, Raftos DA, Pörtner H-O, Ross PM (2015) Persistence of positive carryover effects in the oyster, *Saccostrea glomerata*, following transgenerational exposure to ocean acidification. PLoS One 10:e0132276

Partecke J, Schwabl I, Gwinner E (2006) Stress and the city: urbanization and its effects on the stress physiology in European blackbirds. Ecology 87:1945–1952

Pegado MR, Santos CP, Pimentel M, Cyrne R, Sampaio E, Temporão A, Röckner J, Diniz M, Rosa R (2020) Lack of oxidative damage on temperate juvenile catsharks after a long-term ocean acidification exposure. Mar Biol 167:165

Pereira LF, Goncalves JFM, Peixoto MJ, Domingues A, Fonseca SB, Magalhaes FA, Santos GA, Ozorio ROA (2017) Cross-effects of dietary probiotic supplementation and rearing temperature on antioxidant responses in European seabass *Dicentrarchus labrax* (Linnaeus, 1758) juveniles. J Appl Ichthyol 33:377–385

Pereira LAL, Amanajas RD, de Oliveira AM, de Silva MDP, Val AL (2021) Health of the Amazonian fish tambaqui (*Colossoma macropomum*): effects of prolonged photoperiod and high temperature. Aquacult 541:736836

Petrović TG, Kijanović A, Kolarov Tomašević N, Gavrić JP, Despotović SG, Gavrilović BR, Radovanović TB, Vukov T, Faggio C, Prokić MD (2021) Effects of desiccation on metamorphic climax in *Bombina variegata*: changes in levels and patterns of oxidative stress parameters. Animals 11:953

Pinto D, Pellegrin L, Nitz LF, da Costa ST, Monserrat JM, Garcia L (2019) Haematological and oxidative stress responses in *Piaractus mesopotamicus* under temperature variations in water. Aquac Res 50:3017–3027

Plantivaux A, Furla P, Zoccola D, Garello G, Forcioli D, Richier S, Merle PL, Tambutte E, Tambutte S, Allemand D (2004) Molecular characterization of two CuZn-superoxide dismutases in a sea anemone. Free Radic Biol Med 37:1170–1181

Pöhlmann K, Koenigstein S, Alter K, Abele D, Held C (2011) Heat-shock response and antioxidant defense during air exposure in Patagonian shallow-water limpets from different climatic habitats. Cell Stress Chaper 16:621–632

Pörtner HO, Langenbuch M, Reipschlager A (2004) Biological impact of elevated ocean $CO_2$ concentrations: lessons from animal physiology and earth history. J Ocean 60:705–718

Poyton RO, Castello PR, Ball KA, Woo DK, Pan N (2009) Mitochondria and hypoxic signaling: a new view. Ann N Y Acad Sci 1177:48–56

Radovanović TB, Petrović TG, Gavrilović BR, Despotović SG, Gavrić JP, Kijanović A, Mirč M, Tomašević Kolarov N, Vukov T, Prokić MD (2023) What coloration brings: implications of background adaptation to oxidative stress in anurans. Front Zool 20:6

Rebecchi L (2013) Dry and survive: the role of antioxidant defenses in anhydrobiotic organisms. J Insect Physiol 72:62–72

Regoli F, Cerrano C, Chierici E, Bompadre S, Bavestrello G (2000) Susceptibility to oxidative stress of the Mediterranean domosponge *Petrosia ficiformis*: role of endosymbionts and solar irradiance. Mar Biol 137:453–461

Regoli F, Benedetti M, Krell A, Abele D (2012) Oxidative challenges in polar seas. In: Abele D, Vázquez-Medina JP, Zenteno-Savín T (eds) Oxidative stress in aquatic ecosystems. Wiley-Blackwell, pp 20–40

Reguera S, Zamora-Camacho FJ, Melero E, García-Mesa S, Trenzado CE, Cabrerizo MJ, Sanz A, Moreno-Rueda G (2015) Ultraviolet radiation does not increase oxidative stress in the lizard Psammodromus algirus along an elevational gradient. Comp Biochem Physiol Part A 183:20–26

Reischl E (1986) High sulphydryl content in turtle erythrocytes: is there a relation with resistance to hypoxia? Comp Biochem Physiol B 85:723–726

Reudink MW, Marra PP, Kyser TK, Boag PT, Langin KM, Ratcliffe LM (2009) Non-breeding season events influence sexual selection in a long-distance migratory bird. Proc R Soc Lond B 276:1619–1626

Rey B, Roussel D, Romestaing C, Belouze M, Rouanet JL, Desplanches D, Sibille B, Servais S, Duchamp C (2010) Up-regulation of avian uncoupling protein in cold-acclimated and hyperthyroid ducklings prevents reactive oxygen species production by skeletal muscle mitochondria. BMC Physiol 10:5

Richier S, Merle PL, Furla P, Pigozzi D, Sola F, Allemand D (2003) Characterization of superoxide dismutases in anoxia- and hyperoxia-tolerant symbiotic cnidarians. Biochim Biophys Acta 1621:84–91

Richier S, Furla P, Plantivaux A, Merle PL, Allemand D (2005) Symbiosis-induced adaptation to oxidative stress. J Exp Biol 208:277–285

Ricklefs RE (1976) Growth rates of birds in the humid New World tropics. Ibis 118:179–207

Ritchie DJ, Friesen CR (2015) Thermal effects on oxidative stress in vertebrate ectotherms. Comp Biochem Physiol A 263:111082

Ritola O, Livingstone DR, Peters LD, Lindstrom-Seppa P (2002a) Antioxidant processes are affected in juvenile rainbow trout (*Oncorhynchus mykiss*) exposed to ozone and oxygen-supersaturated water. Aquacult 210:1–19

Ritola O, Peters LD, Livingstone DR, Lindstrom-Seppa P (2002b) Effects of in vitro exposure to ozone and/or hyperoxia on superoxide dismutase, catalase, glutathione and lipid peroxidation in red blood cells and plasma of rainbow trout, *Oncorhynchus mykiss* (Walbaum). Aquac Res 33:165–175

Ritola O, Tossavainen K, Kiuru T, Lindstrom-Seppa P, Molsa H (2002c) Effects of continuous and episodic hyperoxia on stress and hepatic glutathione levels in one-summer-old rainbow trout (*Oncorhynchus mykiss*). J Appl Ichthyol 18:159–164

Rivera-Ingraham GA, Lignot JH (2017) Osmoregulation, bioenergetics and oxidative stress in coastal marine invertebrates: raising the questions for future research. J Exp Biol 220:1749–1760

Rivera-Ingraham GA, Bickmeyer U, Abele D (2013a) The physiological response of the marine platyhelminth *Macrostomum lignano* to different environmental oxygen concentrations. J Exp Biol 216:2741–2751

Rivera-Ingraham GA, Rocchetta I, Meyer S, Abele D (2013b) Oxygen radical formation in anoxic transgression and anoxia-reoxygenation: foe or phantom? Experiments with a hypoxia tolerant bivalve. Mar Environ Res 92:110–119

Rivera-Ingraham GA, Barri K, Boël M, Farcy E, Charles A-L, Geny B, Lignot J-H (2016a) Osmoregulation and salinity-induces oxidative stress: is oxidative adaptation determined by gill function? J Exp Biol 219:80–89

Rivera-Ingraham GA, Nommick A, Blondeau-Bidet E, Ladurner P, Lignot J-H (2016b) Salinity stress from the perspective of the energy-redox axis: lessons from a marine intertidal flatworm. Redox Biol 10:53–64

Rosa R, Pimentel MS, Boavida-Portugal J, Teixeira T, Trübenbach K, Diniz M (2012) Ocean warming enhances malformations, premature hatching, metabolic suppression and oxidative stress in the early life stages of a keystone squid. PLoS One 7:e38282

Rosa R, Ricardo Paula J, Sampaio E, Pimentel M, Lopes AR, Baptista M, Guerreiro M, Santos C, Campos D, Almeida-Val VMF, Calado R, Diniz M, Repolho T (2016) Neuro-oxidative damage and aerobic potential loss of sharks under elevated CO2 and warming. Mar Biol 163:119

Ross SW, Dalton DA, Kramer S, Christensen BL (2001) Physiological (antioxidant) responses of estuarine fishes to variability in dissolved oxygen. Comp Biochem Physiol C 130:289–303

Rossi A, Bacchetta C, Cazenave J (2017) Effect of thermal stress on metabolic and oxidative stress biomarkers of *Hoplosternum littorale* (Teleostei, Callichthyidae). Ecol Indic 79:361–370

Rowan R, Knowlton N, Baker AC, Jara J (1997) Landscape ecology of algal symbionts creates variation in episodes of coral bleaching. Nature 388:265–269

Roychowdhury P, Aftabuddin M, Pati MK (2021) Thermal stress-induced oxidative damages in the liver and associated death in fish, *Labeo rohita*. Fish Physiol Biochem 47:21–32

Sabat P, Narváez C, Peña-Villalobos I, Contreras C, Maldonado K, Sanchez-Hernandez JC, Newsome SD, Nespolo R, Bozinovic F (2017) Coping with salt water habitats: metabolic and oxidative responses to salt intake in the rufous-collared sparrow. Front Physiol 8:654

Sabat P, Bozinovic F, Contreras-Ramos C, Nespolo RF, Newsome SD, Quirici V, Maldonado K, Peña-Villalobos I, Ramirez-Otarola N, Sanchez-Hernandez JC (2019) The interplay between ambient temperature and salt intake affects oxidative status and immune responses in a ubiquitous Neotropical passerine, the rufous-collared sparrow. Comp Biochem Physiol A 234:50–59

Salmón P, Watson H, Nord A, Isaksson C (2018) Effects of the urban environment on oxidative stress in early life: insights from a cross-fostering experiment. Integr Comp Biol 58:986–994

Sampaio E, Lopes AR, Francisco S, Paula JR, Pimentel M, Maulvault AL, Repolho T, Grilo TF, Pousão-Ferreira P, Marques A, Rosa R (2018) Ocean acidification dampens physiological stress response to warming and contamination in a commercially-important fish (*Argyrosomus regius*). Sci Total Environ 618:388–398

Schöttler U (1979) On the anaerobic metabolism of three species of *Nereis* (Annelida). Mar Ecol Prog Ser 1:249–254

Selman C, McLaren JS, Himanka MJ, Speakman JR (2000) Effect of long-term cold exposure on antioxidant enzyme activities in a small mammal. Free Radic Biol Med 28:1279–1285

Selman C, Grune T, Stolzing A, Jakstadt M, McLaren JS, Speakman JR (2002) The consequences of acute cold exposure on protein oxidation and proteasome activity in short-tailed field voles, *Microtus agrestis*. Free Radic Biol Med 33:259–265

Shaklee JB, Christiansen JA, Sidell BD, Prosser CL, Whitt GS (1977) Molecular aspects of temperature-acclimation in fish - contributions of changes in enzyme activities and isoenzyme patterns to metabolic reorganization in Green Sunfish. J Exp Zool 201:1–20

Shama S, Lai C-Y, Antoniazzi JM, Jiang JC, Jazwinski SM (1998) Heat stress-induced life span extension in yeast. Exp Cell Res 245:379–388

Shick JM (1990) Diffusion limitation and hyperoxic enhancement of oxygen consumption in zooxanthellate sea anemones, zoanthids and corals. Biol Bull 179:148–158

Shick JM, Dykens JA (1985) Oxygen detoxification in algal-invertebrate symbioses from the Great Barrier Reef. Oecologia 66:33–41
Shick JM, Lesser MP, Jokiel PL (1996) Effects of ultraviolet radiation on corals and other coral reef organisms. Glob Chang Biol 2:527–545
Shreve SM, Kelty JD, Lee RE (2004) Preservation of reproductive behaviors during modest cooling: rapid cold-hardening fine-tunes organismal response. J Exp Biol 207:1797–1802
Sies H, Stahl W (2007) Carotenoids and flavonoids contribute to nutritional protection against skin damage from sunlight. Mol Biotechnol 37:26–30
Silva CSE, Novais SC, Lemos MFL, Mendes S, Oliveira AP, Gonçalves EJ, Faria AM (2016) Effects of ocean acidification on the swimming ability, development and biochemical responses of sand smelt larvae. Sci Total Environ 563-564:89–98
Simčič T, Jesensek D, Brancelj A (2015) Effects of increased temperature on metabolic activity and oxidative stress in the first life stages of marble trout (*Salmo marmoratus*). Fish Physiol Biochem 41:1005–1014
Singh SP, Ahmad T, Sharma J, Chakrabarti R (2021) Effect of temperature on food consumption, immune system, antioxidant enzymes, and heat shock protein 70 of *Channa punctata* (Bloch, 1793). Fish Physiol Biochem 47:79–91
Sol D, Lapiedra O, González-Lagos C (2013) Behavioural adjustments for a life in the city. Anim Behav 85:1101–1112
Somero GN (1995) Proteins and temperature. Annu Rev Physiol 57:43–68
Somero GN (2004) Adaptation of enzymes to temperature: searching for basic "strategies". Comp Biochem Physiol B 139:321–333
Sørensen JG, Loeschcke V, Kristensen TN (2013) Cellular damage as induced by high temperature is dependent on rate of temperature change - investigating consequences of ramping rates on molecular and organismal phenotypes in *Drosophila melanogaster*. J Exp Biol 216:809–814
Stahlschmidt ZR, French SS, Ahn A, Webb A, Butler MW (2017) A simulated heat wave has diverse effects on immune function and oxidative physiology in the corn snake (*Pantherophis guttatus*). Physiol Biochem Zool 90:434–444
Stark LR, McLetchie DN, Roberts SP (2009) Gender differences and a new adult eukaryotic record for upper thermal tolerance in the desert moss *Syntrichia caninervis*. J Therm Biol 34:131–137
Stier A, Massemin S, Criscuolo F (2014) Chronic mitochondrial uncoupling treatment prevents acute cold-induced oxidative stress in birds. J Comp Physiol B 184:1021–1029
Storey KB, Storey JM (1990) Facultative metabolic rate depression: molecular regulation and biochemical adaptation in anaerobiosis, hibernation, and estivation. Q Rev Biol 65:145–174
Storey KB, Tanino KK (2012) Temperature adaptation in a changing climate: nature at risk. Springer, Berlin
Studds CE, Marra PP (2005) Nonbreeding habitat occupancy and population processes: an upgrade experiment with a migratory bird. Ecology 86:2380–2385
Studds CE, Marra PP (2007) Linking fluctuations in rainfall to non-breeding season performance in a long-distance migratory bird, *Setophaga ruticilla*. Clim Res 35:115–122
Studds CE, Marra PP (2011) Rainfall-induced changes in food availability modify the spring departure programme of a migratory bird. Proc R Soc Lond B 278:3437–3443
Szmant AM, Gassman NJ (1990) The effects of prolonged "bleaching" on the tissue biomass and reproduction of the reef coral *Montastraea annularis*. Coral Reefs 8:217–224
Szymczak R, Waite TD (1988) Generation and decay of hydrogen peroxide in estuarine waters. Aust J Freshwat Res 39:289–299
Takatsu K, Selz OM, Brodersen J (2022) Temperature regime during embryogenesis alters subsequent behavioural phenotypes of juvenile brown trout. Biol Lett 18:20220369
Tanaka M, Mokhtari GK, Terry RD, Balsam LB, Lee KH, Kofidis T, Tsao PS, Robbins RC (2004) Overexpression of human copper/zinc superoxide dismutase (SOD1) suppresses ischemia–reperfusion injury and subsequent development of graft coronary artery disease in murine cardiac grafts. Circulation 110:II200–II206

Tattersall GJ, Sinclair BJ, Withers PC, Fields PA, Seebacher F, Cooper CE, Maloney SK (2012) Coping with thermal challenges: physiological adaptations to environmental temperatures. Compr Physiol 2:2151–2202

Teixeira T, Diniz M, Calado R, Rosa R (2013) Coral physiological adaptations to air exposure: heat shock and oxidative stress responses in *Veretillum cynomorium*. J Exp Mar Biol Ecol 439:35–41

Tökölyi J, Ősz Z, Sebestyén F, Barta Z (2016) Resource allocation and post-reproductive degeneration in the freshwater cnidarian Hydra oligactis (Pallas, 1766). Zoology 120:110–116

Tökölyi J, Kozma B, Sebestyén F, Miklós M, Barta Z (2017) Life history traits and previous exposure predict resistance to UV irradiation in the freshwater cnidarian Hydra oligactis. Inver Biol 136:217–227

Tomanek L, Zuzow MJ, Ivanina AV, Beniash E, Sokolova IM (2011) Proteomic response to elevated $P_{CO2}$ level in eastern oysters, *Crassostrea virginica*: evidence for oxidative stress. J Exp Biol 214:1836–1844

Tremblay N, Zenteno-Savín T, Gómez-Gutiérrez J, Maeda-Martínez AN (2012) Migrating to the oxygen minimum layer: Euphausiids. In: Abele D, Vázquez-Medina JP, Zenteno-Savín T (eds) Oxidative stress in aquatic ecosystems. Wiley-Blackwell, pp 89–98

Uehling JJ, Taff CC, Winkler DW, Vitousek MN (2020) Developmental temperature predicts the adult response to stressors in a free-living passerine. J Anim Ecol 89:842–854

van de Crommenacker J, Komdeur J, Burke T, Richardson DS (2011) Spatio-temporal variation in territory quality and oxidative status: a natural experiment in the Seychelles warbler (Acrocephalus sechellensis). J Anim Ecol 80:668–680

Veselá A, Wilhelm J (2002) The role of carbon dioxide in free radical reactions of the organism. Physiol Res 51:335–339

Vieira HC, Bordalo MD, Rodrigues ACM, Pires SFS, Rocha RJM, Soares A, Rendon-von Osten J, Abreu SN, Morgado F (2021) Water temperature modulates mercury accumulation and oxidative stress status of common goby (*Pomatoschistus microps*). Environ Res 193:110585

Vinagre C, Madeira D, Narciso L, Cabral H, Diniz M (2012) Effect of temperature on oxidative stress in fish: lipid peroxidation and catalase activity in the muscle of juvenile seabass, *Dicentrarchus labrax*. Ecol Indic 23:274–279

Vinagre C, Madeira D, Mendonça V, Dias M, Roma J, Diniz MS (2014) Effect of temperature in multiple biomarkers of oxidative stress in coastal shrimp. J Therm Biol 41:38–42

Visser B, Williams CM, Hahn DA, Short CA, López-Martínez G (2018) Hormetic benefits of prior anoxia exposure in buffering anoxia stress in a soil-pupating insect. J Exp Biol 221:jeb167825

Wang P, Chen H, Qin H, Sankarapandi S, Becher MW, Wong PC, Zweier JL (1998) Overexpression of human copper, zinc-superoxide dismutase (SOD1) prevents postischemic injury. Proc Natl Acad Sci USA 95:4556–4560

Watson H, Videvall E, Andersson MN, Isaksson C (2017) Transcriptome analysis of a wild bird reveals physiological responses to the urban environment. Sci Rep 7:44180

Welker AF, Campos EG, Cardoso LA, Hermes-Lima M (2012) Role of catalase on the hypoxia/reoxygenation stress in the hypoxia-tolerant Nile tilapia. Am J Phys 302:R1111–R1118

Welker AF, Moreira DC, Campos EG, Hermes-Lima M (2013) Role of redox metabolism for adaptation of aquatic animals to drastic changes in oxygen availability. Comp Biochem Physiol A 165:384–404

Wiersma P, Chappell MA, Williams JB (2007a) Cold- and exercise-induced peak metabolic rates in tropical birds. Proc Natl Acad Sci USA 104:20866–20871

Wiersma P, Muñoz-Garcia A, Walker A, Williams JB (2007b) Tropical birds have a slow pace of life. Proc Natl Acad Sci USA 104:9340–9345

Wiersma P, Nowak B, Williams JB (2012) Small organ size contributes to the slow pace of life in tropical birds. J Exp Biol 215:1662–1669

Wikelski M, Spinney L, Schelsky W, Scheuerlein A, Gwinner E (2003) Slow pace of life in tropical sedentary birds: a common-garden experiment on four stonechat populations from different latitudes. Proc R Soc Lond B 270:2383–2388

Wilcoxen TE, Horn DJ, Hogan BM, Hubble CN, Huber SJ, Flamm J, Knott M, Lundstrom L, Salik F, Wassenhove SJ, Wrobel ER (2015) Effects of bird-feeding activities on the health of wild birds. Conserv Physiol 3:cov058

Willmore WG, Storey KB (1997a) Antioxidant systems and anoxia tolerance in a freshwater turtle, *Trachemys scripta elegans*. Mol Cell Biochem 170:177–185

Willmore WG, Storey KB (1997b) Glutathione systems and anoxia tolerance in turtles. Am J Phys 273:R219–R225

Wilson RS, Franklin CE (2002) Testing the beneficial acclimation hypothesis. Trends Ecol Evol 17:66–70

Witas H, Gabryelak T, Matkovics B (1984) Comparative studies on superoxide dismutase and catalase activities in livers of fish and other Antarctic vertebrates. Comp Biochem Physiol C 77:409–411

Wu BS, Lee JK, Thompson KM, Walker VK, Moyes CD, Robertson RM (2002) Anoxia induces thermotolerance in the locust flight system. J Exp Biol 205:815–827

Xie C-Y, Yanchuk AD, Kiss GK (1998) Genetics of interior spruce in British Columbia: performance and variability of open-pollinated families in the east Kootenays. Victoria, Ministry of Forestry

Yahav S, McMurtry JP (2001) Thermotolerance acquisition in broiler chickens by temperature conditioning early in life–the effect of timing and ambient temperature. Poult Sci 80:1662–1666

Yang SY, Yang XX, Li YK, Li DT, Gong Q, Huang XL, Wu JY, Huang AQ, Kong FL, Han XF, Zeng XY, Zhang CY, Du J, Du XG (2021) The multilevel responses of *Acipenser baerii* and its hybrids (*A. baerii* female x *A. schrenckii* male) to chronic heat stress. Aquacult 541:736773

Yang Y, Hao Z, An N, Han Y, Miao W, Storey KB, Lefai E, Liu X, Wang J, Liu S, Xie M, Chang H (2023) Integrated transcriptomics and metabolomics reveal protective effects on heart of hibernating Daurian ground squirrels. J Cell Physiol 238:2724–2748

Yi S, Moore C, Lee R (2007) Rapid cold-hardening protects *Drosophila melanogaster* from cold-induced apoptosis. Apoptosis 12:1183–1193

Yin X, Chen P, Chen H, Jin W, Yan X (2017) Physiological performance of the intertidal Manila clam (*Ruditapes philippinarum*) to long-term daily rhythms of air exposure. Sci Rep 7:41648

Zambonino-Infante JL, Claireaux G, Ernande B, Jolivet A, Quazuguel P, Sévère A, Huelvan C, Mazurais D (2013) Hypoxia tolerance of common sole juveniles depends on dietary regime and temperature at the larval stage: evidence for environmental conditioning. Proc R Soc Lond B 280:20123022

Zepp RG, Shank GC, Stabenau E, Patterson KW, Cyterski MJ, Fisher WS, Bartels E, Anderson SL (2008) Spatial and temporal variability of solar ultraviolet exposure of coral assemblages in the Florida keys: importance of colored dissolved organic matter. Limnol Oceanogr 53:1909–1922

Zhang W, Niu C, Liu Y, Storey KB (2019) Positive or negative? The shell alters the relationship among behavioral defense strategy, energy metabolic levels and antioxidant capacity in freshwater turtles. Front Zool 16:3

Zhang Y, Liang S, He J, Bai Y, Niu Y, Tang X, Li D, Chen Q (2015) Oxidative stress and antioxidant status in a lizard Phrynocephalus vlangalii at different altitudes or acclimated to hypoxia. Comp Biochem Physiol Part A 190:9–14

Zika RG, Moffett JW, Petasne RG, Cooper WJ, Saltzman ES (1985) Spatial and temporal variations of hydrogen peroxide in Gulf of Mexico waters. Geochim Cosmochim Acta 49:1173–1184

# Chapter 6
# From Strenuous Migration to Dormancy and Arousal: How Physical (In)Activity Affects Oxidative Status

**Abstract** Physical effort requires a number of energetic, metabolic, hormonal, and cardiovascular responses that can be highly costly for the organism to sustain. Even physical inactivity, typical of animals during hibernation or aestivation, requires several physiological adjustments of the metabolic machinery, which have also to prepare the organism for the reoxygenation and increased metabolic activity that will characterise the phase of arousal. A number of studies described the many ways in which animals regulate their oxidative status during different phases of physical (in)activity, such as strenuous migratory flight, dormancy, or arousal. Many authors also pointed to a role of exercise-induced reactive oxygen species in mediating hormetic responses to physical effort. In this chapter, I have examined the role of oxidative stress induced by physical effort as cost of and constraint on migration strategies and sibling competition. I have also analysed the oxidative challenges that animals face during periods of inactivity, and the transition from dormancy to arousal due to tissue reoxygenation and consequent burst of reactive oxygen species generation in body tissues.

## 6.1 Introduction

Research has documented a number of factors that influence patterns of physical activity in animals (e.g. Kahurananga and Silkiluwasha 1997; Boldt and Ingold 2005; Hopcraft et al. 2005; Nevoux et al. 2013; Bennie et al. 2014; Récapet et al. 2016). Moving from one place to another one or fighting for a certain resource may carry benefits, such as higher chances of mating or larger availability of food. Increased physical activity, however, requires energy and many types of physiological responses that can be costly for the organism to sustain (e.g. McWilliams et al. 2004; Bonte et al. 2012; Récapet et al. 2016). For example, oxygen uptake and hear rate can increase dramatically just after 6 min of flight in a wind tunnel in pigeons *Columba livia* and 20 min running on a treadmill in emus *Dromaius novaehollandiae* (Butler et al. 1977; Grubb et al. 1983). Increased metabolic activity of

© The Author(s), under exclusive license to Springer Nature Switzerland AG 2024

D. Costantini, *The Role of Organismal Oxidative Stress in the Ecology and Life-History Evolution of Animals*, Fascinating Life Sciences,
https://doi.org/10.1007/978-3-031-65183-0_6

skeletal muscle is a main driver of these physiological costs. But these costs vary across the environments because other factors can also mediate the consequences of muscular effort. For example, flying across vast stretches of windswept ocean requires physiological demands much higher than those required for flying over calm prairies. Diving for fishing also exposes birds and mammals to episodes of hypoxia and reoxygenation (see Chap. 5). Similarly, animals that spend much of their life underground, live at high altitude, or fly across mountainous regions perform their activities under conditions of low partial pressure of oxygen (Ramirez et al. 2007). For example, peroxiredoxin 6, glycosylase 1, glutathione peroxidase, superoxide dismutase 2, copper-zinc superoxide dismutase, and haem oxygenase (all enzymes involved in minimising oxidative damage) were upregulated in high-altitude rufous-collared sparrows *Zonotrichia capensis* compared to low-altitude rufous-collared sparrows (Cheviron et al. 2008). However, resistance against oxidative stress in relation to abiotic conditions connected to altitude varies across taxa and tissues (e.g. Reguera et al. 2015; Zhang et al. 2015).

Physical activity can have important effects on the organismal oxidative status. For example, mitochondria of contractile skeletal muscle cells generate reactive oxygen species during exercise, which may cause oxidative damage, especially when the exercise is intense or prolonged (e.g. Alessio 1993; Ji 1999; Bailey et al. 2003, 2004; Powers and Jackson 2008; Powers et al. 2011; Chen et al. 2022; Davis et al. 2023). Powers and Jackson (2008) reported that the first suggestion that physical exercise results in free radical-mediated damage to tissues had probably appeared in 1978 (Dillard et al. 1978). Since then, there have been a growing number of studies that attempted to elucidate the molecular and cellular mechanisms linking physical effort to oxidative stress. In skeletal muscle, there are several potential subcellular sites and mechanisms that are responsible for the generation of the superoxide free radical, such as mitochondrial matrix and electron transport chain, sarcoplasmic reticulum, or phospholipase $A_2$-dependent processes (Powers and Jackson 2008). Moreover, physical activity can cause inflammatory responses (Shephard and Shek 1994; Woods et al. 1999; Murphy et al. 2008), which in turn may influence the oxidative status (see Chap. 2). For example, animals may redirect macrophages and other phagocytic cells towards muscle fibres that were injured during the contractile activity (McArdle et al. 2004). This process is important for the regeneration of the tissue damaged; however, phagocytic cells release reactive oxygen species that may cause further oxidative damage (Zerba et al. 1990; Powers and Jackson 2008).

Although physical activity may cause an elevation of metabolic rate and energy expenditure, oxidative damage does not necessarily increase proportionally or does not increase at all. One possible reason for this result may lie with the stimulation of antioxidant mechanisms that are upregulated in response to reactive oxygen species, or the reduction in the leakage of protons from the mitochondria that reduces the generation of reactive oxygen species and oxidative damage as a consequence (e.g. Loschen et al. 1971; Herrero and Barja 1997; Gomez-Cabrera et al. 2008; Nanayakkara et al. 2019). Birds may decrease energy expenditure with flight duration (until ca. 200 min) and balance aerobic and anaerobic metabolism in order to make the flight both economic and tolerable and to avoid oxidative stress (Bairlein et al. 2015). A number of studies found that physical activity may result in the hormetic activation of many molecular pathways, involving anti-stress and anti-inflammatory responses or mechanisms of damage repair (e.g. Ji et al. 2006; Gomez-Cabrera et al. 2008; Radak et al. 2008, 2017; Rattan 2008; Nikolaidis et al. 2012a, b; Lesmana et al. 2022). There is, however, an impressive variation in patterns of (in)activity among animal species, which illustrates the difficulties of generating general models, but also offers exciting opportunities for comparative studies.

In this chapter, I have examined the role of oxidative stress induced by physical effort as cost of and constraint on animal performance in the contexts of migration and competition between individuals. I have also analysed the oxidative challenges that animals face during periods of inactivity, such as hibernation, and the transition from dormancy to arousal owing to the rapid tissue reoxygenation and consequent burst of reactive oxygen species generation in body tissues.

## 6.2 Physical Effort and Oxidative Stress

There is considerable evidence across animal species that physical effort (or proxies of it) and oxidative status markers have a non-monotonic relationship. For example, one of the first experiments about the oxidative costs of flight on birds showed that pigeons that flew around 200 km in a few hours had an increase in serum levels of reactive oxygen metabolites and a decrease in serum non-enzymatic antioxidant capacity compared to pigeons that were kept inside lofts or transport crates for the same duration of time. By contrast, serum antioxidants did not change, and serum reactive oxygen metabolites even decreased in pigeons that flew around 60 km for 1 h and a half (Costantini et al. 2008; Costantini and Lipp 2010). In addition, considering only pigeons that flew the same distance, those that came back to the loft earlier had less oxidative damage and saved more antioxidants than those that came back later (Costantini 2014). Prior studies on pigeons that were left to fly in a wind tunnel (Rothe et al. 1987) or in the wild (Schwilch et al. 1996) found that there is a shift to a high and stable lipid-based metabolism within the first 2 h of the flight. As a consequence, prolonged flights caused increased hydrolysis of triglycerides to free fatty acids and glycerol, and β-oxidation of free fatty acids in muscle cells (Schwilch et al. 1996). These metabolic changes may be determined to some degree by the time elapsed since feeding, which is an important determinant of the nutrients that pigeons use to sustain the metabolic and energetic demands of flight (Gannes et al. 2001). Overall, a lipid-based metabolism may have further contributed to increasing oxidative stress in pigeons that flew 200 km. Unsaturated fatty acids are the main forms of lipids stored in avian tissues (McWilliams et al. 2004) and are quite susceptible to free radical-induced oxidative damage compared to other molecules (see Chap. 1). It might also be that an increase in the ratio of unsaturated over saturated fatty acids, which occurs in the blood after exercise, may have further exacerbated the level of oxidation given the higher propensity of unsaturated fatty acids to be oxidised compared to saturated fatty acids (Nikolaidis and Mougios 2004).

In mammals, Vaanholt et al. (2008) found that, although mice *Mus musculus* selectively bred for high wheel-running activity showed higher long-term voluntary exercise than control lines of mice, they had similar levels of the antioxidant enzymes superoxide dismutase and glutathione peroxidase. Similarly, short-tailed field voles *Microtus agrestis* that used running wheels had significantly higher energy expenditure than sedentary voles, but had similar levels of catalase, glutathione peroxidase, and superoxide dismutase in hind- and forelimb skeletal muscle, oxidative DNA damage in lymphocytes and liver, and thiobarbituric acid reactive substances in hindlimb (Selman et al. 2002). By contrast, levels of superoxide dismutase in lung and heart, catalase in liver, and glutathione peroxidase in liver, lung, and heart were increased significantly in rats which performed exercise compared to sedentary rats (Gunduz et al. 2004). The results of these studies show that the effects of physical effort on oxidative status markers can vary across tissues. They also suggest that, when non-experimentally forced to increase their physical effort, rodents might choose their own level of physical activity in order to avoid oxidative stress.

Although devices already exist for quantifying running activity in mice and rats (e.g. running wheels, Gattermann et al. 2004; Judge et al. 2005), these have the disadvantage of enabling individual animals to choose their level of activity, so they may slow down the rate of running if the physiological costs become significant. The use of wind tunnels has provided a powerful experimental tool to analyse the kinematics of flight and the effects of flight effort on the physiological status of birds (Pennycuick et al. 1997; Engel et al. 2010). Many bird species studied so far did not require training. By contrast, some species have appeared to be problematic because they required large amounts of space and extensive training sessions. A more user-friendly system for avian biologists involves the use of automated moveable perches that are positioned on opposite sides of a flight aviary, and that are operated on timers such that they alternately fold away, forcing the birds to fly from one side of the aviary to the other throughout the experimental session (Nudds and Bryant 2000).

Costantini et al. (2012, 2013) used one of these automated systems to test the effects of a short-term increase in flight activity on the oxidative status of zebra finches *Taeniopygia guttata*. Birds were allocated to three experimental groups that differed in the intensity of flight activity, including a control group, a group of birds that were faced with a mild flight effort, and a group of birds that were faced with a strong flight effort. Compared to control birds, those that flew more had a greater increase in plasma levels of reactive oxygen metabolites and protein carbonyls (Fig. 6.1a) and a greater decrease in red blood cell levels of antioxidant thiols (Fig. 6.1b), respectively. Birds that flew more had also greater increases in protein carbonyls and decreases in thiols than those birds that were faced with a mild flight effort (Costantini et al. 2013). The concentration of plasma uric acid in control birds significantly decreased over the experimental trials compared to the other two experimental groups. By contrast, levels of catalase, glutathione peroxidase, superoxide dismutase, and carotenoids did not differ among groups. The multivariate

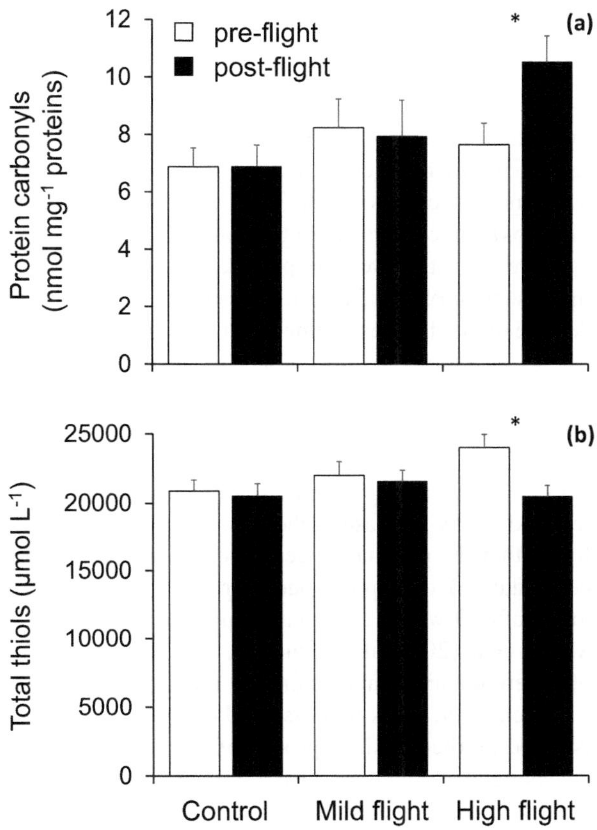


**Fig. 6.1** High flying effort (**a**) increased significantly the concentration of protein carbonyls in red blood cells and (**b**) decreased significantly the concentration of total thiols in red blood cells of zebra finches *Taeniopygia guttata*. Asterisks indicate significant differences between pre- and post-flight values. Values are shown as mean and standard error. Data were collected from Costantini et al. (2013)

analysis of variance further supported that the physiological response of birds differed as a function of the flight effort, resulting in a smaller value of the index of integration (i.e. correlations among oxidative status markers were weaker in these birds compared to the other two experimental groups; see Pavlicev et al. 2009; Haber 2011) in the high-flying effort group. It might be that these birds could have perceived this sudden increase of flight activity as an unpredictable and stressful event. Although in the wild, zebra finches can fly tens of kilometres per day, under captive conditions the physical activity of birds is usually low. Birds that experienced the mild and high-flying effort flew, respectively, around 6 and 16 times the distance of control birds, who were probably as active as all birds had been in their holding cages prior to the experiment (Costantini et al. 2013). Therefore, the flying effort was so great in such a short-time period that birds could not have been flexible enough to rapidly upregulate their antioxidant levels to respond to it (Jones 2006; Sohal and Orr 2012).

Another explanation for the increase in oxidative stress in birds that flew more might lie with the costs incurred for upregulating antioxidant defences. For example, zebra finches could be programmed to sacrifice protection against oxidative stress during periods when allocation of resources to other functions is more relevant. Zebra finches, whose flight activity was experimentally increased (by increasing the number of nestlings that they had to feed), showed a decrease in the levels of superoxide dismutase and glutathione peroxidase in pectoral muscle (Wiersma et al. 2004). Although the zebra finches involved in the flight experiment by Costantini and collaborators were not breeding, it could be that the experimental setting might have created conditions similar to the experiment carried out by Wiersma and collaborators. This would imply that the regulatory mechanisms of blood oxidative status might be similarly expressed across different stages of the life cycle (e.g. nestling-rearing period, migration) that require animals to significantly increase their physical effort. If this were the case, why did not antioxidant enzymes decline in the flight experiment as they did in the reproduction experiment? It might be that the additional effort of nestling-rearing, the younger age of the birds involved in the flight experiment, or the tissue analysed (blood vs. muscle) could have affected the results. The flight for breeding birds was also probably very demanding because the birds did not have training as much as the birds included in the flight experiment. Training is very relevant to prepare the physiological machinery to respond to any increase in the physical effort. For example, Larcombe et al. (2010) found that the plasma concentration of malondialdehyde (an end-product of lipid peroxidation) of captive budgerigars *Melopsittacus undulatus* was significantly lower after 9 weeks of flight training than after a single exercise session. Cooper-Mullin et al. (2019) also found that exercise training might synchronise the response of different antioxidants to energy-intensive activity in zebra finches.

Much of the work that assessed the impact of physical effort on oxidative status is based on the classic approach of increasing the physical activity of animals and analysing the effects at physiological level. We have also relevant studies that addressed the link between physical effort and oxidative status by reducing somehow the activity of animals. For example, prevention of flight activity of male houseflies *Musca domestica* decreased their rate of oxygen consumption and almost tripled their lifespan compared to those houseflies that were free to fly (Yan and Sohal 2000). Low activity houseflies had a lower generation of mitochondrial hydrogen peroxide at various ages than high activity houseflies. Finally, low activity houseflies had lower levels of oxidative damage to mitochondrial proteins and a slower loss in mitochondrial functionality with age than high activity houseflies (Yan and Sohal 2000). Honey bees *Apis mellifera* with much flight experience exhibited elevated levels of 8-hydroxy-2′-deoxyguanosine in flight muscles relative to bees with less flight experience. Bees that had much flight experience also showed increased levels of superoxide free radical and hydrogen peroxide, and decreased or unchanged levels of superoxide dismutase and catalase (Margotta et al. 2018).

Things are not always what they seem. Although the results of these studies suggested that increased generation of reactive oxygen species and oxidative damage are two main consequences of increased physical effort, we have also empirical evidence that supports an opposite scenario. For example, fruit flies *Drosophila melanogaster* that were allowed to fly throughout their life had higher oxidative damage than control fruit flies, but their generation of mitochondrial reactive oxygen species was similar to that of control fruit flies (Magwere et al. 2006). A number of studies also found that the low levels of reactive oxygen species generated during a moderate exercise can stimulate the upregulation of antioxidant enzymes, or of those molecules that can repair damage to proteins or nucleic acids, resulting in high protection against oxidative stress as a consequence (e.g. Davies 1986; Hollander et al. 2001; Radak et al. 2003, 2008, 2017; Alessio and Hagerman 2006; Ji et al. 2006; Powers et al. 2011; Nikolaidis et al. 2012a, b; Lesmana et al. 2022). For example, in birds, factors such as the mode of flying (e.g. short with repeated take-offs, steady-state), duration of flight, or intensity of flight determine whether the flight effort is either stimulatory or detrimental in terms of energy expenditure or stress level (e.g. Costantini et al. 2008). Similarly, in mammals, moderate physical activity reduced levels of oxidative damage to nuclear and mitochondrial nucleic acids in rats (Nakamoto et al. 2007), while a high physical effort increased levels of isoprostanes, thiobarbituric acid reactive substances, protein carbonyls, oxidised glutathione, catalase, and glutathione peroxidase, and decreased levels of reduced glutathione and non-enzymatic antioxidant capacity in humans (Margonis et al. 2007).

Reactive oxygen species appear to have a key role in triggering the hormetic stimulation of the antioxidant and redox regulation machinery, probably through their action as cellular signals. Supplementation with antioxidants can preclude the exercise-induced activation of health-promoting mechanisms in rats or humans, probably through the suppression of the signalling activity of reactive oxygen species (e.g. Gomez-Cabrera et al. 2005; Ristow et al. 2009). However, other studies did not find any detrimental effects of antioxidant supplementation on the positive effects of exercise in mammals (e.g. Theodorou et al. 2011; Nikolaidis et al. 2012a). Studies on birds also did not find any evidence for these detrimental effects of antioxidant supplementation. For example, male budgerigars that were on a diet rich in antioxidants during 2 months of flight training had a greater performance in repeated escape flights than males that were on a diet poor in antioxidants (Arnold et al. 2010). Moreover, budgerigars that were experimentally stimulated to perform short flights mimicking escapes from predators had increased levels of oxidative DNA damage and malondialdehyde when on the antioxidant-poor diet, but not when on the enhanced, antioxidant-rich diet (Larcombe et al. 2008).

## 6.3 Costs of Migration and Strategies to Reduce Them: From Muscle Damage to Cognition

Every year, millions of animals engage in strenuous journeys over long distances from the breeding to the wintering grounds and back. Migration is a ubiquitous phenomenon, described in many animal groups as diverse as insects, crustaceans, fish, amphibians, reptiles, birds, and mammals (Aidley 1981; Milner-Gulland et al. 2011). There are many benefits for migrating animals, such as moving towards environments with better climatic conditions, higher availability of food, or other better conditions for successfully reproducing. However, migration imposes a number of physiological challenges, such as a high energetic turnover, fasting, and reduction in organ sizes (Gwinner 1990; McWilliams et al. 2004). For example, birds fuel their migratory flight deriving most of the energy from fat and protein stores accumulated in the pre-migratory phase (Berthold 2001). Accumulating fat and protein stores before migrating and taking stops during the migration to refuel are fundamental for the timing and success of migration in birds (Alerstam and Lindstrom 1990; Jenni et al. 2000). However, we have much evidence that a high caloric intake can increase the generation of reactive oxygen species (see Chap. 3). Moreover, these super-athletes of the animal kingdom start their migratory journey with much body fat, which increases their wing loading. The question then is how birds balance the benefits of accumulating and metabolising large amounts of fats with the potential costs of increased generation of reactive oxygen species.

Marasco et al. (2021) examined whether the emergence of the migratory phenotype, primarily signalled by increased food intake and fuelling, is associated with changes of the oxidative status. Relying on an experimental induction of autumn migration followed by a non-migratory wintering phase, they compared levels of oxidative status markers between common quails *Coturnix coturnix* sampled during the migratory and the non-migratory phases, respectively. The emergence of the migratory phenotype was associated with (1) higher levels of thiobarbituric acid reactive substances in the liver, (2) lower levels of superoxide dismutase in red blood cells and, marginally, in the liver, and (3) higher levels of glutathione peroxidase in the pectoral muscle. There was no link between food intake and variation in markers of oxidative status in any of the tissues examined, despite food intake was higher in the migratory quails. The increase in body mass was, however, positively correlated with muscle levels of glutathione peroxidase as birds entered the pre-migratory fattening phase, while the amount of decrease in body mass was negatively correlated with muscle levels of glutathione peroxidase as birds transitioned to the non-migratory phase (Marasco et al. 2021). The results of this experimental study suggested that, during the emergence of the migratory phenotype, birds might strategically displace oxidative costs on the liver in order to safeguard the pectoral muscles, which play a fundamental role in achieving a successful migratory flight. This interpretation would suggest that birds might modify their physiological status in order to be prepared to cope with the oxidative challenges of the migration flight (*preparation for the oxidative costs of migration hypothesis*). Similarly, Gutiérrez et al. (2019) found that Hudsonian godwit *Limosa haemastica* increased plasma non-enzymatic antioxidant capacity and reduced thiobarbituric acid reactive substances and lipid hydroperoxides in red blood cells as the pre-migratory season progressed, despite higher basal metabolic rates before departure, which might represent an adaptation to diminish the physiological costs of long-distance migration. Photo-stimulation of male red-headed bunting *Emberiza bruniceps* (used to trigger their pre-migratory phenotype) increased expression of the genes GPX1 and SOD1 in spleen (Sur et al. 2020) and reduced the generation of reactive oxygen species in erythrocytes (Bhardwaj et al. 2023).

The active migratory flight is the next phase during which the oxidative status homeostasis of birds can be severely challenged. Birds maintain an intense and prolonged metabolic activity required to sustain their flight, experience depletion of nutrients and dietary antioxidants because of the increased metabolic rate and prolonged fasting, and suffer tissue inflammation. Moreover, migratory animals may be relatively sedentary outside of the period of migration, leading to marked contrasts in the intensity of flight activity required during the year. Bird species also differ in their migratory strategies, which might build up further variation in the physiological performance (McWilliams et al. 2004; Falsone et al. 2009). For example, while some species migrate during the daylight, other species fly during the night and rest and forage during the daylight. The time budget dedicated to foraging or, in general, to stopover is influenced by the amount of energy stores that are still available: fat individuals leave the site sooner than lean individuals (Fusani et al. 2009; Goymann et al. 2010). The environmental conditions at the stopover site, where birds stop to rest and/or feed, may also influence the capability of birds of coping with oxidative stress. Good stopover sites would provide food and rest that birds need in order to recover from the strenuous flight, but inhospitable places would probably reduce the chances of obtaining enough resources to sustain their endogenous functions. This scenario is supported by studies carried out on migrating songbirds at stopover sites, which appear to prepare for, and recover from, oxidative stress posed by long-distance flights (Skrip et al. 2015; Eikenaar et al. 2020a, 2023), for example, through replenishing of dietary antioxidant stores (Eikenaar et al. 2016; Oguchi et al. 2017; Cooper-Mullin and McWilliams 2022).

Stopover sites are also probably important for migratory mammals. Costantini et al. (2019) tested this hypothesis by comparing values of blood-based markers of oxidative status between Nathusius' bats *Pipistrellus nathusii* that were caught during migration flights with those measured in conspecific individuals that, after capture, were left resting into cages for 18 or 24 h. Flying Nathusius' bats had higher oxidative damage and different levels of both non-enzymatic and enzymatic antioxidants (glutathione peroxidase) compared to resting bats (Fig. 6.2). The differences in oxidative status markers varied between sampling years and were independent from individual body condition or sex. These results provided evidence that migratory flight might impose acute oxidative stress to bats, and that resting helped bats to recover from oxidative damage accrued *en route*. Although birds and bats have disparate adaptations with respect to their physiology during migration, the results of these studies suggest that migrating birds and bats might share similar strategies of mitigating and recovering from oxidative stress. Further work on bats found that oxidative damage was higher in migrating than in pre-migrating Nathusius' pipistrelles (Fig. 6.3) (Voigt et al. 2020). However, injection of the antigen lipopolysaccharide did not affect plasma oxidative status markers in both migrating and pre-migrating Nathusius' pipistrelles. Inflammatory markers were also similar between migrating and pre-migrating animals. These results supported the hypothesis that the intense physical exercise of migratory flight results in high levels of oxidative stress, most likely caused by intense metabolic activity rather than by inflammatory processes.

Migratory species have evolved a number of physiological and morphological characteristics that make them better at coping with flight-induced oxidative stress

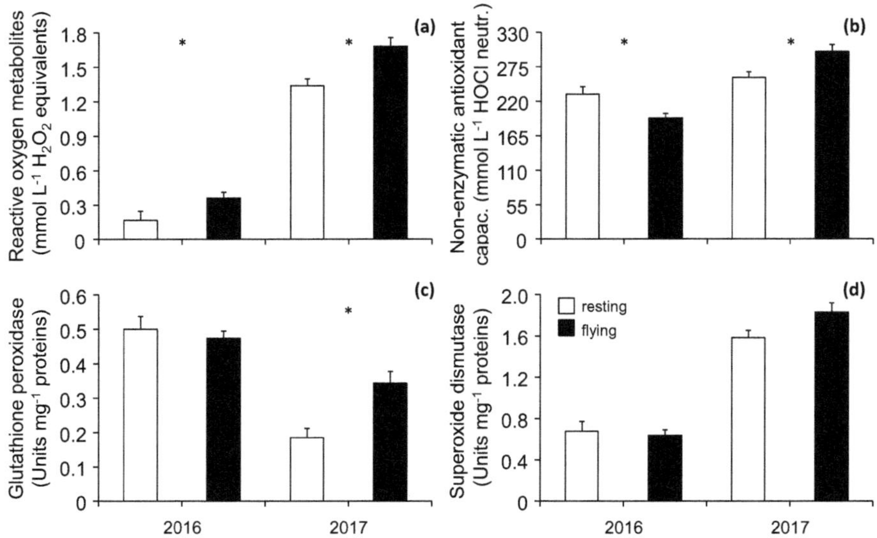


**Fig. 6.2** Differences between flying and resting bats in the sampling years 2016 and 2017 for (**a**) plasma reactive oxygen metabolites, (**b**) plasma non-enzymatic antioxidant capacity, (**c**) erythrocyte glutathione peroxidase, and (**d**) erythrocyte superoxide dismutase. Asterisks indicate significant differences between flying and resting bats. Values are shown as least square mean and standard error. Reprinted with minor modifications from Costantini et al. (2019) with permission of Oxford University Press

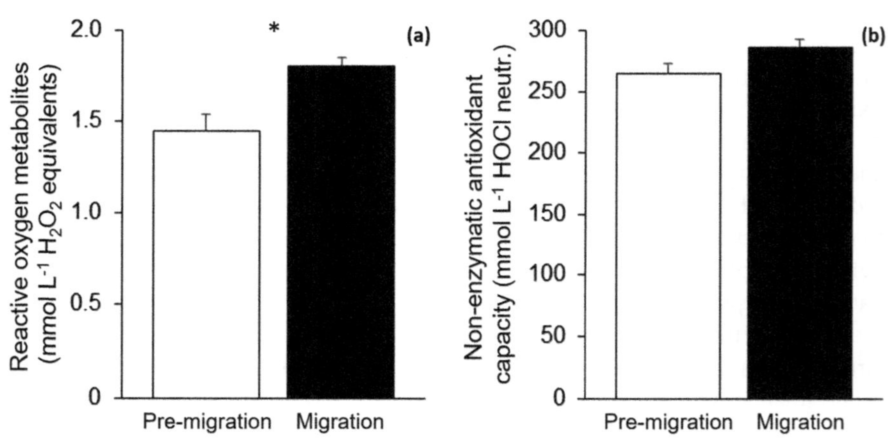


**Fig. 6.3** Differences between pre-migration and migration seasons in Nathusius' pipistrelles *Pipistrellus nathusii* for (**a**) plasma reactive oxygen metabolites and (**b**) plasma non-enzymatic antioxidant capacity. Asterisks indicate significant differences between pre-migration and migration seasons. Values are shown as least square mean and standard error. Redrawn from Voigt et al. (2020). © 2020, The Author(s); distributed under the terms of the Creative Commons Attribution License (CC BY)

than non-migratory species. Migratory endurance flights are fuelled mainly with fatty acids, and the risk of their peroxidation (resulting in oxidative damage to lipids) increases with the number of double bonds (see Chaps 1 and 3). Arctic terns *Sterna paradisaea* migrate annually between the northern and southern polar areas. That they might make this demanding trip dozens of times during their long life shows that they have evolved appropriate mechanisms for this intermittent high exercise level. Seabirds, like the Arctic tern, have cell membranes particularly resistant to oxidative damage. For example, the analysis of myocardial phospholipids showed that *Procellariiformes* have cell membranes that are more resistant against peroxidation than those of *Galliformes*. The reason for this difference likely lies with the cell membrane composition, because seabirds have proportionately higher concentrations of monounsaturated fatty acids and lower concentrations of polyunsaturated (particularly n-6) fatty acids in their cell membranes than fowl, respectively (Buttemer et al. 2008). Similarly, migratory common blackbirds *Turdus merula* appeared to change the composition of key fatty acids in order to retain an unsaturation index constant and similar to resident blackbirds, despite the increased levels of peroxidisable polyunsaturated fatty acids (Jensen et al. 2020). Other authors found evidence that, by fuelling their flights with relatively few polyunsaturated fatty acids, migratory animals could limit oxidative damage to lipids (McWilliams et al. 2020; Eikenaar et al. 2022). However, Dick and Guglielmo (2019a, b) did not find any significant effects of dietary long-chain polyunsaturated fatty acids on levels of protein carbonyls, catalase, superoxide dismutase, and glutathione in flight muscle at rest or after endurance flight, nor any effect on flight performance.

We have also evidence that conspecific migratory individuals or different species do not all perform as good as we might expect. Costantini et al. (2007) quantified plasma levels of reactive oxygen metabolites and non-enzymatic antioxidant capacity in garden warblers *Sylvia borin* and barn swallows *Hirundo rustica* caught by mist-netting at a stopover site on a small island located around 500 km off the African coast from which birds departed for the migratory flight towards the breeding European grounds. Estimates made on a neighbouring island suggested that birds landing in these islands had completed a 14–16 h non-stop flight (Pilastro et al. 1995; Schwilch et al. 2002). The results of this correlative work showed that garden warblers that landed in better condition (i.e. with higher fat stores) had higher levels of plasma non-enzymatic antioxidant capacity than garden warblers that landed in worst condition (Costantini et al. 2007). By contrast, the correlation between fat stores and plasma non-enzymatic antioxidant capacity was small in barn swallows. Interestingly, barn swallows have relatively low metabolic activity compared to other birds (Hails 1979; Pennycuick 1990; Bruderer et al. 2001). It has been estimated that the metabolic rate of flying hirundines, like barn swallows and swifts, is around 50–70% lower than the metabolic rate of other birds of similar body size (Hails 1979). Moreover, unlike most other passerines, hirundines rarely show regular flapping flight (Bruderer et al. 2001), a type of flight that requires one of the highest metabolic demands of any form of endurance locomotion in vertebrates (Pennycuick 1990). Similar correlations between fat stores and markers of oxidative status were also documented in other songbirds. For example, Eikenaar et al. (2020b) found that northern wheatear *Oenanthe oenanthe* with larger energy stores at a stopover during autumn migration had lower plasma levels of malondialdehyde, a secondary product of peroxidation of polyunsaturated fatty acids (marker of oxidative damage to lipids). However, Ferretti et al. (2020) did not find any significant correlations between oxidative status markers and energy stores at stopover site in both garden warblers and whitethroats *Sylvia communis*. Although the results of these studies did not provide any proof that birds suffered oxidative stress while migrating, levels of fat stores are a significant proxy of migration effort, and a predictor of the likelihood to complete endurance flights. Thus, if oxidative stress accrued during migration reduces future chances of reproducing successfully, saving antioxidants during migration might carry benefits for reproduction. Oxidative stress might therefore be one important mediator of the trade-off between migratory effort and expression of those traits that prepare the organism for breeding. The hypothesis that oxidative costs owing to the strenuous physical efforts of migration might impinge on subsequent reproduction is supported by an experimental study that found an effect of flight effort on the deposition of lutein (a type of carotenoid) into eggs in captive zebra finches (Skrip et al. 2016). This is a topic that definitely warrants more research in order to get more insight about the significance of variation in migration strategies.

Further studies provided evidence in favour of a link between migration and oxidative stress. For example, Ninni et al. (2004) found that late arriving male barn swallows had higher levels of circulating carotenoids, but they depleted a larger quantity of carotenoids shortly after settling in the breeding area compared to early arriving male barn swallows. Carotenoids are stored in fat and used in times of high physiological demands as shown in garden warblers before the autumn migration (Metzger and Bairlein 2011). Jenni-Eiermann et al. (2014) found that protein carbonyls and glutathione peroxidase were higher in European robin *Erithacus rubecula* caught out of their nocturnal flight than in conspecifics caught during the day while resting, and, particularly, in those individuals that had reduced flight muscles. Eikenaar et al. (2017) found that migrating blackbirds had higher plasma non-enzymatic antioxidant capacity than sedentary blackbirds. After flying for up to a maximum of 6 h in a wind tunnel, yellow-rumped warbler *Setophaga coronata coronata* had increased activity of superoxide dismutase in flight muscle, but also evidence of increased oxidation of glutathione (GSSG) and levels of protein carbonyls (Dick and Guglielmo 2019a, b). Furthermore, protein damage was strongly linked to energy expenditure and flight efficiency at the individual level (Dick and Guglielmo 2019a, b). There is also indirect evidence that migratory flight may cause oxidative stress. For example, experimentally increased foraging effort augmented the levels of reactive oxygen metabolites in captive zebra finches (Yap et al. 2017). Correlative data collected from GPS/accelerometer-tagged streaked shearwaters *Calonectris leucomelas* and Cory's shearwaters *Calonectris borealis* or geolocator-tagged long-tailed jaegers *Stercorarius longicaudus* and Arctic terns also suggested a link between oxidative stress and increased foraging activity or flight distance, respectively (Koyama et al. 2021; Laranjeiro et al. 2021; Fowler et al. 2023). In fish, adult sea-run brown trout *Salmo trutta* had higher values of the GSSG/GSH ratio than resident adult brown trout (Peiman et al. 2017). By contrast, migratory juvenile brown trout had higher levels of reduced glutathione and non-enzymatic antioxidant capacity in red blood cells than resident brown trout, while the concentration of oxidised glutathione in red blood cells did not differ between them (Birnie-Gauvin et al. 2017).

Most of this work focussed on oxidative stress as a cost of migration. But can oxidative stress accrued during the migration constrain the flight performance? For example, oxidative stress could affect the contractile function of skeletal muscle. Many studies provided compelling evidence that prolonged muscular contractions cause accumulation of reactive oxygen species that contribute to muscle fatigue (e.g. Powers and Jackson 2008; Reid 2008; Powers et al. 2011; Westerblad and Allen 2011; Debold 2015). For example, induction of oxidative stress by means of the experimental knockout of manganese superoxide dismutase impaired the activity of mitochondria and lowered endurance capacity in energetically demanding activities (e.g. Williams et al. 1998; Kinugawa et al. 2005; Lustgarten et al. 2009, 2011). Similarly, oxidative stress limited prolonged flight performance in migratory locust *Locusta migratoria* (Du et al. 2022). The involvement of reactive oxygen species in endurance to muscle effort is further supported by studies that provided antioxidants to experimental subjects. For example, supplementation of the dietary antioxidant vitamin E prolonged endurance to muscle fatigue of mice submitted to a swimming test (Novelli et al. 1990). Similarly, intravenous infusions of the antioxidant N-acetylcysteine significantly increased the time to fatigue in human subjects that exercised to fatigue on a cycle ergometer (McKenna et al. 2006).

Oxidative stress could also impair the migratory performance through its potentially negative effects on cognitive function. Many studies showed that oxidative stress may cause memory impairment, decreased learning, or reduced homing abilities (e.g. Cram 2022). For example, Liu et al. (2003) found that the dramatic decline in learning and memory taking place in early middle-aged mice was associated with a large increase in brain oxidative stress, and that chronic treatment with low doses of mimetics of superoxide dismutase and catalase almost completely reversed the behavioural and the biochemical changes. The results of this experimental study supported the role of reactive oxygen species in age-related learning impairment. Similarly, Rosa et al. (2007) investigated the effects of intense and exhaustive exercise on both cognition (passive avoidance task, contextual fear conditioning, and tone fear conditioning, performed 24 h after the last exercise session) and brain oxidative stress in male mice. There was a remarkable memory reduction of exercised mice in comparison with control mice, and this memory reduction was associated with an increase in the levels of thiobarbituric acid reactive substances and protein carbonyls in brain. Both effects of exercise on memory and oxidative damage were prevented by concurrent supplementation of vitamins C and E (Rosa et al. 2007).

Exposure to high levels of reactive oxygen species is implicated in brain ageing and in the pathogenesis of a number of neurodegenerative diseases. However, it is increasingly recognised that exposure to mild doses of reactive oxygen species may enhance some brain functions (e.g. spatial learning, memory) and neurogenesis, possibly through hormetic effects (e.g. Anderson et al. 2000; Ang et al. 2006; Pereira et al. 2007; Calabrese et al. 2010; Mao and Franke 2013; Tauffenberger and Magistretti 2021). Consistent with the concept of mitohormesis, there is evidence that transiently increased levels of exercise-induced reactive oxygen species stimulate a number of physiological responses that promote the maintenance of organismal homeostasis (Ristow et al. 2009; Mattson and Calabrese 2010). There might be a potential for hormesis to play a role as a mediator of physiological adjustments to the migratory effort, because migration can be performed under very variable conditions that might expose individuals to different amounts of reactive oxygen species. Empirical research on this topic is an important area for future research.

Another important topic related to the physiological demands of migration concerns the need of antioxidants that would be consumed to buffer flight-induced oxidative stress. Animals might mitigate oxidative costs of migration or attenuate costs linked to other physiological pathways by consuming prophylactically (before departure) or therapeutically (while *en route* at stopover sites) food rich in antioxidants, such as polyphenols (Bolser et al. 2013; Skrip et al. 2015; Cooper-Mullin and McWilliams 2016; Casagrande et al. 2020; Frawley et al. 2021). However, other studies did not find any support for a role of antioxidants as enhancers of performance of flying birds (Carter et al. 2020). Research on laboratory rodents found that polyphenols may enhance both cognitive abilities and resistance against oxidative stress because polyphenolic compounds can cross the blood brain-barrier and localise in those regions of brain that regulate learning and memory (e.g. Andres-Lacueva et al. 2005; Shukitt-Hale et al. 2005; Papandreou et al. 2009). For example, administration of polyphenol extracts to rats increased their muscle strength and balance, or enhanced neuronal functioning and restored the capability of brain to generate a neuroprotective response to stress (Fig. 6.4) (Shukitt-Hale et al. 2005). Similarly, administration of a polyphenol-rich extract of wild blueberries *Vaccinium angustifolium* to mice increased their learning and memory and reduced their levels of thiobarbituric acid reactive substances in brain, respectively (Papandreou et al. 2009). Grape powder (rich in polyphenols, such as quercetin and resveratrol) treatment of laboratory rats (1) prevented and reversed social defeat-induced behavioural and cognitive deficits, (2) decreased social defeat-induced increase in plasma 8-isoprostanes, (3) replenished social defeat-induced depleted pool of key antioxidant enzymes, and (4) protected hippocampal cells against damage (Solanki et al. 2017). Given that polyphenols were detected in the avian brain (greenfinch *Carduelis chloris*, Mullen et al. 2010), there is also potential for polyphenols to play a relevant role as protectors of brain functions in migrating birds. However, experiments of supplementation of dietary polyphenols found contrasting evidence about their antioxidant role in birds (Cecere et al. 2016; Sebastiano et al. 2018). Further research is definitely required to get insight into the circumstances under which polyphenols provide benefits for individual physical performance.

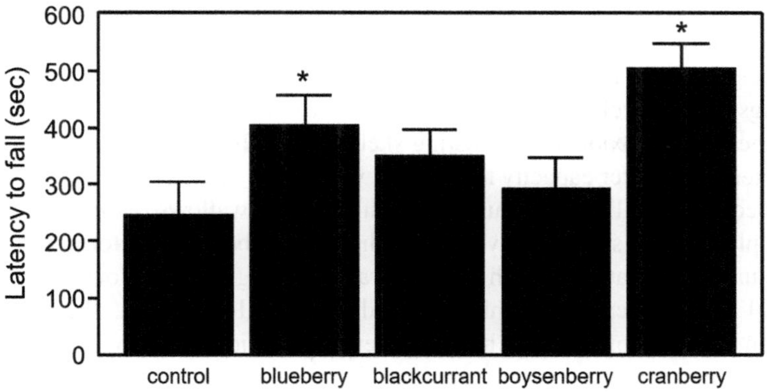


**Fig. 6.4** Latency to fall of rats on the inclined screen test for the control, blueberry, blackcurrant, boysenberry, and cranberry diet groups. The latency to fall is used as a quantitative metric of motor function: the more neurological impairment the animals have, the sooner they fall off the rod. Asterisks indicate significant differences from the control group. Values are shown as mean and standard error. Reprinted with minor modifications from Shukitt-Hale et al. (2005) with permission of Springer

## 6.4 Oxidative Stress and Physical Effort: Lessons from Sibling Rivalry and Quarrels

The physiological demands of physical activity are also an important proximate determinant of the extent and form of competition (e.g. begging, fights) among siblings over access to limited parental resources. In species with parental care, offspring have evolved behaviours to solicit their parents to provide food and care, meaning that more resources are given to those siblings that perform better in competition (e.g. Rodríguez-Gironés et al. 1996; Godfray 1991; Caro et al. 2016). Life-history theory predicts, however, that such behaviours carry costs for the offspring, and these costs would prevent a runaway escalation of solicitation behaviours (e.g. Macnair and Parker 1979; Parker et al. 2002). The *oxidative cost of sibling competition hypothesis* would predict that oxidative stress might be one of these endogenous costs.

In many altricial bird species, hatching asynchrony (time span between the hatching of the first and last egg within the same clutch) generates different casts of offspring (structured family), with core and marginal offspring being the earlier and later hatched, respectively. Consequently, the younger and smaller marginal offspring may be at a disadvantage compared to their older and larger siblings. Being born in a large family might also increase the cost of competition compared to being born in a small family because of the higher number of siblings to compete with. In addition, competition between siblings could also affect indirectly the organismal oxidative status through an effect on the growth rate. Hall et al. (2010) found that an experimental increase of the availability of antioxidants could increase the growth rate of nestling red-winged blackbirds *Agelaius phoeniceus*, irrespective of hatching order, but it did not influence their begging behaviour. Furthermore, the supplementation of antioxidants reduced the plasma concentration of malondialdehyde in marginal nestlings and increased the plasma concentration of malondialdehyde in core nestlings, respectively. One explanation for this result might lie with the higher investment of antioxidants in boosting skeletal growth because this strategy would pay in terms of better capacity to compete with larger siblings.

Boncoraglio et al. (2012) found that nestling barn swallows *Hirundo rustica* with higher plasma levels of oxidative damage had poorer begging performance in terms of posture and duration, which are two traits of begging behaviour that influence parental food allocation in this songbird species. In addition, the relationship between duration of begging bout and levels of oxidative damage was stronger when nestlings were experimentally food-deprived, pointing to nutritional stress as an important factor that might drive this relationship. Moreno-Rueda et al. (2012) also found that begging may entail some oxidative costs in birds. Nestling magpies *Pica pica* that were experimentally solicited to spend more time begging had a higher plasma concentration of thiobarbituric acid reacting substances. Nestlings that begged for longer bouts also reduced their growth and downregulated their immune response, possibly to maintain the homeostasis of their oxidative status. In contrast to these studies, Maronde and Richner (2015) found poor support for a role of oxidative stress as a cost of begging behaviour in great tits *Parus major*. Fuertes-Recuero et al. (2023) also found that oxidative costs of begging were minor in nestling barn swallows.

We have also indirect evidence in support of an oxidative cost of competition. This evidence is based on studies that compared markers of oxidative status between siblings, different broods in relation to their size (i.e. number of siblings), or broods whose size was experimentally enlarged or reduced. For example, Costantini et al. (2006) found that, irrespective of their body mass, nestlings from larger broods had higher levels of reactive oxygen metabolites than nestlings from smaller broods in the common kestrel *Falco tinnunculus* (Costantini et al. 2006). Similarly, Costantini et al. (2010) and Bourgeon et al. (2011) found that nestlings from larger broods had higher levels of oxidative stress than nestlings from smaller broods in the common starling *Sturnus vulgaris*. Several studies also found that circulating levels of non-enzymatic antioxidants were lower in nestlings from larger broods than nestlings from smaller broods (Alonso-Alvarez et al. 2006; Costantini et al. 2010; Bourgeon et al. 2011).

Comparisons between siblings provided further evidence for an oxidative cost of competition. For example, last-hatched nestlings of great tits had higher oxidative damage and ended growth with shorter telomeres than first-hatched nestlings (Stier et al. 2015). Nestling white-throated sparrows *Zonotrichia albicollis* that were smaller than nest mates had higher oxidative stress, and nestlings with higher oxidative stress and faster growth rates displayed shorter telomeres (Grunst et al. 2019). DNA integrity was higher in core than in marginal nestlings of the spotless starling *Sturnus unicolor* (Montoya et al. 2020). In Adelie penguins *Pygoscelis adeliae*, second-hatched chicks from nests with two siblings had higher levels of oxidative damage than first-hatched chicks and single chicks (Fig. 6.5) (Marciau et al. 2023).

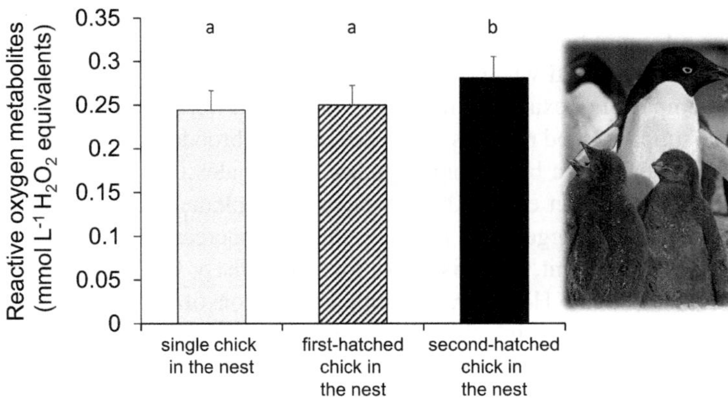


**Fig. 6.5** Second-hatched chicks of Adélie penguins *Pygoscelis adeliae* from nests with two siblings had higher levels of reactive oxygen metabolites than single chicks and first-hatched chicks from nests with two siblings at 32 days of age (Marciau et al. 2023). Shared letters indicate that the two groups did not differ significantly. Values are shown as mean and standard error. Image credit: Peter West, National Science Foundation, under Antarctic Conservation Act (ACA) permit West 2014–009

In mammals, the concentration of protein thiols was lower at weaning in the liver and kidneys of house mice *Mus musculus domesticus* from enlarged litters (Gibson et al. 2015), and levels of oxidative damage to DNA (8-OHdG) in the liver were higher in mound-building mice *Mus spicilegus* stemming from larger litters than those from smaller litters (Rödel et al. 2022). In some cases, the variation in oxidative status markers between siblings could be induced indirectly by different growth rates rather than by direct fights. For example, Smith (2016) found that, when considering within-brood growth differences, the concentration of protein carbonyls in blood was greater in nestling blue tits *Cyanistes caeruleus* that had grown faster than their siblings.

We have also studies that did not find support for the oxidative cost of sibling competition hypothesis. For example, levels of oxidative damage were lower in nestlings from larger than smaller broods in great tits (Costantini et al. 2010), and an experimental increase or decrease of brood size did not influence any markers of oxidative status in nestling red-winged blackbirds (Hall et al. 2010) or nestling great tits (Losdat et al. 2010), respectively. Levels of malondialdehyde were not affected by an experimental enlargement of the brood size in spotless starlings *Sturnus unicolor*; however, the levels of the antioxidant glutathione decreased with increasing brood size (Gil et al. 2019). Nestling zebra finches from reduced broods grew faster, but without any apparent deleterious impact on blood oxidative status markers and telomere length, compared to nestlings from control broods, and nestlings from enlarged broods grew at the same rates of control chicks, but they suffered from greater oxidative damage and telomere loss than control broods (Reichert et al. 2015). Interestingly, fledglings near independence that grew up with siblings whose corticosterone level was experimentally increased had a smaller body size, higher levels of oxidative damage to lipids and proteins, and a fragile juvenile plumage, suggesting a form of horizontal transfer of stress (Noguera et al. 2017).

Further studies tested whether oxidative stress constrains the capability of offspring to compete. For example, nestling great tits that were born from eggs laid by carotenoid-supplemented mothers and raised in small broods begged more intensely than nestlings that were born from eggs laid by females that were not given any carotenoids (Helfenstein et al. 2008). Similarly, supplementation of vitamin E to chicks of the yellow-legged gull *Larus michahellis* increased the total number of calls that they could emit, and this effect was particularly strong in smaller chicks (Noguera et al. 2010). However, the supplementation of a cocktail of vitamins (including C and E) and minerals did not influence the begging behaviour of nestling red-winged blackbirds (Hall et al. 2010).

## 6.5 The Hidden Risks of Physical Inactivity

Oxidative stress can also threaten organismal functions when the organism passes from a metabolically depressed dormant state to an active one, such as in the case of those species that wake up from hibernation, aestivation, or short-term torpor. Many species have evolved the capability to suppress their basal metabolic rate in order to enter a hypometabolic or dormant state during periods when food occurs in very limited supply and/or environmental conditions are very energetically and physiologically demanding. For example, this bunch of species includes those that live in seasonal environments characterised by cold or hot periods and dryness, or those that live in intertidal zones of rocky shorelines where they experience daily a transition from an active and fully aerobic state during high tides to an inactive and metabolically depressed state during low tides, respectively. Many species may also enter short-term periods of metabolic suppression, such as during torpor in birds or mammals. Metabolic suppression can, therefore, come in varying degrees, ranging from a 20–30% reduction in metabolic rate for a few hours during nightly torpor in small birds and mammals, to over a 95% reduction in metabolic rate for many weeks during seasonal hibernation in mammals or diapause in insects, through to a virtually ametabolic state that enables desiccated seeds, spores, and cysts of many plant and animal species to remain viable for many years (e.g. Storey and Storey 1990; Hand and Hardewig 1996; Storey 2001; Staples and Brown 2008).

All of these species have evolved a number of ways of coping with the demands of being in a dormant state or of arousal from this dormant state. For example, many species that occur in regions that are seasonally or perpetually cold go into hibernation. During the period of dormancy, these animals cope with temperatures that fall substantially below 0 °C. Lipids are the primary substrates for energy production during hibernation (Heldmaier et al. 1999). As a consequence, they rely on their lipid reserves, meaning that a larger number of potential substrates of lipid peroxidation are available in the tissues. Importantly, accumulation of oxidative damage to tissues may jeopardise anti-freezing mechanisms that enable hibernating species to survive the harsh season (Costanzo and Lee 2013). Similarly, species that aestivate in warm environments have to cope with heat and hydric stress. During the transition from a dormant to an active state, cells are reoxygenated and the metabolic rate increases as a consequence. This burst of oxygen in tissues also increases the generation of reactive oxygen species and, possibly, oxidative damage. Many studies suggested that antioxidant defences are programmed endogenously to increase during the dormant period or at arousal (see Chap. 5) in a diverse group of species, from terrestrial snails, frogs, to snakes and squirrels to protect against oxidative damage that can be potentially generated (Fig. 6.6, 6.7, and 6.8) (e.g. Hermes-Lima and Storey 1993; Hermes-Lima et al. 1998; Carey et al. 2000; Chauhan et al. 2002; Drew et al. 2002; Hermes-Lima and Zenteno-Savín 2002; Bagnyukova et al. 2003; Okamoto et al. 2006; Voituron et al. 2006; Baker et al. 2007; Orr et al. 2009; Salway et al. 2010; Wang et al. 2018; Jefimow et al. 2020; Patnaik and Sahoo 2021). For example, Bertuglia and Giusti (2003) found that plasma levels of organic

**Fig. 6.6** Levels of thiobarbituric acid reactive substances (TBARS) in (**a**) plasma and (**b**) erythrocytes in hibernating and active American black bears *Ursus americanus*, respectively. Values are shown as mean and standard error. Reprinted with minor modifications from Chauhan et al. (2002) with permission of Elsevier

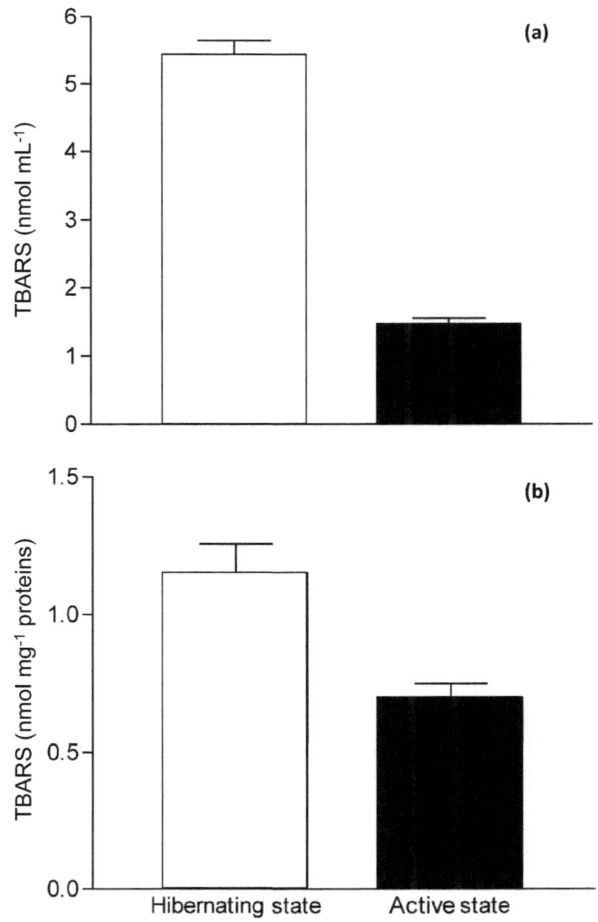

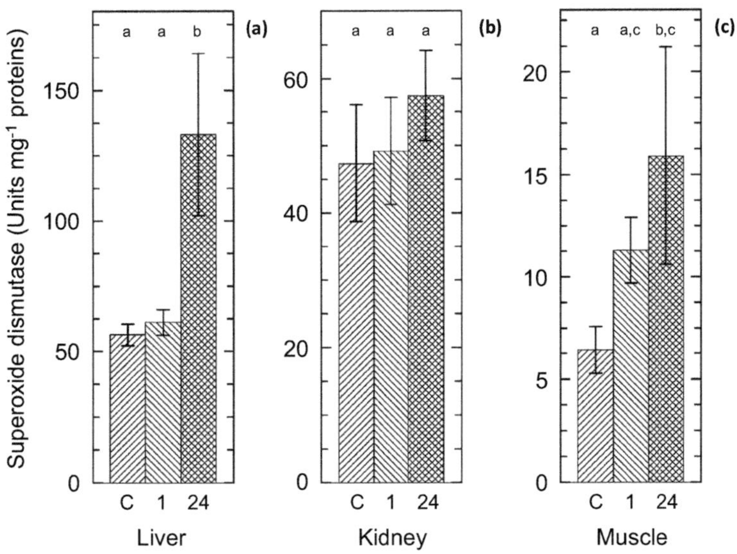


**Fig. 6.7** Activity of superoxide dismutase in (**a**) liver, (**b**) kidney, and (**c**) skeletal muscle of frogs *Rana ridibunda* after 2 months of hibernation at 5 °C, and 1 or 24 h of recovery from hibernation at 20 °C, respectively. Outcomes of contrasts between groups are shown as letters: shared letters (e.g. both with **a**) indicate that the two groups did not differ significantly. Values are shown as mean and standard error. Reprinted with minor modifications from Bagnyukova et al. (2003) with permission of Elsevier

hydroperoxides did not change during a period of hypoxia, while they increased by 90 and 72% within 5 and 15 min of reperfusion in male Syrian hamsters *Mesocricetus auratus*, respectively. After 15 min from the start of reperfusion, levels of organic hydroperoxides decreased with time significantly. This increase in plasma organic hydroperoxides was, however, not observed in those hamsters that were injected a bolus of superoxide dismutase half an hour before the reoxygenation took place, a finding that highlights the relevance of keeping antioxidant defences high before arousal (Bertuglia and Giusti 2003).

Some species are quite flexible in terms of aestivation behaviour, being able to emerge rapidly from their subterranean burrows, thereby permitting opportunistic feeding and breeding when the availability of water and food increase. To do so, these species give priority to the protection of skeletal muscle to avoid atrophy, which would compromise their capability of rapidly coming out of aestivation and resume their activities. For example, green-striped burrowing frogs *Cyclorana alboguttata* keep their mitochondria quiescent during aestivation in order to reduce the generation of reactive oxygen species, maintain active their uncoupling proteins (which reduce mitochondrial generation of reactive oxygen species), upregulate levels of the mitochondrially encoded antioxidant superoxide dismutase type 2, and preferentially combust lipids consistent with the sparing of lean muscle mass (Hudson et al. 2006; Reilly et al. 2014). There is also evidence that even skeletal muscles that perform different functions are protected in different ways. Temperature-induced increases in

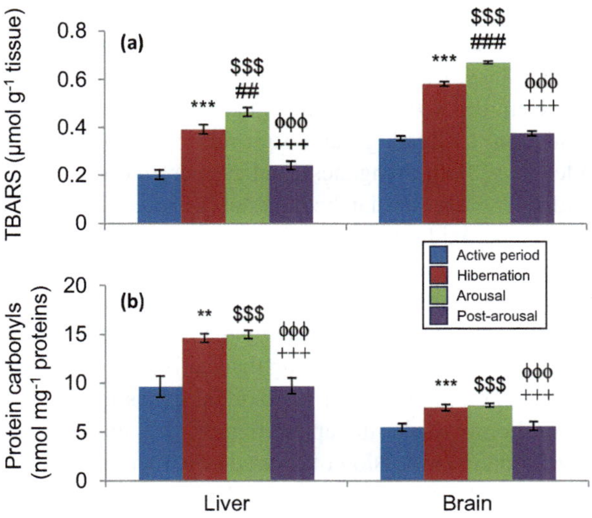


**Fig. 6.8** Effects of hibernation, arousal, and post-arousal on the level of (**a**) thiobarbituric acid reactive substances and (**b**) protein carbonyls in liver and brain tissues of male common Asian toads *Duttaphrynus melanostictus*. Asterisks indicate significant differences among groups: ** $P < 0.01$ compared with active period; *** $P < 0.001$ compared with active period; ## $P < 0.01$ compared with hibernation; ### $P < 0.001$ compared with hibernation; $$$ $P < 0.001$ compared with active period; +++ $P < 0.001$ compared with arousal; φφφ $P < 0.001$ compared with hibernation. Values are shown as mean and standard error. Reprinted with minor modifications from Patnaik and Sahoo (2021). © 2021, published by The Company of Biologists Ltd.; distributed under the terms of the Creative Commons Attribution License (CC BY)

oxygen consumption rate occur in some muscles of ectotherms undergoing prolonged muscular disuse (atrophy) during aestivation. However, the extent of atrophy in jumping muscles is not enhanced at higher temperatures. A study on green-striped burrowing frogs found that (1) muscles differed in the expression of biochemical regulation mechanisms to avoid increased damage at higher temperatures during aestivation and (2) priority was given to the protection of jumping muscles, which have important post-aestivation locomotor functions (Young et al. 2013).

Many studies showed that only certain antioxidant mechanisms are upregulated during the whole or a part of the dormancy period, and only in certain tissues. For example, a study on Couch's spadefoot toads *Scaphiopus couchii* found that (1) levels of glutathione were higher in active than aestivating toads in leg muscle, liver, and lung, but were similar between them in heart, kidney, and gut, (2) levels of glutathione-S-transferase were higher in active than aestivating toads in gut, heart, liver, lung, and kidney, but were similar between them in leg muscle, (3) levels of glutathione reductase were higher in active than aestivating toads in heart, liver, and kidney, but were similar between them in gut, leg muscle, and lung, (4) levels of selenium-dependent glutathione peroxidase were higher in active than aestivating toads in gut and liver, were lower in active than aestivating toads in kidney, and were similar between them in heart, leg muscle, and lung, (5) levels of catalase were higher in active than aestivating toads in heart and liver, were lower in active than aestivating toads in kidney, and were similar between them in gut, leg muscle, and lung, and (6) levels of superoxide dismutase were higher in active than aestivating toads in heart and kidney, were lower in active than aestivating toads in leg muscle and liver, and were similar between them in gut and lung (Grundy and Storey 1998). Similarly, a study on 13-lined ground squirrels *Spermophilus tridecemlineatus* found that (1) levels of both manganese and copper-zinc superoxide dismutase in brain, heart, and liver were similar between active and hibernating squirrels, (2) levels of glutathione peroxidase were similar between active and hibernating squirrels in brain and heart, and higher in hibernating than active squirrels in liver, (3) levels of glutathione reductase were similar between active and hibernating squirrels in heart and liver, and lower in hibernating than active squirrels in brain, and (4) levels of catalase were higher in hibernating than active squirrels in brain, heart, and liver (Page et al. 2009). Expression of antioxidant genes may also vary across tissues during torpor. For example, in captive female gray mouse lemurs *Microcebus murinus*, levels of protein expression of thioredoxin 1 were elevated to 2.19-fold in white adipose tissue, and copper-zinc superoxide dismutase 1 levels rose to 1.1-fold in skeletal muscle in torpid compared to active lemurs, respectively (Fig. 6.9) (Wu et al. 2015). Additionally, non-enzymatic antioxidant capacity was increased to 1.6-fold in liver but not in other tissues in torpid compared to active lemurs (Wu et al. 2015). Further work found that markers of lipid peroxidation (conjugated dienes and lipid hydroperoxides) were higher in tissues of hibernating squirrels or aestivating toads than in those of active individuals, suggesting that there is accumulation of oxidative damage to lipids during the dormancy period (e.g. Grundy and Storey 1998; Carey et al. 2000). Thus, upregulation of antioxidant defences as an anticipatory response to the awakening from the dormancy period is not common across all tissues and species. For example, red-eared slider turtles *Trachemys scripta elegans* that experienced 20 h of anoxic conditions had either suppressed or increased levels of certain antioxidant enzymes across some tissues compared to control turtles that minimised or prevented damage to lipids (levels of lipid hydroperoxides and thiobarbituric acid reactive substances) during anoxia and the reoxygenation of organs after anoxic submergence (Willmore and Storey 1997a, b). One explanation for these results might lie with the high constitutive antioxidant levels that red-eared slider turtles have compared to other non-mammalian species (Hermes-Lima et al. 2001; Hermes-Lima and Zenteno-Savín 2002). Thus, turtles would not need to upregulate all their antioxidant defences. This characteristic of turtles might be explained to some extent by their diving habits, which make turtles exposed daily to repeated periods of anoxia–reoxygenation that require high constitutive antioxidant defences if compared with non-diving species. Another explanation might lie with the duration of the dormancy period. Toads were aestivating for 2 months, while turtles were dormant for only 1 day. It might be that long-term dormancy is costly and demands a large consumption of antioxidants compared to short-term dormancy. The mechanisms underlying this differential expression of antioxidants across time and tissues, and its functional consequences are far from being elucidated.

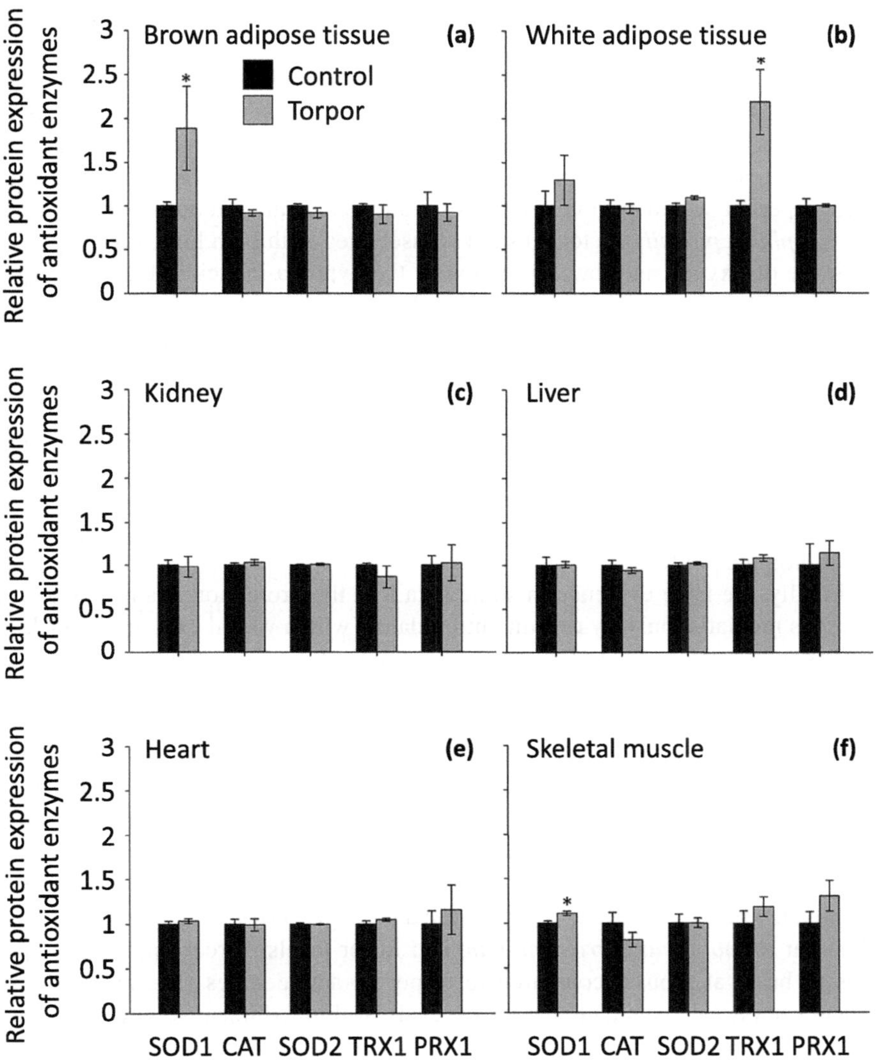


**Fig. 6.9** Protein expression of Cu/Zn-superoxide dismutase (SOD1), catalase (CAT), Mn-superoxide dismutase (SOD2), thioredoxin 1 (TRX1), and peroxiredoxin 2 (PRX2) in (**a**) brown adipose tissue, (**b**) white adipose tissue, (**c**) kidney, (**d**) liver, (**e**) heart, and (**f**) skeletal muscle from control and torpid gray mouse lemurs *Microcebus murinus*, respectively. Asterisks indicate significant differences in comparison with values of controls. Values are shown as median and standard error. Reprinted with minor modifications from Wu et al. (2015). Copyright © 2015, published by Elsevier B.V.; distributed under the terms of the Creative Commons Attribution License (CC BY)

The suppression of antioxidant defences observed in turtles does not seem to support a possible hormetic-like stimulatory effect of the anoxic period that would prepare turtles for the reoxygenation-induced burst of reactive oxygen species. Thus, it might be that some species have evolved other mechanisms that protect them against oxidative stress at arousal. For example, it has been suggested that some species could simply avoid profound tissue hypoxia in order to make the transition from dormancy to arousal as little stressful as possible (Page et al. 2009). This strategy of avoiding profound tissue hypoxia would make levels of reactive oxygen species higher than in normal hypoxia, so that the increase in levels of reactive oxygen species at arousal compared to those during the dormancy period would be mitigated. For example, in the Arctic ground squirrel *Spermophilus parryii*, the torpid state is associated with both high arterial partial pressure of oxygen and low expression of the hypoxia-inducible factor 1-α (Ma et al. 2005). It has also been suggested that some species might rely on repair systems that would be activated in order to detoxify the body from the accumulation of oxidative damage generated at arousal (Hermes-Lima and Zenteno-Savín 2002). By contrast, the high tolerance of anoxia of the proteus *Proteus anguinus* appears to be determined by a strong upregulation of an uncoupling protein during reoxygenation rather than by the classic protective role of antioxidant enzymes (Issartel et al. 2009). The uncoupling proteins would enable the proteus to decrease the mitochondrial generation of reactive oxygen species through the dissipation of the proton motive force.

Finally, we have evidence in some taxa that the protection against oxidative stress is mediated only by certain antioxidants, which would explain the lack of a consistent upregulation of all antioxidant molecules. For example, channelled apple snails *Pomacea canaliculata* had increased levels of uric acid during aestivation (Fig. 6.10) and increased levels of allantoin (the product of the non-enzymatic oxidation of uric acid) 1 day after arousal, respectively (Giraud-Billoud et al. 2011, 2013; Rodriguez et al. 2023). Urate is certainly a product of nitrogen metabolism, but it has also antioxidant properties (see Chap. 1). Hence, accumulation of uric acid during the aestivation might be favoured by selection because it would be advantageous in terms of better protection against oxidative stress. Finally, the bat species Rickett's big-footed bat *Myotis ricketti* and greater horseshoe bat *Rhinolophus ferrumequinum* had lower levels of reactive oxygen species in brain at arousal compared to torpor or active states (Yin et al. 2016). Greater horseshoe bats had also lower levels of thiobarbituric acid reactive substances in brain at arousal compared to torpor or active states (Yin et al. 2016). This is surprising given the strong increase in oxygen consumption that occurs at arousal (Lee et al. 2002). Further data suggested that Rickett's big-footed bats use to some degree antioxidant enzymes to control the generation of reactive oxygen species, while greater horseshoe bats seem to rely more on low molecular weight antioxidants, like glutathione, to cope with reactive oxygen species during arousal (Yin et al. 2016). However, other mechanisms are probably involved in the control of reactive oxygen species because the expression of antioxidants was not strongly associated with the patterns of reactive oxygen species generation.

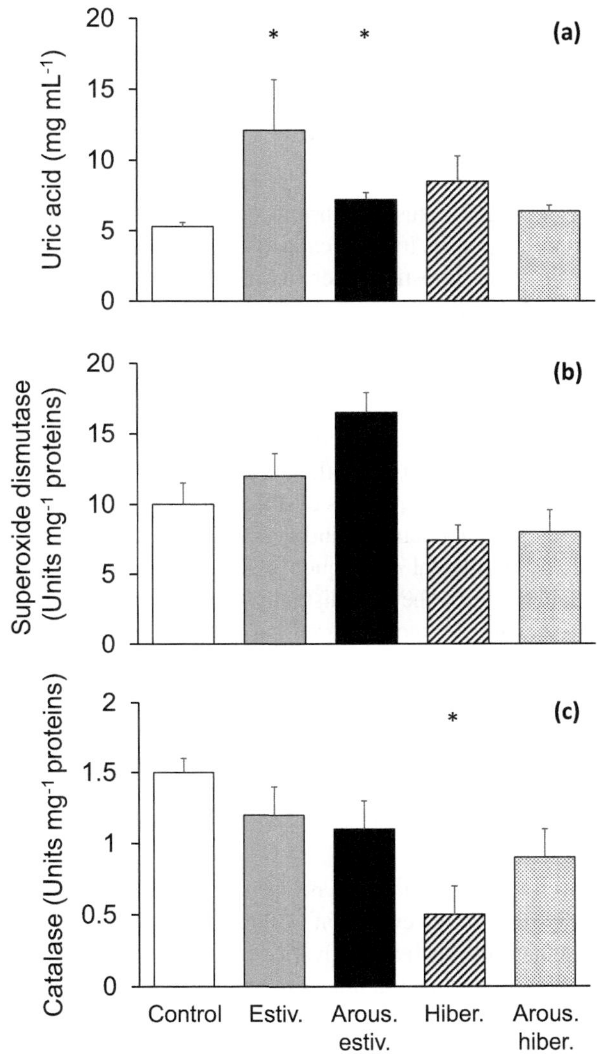


**Fig. 6.10** Effects of aestivation (Estiv.), arousal from aestivation (Arous. estiv.), hibernation (Hiber.), and arousal from hibernation (Arous. hiber.) on (**a**) uric acid, (**b**) superoxide dismutase, and (**c**) catalase in apple snails *Pomacea canaliculata*. Asterisks indicate significant differences in comparison with values of controls (Control). Values are shown as mean and standard error. Data were collected from the Table 1 in Rodriguez et al. (2023)

## 6.6 Conclusions and Future Directions

There is ample evidence to suggest that a strong increase in physical effort can reduce survival in free-living animals, but, as highlighted by Soulsbury and Halsey (2018) the term effort refers to different processes in different studies and is rarely clarified or quantified. Also, to understand the role of activity in functional senescence, studies should better describe the activity of free-ranging animals in terms of intensity, frequency, and duration, coupled with records of resultant oxidative status markers (Soulsbury and Halsey 2018). Finally, it would be important to describe the type of damage to skeletal muscle that a given physical effort can generate.

There is also strong evidence supporting a link among physical effort, regulation of oxidative status, and hormesis. Studies on laboratory animals showed that hormetic responses to physical activity can occur in a number of tissues, such as bone, brain, heart, intestine, liver, or skeletal muscle (e.g. Caillaud et al. 1999; Ogonovszky et al. 2005; Pereira et al. 2007; de Lira et al. 2008; Radak et al. 2008, 2017; Rattan 2008; Rosa et al. 2008; Stranahan and Mattson 2010; Nikolaidis et al. 2012a, b; Lesmana et al. 2022). However, we do not know if these paradigms are relevant for natural animal populations. There is much scope for novel experimental work to assess the roles that reactive oxygen species, antioxidant machinery, and oxidative stress have in moulding the adaptations to physical (in)activity, and in determining causes and consequences of different levels in physical effort and performance. The merger of biochemical and genomic tools would greatly further our understanding of the biological consequences of physical activity. For example, whole-genome sequencing of the two distantly related bat species, the black flying fox *Pteropus alecto* and the mouse-eared bat *Myotis davidii*, showed that a high proportion of genes regulating the DNA damage checkpoint–DNA repair pathway was under positive selection in the bat ancestor (Zhang et al. 2013). Similarly, longitudinal comparative transcriptomics revealed that bats exhibited an age-related gene expression pattern associated with DNA repair (Huang et al. 2019). These results might indicate that the evolution of powered flight required selection of those mechanisms that control levels of damage to DNA induced by reactive oxygen species (Zhang et al. 2013).

Finally, this chapter has also shown that oxidative stress has likely played a relevant role in the evolution of those behavioural strategies that require long periods of inactivity, such as aestivation or hibernation. We have evidence that the upregulation of antioxidant defences during the metabolically depressed dormancy period is important to control the small, but ongoing generation of reactive oxygen species and, in particular, to prepare the organism to the burst of reactive oxygen species that will be caused by the reoxygenation of tissues at arousal. However, many studies found that other mechanisms are also or even more important than keeping constitutive antioxidant levels high during the dormancy period. We have also evidence that protection against oxidative stress might vary across tissues, indicating that animals might strategically protect those tissues that are more relevant to move the first steps after arousal. It will be important in future work to assess in more detail how the expression of these protective mechanisms varies as a function of species, population, sex, age, genotype, life history, season, or abiotic conditions of the development and adult environments. It will also be important to elucidate whether protection of certain tissues is really prioritised above others, or rather the observed differences in antioxidant levels among tissues is owing to limitations of the sampling design.

## References

Aidley DJ (1981) Animal migration. Cambridge University Press, Cambridge

Alerstam T, Lindstrom A (1990) Optimal bird migration: the relative importance of time, energy, and safety. In: Gwinner E (ed) Bird migration: physiology and ecophysiology. Springer-Verlag, Berlin, pp 331–351

Alessio HM (1993) Exercise-induced oxidative stress. Med Sci Sports Exer 25:218–224

Alessio HM, Hagerman AE (2006) Oxidative stress, exercise and aging. Imperial College Press, London

Alonso-Alvarez C, Bertrand S, Devevey G, Prost J, Faivre B, Chastel O, Sorci G (2006) An experimental manipulation of life history trajectories and resistance to oxidative stress. Evolution 60:1913–1924

Anderson BJ, Rapp DN, Baek DH, McCloskey DP, Coburn-Litvak PS, Robinson JK (2000) Exercise influences spatial learning in the radial arm maze. Physiol Behav 70:425–429

Andres-Lacueva C, Shukitt-Hale B, Galli RL, Jauregui O, Lamuela-Raventos RM, Joseph JA (2005) Anthocyanins in aged blueberry-fed rats are found centrally and may enhance memory. Nutr Neurosci 8:111–120

Ang ET, Dawe GS, Wong PT, Moochhala S, Ng YK (2006) Alterations in spatial learning and memory after forced exercise. Brain Res 1113:186–193

Arnold KE, Larcombe SD, Ducaroir L, Alexander L (2010) Antioxidant status, flight performance and sexual signalling in wild-type parrots. Behav Ecol Sociobiol 64:1857–1866

Bagnyukova TV, Storey KB, Lushchak VI (2003) Induction of oxidative stress in *Rana ridibunda* during recovery from winter hibernation. J Therm Biol 28:21–28

Bailey DM, Davies B, Young IS, Jackson MJ, Davison GW, Isaacson R, Richardson RS (2003) EPR spectroscopic detection of free radical outflow from an isolated muscle bed in exercising humans. J Appl Physiol 94:1714–1718

Bailey DM, Young IS, McEneny J, Lawrenson L, Kim J, Barden J, Richardson RS (2004) Regulation of free radical outflow from an isolated muscle bed in exercising humans. Am J Physiol Heart Circ Physiol 287:H1689–H1699

Bairlein F, Fritz J, Scope A, Schwendenwein I, Stanclova G, van Dijk G, Meijer HA, Verhulst S, Dittami J (2015) Energy expenditure and metabolic changes of free-flying migrating northern bald ibis. PLoS One 10:e0134433

Baker PJ, Costanzo JP, Lee RE Jr (2007) Oxidative stress and antioxidant capacity of a terrestrially hibernating hatchling turtle. J Comp Physiol B 177:875–883

Bennie JJ, Duffy JP, Inger R, Gaston KJ (2014) Biogeography of time partitioning in mammals. Proc Natl Acad Sci USA 111:13727–13732

Berthold P (2001) Bird migration: a general survey. Oxford University Press, Oxford

Bertuglia S, Giusti A (2003) Microvascular oxygenation, oxidative stress, NO suppression and superoxide dismutase during postischemic reperfusion. Am J Phys 285:H1064–H1071

Bhardwaj N, Kumar A, Gupta NJ (2023) Altered dynamics of mitochondria and reactive oxygen species in the erythrocytes of migrating red-headed buntings. Front Physiol 14:1111490

Birnie-Gauvin K, Peiman KS, Larsen MH, Baktoft H, Aarestrup K, Willmore WG, Cooke SJ (2017) Oxidative stress and partial migration in brown trout (Salmo trutta). Can J Zool 95:829–835

Boldt A, Ingold P (2005) Effects of air traffic, snow cover and weather on altitudinal short-term and medium-term movements of female Alpine chamois *Rupicapra rupicapra* in winter. Wildlife Biol 11:351–362

Bolser JA, Alan RR, Smith AD, Li L, Seeram NP, McWilliams SR (2013) Birds select fruits with more anthocyanins and phenolic compounds during autumn migration. Wilson J Ornithol 125:97–108

Boncoraglio G, Caprioli M, Saino N (2012) Solicitation displays reliably reflect oxidative damage in barn swallow nestlings. Behav Ecol Sociobiol 66:539–546

Bonte D, Van Dyck H, Bullock JM, Coulon A, Delgado M, Gibbs M, Lehouck V, Matthysen E, Mustin K, Saastamoinen M, Schtickzelle N, Stevens VM, Vandewoestijne S, Baguette M,

Barton K, Benton TG, Chaput-Bardy A, Clobert J, Dytham C, Hovestadt T, Meier CM, Palmer SC, Turlure C, Travis JM (2012) Costs of dispersal. Biol Rev 87:290–312

Bourgeon S, Guindre-Parker S, Williams TD (2011) Effects of sibling competition on growth, oxidative stress, and humoral immunity: a two-year brood-size manipulation. Physiol Biochem Zool 84:429–437

Bruderer L, Liechti F, Bilo D (2001) Flexibility in flight behaviour of barn swallows (*Hirundo rustica*) and house martins (*Delichon urbica*) tested in a wind tunnel. J Exp Biol 204:1473–1484

Butler PJ, West NH, Jones DR (1977) Respiratory and cardiovascular responses of the pigeon to sustained, level flight in a wind-tunnel. J Exp Biol 71:7–26

Buttemer W, Battam H, Hulbert AJ (2008) Fowl play and the price of petrel: long-living Procellariformes have peroxidation resistant membrane composition compared with short-living Galliformes. Biol Lett 4:351–354

Caillaud C, Py G, Eydoux N, Legros P, Prefaut C, Mercier J (1999) Antioxidants and mitochondrial respiration in lung, diaphragm, and locomotor muscles: effect of exercise. Free Radic Biol Med 26:1292–1299

Calabrese V, Cornelius C, Dinkova-Kostova AT, Calabrese EJ, Mattson MP (2010) Cellular stress responses, the hormesis paradigm, and vitagenes: novel targets for therapeutic intervention in neurodegenerative disorders. Antioxid Redox Signal 13:1763–1811

Carey HV, Frank CL, Seifert JP (2000) Hibernation induces oxidative stress and activation of NF-kappa B in ground squirrel intestine. J Comp Physiol B 170:551–559

Caro SM, West SA, Griffin AS (2016) Sibling conflict and dishonest signaling in birds. Proc Natl Acad Sci USA 113:13803–13808

Carter WA, DeMoranville KJ, Pierce BJ, McWilliams SR (2020) The effects of dietary linoleic acid and hydrophilic antioxidants on basal, peak, and sustained metabolism in flight-trained European starlings. Ecol Evol 10:1552–1566

Casagrande S, DeMoranville KJ, Trost L, Pierce B, Bryła A, Dzialo M, Sadowska ET, Bauchinger U, McWilliams SR (2020) Dietary antioxidants attenuate the endocrine stress response during long-duration flight of a migratory bird. Proc R Soc Lond B 287:20200744

Cecere JG, Caprioli M, Carnevali C, Colombo G, Dalle-Donne I, Mancuso E, Milzani A, Parolini M, Portanova A, Saino N, Serra L, Rubolini D (2016) Dietary flavonoids advance timing of moult but do not affect redox status of juvenile blackbirds (Turdus merula). J Exp Biol 219:3155–3162

Chauhan VP, Tsiouris JA, Chauhan A, Sheikh AM, Brown WT, Vaughan M (2002) Increased oxidative stress and decreased activities of Ca(2+)/Mg(2+)-ATPase and Na(+)/K(+)-ATPase in the red blood cells of the hibernating black bear. Life Sci 71:153–161

Chen MM, Li Y, Deng SL, Zhao Y, Lian ZX, Yu K (2022) Mitochondrial function and reactive oxygen/nitrogen species in skeletal muscle. Front Cell Dev Biol 10:826981

Cheviron ZA, Whitehead A, Brumfield RT (2008) Transcriptomic variation and plasticity in rufous-collared sparrows (*Zonotrichia capensis*) along an altitudinal gradient. Mol Ecol 17:4556–4669

Cooper-Mullin C, McWilliams SR (2016) The role of the antioxidant system during intense endurance exercise: lessons from migrating birds. J Exp Biol 219:3684–3695

Cooper-Mullin C, McWilliams SR (2022) Fat stores and antioxidant capacity affect stopover decisions in three of four species of migratory passerines with different migration strategies: an experimental approach. Front Ecol Evol 10:762146

Cooper-Mullin C, Carter WA, McWilliams SR (2019) Acute effects of intense exercise on the antioxidant system in birds: does exercise training help? J Exp Biol 222:jeb210443

Costantini D (2014) Oxidative stress and hormesis in evolutionary ecology and physiology: a marriage between mechanistic and evolutionary approaches. Springer-Verlag, Berlin, p 348

Costantini D, Lipp H-P (2010) Short restraint time does not influence markers of serum oxidative stress in homing pigeons (*Columba livia*). J Anim Physiol Anim Nutr 94:24–28

Costantini D, Casagrande S, De Filippis S, Brambilla G, Fanfani A, Tagliavini J, Dell'Omo G (2006) Correlates of oxidative stress in wild kestrel nestlings (*Falco tinnunculus*). J Comp Physiol B 176:329–337

Costantini D, Cardinale M, Carere C (2007) Oxidative damage and anti-oxidant capacity in two migratory bird species at a stop-over site. Comp Biochem Physiol Part C 144:363–337

Costantini D, Dell'Ariccia G, Lipp H-P (2008) Long flights and age affect oxidative status of homing pigeons (*Columba livia*). J Exp Biol 211:377–381

Costantini D, Carello L, Fanfani A (2010) Relationships among oxidative status, breeding conditions and life-history traits in free-living great tits *Parus major* and common starlings *Sturnus vulgaris*. Ibis 152:793–802

Costantini D, Mirzai N, Metcalfe N (2012) An automated system to control and manipulate the flight activity of captive birds. Behav Ecol Sociobiol 66:1195–1199

Costantini D, Monaghan P, Metcalfe N (2013) Loss of integration is associated with reduced resistance to oxidative stress. J Exp Biol 216:2213–2220

Costantini D, Lindecke O, Pētersons G, Voigt CC (2019) Migratory flight imposes oxidative stress in bats. Curr Zool 65:147–153

Costanzo JP, Lee RE Jr (2013) Avoidance and tolerance of freezing in ectothermic vertebrates. J Exp Biol 216:1961–1967

Cram DL (2022) Oxidative stress and cognition in ecology. J Zool 318:229–240

Davies KJ (1986) Intracellular proteolytic systems may function as secondary antioxidant defenses: an hypothesis. J Free Rad Biol Med 2:155–173

Davis MS, Bayly WM, Hansen CM, Barrett MR, Blake CA (2023) Effects of hyperthermia and acidosis on mitochondrial production of reactive oxygen species. Am J Phys 325:R725–R734

Debold EP (2015) Potential molecular mechanisms underlying muscle fatigue mediated by reactive oxygen and nitrogen species. Front Physiol 6:239

de Lira CA, Vancini RL, Ihara SS, da Silva AC, Aboulafia J, Nouailhetas VL (2008) Aerobic exercise affects C57BL/6 murine intestinal contractile function. Eur J Appl Physiol 103:215–223

Dick MF, Guglielmo CG (2019a) Dietary polyunsaturated fatty acids influence flight muscle oxidative capacity but not endurance flight performance in a migratory songbird. Am J Phys 316:R362–R375

Dick MF, Guglielmo CG (2019b) Flight muscle protein damage during endurance flight is related to energy expenditure but not dietary polyunsaturated fatty acids in a migratory bird. J Exp Biol 222:jeb187708

Dillard CJ, Litov RE, Savin WM, Dumelin EE, Tappel AL (1978) Effects of exercise, vitamin E, and ozone on pulmonary function and lipid peroxidation. J Appl Physiol 45:927–932

Drew KL, Tøien Ø, Rivera PM, Smith MA, Perry G, Rice ME (2002) Role of the antioxidant ascorbate in hibernation and warming from hibernation. Comp Biochem Physiol C 133:483–492

Du B, Ding D, Ma C, Guo W, Kang L (2022) Locust density shapes energy metabolism and oxidative stress resulting in divergence of flight traits. Proc Natl Acad Sci USA 119:e2115753118

Eikenaar C, Jönsson J, Fritzsch A, Wang HL, Isaksson C (2016) Migratory refueling affects non-enzymatic antioxidant capacity, but does not increase lipid peroxidation. Physiol Behav 158:26–32

Eikenaar C, Källstig E, Andersson MN, Herrera-Dueñas A, Isaksson C (2017) Oxidative challenges of avian migration: a comparative field study on a partial migrant. Physiol Biochem Zool 90:223–229

Eikenaar C, Winslott E, Hessler S, Isaksson C (2020a) Oxidative damage to lipids is rapidly reduced during migratory stopovers. Funct Ecol 34:1215–1222

Eikenaar C, Hegemann A, Packmor F, Kleudgen I, Isaksson C (2020b) Not just fuel: energy stores are correlated with immune function and oxidative damage in a long-distance migrant. Curr Zool 66:21–28

Eikenaar C, Winslott E, Schmaljohann H, Wang HL, Isaksson C (2022) Can differential fatty acid composition help migrating birds to limit oxidative lipid damage? Physiol Behav 249:113768

Eikenaar C, Ostolani A, Hessler S, Ye EY, Karwinkel T, Isaksson C (2023) Stopovers serve physiological recovery in migratory songbirds. Physiol Biochem Zool 96:378–389

Engel S, Bowlin MS, Hedenström A (2010) The role of wind-tunnel studies in integrative research on migration biology. Integr Comp Biol 50:323–335

Falsone K, Jenni-Eiermann S, Jenni L (2009) Corticosterone in migrating songbirds during endurance flight. Horm Behav 56:548–556

Ferretti A, McWilliams SR, Rattenborg NC, Maggini I, Cardinale M, Fusani L (2020) Energy stores, oxidative balance, and sleep in migratory garden warblers (Sylvia borin) and whitethroats (Sylvia communis) at a spring stopover site. Integr Org Biol 2:obaa010

Fowler MA, Wong JB, Harrison AL (2023) Oxidative physiology of two small and highly migratory Arctic seabirds: Arctic terns (Sterna paradisaea) and long-tailed jaegers (Stercorarius longicaudus). Conserv Physiol 11:coad060

Frawley AE, DeMoranville KJ, Carbeck KM, Trost L, Bryła A, Działo M, Sadowska ET, Bauchinger U, Pierce BJ, McWilliams SR (2021) Flight training and dietary antioxidants have mixed effects on the oxidative status of multiple tissues in a female migratory songbird. J Exp Biol 224:jeb243158

Fuertes-Recuero M, Cantarero A, Redondo AJ, Sánchez-Sánchez N, Redondo T (2023) Do barn swallow nestlings incur an oxidative cost of begging? J Ornithol 165(1):231–239

Fusani L, Cardinale M, Carere C, Goymann W (2009) Stopover decision during migration: physiological conditions predict nocturnal restlessness in wild passerines. Biol Lett 5:302–305

Gannes LZ, Hatch KA, Pinshow B (2001) How does time since feeding affect the fuels pigeons use during flight? Physiol Biochem Zool 74:1–10

Gattermann R, Weinandy R, Fritzsche P (2004) Running-wheel activity and body composition in golden hamsters (*Mesocricetus auratus*). Physiol Behav 82:541–544

Gibson AB, Garratt M, Brooks RC (2015) Experimental evidence that litter size imposes an oxidative challenge to offspring. J Exp Biol 218:3911–3918

Gil D, Alfonso-Iñiguez S, Pérez-Rodríguez L, Muriel J, Monclús R (2019) Harsh conditions during early development influence telomere length in an altricial passerine: links with oxidative stress and corticosteroids. J Evol Biol 32:111–125

Giraud-Billoud M, Abud MA, Cueto JA, Vega IA, Castro-Vazquez A (2011) Uric acid deposits and estivation in the invasive apple-snail, *Pomacea canaliculata*. Comp Biochem Physiol A 158:506–512

Giraud-Billoud M, Vega IA, Tosi ME, Abud MA, Calderón ML, Castro-Vazquez A (2013) Antioxidant and molecular chaperone defences during estivation and arousal in the South American apple snail *Pomacea canaliculata*. J Exp Biol 216:614–622

Godfray HCJ (1991) Signalling of need by offspring to their parents. Nature 352:328–330

Gomez-Cabrera MC, Borrás C, Pallardó FV, Sastre J, Ji LL, Viña J (2005) Decreasing xanthine oxidase-mediated oxidative stress prevents useful cellular adaptations to exercise in rats. J Physiol 567:113–120

Gomez-Cabrera MC, Domenech E, Viña J (2008) Moderate exercise is an antioxidant: upregulation of antioxidant genes by training. Free Rad Biol Med 44:126–131

Goymann W, Spina F, Ferri A, Fusani L (2010) Body fat influences departure from stopover sites in migratory birds: evidence from whole-Island telemetry. Biol Lett 6:478–481

Grubb B, Jorgensen DD, Conner M (1983) Cardiovascular changes in the exercising emu. J Exp Biol 104:193–201

Grundy JE, Storey KB (1998) Antioxidant defenses and lipid peroxidation damage in estivating toads, *Scaphiopus couchii*. J Comp Physiol B 169:132–142

Grunst AS, Grunst ML, Gonser RA, Tuttle EM (2019) Developmental stress and telomere dynamics in a genetically polymorphic species. J Evol Biol 32:134–143

Gunduz F, Senturk UK, Kuru O, Aktekin B, Aktekin MR (2004) The effect of one year's swimming exercise on oxidant stress and antioxidant capacity in aged rats. Physiol Res 53:171–176

Gutiérrez JS, Sabat P, Castañeda LE, Contreras C, Navarrete L, Peña-Villalobos I, Navedo JG (2019) Oxidative status and metabolic profile in a long-lived bird preparing for extreme endurance migration. Sci Rep 9:17616

Gwinner E (1990) Bird migration: physiology and ecophysiology. Springer-Verlag, Berlin

Haber A (2011) A comparative analysis of integration indices. Evol Biol 38:476–488

Hails CJ (1979) A comparison of flight energetics in hirundines and other birds. Comp Biochem Physiol A 63:581–585

Hall ME, Blount JD, Forbes S, Royle NJ (2010) Does oxidative stress mediate the trade-off between growth and self-maintenance in structured families? Funct Ecol 24:365–373

Hand SC, Hardewig I (1996) Downregulation of cellular metabolism during environmental stress: mechanisms and implications. Annu Rev Physiol 58:539–563
Heldmaier G, Klingenspor M, Werneyer M, Lampi BJ, Brooks SP, Storey KB (1999) Metabolic adjustments during daily torpor in the Djungarian hamster. Am J Phys 276:896–906
Helfenstein F, Berthouly A, Tanner M, Karadas F, Richner H (2008) Nestling begging intensity and parental effort in relation to prelaying carotenoid availability. Behav Ecol 19:108–115
Hermes-Lima M, Storey KB (1993) Role of antioxidants in the tolerance of freezing and anoxia by garter snakes. Am J Phys 265:646–652
Hermes-Lima M, Zenteno-Savín T (2002) Animal response to drastic changes in oxygen availability and physiological oxidative stress. Comp Biochem Physiol C 133:537–556
Hermes-Lima M, Storey JM, Storey KB (1998) Antioxidant defenses and metabolic depression. The hypothesis of preparation for oxidative stress in land snails. Comp Biochem Physiol B 120:437–448
Hermes-Lima M, Storey JM, Storey KB (2001) Antioxidant defenses and animal adaptation to oxygen availability during environmental stress. In: Storey KB, Storey JM (eds) Cell and molecular responses to stress. Elsevier, Amsterdam, pp 263–287
Herrero A, Barja G (1997) ADP-regulation of mitochondrial free radical production is different with complex I- or complex II-linked substrates: implications for the exercise paradox and brain hypermetabolism. J Bioenerg Biomembr 29:241–249
Hollander J, Fiebig R, Gore M, Ookawara T, Ohno H, Ji LL (2001) Superoxide dismutase gene expression is activated by a single bout of exercise. Pflug Arch (Eur J Physiol) 442:426–434
Hopcraft JGC, Sinclair ARE, Packer C (2005) Planning for success: Serengeti lions seek prey accessibility rather than abundance. J Anim Ecol 74:559–566
Huang Z, Whelan CV, Foley NM, Jebb D, Touzalin F, Petit EJ, Puechmaille SJ, Teeling EC (2019) Longitudinal comparative transcriptomics reveals unique mechanisms underlying extended healthspan in bats. Nat Ecol Evol 3:1110–1120
Hudson NJ, Lehnert SA, Ingham AB, Symonds B, Franklin CE, Harper GS (2006) Lessons from an estivating frog: sparing muscle protein despite starvation and disuse. Am J Phys 290:836–843
Issartel J, Hervant F, de Fraipont M, Clobert J, Voituron Y (2009) High anoxia tolerance in the subterranean salamander *Proteus anguinus* without oxidative stress nor activation of antioxidant defenses during reoxygenation. J Comp Physiol B 179:543–551
Jefimow M, Przybylska-Piech AS, Wojciechowski MS (2020) Predictive and reactive changes in antioxidant defence system in a heterothermic rodent. J Comp Physiol B 190:479–492
Jenni L, Jenni-Eiermann SJ, Spina F, Schwabl H (2000) Regulation of protein breakdown and adrenocortical response to stress in birds during migratory flight. Am J Phys 278:1182–1189
Jenni-Eiermann S, Jenni L, Smith S, Costantini D (2014) Oxidative stress in endurance flight: an unconsidered factor in bird migration. PLoS One 9:e97650
Jensen JK, Isaksson C, Eikenaar C, Andersson MN (2020) Migrant blackbirds, Turdus merula, have higher plasma levels of polyunsaturated fatty acids compared to residents, but not enhanced fatty acid unsaturation index. Ecol Evol 10:10196–10206
Ji LL (1999) Antioxidants and oxidative stress in exercise. Proc Soc Exp Biol Med 222:283–292
Ji LL, Gomez-Cabrera M-C, Vina J (2006) Exercise and hormesis: activation of cellular antioxidant signaling pathway. Ann N Y Acad Sci 1067:425–435
Jones DP (2006) Redefining oxidative stress. Antiox Redox Signals 8:1865–1879
Judge S, Jang YM, Smith A, Selman C, Phillips T, Speakman JR, Hagen T, Leeuwenburgh C (2005) Exercise by lifelong voluntary wheel running reduces subsarcolemmal and interfibrillar mitochondrial hydrogen peroxide production in the heart. Am J Phys 289:1564–1572
Kahurananga J, Silkiluwasha F (1997) The migration of zebra and wildebeest between Tarangire National Park and Simanjiro Plains, northern Tanzania, in 1972 and recent trends. Afr J Ecol 35:179–185
Kinugawa S, Wang ZP, Kaminski PM, Wolin MS, Edwards JG, Kaley G, Hintze TH (2005) Limited exercise capacity in heterozygous manganese superoxide dismutase gene-knockout mice: roles of superoxide anion and nitric oxide. Circulation 111:1480–1486

Koyama S, Mizutani Y, Yoda K (2021) Exhausted with foraging: foraging behavior is related to oxidative stress in chick-rearing seabirds. Comp Biochem Physiol A 258:110984

Laranjeiro MI, Alves LMF, da Silva JM, Pereira JM, Norte AC, Paiva VH, Lemos MFL, Ramos JA, Novais SC, Ceia FR (2021) Year-round element quantification of a wide-ranging seabird and their relationships with oxidative stress, trophic ecology, and foraging patterns. Environ Pollut 284:117502

Larcombe SD, Tregaskes CA, Coffey JS, Stevenson AE, Alexander L, Arnold KE (2008) The effects of short-term antioxidant supplementation on oxidative stress and flight performance in adult budgerigars *Melopsittacus undulatus*. J Exp Biol 211:2859–2864

Larcombe SD, Coffey JS, Bann D, Alexander L, Arnold KE (2010) Impacts of dietary antioxidants and flight training on post-exercise oxidative damage in adult parrots. Comp Biochem Physiol B 155:49–53

Lee M, Choi I, Park K (2002) Activation of stress signaling molecules in bat brain during arousal from hibernation. J Neurochem 82:867–873

Lesmana R, Parameswari C, Mandagi GF, Wahyudi JF, Permana NJ, Radhiyanti PT, Gunadi JW (2022) The role of exercise-induced reactive oxygen species (ROS) hormesis in aging: friend or foe. Cell Physiol Biochem 56:692–706

Liu R, Liu IY, Bi X, Thompson RF, Doctrow SR, Malfroy B, Baudry M (2003) Reversal of age-related learning deficits and brain oxidative stress in mice with superoxide dismutase/catalase mimetics. Proc Natl Acad Sci USA 100:8526–8531

Loschen G, Flohe L, Chance B (1971) Respiratory chain linked $H_2O_2$ production in pigeon heart mitochondria. FEBS Lett 18:261–264

Losdat S, Helfenstein F, Gaude B, Richner H (2010) Effect of sibling competition and male carotenoid supply on offspring condition and oxidative stress. Behav Ecol 21:1271–1277

Lustgarten MS, Jang YC, Liu Y, Muller FL, Qi W, Steinhelper M, Brooks SV, Larkin L, Shimizu T, Shirasawa T, McManus LM, Bhattacharya A, Richardson A, Van Remmen H (2009) Conditional knockout of Mn-SOD targeted to type IIB skeletal muscle fibers increases oxidative stress and is sufficient to alter aerobic exercise capacity. Am J Phys 297:C1520–C1532

Lustgarten MS, Jang YC, Liu Y, Qi W, Qin Y, Dahia PL, Shi Y, Bhattacharya A, Muller FL, Shimizu T, Shirasawa T, Richardson A, Van Remmen H (2011) MnSOD deficiency results in elevated oxidative stress and decreased mitochondrial function but does not lead to muscle atrophy during aging. Aging Cell 10:493–505

Ma LY, Zhu X, Rivera PM, Tøien Ø, Barnes BM, LaManna JC, Smith MA, Drew KL (2005) Absence of cellular stress in brain after hypoxia induced by arousal from hibernation in Arctic ground squirrels. Am J Phys 289:R1297–R1306

Macnair M, Parker GA (1979) Models of parent–offspring conflict. III. Intrabrood conflict. Anim Behav 27:1202–1209

Magwere T, Pamplona R, Miwa S, Martinez-Diaz P, Portero-Otin M, Brand MD, Partridge L (2006) Flight activity, mortality rates, and lipoxidative damage in *Drosophila*. J Geront Ser A 61:136–145

Mao L, Franke J (2013) Hormesis in aging and neurodegeneration-a prodigy awaiting dissection. Int J Mol Sci 14:13109–13128

Marasco V, Sebastiano M, Costantini D, Pola G, Fusani L (2021) Controlled expression of the migratory phenotype affects oxidative status in birds. J Exp Biol 224:jeb233486

Marciau M, Costantini D, Bestley S, Hicks O, Hindell MA, Kato A, Raclot T, Ribout C, Ropert-Coudert Y, Angelier F (2023) Environmental drivers of growth and oxidative status during early life in a long-lived Antarctic seabird, the Adélie Penguin. Physiol Biochem Zool 96:177–191

Margonis K, Fatouros IG, Jamurtas AZ, Nikolaidis MG, Douroudos I, Chatzinikolaou A, Mitrakou A, Mastorakos G, Papassotiriou I, Taxildaris K, Kouretas D (2007) Oxidative stress biomarkers responses to physical overtraining: implications for diagnosis. Free Radic Biol Med 43:901–910

Margotta JW, Roberts SP, Elekonich MM (2018) Effects of flight activity and age on oxidative damage in the honey bee, *Apis mellifera*. J Exp Biol 221:jeb183228

Maronde L, Richner H (2015) Effects of increased begging and vitamin E supplements on oxidative stress and fledging probability. Behav Ecol 26:465–471
Mattson MP, Calabrese EJ (2010) Hormesis: a revolution in biology, toxicology and medicine. Springer, New York and London
McArdle F, Spiers S, Aldemir H, Vasilaki A, Beaver A, Iwanejko L, McArdle A, Jackson MJ (2004) Preconditioning of skeletal muscle against contraction-induced damage: the role of adaptations to oxidants in mice. J Physiol 561:233–244
McKenna MJ, Medved I, Goodman CA, Brown MJ, Bjorksten AR, Murphy KT, Petersen AC, Sostaric S, Gong X (2006) N-acetylcysteine attenuates the decline in muscle Na+,K+-pump activity and delays fatigue during prolonged exercise in humans. J Physiol 576:279–288
McWilliams SR, Guglielmo C, Pierce B, Klaassen M (2004) Flying, fasting, and feeding in birds during migration: a nutritional and physiological ecology perspective. J Avian Biol 35:377–393
McWilliams S, Pierce B, Wittenzellner A, Langlois L, Engel S, Speakman JR, Fatica O, DeMoranville K, Goymann W, Trost L, Bryla A, Dzialo M, Sadowska E, Bauchinger U (2020) The energy savings-oxidative cost trade-off for migratory birds during endurance flight. eLife 9:e60626
Metzger BJ, Bairlein F (2011) Fat stores in a migratory bird: a reservoir of carotenoid pigments for times of need? J Comp Physiol B 181:269–275
Milner-Gulland EJ, Fryxell JM, Sinclair ARE (2011) Animal migration: a synthesis. Oxford University Press, Oxford
Montoya B, Gil D, Valverde M, Rojas E, Pérez-Rodríguez L (2020) DNA integrity estimated via the comet assay reflects oxidative stress and competitive disadvantage in developing birds. Physiol Biochem Zool 93:384–395
Moreno-Rueda G, Redondo T, Trenzado CE, Sanz A, Zúñiga JM (2012) Oxidative stress mediates physiological costs of begging in magpie (*Pica pica*) nestlings. PLoS One 7:e40367
Mullen W, Larcombe S, Arnold K, Welchman H, Crozier A (2010) Use of accurate mass full scan mass spectrometry for the analysis of anthocyanins in berries and berry-fed tissues. J Agric Food Chem 58:3910–3915
Murphy EA, Davis JM, Carmichael MD, Gangemi JD, Ghaffar A, Mayer EP (2008) Exercise stress increases susceptibility to influenza infection. Brain Behav Immun 22:1152–1155
Nakamoto H, Kaneko T, Tahara S, Hayashi E, Naito H, Radak Z, Goto S (2007) Regular exercise reduces 8-oxodG in the nuclear and mitochondrial DNA and modulates the DNA repair activity in the liver of old rats. Exp Geront 42:287–295
Nanayakkara GK, Wang H, Yang X (2019) Proton leak regulates mitochondrial reactive oxygen species generation in endothelial cell activation and inflammation - a novel concept. Arch Biochem Biophys 662:68–74
Nevoux M, Arlt D, Nicoll M, Jones C, Norris K (2013) The short- and long-term fitness consequences of natal dispersal in a wild bird population. Ecol Lett 16:438–445
Nikolaidis MG, Mougios V (2004) Effects of exercise on the fatty-acid composition of blood and tissue lipids. Sports Med 34:1051–1076
Nikolaidis MG, Kerksick CM, Lamprecht M, McAnulty SR (2012a) Does vitamin C and E supplementation impair the favorable adaptations of regular exercise? Oxidative Med Cell Longev 2012:707941
Nikolaidis MG, Kyparos A, Spanou C, Paschalis V, Theodorou AA, Vrabas IS (2012b) Redox biology of exercise: an integrative and comparative consideration of some overlooked issues. J Exp Biol 215:1615–1625
Ninni P, de Lope F, Saino N, Haussy C, Møller AP (2004) Antioxidants and condition dependence of arrival date in a migratory passerine. Oikos 105:55–64
Noguera JC, Morales J, Pérez C, Velando A (2010) On the oxidative cost of begging: antioxidants enhance vocalizations in gull chicks. Behav Ecol 21:479–484
Noguera JC, Kim SY, Velando A (2017) Family-transmitted stress in a wild bird. Proc Natl Acad Sci USA 114:6794–6799
Novelli GP, Bracciotti G, Falsini S (1990) Spin-trappers and vitamin E prolong endurance to muscle fatigue in mice. Free Rad Biol Med 8:9–13

Nudds RL, Bryant DM (2000) The energetic cost of short flights in birds. J Exp Biol 203:1561–1572
Ogonovszky H, Sasvári M, Dosek A, Berkes I, Kaneko T, Tahara S, Nakamoto H, Goto S, Radák Z (2005) The effects of moderate, strenuous, and overtraining on oxidative stress markers and DNA repair in rat liver. Can J Appl Physiol 30:186–195
Oguchi Y, Smith RJ, Owen JC (2017) Fruits and migrant health: consequences of stopping over in exotic- vs. native-dominated shrublands on immune and antioxidant status of Swainson's Thrushes and Gray Catbirds. Condor 119:800–816
Okamoto I, Kayano T, Hanaya T, Arai S, Ikeda M, Kurimoto M (2006) Up-regulation of an extra-cellular superoxide dismutase-like activity in hibernating hamsters subjected to oxidative stress in mid- to late arousal from torpor. Comp Biochem Physiol C 144:47–56
Orr AL, Lohse LA, Drew KL, Hermes-Lima M (2009) Physiological oxidative stress after arousal from hibernation in Arctic ground squirrel. Comp Biochem Physiol A 153:213–221
Page MM, Peters CW, Staples JF, Stuart JA (2009) Intracellular antioxidant enzymes are not globally upregulated during hibernation in the major oxidative tissues of the 13-lined ground squirrel *Spermophilus tridecemlineatus*. Comp Biochem Physiol A 152:115–122
Papandreou MA, Dimakopoulou A, Linardaki ZI, Cordopatis P, Klimis-Zacas D, Margarity M, Lamari FN (2009) Effect of a polyphenol-rich wild blueberry extract on cognitive performance of mice, brain antioxidant markers and acetylcholinesterase activity. Behav Brain Res 198:352–358
Parker GA, Royle NJ, Hartley IR (2002) Begging scrambles with unequal chicks: interactions between need and competitive ability. Ecol Lett 5:206–215
Patnaik P, Sahoo DD (2021) Variations in oxidative stress and antioxidant defense level during different phases of hibernation in common Asian toad, *Duttaphrynus melanostictus*. Biol Open 10:bio058567
Pavlicev M, Cheverud JM, Wagner GP (2009) Measuring morphological integration using eigenvalue variance. Evol Biol 36:157–170
Peiman KS, Birnie-Gauvin K, Larsen MH, Colborne SF, Gilmour KM, Aarestrup K, Willmore WG, Cooke SJ (2017) Morphological, physiological and dietary covariation in migratory and resident adult brown trout (Salmo trutta). Zoology 123:79–90
Pennycuick CJ (1990) Predicting wingbeat frequency and wavelength of birds. J Exp Biol 150:171–185
Pennycuick CJ, Alerstam T, Hedenström A (1997) A new low-turbulence wind tunnel for bird flight experiments at Lund University, Sweden. J Exp Biol 200:1441–1449
Pereira AC, Huddleston DE, Brickman AM, Sosunov AA, Hen R, McKhann GM, Sloan R, Gage FH, Brown TR, Small SA (2007) An in vivo correlate of exercise-induced neurogenesis in the adult dentate gyrus. Proc Natl Acad Sci USA 104:5638–5643
Pilastro A, Baccetti N, Massi A, Montemaggiori A, Roselli A, Spina F (1995) Stima della direzione di migrazione e del consumo di grasso per ora di volo nel beccafico (*Sylvia borin*) durante la migrazione primaverile. In: Pandolfi M, Foschi UF (eds) Atti del VII Convegno Italiano di Ornitologia, Supplemento alle Ricerche di Biologia della Selvaggina, vol 22. Bologna, Istituto Nazionale per la Fauna Selvatica, pp 453–463
Powers SK, Jackson MJ (2008) Exercise-induced oxidative stress: cellular mechanisms and impact on muscle force production. Physiol Rev 88:1243–1276
Powers SK, Ji LL, Kavazis AN, Jackson MJ (2011) Reactive oxygen species: impact on skeletal muscle. Compr Physiol 1:941–969
Radak Z, Apor P, Pucsok J, Berkes I, Ogonovszky H, Pavlik G, Nakamoto H, Goto S (2003) Marathon running alters the DNA base excision repair in human skeletal muscle. Life Sci 72:1627–1633
Radak Z, Chung HY, Koltai E, Taylor AW, Goto S (2008) Exercise, oxidative stress and hormesis. Ageing Res Rev 7:34–42
Radak Z, Ishihara K, Tekus E, Varga C, Posa A, Balogh L, Boldogh I, Koltai E (2017) Exercise, oxidants, and antioxidants change the shape of the bell-shaped hormesis curve. Redox Biol 12:285–290
Ramirez JM, Folkow LP, Blix AS (2007) Hypoxia tolerance in mammals and birds: from the wilderness to the clinic. Annu Rev Physiol 69:113–143

Rattan SIS (2008) Hormesis in aging. Ageing Res Rev 7:63–78

Récapet C, Zahariev A, Blanc S, Arrivé M, Criscuolo F, Bize P, Doligez B (2016) Differences in the oxidative balance of dispersing and non-dispersing individuals: an experimental approach in a passerine bird. BMC Evol Biol 16:125

Reguera S, Zamora-Camacho FJ, Melero E, García-Mesa S, Trenzado CE, Cabrerizo MJ, Sanz A, Moreno-Rueda G (2015) Ultraviolet radiation does not increase oxidative stress in the lizard *Psammodromus algirus* along an elevational gradient. Comp Biochem Physiol A 183:20–26

Reichert S, Criscuolo F, Zahn S, Arrivé M, Bize P, Massemin S (2015) Immediate and delayed effects of growth conditions on ageing parameters in nestling zebra finches. J Exp Biol 218:491–499

Reid MB (2008) Free radicals and muscle fatigue: of ROS, canaries, and the IOC. Free Radic Biol Med 44:169–179

Reilly BD, Hickey AJR, Cramp RL, Franklin CE (2014) Decreased hydrogen peroxide production and mitochondrial respiration in skeletal muscle but not cardiac muscle of the green-striped burrowing frog, a natural model of muscle disuse. J Exp Biol 217:1087–1093

Ristow M, Zarse K, Oberbach A, Kloting N, Birringer M, Kiehntopf M, Stumvoll M, Kahn CR, Bluher M (2009) Antioxidants prevent health-promoting effects of physical exercise in humans. Proc Natl Acad Sci USA 106:8665–8670

Rödel HG, Jardim V, Rangassamy M, Jaravel L, Jacquet D, Monclús R, Féron C, Costantini D (2022) Early life parameters and personality affect oxidative status during adulthood in an altricial rodent. Physiol Rep 10:e15427

Rodriguez C, Campoy-Diaz AD, Giraud-Billoud M (2023) Short-term estivation and hibernation induce changes in the blood and circulating hemocytes of the apple snail Pomacea canaliculata. Meta 13:289

Rodríguez-Gironés MA, Cotton PA, Kacelnik A (1996) The evolution of begging: signaling and sibling competition. Proc Natl Acad Sci USA 93:14637–14641

Rosa EF, Takahashi S, Aboulafia J, Nouailhetas VL, Oliveira MG (2007) Oxidative stress induced by intense and exhaustive exercise impairs murine cognitive function. J Neurophysiol 98:1820–1826

Rosa EF, Freymüller E, Ihara SS, Aboulafia J, Nouailhetas VL (2008) Damaging effects of intense repetitive treadmill running on murine intestinal musculature. J Appl Physiol 104:1410–1417

Rothe HJ, Biesel W, Natchigall W (1987) Pigeon flight in a wind tunnel. II. Gas exchange and power requirements. J Comp Physiol B 157:99–109

Salway KD, Tattersall GJ, Stuart JA (2010) Rapid upregulation of heart antioxidant enzymes during arousal from estivation in the Giant African snail (*Achatina fulica*). Comp Biochem Physiol A 157:229–236

Schwilch R, Jenni L, Jenni-Eiermann SJ (1996) Metabolic responses of homing pigeons to flight and subsequent recovery. J Comp Physiol B 166:77–87

Schwilch R, Grattarola A, Spina F, Jenni L (2002) Protein loss during long distance migratory flight in passerine birds: adaptation and constraint. J Exp Biol 205:687–695

Sebastiano M, Eens M, Messina S, AbdElgawad H, Pineau K, Gerrit B, Chastel O, Costantini D (2018) Resveratrol supplementation reduces oxidative stress and modulates the immune response in free-living animals during a viral infection. Funct Ecol 32:2509–2519

Selman C, McLaren JS, Collins AR, Duthie GG, Speakman JR (2002) Antioxidant enzyme activities, lipid peroxidation, and DNA oxidative damage: the effects of short-term voluntary wheel running. Arch Biochem Biophys 401:225–261

Shephard RJ, Shek PN (1994) Potential impact of physical activity and sport on the immune system - a brief review. Br J Sports Med 28:247–255

Shukitt-Hale B, Galli RL, Meterko V, Carey A, Bielinski DF, McGhie T, Joseph JA (2005) Dietary supplementation with fruit polyphenolics ameliorates age-related deficits in behavior and neuronal markers of inflammation and oxidative stress. Age 27:49–57

Skrip MM, Bauchinger U, Goymann W, Fusani L, Cardinale M, Alan RR, McWilliams SR (2015) Migrating songbirds on stopover prepare for, and recover from, oxidative challenges posed by long-distance flight. Ecol Evol 5:3198–3209
Skrip MM, Seeram NP, Yuan T, Ma H, McWilliams SR (2016) Dietary antioxidants and flight exercise in female birds affect allocation of nutrients to eggs: how carry-over effects work. J Exp Biol 219:2716–2725
Smith SM (2016) The role of oxidative stress as an underlying mechanism in life-history trade-offs. PhD thesis, University of Glasgow
Sohal RS, Orr WC (2012) The redox stress hypothesis of aging. Free Rad Biol Med 52:539–555
Solanki N, Salvi A, Patki G, Salim S (2017) Modulating oxidative stress relieves stress-induced behavioral and cognitive impairments in rats. Int J Neuropsychopharmacol 20:550–561
Soulsbury CD, Halsey LG (2018) Does physical activity age wild animals? Front Ecol Evol 6:222
Staples JF, Brown JC (2008) Mitochondrial metabolism in hibernation and daily torpor: a review. J Comp Physiol B 178:811–827
Stier A, Massemin S, Zahn S, Tissier ML, Criscuolo F (2015) Starting with a handicap: effects of asynchronous hatching on growth rate, oxidative stress and telomere dynamics in free-living great tits. Oecologia 179:999–1010
Storey KB (2001) Turning down the fires of life: metabolic regulation of hibernation and estivation. In: Storey KB (ed) Molecular mechanisms of metabolic arrest: life in limbo. BIOS Scientific Publishers, Oxford, pp 1–21
Storey KB, Storey JM (1990) Facultative metabolic rate depression: molecular regulation and biochemical adaptation in anaerobiosis, hibernation, and estivation. Q Rev Biol 65:145–174
Stranahan AM, Mattson MP (2010) Exercise-induced hormesis. In: Mattson MP, Calabrese EJ (eds) Hormesis: a revolution in biology, toxicology and medicine. Springer, New York, pp 109–122
Sur S, Sharma A, Bhardwaj SK, Kumar V (2020) Involvement of steroid and antioxidant pathways in spleen-mediated immunity in migratory birds. Comp Biochem Physiol A 250:110790
Tauffenberger A, Magistretti PJ (2021) Reactive oxygen species: beyond their reactive behavior. Neurochem Res 46:77–87
Theodorou AA, Nikolaidis MG, Paschalis V, Koutsias S, Panayiotou G, Fatouros IG, Koutedakis Y, Jamurtas AZ (2011) No effect of antioxidant supplementation on muscle performance and blood redox status adaptations to eccentric training. Am J Clin Nutr 93:1373–1383
Vaanholt LM, Speakman JR, Garland T Jr, Lobley GE, Visser GH (2008) Protein synthesis and antioxidant capacity in aging mice: effects of long-term voluntary exercise. Physiol Biochem Zool 81:148–157
Voigt CC, Fritze M, Lindecke O, Costantini D, Pētersons G, Czirjak GA (2020) The immune response of bats differs between pre-migration and migration seasons. Sci Rep 10:17384
Voituron Y, Servais S, Romestaing C, Douki T, Barré H (2006) Oxidative DNA damage and antioxidant defenses in the European common lizard (*Lacerta vivipara*) in supercooled and frozen states. Cryobiology 52:74–82
Wang J, Shi H, Guo Q, Liu F, Yan S, Dai D, Wu M (2018) The impact of hibernation and arousal on energy metabolism and antioxidant defenses in leech (*Whitmania pigra*). Aquac Res 49:188
Westerblad H, Allen DG (2011) Emerging roles of ROS/RNS in muscle function and fatigue. Antioxid Redox Signal 15:2487–2499
Wiersma P, Selman C, Speakman JR, Verhulst S (2004) Birds sacrifice oxidative protection for reproduction. Biol Lett 271:360–363
Williams MD, Van Remmen H, Conrad CC, Huang TT, Epstein CJ, Richardson A (1998) Increased oxidative damage is correlated to altered mitochondrial function in heterozygous manganese superoxide dismutase knockout mice. J Biol Chem 273:28510–28515
Willmore WG, Storey KB (1997a) Antioxidant systems and anoxia tolerance in a freshwater turtle, *Trachemys scripta elegans*. Mol Cell Biochem 170:177–185
Willmore WG, Storey KB (1997b) Glutathione systems and anoxia tolerance in turtles. Am J Phys 273:R219–R225
Woods JA, Davis JM, Smith JA, Nieman DC (1999) Exercise and cellular innate immune function. Med Sci Sports Exerc 31:57–66

Wu CW, Biggar KK, Zhang J, Tessier SN, Pifferi F, Perret M, Storey KB (2015) Induction of antioxidant and heat shock protein responses during torpor in the gray mouse lemur, *Microcebus murinus*. Genom Prot Bioinform 13:119–126

Yan L-J, Sohal RS (2000) Prevention of flight activity prolongs the life span of the housefly, Musca domestica, and attenuates the age-associated oxidative damage to specific mitochondrial proteins. Free Radic Biol Med 29:1143–1150

Yap KN, Kim OR, Harris KC, Williams TD (2017) Physiological effects of increased foraging effort in a small passerine. J Exp Biol 220:4282–4291

Yin Q, Ge H, Liao C-C, Liu D, Zhang S, Pan Y-H (2016) Antioxidant defenses in the brains of bats during hibernation. PLoS One 11:e0152135

Young KM, Cramp RL, Franklin CE (2013) Each to their own: skeletal muscles of different function use different biochemical strategies during aestivation at high temperature. J Exp Biol 216:1012–1124

Zerba E, Komorowski TE, Faulkner JA (1990) Free radical injury to skeletal muscles of young, adult, and old mice. Am J Phys 258:429–435

Zhang G, Cowled C, Shi Z, Huang Z, Bishop-Lilly KA, Fang X, Wynne JW, Xiong Z, Baker ML, Zhao W, Tachedjian M, Zhu Y, Zhou P, Jiang X, Ng J, Yang L, Wu L, Xiao J, Feng Y, Chen Y, Sun X, Zhang Y, Marsh GA, Crameri G, Broder CC, Frey KG, Wang LF, Wang J (2013) Comparative analysis of bat genomes provides insight into the evolution of flight and immunity. Science 339:456–460

Zhang Y, Liang S, He J, Bai Y, Niu Y, Tang X, Li D, Chen Q (2015) Oxidative stress and antioxidant status in a lizard *Phrynocephalus vlangalii* at different altitudes or acclimated to hypoxia. Comp Biochem Physiol A 190:9–14

# Chapter 7
# For Better or Worse: How Early Life Oxidative Stress Moulds the Phenotype

**Abstract** The early environment in which an individual grows up and develops has profound, long-lasting, and often irreversible consequences for the phenotype throughout the individual lifetime. Identifying the physiological mechanisms that orchestrate the phenotypic development and mediate the trade-offs that occur early in life is critical for our understanding of how the diversity in growth and developmental strategies evolves. It is also important to explain why a strategy may be more successful under certain environmental circumstances than others, and also to determine how the individual oxidative status itself is moulded by early experiences. In this chapter, I have discussed (1) the role of maternal effects in moulding the oxidative status of their progeny and (2) how reactive oxygen species organise the phenotypic development. I have also illustrated the role of oxidative stress as both a constraint on and a cost of growth, and how the passive antioxidant protection transferred by mothers to their progeny through the egg, the placenta, or the milk interacts with the maturation of the endogenous antioxidant machinery in determining the organismal capacity to cope with oxidative stress. Finally, I have examined how hormetic priming early in life contributes to build phenotypic resistance to the many challenges of life.

## 7.1 Introduction

Early life experiences have profound effects on life (Costantini and Marasco 2022). The adult phenotype is determined by complex interactions between genes and environmental stimuli that are received from conception to sexual maturity and beyond. Molecular regulators of development do not all reside only within the embryo or the egg, but also play relevant roles after birth. Exposure to post-natal environmental stimuli contributes significantly to build up phenotypic variation, sometimes amplifying or mitigating the effects of pre-natal experiences (Gilbert 2001; Metcalfe and Monaghan 2001a, b; West-Eberhard 2003; Monaghan 2008; Gilbert and Epel 2009; Costantini and Marasco 2022). Developmental systems are

© The Author(s), under exclusive license to Springer Nature Switzerland AG 2024
D. Costantini, *The Role of Organismal Oxidative Stress in the Ecology and Life-History Evolution of Animals*, Fascinating Life Sciences,
https://doi.org/10.1007/978-3-031-65183-0_7

also guided by processes other than genetic control or natural selection, such as self-organisation properties of components of a developing organism, neutral evolution, epigenetic modifications of the genome, or compensatory rearrangements of anatomical traits and metabolic networks (e.g. Szyf et al. 2005; Gerhart and Kirschner 2007; Badyaev 2011; Curley and Branchi 2013).

Adult body size has a significant impact on reproductive fitness in many animal species because early attainment of a large body size may confer selective advantages, such as in the competition among males to attain a dominant position in the group, in laying large quantities of eggs, or in the reduction of predation risk (Metcalfe and Monaghan 2001a, b). Thus, we might expect that individuals grow as fast as possible in order to achieve an adult body size that would bring benefits for reproduction. However, growth rates are usually slower than the maximum achievable. This implies that there are constraints on rapid or accelerated growth through a trade-off in the allocation of resources (e.g. energy, nutrients), where energetically expensive growth causes resources to be diverted away from other processes, such as physiological development (Metcalfe and Monaghan 2001a, b). If individuals are not able to completely adjust that trade-off, it may result in costs (e.g. cellular damage, reduced lifespan) induced by a rapid growth that reduce the performance of self-maintenance mechanisms. Identifying the proximate causes that mediate early in life trade-offs is very important to further our understanding of how evolution has generated the impressive diversity in growth and developmental strategies, and the conditions that make a certain strategy more evolutionary successful than others. This continuous complex interplay among different proximate and ultimate processes, and the benefits and costs associated has indeed determined the evolution of a variety of growth and developmental strategies, such as metamorphosis, altricial–precocial continuum, indeterminate and determinate growth (Stearns 1992; Lindström 1999; Zera and Harshman 2001; Ricklefs and Wikelski 2002; Monaghan 2008; Smith et al. 2016; Costantini and Marasco 2022).

In this chapter, I have discussed the roles that (1) maternal hormones play in moulding the oxidative status of the progeny (see also Chap. 3 for the role of maternal nutrition) and (2) reactive oxygen species have in the organisation of phenotypic development. I have also illustrated the role of oxidative stress as both a constraint on and a cost of growth and development. Finally, I have examined how early in life hormetic priming contributes to mould a phenotype that is more resistant to the many challenges of life.

## 7.2 A Mother's Legacy: How Maternal Hormones Mould the Oxidative Status of the Progeny

Mothers pass a number of molecules (e.g. hormones, antioxidants) along to the offspring that are able of moulding many of their phenotypic traits for life (Moore et al. 2019). In particular, the synthesis of hormones and their secretion are sensitive to changes in environmental conditions, so that maternal hormones are considered to be important molecular carriers for the transfer of environmental cues to the next generation. So far, most of the studies on the link between maternal hormones and oxidative status of offspring were focussed on androgens and glucocorticoids (e.g. Henriksen et al. 2011; Sheriff and Love 2013; Groothuis et al. 2005, 2019). Surprisingly, other hormones with clear effects on metabolism (e.g. thyroid hormones, Hsu et al. 2019; Sarraude et al. 2020) have received comparatively little experimental investigation.

Sex hormones play a relevant role as modulators of offspring growth and development (e.g. Schwabl 1993; Groothuis et al. 2005, 2019; Carere and Balthazart 2007; Groothuis and Schwabl 2008). In birds, the importance of maternal sex hormones was first highlighted in the 1990s, when Schwabl (1993, 1996) found that female canaries *Serinus canaria* deposited testosterone into their eggs, and that this deposition was not carried out randomly across the laying sequence. Indeed, earlier studies suggested that the avian embryo was the main source of sex hormones because the embryo develops in the egg outside the mother. However, given the lipophilic nature of sex hormones, mothers can pass them into the lipoprotein matrix of yolk during vitellogenesis. Although yolk testosterone has a number of potentially beneficial effects on offspring (Schwabl 1993), it can also carry costs, such as suppression of immune function, oxidative stress, and mortality (Sockman and Schwabl 2000).

The question then is whether yolk testosterone has indeed relevant effects on the embryo. To tackle this question, Tobler and Sandell (2009) injected testosterone into the eggs. They found that the injection of 0.5 ng of testosterone on the third day of incubation had short-term effects on male zebra finches *Taeniopygia guttata*. Compared to control males, plasma levels of non-enzymatic antioxidant capacity were reduced in males that were born from eggs enriched with testosterone when they were 10 days old. However, control and treated birds had similar levels of antioxidants at 34 days of age. In a similar experiment, however, Tobler et al. (2013) found that males from testosterone-injected eggs had higher levels of antioxidants than control males, and this difference was still evident when birds were seven months old.

Experiments on other bird species also found contrasting results. Galvan and Alonso-Alvarez (2010) did not find any effects of an injection of 26 ng of testosterone into the eggs on the red blood cell concentration of glutathione and plasma levels of non-enzymatic antioxidant capacity in nestling great tits *Parus major* at 15 days of age. By contrast, Noguera et al. (2011b) found that injecting third-laid eggs with 261 ng of testosterone on the same day in which the egg was laid increased temporarily the plasma non-enzymatic antioxidant capacity and reduced the plasma concentration of malondialdehyde in chick yellow-legged gulls *Larus michahellis* during the first 9 days of life, respectively. Treidel et al. (2013) found that the injection of 5 ng of testosterone into the eggs prior to the start of incubation did not affect levels of reactive oxygen metabolites or DNA damage, but impaired the DNA damage repair efficiency of 17–18 days old chickens *Gallus gallus*, in comparison with the control birds. Finally, Parolini et al. (2017, 2018a) injected different amounts of testosterone, ranging from 42 to 95 ng, into the eggs in order to increase its concentration by 1 standard deviation of the concentration recorded in the yolk of yellow-legged gull eggs from the same colony. The injection of testosterone (1) caused a reduction in the concentration of pro-oxidants in liver, but not in brain, of embryos, (2) did not affect the non-enzymatic antioxidant capacity in brain and in liver of embryos, and (3) did not affect the concentration of pro-oxidants, levels of non-enzymatic antioxidant capacity, and telomere length in blood of hatchlings.

The results of these studies did not provide clear evidence about how testosterone would affect the oxidative status of offspring. Why are results of these experiments so contrasting? One reason might lie in the dose of the testosterone injected into the egg if it falls in the stimulatory range for yellow-legged gulls and in the inhibitory/toxic range for zebra finches and chickens, respectively (Fig. 7.1). Another reason might be that the sensitive window of development (when the organism is responsive to stimuli) during which egg testosterone was manipulated differed among studies. Phenotypic traits, after an initial hypo- or hyper-responsive period, may later transition into a hyper- or hypo-responsive period, respectively. Thus, the emergence of an effect or lack thereof might depend on the development stage because it determines the responsiveness of the organism. A third explanation might lie with the pace of life because species with higher paces of life tend to invest less into antioxidant protection. Zebra finches and chickens have a higher pace of life than gulls; however, testosterone treatment did not affect antioxidants in great tits, which have also a high pace of life. In addition, the response to a same dose of testosterone even differed between studies carried out on the same species. Fourth, studies differed in the age of birds at which the effects of testosterone treatment were quantified. This could explain differences between studies if the physiological effects of testosterone were transient or long-lasting. Fifth, it is unknown if the distribution of the injected testosterone across the yolk layers was similar between studies; if not, it would mean that embryos were exposed to doses that differed from those that were injected (e.g. Lipar et al. 1999). Finally, species could differ in the way embryos metabolise the injected testosterone. Testosterone is a precursor of the androgen dihydrotestosterone and of the oestrogen estradiol (aromatisation: biosynthesis of oestrogens from testosterone catalysed by the enzyme aromatase), which might differ from testosterone in how they affect the organismal oxidative status (e.g. Casagrande et al. 2012a, b).

Glucocorticoids are another relevant group of hormones that mothers use to modulate growth and development of their offspring. The *developmental programming hypothesis* highlights that exposure to certain stimuli during prenatal or early post-natal development can cause phenotypic changes through the action of glucocorticoids, which subsequently will affect health and risk of disease (e.g. Barker et al. 1993; Edwards et al. 1993; Barker 1998; Seckl 1998; Waterland and Garza 1999; Armitage et al. 2008). This hypothesis was proposed to explain the complex interplay between stress exposure of the mammalian embryo, patterns of foetal growth and development, and patho-physiological consequences in adulthood (Barker et al. 1993; Edwards et al. 1993; Barker 1998; Seckl 1998; Waterland and Garza 1999; Armitage et al. 2008; Laviola and Macrì 2013). In particular, many authors suggested that overexposure of foetus

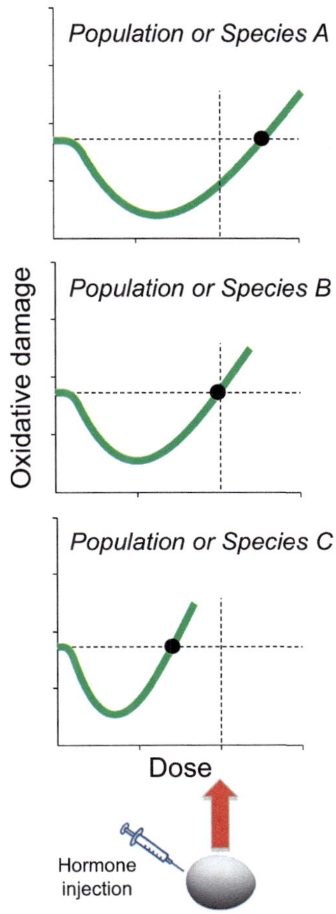


**Fig. 7.1** The range of hormonal doses injected into an egg at which the physiological response to stressors is stimulated likely varies between conspecific populations and species. These differences suggest that it is fundamentally important to know the hormetic zone (i.e. the area under the curve and above the zero-equivalent point that indicates the stimulatory range) in order to be able to predict the response of an individual from a certain population or a species to an environmental stressor. In this illustrated example, the zero-equivalent point (black circle) is the point of the hormetic curve where oxidative damage equals that of the control (i.e. zero exposure), which is represented by a horizontal dashed line. Reprinted with minor modifications from Costantini (2013) with kind permission of Springer Nature

to maternal glucocorticoids provides offspring with environmental cues that would guide the development in order to programme the phenotype to cope with the future post-natal environmental conditions. This is because vertebrates translate stressful environmental and social stimulations into secretions of glucocorticoid hormones, which activate a number of physiological responses to maintain organismal homeostasis (Wingfield et al. 1998; Sapolsky et al. 2000; Romero 2004). Studies carried out on birds supported a role of maternal glucocorticoids in generating adaptive phenotypic plasticity, including plasticity of the hypothalamic–pituitary–adrenal axis that regulates the organismal stress response (Hayward and Wingfield 2004; Love and Williams 2008a, b). However, other authors criticised this hypothesis because the information gathered by glucocorticoids would be maladaptive when early in life and adult environmental conditions differ (e.g. Bateson et al. 2004). These criticisms are based on all-or-nothing scenarios, which do not take into account the continuous nature of stressors nor the importance of their intensity. In this context, hormesis could provide a useful conceptual and mechanistic complementary model to further our understanding of the (mal)adaptive effects of maternal glucocorticoids. For example, several studies found that early in life exposure to mild stressful conditions may stimulate growth (e.g. house crickets *Acheta domesticus*, Luckey 1968; tobacco budworms *Heliothis virescens*, Stipanovic et al. 1986; cotton bollworms *Helicoverpa armigera*, Celorio-Mancera et al. 2011; western fence lizards *Sceloporus occidentalis*, McFarland et al. 2012). Early in life exposure to mild stressful conditions might also increase the stress tolerance or resistance in adulthood (homeodynamics *sensu* Yates 1994; Lloyd et al. 2001; Stebbing 2009).

Some experimental studies provided evidence in support of a role of maternal glucocorticoids in moulding the oxidative status of progeny (e.g. Haussmann et al. 2012; Marasco et al. 2013; Romano et al. 2021). For example, Haussmann et al. (2012) injected two doses of corticosterone (5 or 10 ng $g^{-1}$ yolk) into eggs of chicken in order to elevate it to levels that chickens can have under different stressful conditions. At 25 days of age, baseline levels of corticosterone were similar between the groups. By contrast, levels of corticosterone in the high-dose group increased across the whole duration of an induced stress response, whereas levels in the control and low-dose groups did not. Chicks born from the corticosterone-injected eggs had higher baseline levels of reactive oxygen metabolites than control chicks, and those chicks born from the high-dose group had also an over-representation of short telomeres compared to control and low-dose chicks (Haussmann et al. 2012).

Further experimental work showed that any effects of embryonic exposure to glucocorticoids may depend on the post-natal stressful experiences. Marasco et al. (2012, 2013) used an experimental setting including Japanese quails *Coturnix japonica* that were not exposed to corticosterone, or that were exposed to corticosterone (8.5 ng) during (1) the embryonic development, (2) the post-natal period (between days 5 and 19 of age; 45 μg between 5 and 15 days or 90 μg between 16 and 19 days), or (3) both the embryonic and post-natal periods. At 22 days of age, female chicks that were exposed to corticosterone during the post-natal period, regardless of pre-natal experience, had shorter stress responses in comparison with controls and chicks that were exposed only during the embryonic period, respectively. At 64 days of age, those birds that were previously stressed as embryos showed higher corticosterone concentrations over the stress response compared with controls. This effect was not evident in birds subjected to either post-natal treatment or the combined treatments (Marasco et al. 2012). Corticosterone exposure did not affect any marker of oxidative status in midbrain, and levels of superoxide dismutase and protein carbonyls in both red blood cells and cerebellum (Marasco et al. 2013). By contrast, levels of glutathione peroxidase in red blood cells were higher in all chicks that were exposed to corticosterone, with an additive effect in those chicks that were subjected to the combined pre- and post-natal exposures. In addition, the non-enzymatic antioxidant capacity in red blood cells was lower in all chicks that were exposed to corticosterone at some point during development compared to control chicks. Levels of glutathione peroxidase in cerebellum tended to be higher in all chicks that were exposed to corticosterone compared to control chicks. In addition, the non-enzymatic antioxidant capacity was significantly lower only in those chicks that were subjected to the combined pre- and post-natal exposures. Finally, the expression of the gene superoxide dismutase 3 (extracellular superoxide dismutase) was downregulated in the hippocampus of all chicks that were exposed to corticosterone at some point during development compared to control chicks (Marasco et al. 2016). Thus, corticosterone had significant tissue-dependent effects on the oxidative status, and only in a few cases the effect of embryonic exposure was modulated by the post-natal exposure, suggesting that only certain components of the oxidative status have a prolonged responsiveness (or plasticity) to environmental stressful stimuli.

In fish, exposure of mothers to a source of stress can also increase circulating and egg cortisol levels. Thus, it might be expected that maternal stress would impact on the oxidative status of their offspring. A study on sockeye salmons *Oncorhynchus nerka* did not find any correlation between maternal stress and glutathione or oxygen radical absorptive capacity in their offspring (Taylor et al. 2016). Moreover, the experimental increase of cortisol into the eggs had small effects on the antioxidant status of fish (Taylor et al. 2016).

Being egg-laying taxa, birds and fish offer the possibility to experimentally tease apart pre- versus post-hatching effects. In addition, the reduced physiological intimacy between the developing avian or fish embryo and the mother compared to mammals enables minimising potential compensatory/buffering effects of the parents on the growing offspring. Indeed, in mammals, glucocorticoids can cross the placenta, so that they can pass from the mother to the foetus (e.g. Ward 1972; Dahlöf et al. 1977; Barker et al. 1993; Seckl 1998; Hauser 2013). Some studies showed that (1) the embryonic exposure to glucocorticoids can also have short- or long-lasting impacts on the oxidative status and (2) the effects of glucocorticoids depend on the pregnancy window during which the embryo becomes exposed to stress hormones (e.g. Mendez-Armenta et al. 2003; Atanasova et al. 2009; Bartho et al. 2019; Dantzer et al. 2020). For example, Atanasova et al. (2009) carried out an oral treatment with dexamethasone (a synthetic glucocorticoid) for 1 week on a group of common marmosets *Callithrix jacchus* during an early period of gestation (42–48 days), and on another group during a late period of gestation (90–96 days), respectively. The results of this experiment showed that, at 2 years of age, the offspring of mothers that were given dexamethasone during late pregnancy had higher expression of the genes glutathione peroxidase 1, superoxide dismutase 2, glutamate cysteine ligase, and glutathione peroxidase

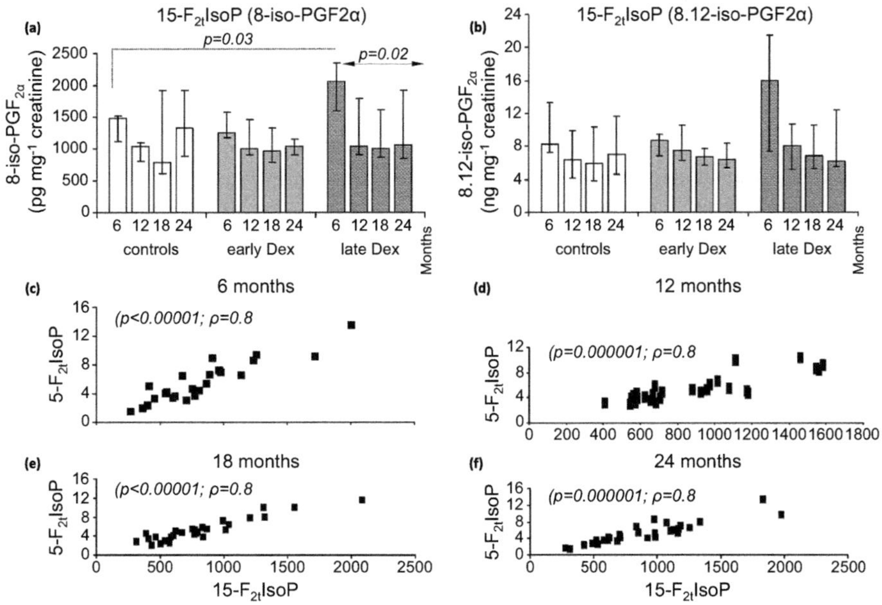


**Fig. 7.2** F2-isoprostane levels measured in offspring urine samples of marmoset monkeys *Callithrix jacchus* at different ages. Panels (**a**) and (**b**) show F2-isoprostanes in controls (oral dexamethasone vehicle, pregnancy days 42–48 and 90–96), early Dex (oral dexamethasone pregnancy days 42–48), and late Dex (oral dexamethasone pregnancy days 90–96). Values are shown as median and 84th and 16th percentiles (shown as error bars). Panels (**c**) to (**f**) show correlations between the 15-F2tIsoP (pg $mg^{-1}$ creatinine) and 5-F2tIsoP (ng $mg^{-1}$ creatinine) urine levels for all offspring at 6, 12, 18, and 24 months of age. Reprinted with minor modifications from Atanasova et al. (2009) with permission of Taylor & Francis

4 in the aorta, and higher urinary levels of F2-isoprostanes than control offspring (Fig. 7.2). By contrast, offspring of mothers that were given dexamethasone during early pregnancy had only higher expression of the gene glutathione reductase in the aorta than control offspring (Fig. 7.2).

In another study, Bartho et al. (2019) implanted C57BL/6 mice with osmotic minipumps containing corticosterone (33 μg $kg^{-1}$ $h^{-1}$) when the embryo had 12.5 days, and their placentas were collected 2 days later. Corticosterone exposure reduced placental levels of hydrogen peroxide and 8-hydroxy-2′-deoxyguanosine, but increased levels of protein carbonyls, advanced glycation end products, and thioredoxin reductase in placentas of female foetuses, but not male foetuses (Fig. 7.3). This study demonstrated that placentae of female foetuses responded differently to maternal corticosterone exposure, and highlighted an important role of reactive oxygen species, mitochondrial adaptations, and antioxidant responses in glucocorticoid-mediated programming of health and disease (Bartho et al. 2019).

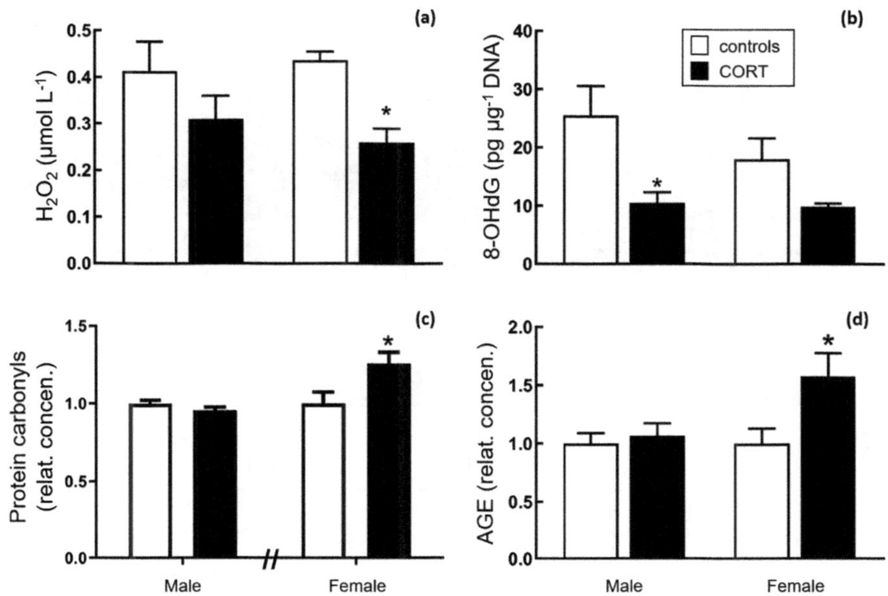


**Fig. 7.3** Effects of maternal corticosterone exposure (black histograms) on (**a**) hydrogen peroxide, (**b**) 8-hydroxy-2′-deoxyguanosine, (**c**) relative protein carbonyl concentrations, and (**d**) relative advanced glycation end product (AGE) concentrations in placental tissues from female and male foetuses compared with untreated controls (open histograms). Asterisks indicate significant differences between groups. Values are shown as mean and standard error. Reprinted with minor modifications from Bartho et al. (2019) with permission of John Wiley and Sons

## 7.3 Organisational Effects of Reactive Oxygen Species

One route through which maternal hormones drive the development of oxidative status might lie with the organisational effects of reactive oxygen species. In the early 1900s, Otto Warburg found that fertilisation is accompanied by an increase of oxygen consumption in sea urchin embryos (Warburg 1908; Shearer 1922). Some decades later, Foerder et al. (1978) found that this dramatic increase in oxygen consumption was linked to a burst of reactive oxygen species generation. We now know that reactive oxygen species can regulate development and even regeneration of tissues. For example, Han et al. (2018) found in oocytes obtained from *Xenopus laevis* that (1) the fertilisation-induced calcium wave induced reactive oxygen species generation in mitochondria, (2) inhibition of this generation of reactive oxygen species resulted in cell-cycle arrest, and (3) removal of inhibitors restored cell division and generation of reactive oxygen species (Figs. 7.4 and 7.5).

Reactive oxygen species are versatile signalling molecules in both animals and plants which work as regulators of or constraints on growth and development (e.g. Xia et al. 2015; Smith et al. 2016). Oxidative stress may induce apoptosis of progenitor cells that are crucial for tissue development, and reactive oxygen species can induce post-translational modifications on signalling proteins or transcriptional

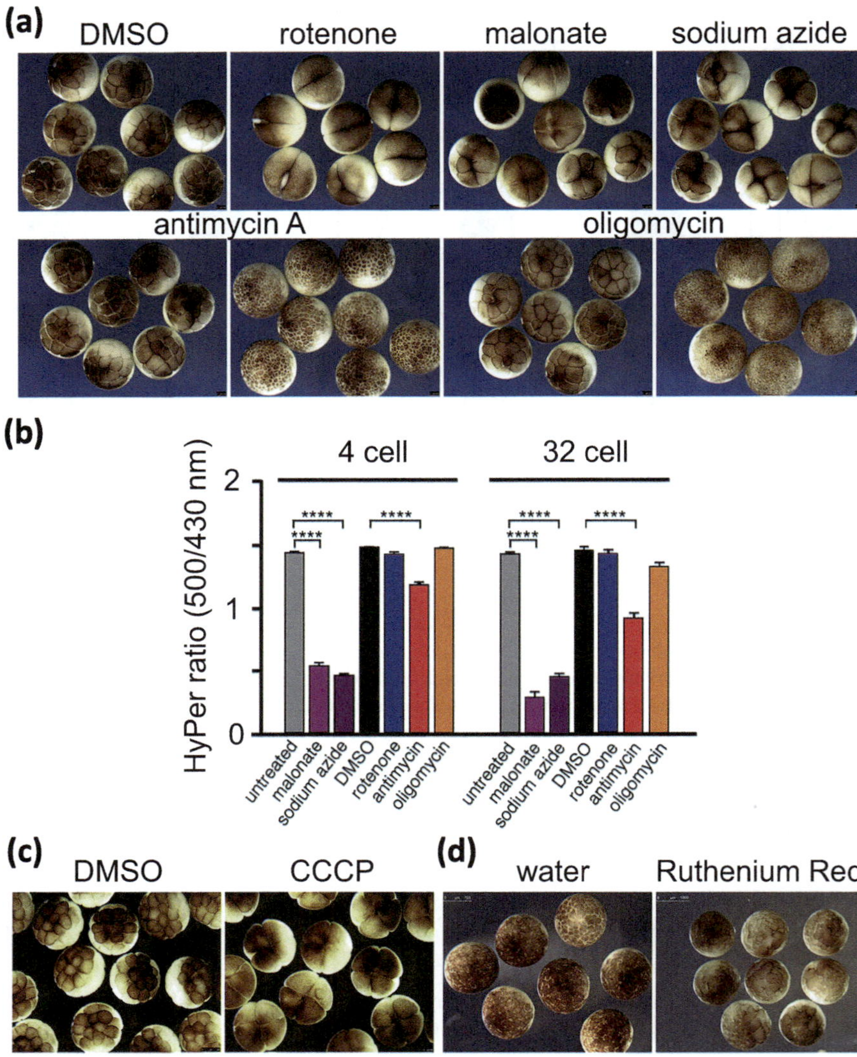


**Fig. 7.4** Inhibition of the generation of reactive oxygen species in mitochondria induced cell division arrest in *Xenopus* early development. (**a**) Embryos were treated at the one-cell stage with 0.1% DMSO, 1 mmol $L^{-1}$ rotenone, 10 mmol $L^{-1}$ malonate, 3 mmol $L^{-1}$ sodium azide, 10 mmol $L^{-1}$ antimycin A, and 6 mmol $L^{-1}$ oligomycin, respectively. Embryos treated with rotenone, malonate, and sodium azide were arrested at the two- to eight-cell stage. By the blastula stage, embryos treated with antimycin A were arrested. (**b**) One-cell stage embryos obtained from females expressing HyPer (hydrogen peroxide sensor) were treated with mitochondrial inhibitors, and HyPer ratio was measured at the 4-cell and 32-cell stages. Asterisks indicate significant differences between groups. (**c**) Embryos were treated with 0.1% DMSO or 2 mmol $L^{-1}$ CCCP. (**d**) Embryos were injected with water as a control and 50 pmol ruthenium red into one cell at the two-cell stage. Injection of ruthenium red induced cell-cycle arrest. Reprinted from Han et al. (2018). © 2017 The Authors; distributed under the terms of the Creative Commons Attribution License (CC BY)

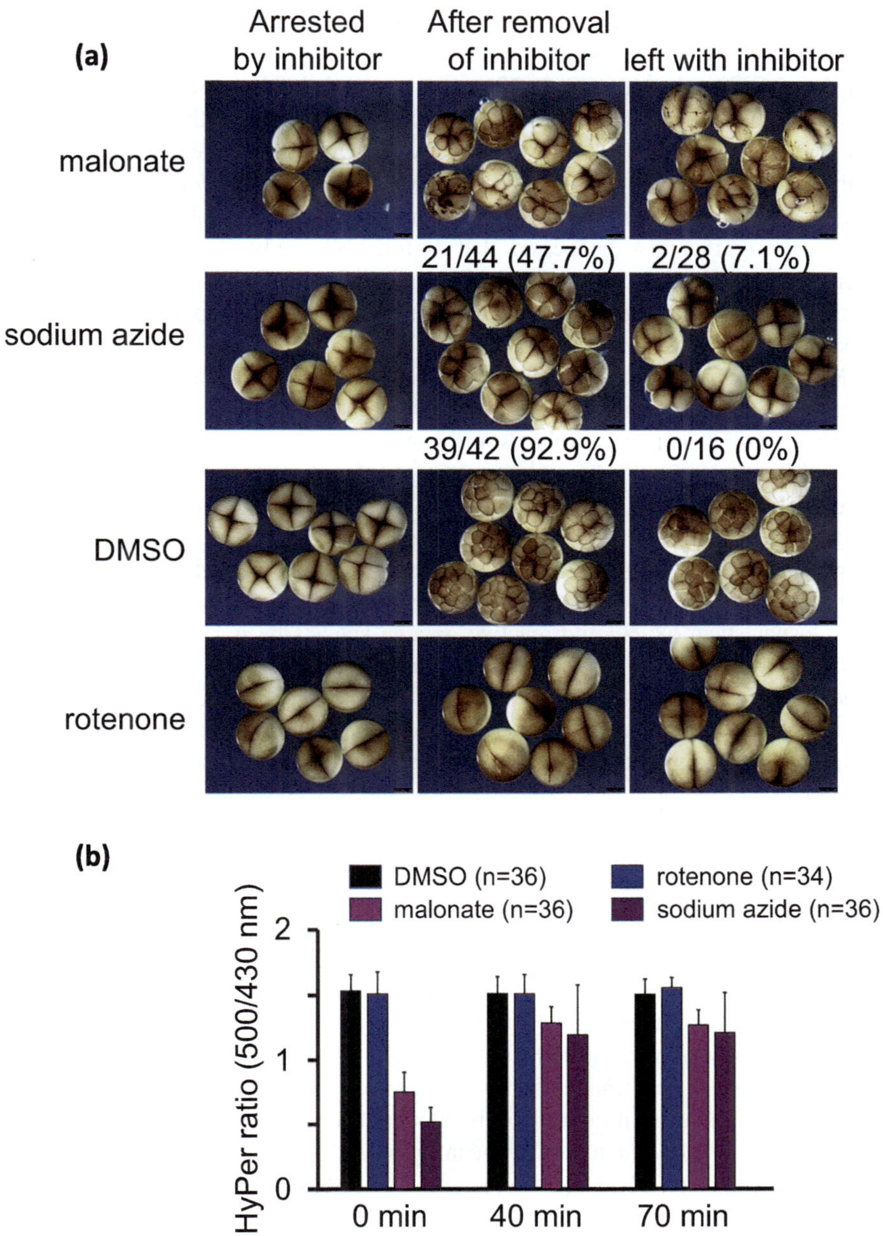


**Fig. 7.5** Removal of inhibitors restored cell division and reactive oxygen species generation in *Xenopus*. (**a**) Embryos were treated with inhibitors, 10 mmol $L^{-1}$ malonate, 3 mmol $L^{-1}$ sodium azide, and 1 mmol $L^{-1}$ rotenone, respectively, for 40 min until cells stopped dividing (left column), and then they were transferred to a dish containing fresh medium without inhibitors. Embryos transferred after treatment with malonate and azide retrieved cell division and divided to 8- to 16-cell stages (middle), but not in medium with inhibitors (right). (**b**) Fertilised transgenic embryos expressing HyPer (hydrogen peroxide sensor) were treated with inhibitors for 40 min and imaged shown as 0 min. Then embryos were transferred to fresh medium without inhibitors, and imaged at 40 and 70 min. Values are shown as mean and standard deviation. Reprinted from Han et al. (2018). © 2017 The Authors; distributed under the terms of the Creative Commons Attribution License (CC BY)

activation of specific genes that play a role in developmental processes (e.g. Covarrubias et al. 2008; Bräutigam et al. 2011; Smith et al. 2013). In sea urchin embryos, Coffman et al. (2009, 2014) found that mitochondrial reactive oxygen species regulate the oral–aboral axis specification by activating the nodal signalling pathway. Han et al. (2018) found that reactive oxygen species are essential for early embryonic development in the African clawed frog *Xenopus laevis*. Apirajkamol et al. (2020) found that oxidative stress induced experimentally using paraquat increased significantly the period that cotton bollworms *Helicoverpa armigera* spent as caterpillars compared to controls, and upregulated clusters of genes related to cuticle and chitin development, glycine metabolism, and oxidation-reduction processes.

Metamorphosis is another relevant biological process, where reactive oxygen species act as important regulators of development and tissue remodelling. For example, studies on metamorphosis in Japanese wrinkled frogs *Glandirana rugosa* (Hanada et al. 1997) or African clawed frogs (Menon and Rozman 2007; Johnson et al. 2013) found that the cellular environments in the intestine and tail become progressively more oxidising during their remodelling and regression, respectively. However, Gavrić et al. (2021) found that newly metamorphosed crested newts had higher antioxidants and lower thiobarbituric acid reactive substances than larvae at late Gosner stage 62. Should reactive oxygen species be important regulators of metamorphosis, we might expect that increased intake of antioxidants could be detrimental. This is because antioxidants would remove reactive oxygen species, causing a disruption of their signalling action. This hypothesis has received experimental support. For example, increased intake of carotenoids delayed completion of metamorphosis in gray treefrogs *Hyla versicolor* and in wood frogs *Lithobates sylvaticus*, and caused lower survival to metamorphosis in wood frogs (Fig. 7.6) (Cothran et al. 2015).

The effects of the exposure of the embryo to reactive oxygen species during development can also be long-lasting. For example, Romero-Haro and Alonso-Alvarez (2015) found that short-term inhibition of glutathione synthesis during development may induce deep modifications in the expression of phenotypic traits in adulthood, such as oxidative status and sexual signals. Additionally, Romero-Haro and Alonso-Alvarez (2020) found that offspring of zebra finches, whose glutathione synthesis was transitorily inhibited when they were nestlings, differed in several body traits from control birds. Obata et al. (2018) showed that transient exposure of *Drosophila* to low concentrations of oxidants during development caused an extension of lifespan of adults, acting in an antibiotic-like manner to selectively deplete the microbiome of *Acetobacter* proteobacteria that cause age-related gut dysfunction and shorten lifespan.

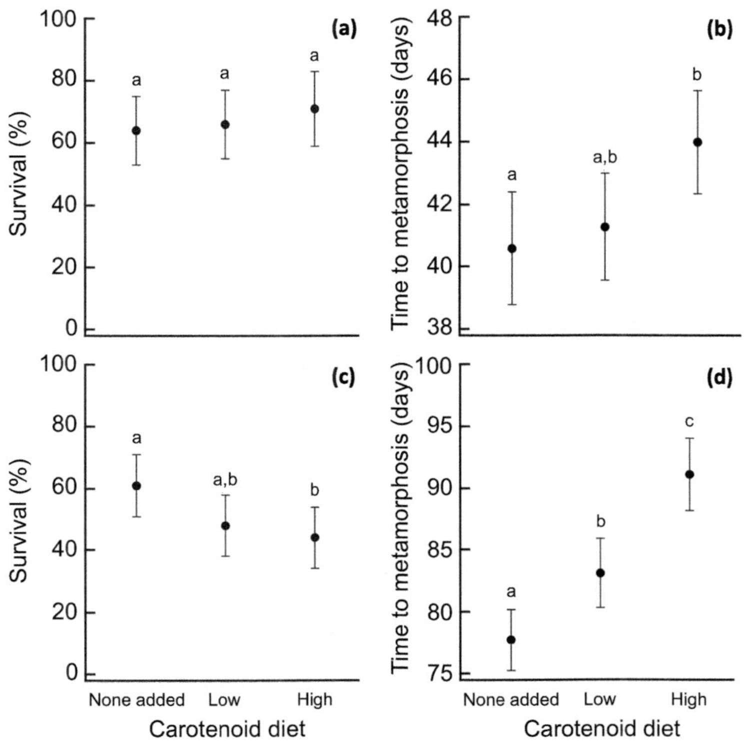


**Fig. 7.6** Effects of diets with different carotenoid contents (control vs. low or high content) provided during metamorphosis on (**a**) survival and (**b**) time to metamorphosis of wood frogs *Lithobates sylvaticus* (*Rana sylvatica*), and (**c**) survival and (**d**) time to metamorphosis of gray treefrogs *Dryophytes versicolor*. Groups that do not share the same superscript letter differ significantly. Values are shown as mean and 95% confidence intervals. Reprinted with minor modifications from Cothran et al. (2015). © The Author 2015. Published by Oxford University Press and the Society for Experimental Biology; distributed under the terms of the Creative Commons Attribution License (CC BY)

Indirect evidence for a role of reactive oxygen species as modulators of tissue development comes from studies on tissue regeneration. For example, Love et al. (2013) found that *Xenopus tropicalis* tadpole tail amputation induced a sustained generation of reactive oxygen species during tail regeneration. Lowering reactive oxygen species levels, using pharmacological or genetic approaches, reduced the level of cell proliferation and impaired tail regeneration (Figs. 7.7 and 7.8).

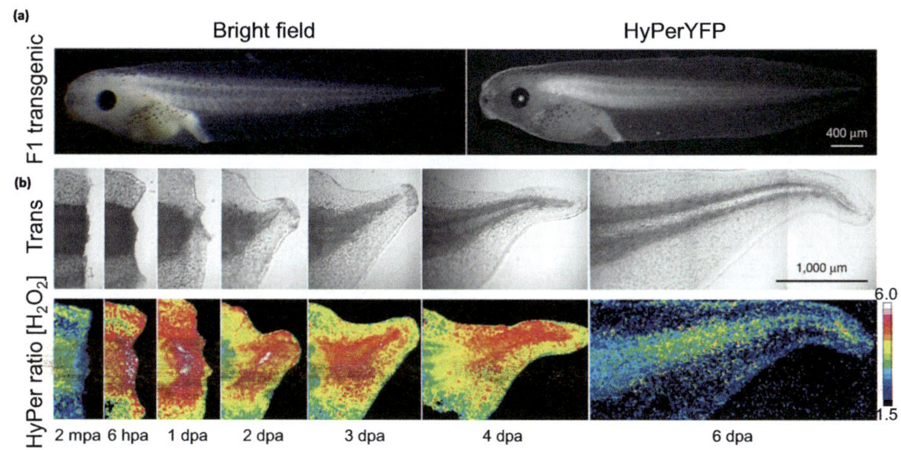


**Fig. 7.7** Generation of reactive oxygen species during tail regeneration in *Xenopus* tadpoles. (**a**) Bright-field and fluorescence micrographs of a tadpole derived from the F1 generation of a transgenic *X. laevis* line that expresses the hydrogen peroxide sensor HyPerYFP ubiquitously. (**b**) HyPerYFP transillumination (Trans) imaging of a representative regenerating tadpole tail. Hydrogen peroxide is derived from the excitation ratio of HyPerYFP 490 nm/HyPerYFP 402 nm. mpa, minutes post-amputation; hpa, hours post-amputation; dpa, days post-amputation. Owing to the size of the regenerated tail, the 6 dpa time point panels are derived from the merging of three images. Reprinted from Love et al. (2013) with permission of Springer Nature

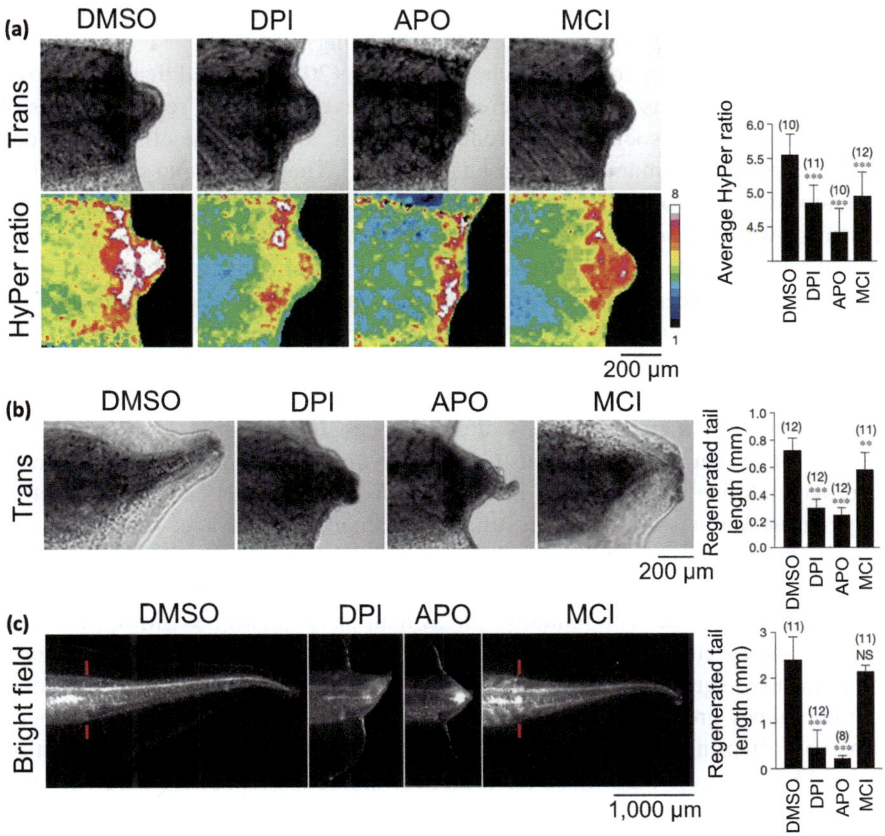


**Fig. 7.8** Pharmacologically lowering levels of reactive oxygen species impaired tail regeneration in *Xenopus* tadpoles. (**a**) Representative transillumination (Trans) and HyPerYFP imaging and quantification of tadpole tails treated with DMSO (dimethylsulphoxide), 2 μmol $L^{-1}$ DPI (diphenyleneiodonium), 200 μmol $L^{-1}$ APO (apocynin), or 200 μmol $L^{-1}$ MCI-186 (Edaravone) treatments at 12 h after amputation. (**b**) Representative tails and quantification of regenerated tail length at 72 h after amputation following the indicated inhibitor treatments. (**c**) Images and quantification of tadpoles that were exposed to DMSO, DPI, APO, or MCI-186 (the same doses as above) from 0 to 72 hpa, and then cultured in normal media until 7 days post-amputation. The red lines indicate initial point of amputation. Asterisks indicate significant differences in comparison with values of DMSO controls; ** $P < 0.01$, *** $P < 0.001$. Values are shown as mean and standard error. Reprinted from Love et al. (2013) with permission of Springer Nature

## 7.4 Oxidative Stress as a Cost of and Constraint on Growth and Development

Rates of growth and development are usually slower than the maximum achievable, implying that there are long-term costs of rapid growth, but also constraints through a trade-off in the allocation of resources between growth and other expensive processes. In particular, a number of long-term costs were documented in individuals that experienced periods of compensatory or catch-up growth, which refer to an accelerated growth of an organism following a period of slowed development (e.g. prolonged nutrient deprivation followed by refeeding).

Using a meta-analytical approach, Smith et al. (2016) showed that greater oxidative stress, in terms of either increased oxidative damage or reduced levels of antioxidants, was associated with reduced growth, although the effect depended on the experimental manipulation used (Fig. 7.9a). They also found support for an oxidative cost of growth, at least in terms of increased oxidative damage, although faster growth was not associated with a change in antioxidant levels (Fig. 7.9b). These findings that oxidative stress can act as a constraint on growth support theoretical links between oxidative stress and animal life histories and provide evidence for a trade-off between growth and self-maintenance. Furthermore, the apparent oxidative costs of growth imply individuals cannot alter this trade-off when faced with enhanced growth. However, it is unclear whether the oxidative costs of growth emerge under any circumstances. For example, Costantini et al. (2010) found that nestling pigeons *Columba livia* that attained a larger body mass and skeletal size after a period of 30 days had higher levels of serum reactive oxygen metabolites than nestlings that were smaller at that age, while they had also higher non-enzymatic antioxidant capacity. Harrison et al. (2013) found that tobacco hornworms *Manduca sexta* that grew faster as larvae had lower survival and capacity to cope with oxidative stress than larvae that grew slower. By contrast, Dantzer et al. (2020) found that offspring from mothers of the North American red squirrel *Tamiasciurus hudsonicus* treated with glucocorticoids during pregnancy grew (in body mass) 17% faster than those from control mothers, whereas offspring from mothers treated with glucocorticoids during lactation grew 34.8% slower than those from control mothers, respectively. Despite this alteration of growth rate, the treatment did not have any impact on the oxidative stress level or telomere length of offspring, and fast-growing offspring did not have higher oxidative stress levels or shorter telomere lengths than slow-growing offspring. Similarly, Tüzün et al. (2020) found that individuals of the damselfly *Ischnura elegans* with a higher growth rate after a transient food shortage did not have shorter adult lifespans or higher levels of thiobarbituric acid reactive substances. Rather, they showed higher levels of superoxide dismutase and catalase. Meniri et al. (2020) found that higher growth rate of Seba's short-tailed bats *Carollia perspicillata* was associated with elevated oxidative damage, suggesting an oxidative cost of growth. However, they found no clear evidence for physiological costs specific to the catch-up growth induced by ad libitum feeding following a period of food restriction.

It is unclear the extent to which tissue development contributes to the physiological costs associated with growth. For example, Gomez-Mestre et al. (2013) found that *Pelobates cultripes* tadpoles accelerated their development in response to reduced water levels, but this acceleration did not influence levels of thiobarbituric acid reactive substances, possibly due to the induced upregulation of the major antioxidant enzymes catalase, superoxide dismutase, and glutathione peroxidase. By contrast, Janssens and Stoks (2018) found that larvae of the damselfly *Lestes viridis* that accelerated development (time elapsed from egg hatching until metamorphosis)

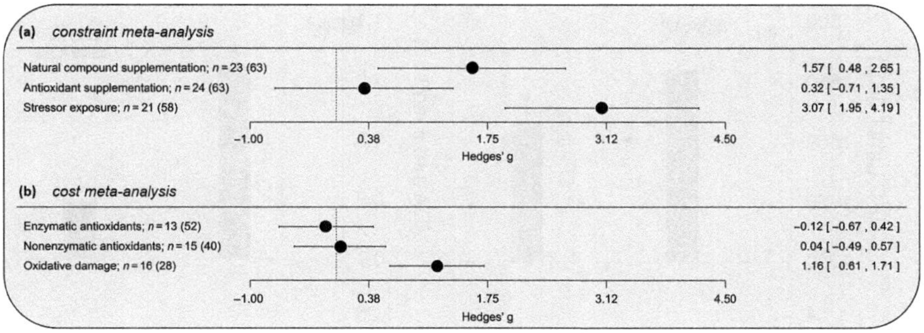


**Fig. 7.9** Predicted effect sizes (mean and 95% confidence interval, CI, at right) for (**a**) constraint meta-analysis (slower growth in groups with experimentally increased oxidative damage and/or reduced antioxidant defences) and (**b**) cost meta-analysis (greater oxidative damage and/or lower antioxidant defences in the experimental group with faster growth). (**a**) A positive effect size indicates increased oxidative stress is associated with reduced growth. Separate effect sizes are given for each experimental manipulation, as the effect size differed significantly between these. (**b**) A positive effect size indicates increased growth is associated with increased oxidative stress. As the effect size differed significantly between markers of oxidative status, separate effect sizes are given for each marker. When the CI does not include zero, the effect size is significant. "*n*" is number of effect sizes. The dataset included the following studies: Kato et al. (1981), Ohtsuka et al. (1998), Taniguchi et al. (1999), Díaz-Cruz et al. (2003), Lee and Dabrowski (2003), Huang and Huang (2004), Huang and Lin (2004), Huang et al. (2004), Chien and Shiau (2005), Filho et al. (2005), Onderci et al. (2005), Sitjà-Bobadilla et al. (2005), de Ayala et al. (2006), Fontagné et al. (2006, 2008), Langley-Evans and Sculley (2006), Wang et al. (2002, 2006, 2011, 2012a, b, 2016), Alonso-Alvarez et al. (2007), De Block and Stoks (2008), Ma et al. (2008), Nain et al. (2008), Savary-Auzeloux et al. (2008), Tarry-Adkins et al. (2008), Yurtseven et al. (2008), Djuric et al. (2009), Li et al. (2009), Peng et al. (2009), Roark et al. (2009), Salomons (2009), Zhang et al. (2009, 2010, 2013a, b), Zheng et al. (2009, 2012, 2013), Costantini (2010), Kader et al. (2010), Kilgas et al. (2010), Liu et al. (2010), Saïdi et al. (2010), Shen et al. (2010), Yang et al. (2010), Mustafa et al. (2011), Noguera et al. (2011a), Özlüer-Hunt et al. (2011), Can et al. (2012), Chen et al. (2012, 2013), Gao et al. (2012, 2013a, b, 2014a, b), Geiger et al. (2012), Guerra et al. (2012), Hargitai et al. (2012), Hisar et al. (2012), Kütter et al. (2012), Luo et al. (2012), Niu et al. (2012, 2013), Salin et al. (2012), Simitzis et al. (2012), Yi et al. (2012), Aziza et al. (2013), Dong et al. (2013), Gomez-Mestre et al. (2013), Ijiri et al. (2013), Keleştemur and Seven (2013), Khempaka et al. (2013), Lee et al. (2013a, b), Radhakrishnan et al. (2013), Sun et al. (2013), Yengkokpam et al. (2013), Zhou et al. (2013), Cinar et al. (2014), Lu et al. (2014), Marri and Richner (2014), Stier et al. (2014), Thiamhirunsopit et al. (2014), Xu and Pan (2014). Reprinted from Smith et al. (2016)

without changing their growth rate (body mass) had a shorter lifespan associated with an increase in malondialdehyde, protein carbonyls, and 8-hydroxy-2′-deoxyguanosine (Fig. 7.10). A study on spiny toads *Bufo spinosus* found a (1) negative covariation between the duration of embryonic development and levels of superoxide dismutase or thiols and (2) negative covariation between the occurrence of malformed hatchlings and thiols, and a positive covariation between the occurrence of malformed hatchlings and catalase levels (Fig. 7.11) (Cheron et al. 2022). These results suggested that a faster development was associated with upregulation of antioxidants, and that oxidative stress might have caused teratogenic effects on embryos.

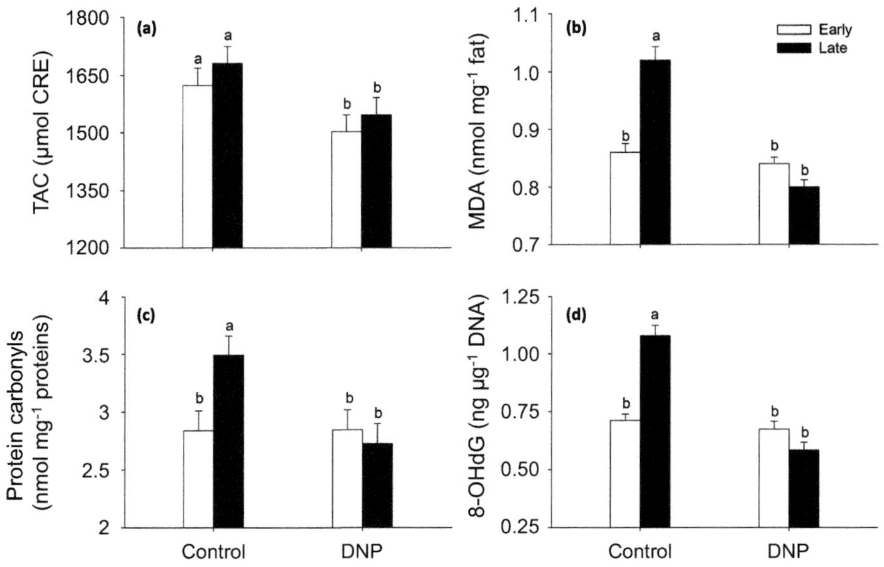


**Fig. 7.10** Effect of photoperiod and DNP (mitochondrial uncoupler) exposure on (**a**) antioxidant capacity as measured using the TAC assay, (**b**) malondialdehyde (MDA), (**c**) protein carbonyls, and (**d**) 8-hydroxy-2′-deoxyguanosine (8-OHdG) of *Lestes viridis* adults. Early and late refer to periods in the larval growth season. Groups that do not share the same superscript letter differ significantly. Values are shown as mean and standard error. Reprinted with minor modifications from Janssens and Stoks (2018) with permission of John Wiley and Sons

There is some evidence that increased intake of antioxidants might boost growth without generating significant changes of the organismal oxidative status. For example, supplementation of diet with vitamin E had stimulatory effects on the growth and protective effects against oxidative stress in pied flycatchers *Ficedula hypoleuca* (Pérez-Rodríguez et al. 2019), and in western bluebirds *Sialia mexicana* (Montoya et al. 2020). However, other studies found little evidence for this protective role of dietary antioxidants. For example, vitamin E boosted growth of European seabasses *Dicentrarchus labrax* during a period of compensatory growth following a prolonged food restriction (Costantini et al. 2018), or of embryos of yellow-legged gulls *Larus michahellis* soon before hatching (Parolini et al. 2018b), but it did not protect against oxidative stress. Moreover, the effects of antioxidants on growth might also depend on the position of the offspring in the sibling hierarchy (e.g. vitamin E in the yellow-legged gull, Parolini et al. 2015, and in the collared flycatcher, Matrková and Remeš 2014).

Investigating the oxidative costs of compensatory growth is particularly important in those contexts where animals need to rapidly modify their growth and development rates in order to respond to the rapid changes that occur in their environments. However, it might be sometimes more convenient to delay a life-history event rather than to pay the costs of trying to being on a time schedule. Under these circumstances, compensatory growth is achieved by extending the growth period, which

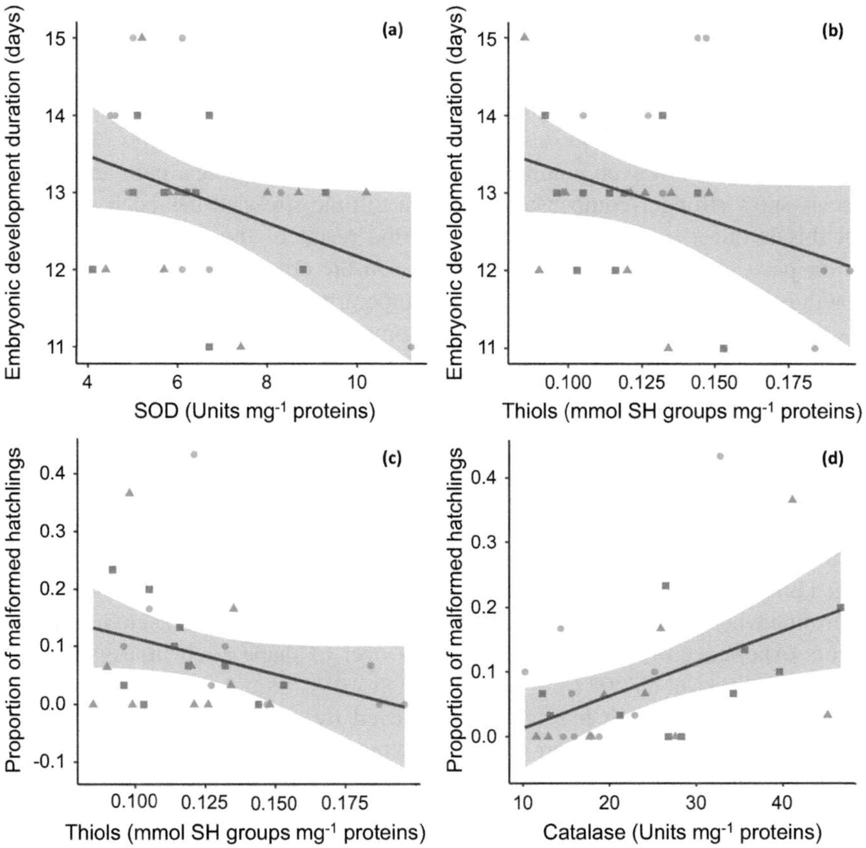


**Fig. 7.11** Oxidative stress, duration of embryonic development, and malformations (axial malformation, oedema, scoliosis of the tail) in tadpoles of spiny toads *Bufo spinosus*. Markers were measured in homogenates of tadpoles. (**a**) Linear regression between duration of embryonic development and superoxide dismutase (SOD); (**b**) linear regression between duration of embryonic development and levels of thiols; (**c**) linear regression between proportion of malformed hatchlings and thiols; (**d**) linear regression between proportion of malformed hatchlings and catalase activity. Colour and shape represent different experimental groups. Reprinted with minor modifications from Cheron et al. (2022) with permission of Elsevier

delays age at emergence or sexual maturity, but also comes at a smaller physiological cost for the individual. Damselfly species belonging to the genus *Lestes* that live in temporary or permanent water bodies express different life-history strategies (Stoks and McPeek 2006). In contrast to permanent water bodies, temporary water bodies may dry out completely in summer, so that *Lestes* species living in this environment lay drought resistant eggs that remain in a diapause state across autumn and winter. After hatching, which occurs in the following spring, individuals spend a few months in an aquatic larval stage and then emerge as adults in summer. Thus, larvae of these species face strong time constraints (time stress) because they must emerge before their pond dries out (De Block et al. 2008). By contrast, eggs laid by species living in permanent ponds do not go through a diapause state and hatch in late summer. The larval phase lasts around 10 months through winter and to the following spring, when individuals emerge as aerial adults in early summer (De Block et al. 2008). Consequently, normal growth rates under non-stressed conditions are higher in temporary-pond *Lestes* compared to permanent-pond *Lestes* (Stoks and McPeek 2003). Moreover, compared to temporary-pond species, permanent-pond species show stronger compensatory growth to time stress. It has been suggested that this stronger compensatory growth is the result of the co-evolution with a slower pace of life (De Block et al. 2008). On the other hand, the shorter larval growth period and higher growth rates in temporary-pond species might give them less opportunity to further accelerate development and growth as they may be near their physiological limits (De Block et al. 2008). Hence, they might suffer higher physiological costs to compensate and, as shown by De Block and Stoks (2008), these costs can carry over from larvae to adults, such as those related to the upregulation of enzymatic antioxidant defences in the temporary-pond *Lestes viridis*. The extent to which metamorphosis is costly can also depend on the adaptations evolved by a particular species (e.g. bioluminescent reactions that can provide auxiliary protection against oxidative damage, Barros and Bechara 1998) or the developmental stage (Burraco et al. 2020). The interplay between these multiple factors will likely determine whether metamorphosis is a new beginning or instead that environmental factors experienced in the larval stage carry over to shape adult fitness (Koyama et al. 2022; Stoks et al. 2022).

Compensatory growth may also be achieved by accelerating the growth rate without changing the duration of the growth period (Metcalfe and Monaghan 2001a,b; Steinberg 2018). This second case of compensatory growth (known also as catch-up growth) is particularly relevant because a number of studies found significant long-term effects on individual fitness. Although catch-up growth in the short-term enables the individual to achieve the body size that is supposed to be normal for its life stage (e.g. adequate body size for leaving the nest in birds), it can carry relevant costs that emerge later in life (Metcalfe and Monaghan 2001a, b; Steinberg 2018).

Many studies tested the role of oxidative stress as a mechanistic link between catch-up growth and fitness. Alonso-Alvarez et al. (2007) enlarged (six nestlings) or reduced (two nestlings) the brood size of zebra finches in order to manipulate the access to food and consequently early growth. As expected, nestlings in the enlarged broods grew slower during their first 20 days of age, but the pattern was reversed during the following growth period (between 21 and 40 days of age), with birds from enlarged broods growing at a faster rate than nestlings from reduced broods. In the next 20 days, the growth rate was similar between the two groups. When nestlings were 60 days old, their red blood cell resistance to an in vitro exposure to free radicals was negatively correlated only with the growth rate during the 21–40 days of age period, nestlings that grew faster being those with the weakest capacity to resist a free radical attack (Alonso-Alvarez et al. 2007). Interestingly, the correlation between red blood cell resistance to free radicals and growth rate was statistically significant for enlarged-brood individuals but not for reduced-brood birds. It might be that an elevated metabolic rate needed to sustain the accelerated growth caused increased generation of reactive oxygen species that weakened the membranes of red blood cells. However, it is complicated to tease apart the effect of growth rate from that of food intake. Nestlings could have accumulated low amounts of dietary antioxidants during the period of food restriction, and this could have affected their capability of coping with the metabolic demands of subsequent fast growth. Other studies on zebra finches actually found that nestlings that were grown on a diet poor in antioxidants had lower levels of vitamins A and E (Blount et al. 2003) and carotenoids (Alonso-Alvarez et al. 2006) later in life.

Compensatory growth was also documented for slow-growing species. Baby king penguins *Aptenodytes patagonicus* face alone the Subantarctic winter because parents are away for foraging in open sea. Over this period, chicks rely on their own energy reserves and suffer high mortality (Weimerskirch et al. 1992). Chicks slow down their growth so much that they may need around 11 months to achieve their complete growth, which is one of the longest growing periods observed in birds (Geiger et al. 2012). At the end of winter, when food is again available, baby penguins can boost their growth rate, but this results in higher levels of reactive oxygen metabolites and telomere shortening (Geiger et al. 2012).

In conclusion, the results of these studies show that compensatory growth determines changes to the organismal oxidative status across a wide range of taxa. These results also suggest that oxidative stress might be a candidate cellular mechanism underlying the negative effect that the compensatory growth has on lifespan (Metcalfe and Monaghan 2001a, b; Rollo 2002; Ricklefs 2006). A major limitation of many studies is that, in order to manipulate the growth rate, young are subjected to periods of food restriction. Thus, reduced nutrient intake or even malnutrition might also contribute to the long-term impact on lifespan (see Chap. 3). To overcome this limitation, other studies manipulated the ambient temperature because it can also influence growth, particularly in ectotherms (e.g. Lee et al. 2013b). However, even manipulations of temperature are not truly risk-free because they can induce a number of metabolic alternations in the organism, including thermal stress.

The role that oxidative stress plays as a constraint on growth and development is better exemplified in studies that tested the effects of experimentally increased oxidative stress or antioxidant administration on the growth rate. For example, many experimental reports found that induction of oxidative stress through supplementation of diet with dietary oxidised lipids impaired growth and survival in various fish species, such as the rainbow trout *Oncorhynchus mykiss* (Hung et al. 1981), the Atlantic salmon *Salmo salar* (Koshio et al. 1994), or the Siberian sturgeon *Acipenser baerii* (Fontagné et al. 2006). Supplementation of diet with natural compounds (e.g. plant extracts, probiotics, herbs that have potential antioxidant properties), but not with specific antioxidants, produced a significant positive effect on growth, supporting the hypothesis that increased oxidative stress constrains growth (e.g. Zhang et al. 2010, 2013a, b; Gao et al. 2013a, b; Lee et al. 2013a; Radhakrishnan et al. 2013; Smith et al. 2016).

## 7.5 Passive Antioxidant Protection and Maturation of Antioxidant Machinery

Animals are born with a naive antioxidant machinery, so that they initially rely on a passive antioxidant protection, which includes the antioxidants that mothers deposited into the egg, or that pass to their foetus through the placenta or to the newborn through the milk.

In mammals, maturation of the antioxidant machinery occurs mainly after birth (e.g. Allen and Venkatraj 1992; Fantel 1996). In a cross-sectional study on hooded seals *Cystophora cristata*, Vázquez-Medina et al. (2011b) found that the generation of the superoxide free radical in the longissimus dorsi muscle was higher in adults than in neonatal pups, while it was intermediate in weaned pups. Importantly, Vázquez-Medina et al. (2011b) also found a similar pattern for the levels of manganese superoxide dismutase and copper-zinc superoxide dismutase, indicating a possible cross-talk between the generation of the superoxide free radical and the expression of these particular antioxidant enzymes during maturation. Finally, neonatal pups, weaned pups, and adults had similar levels of thiobarbituric acid reactive substances, protein carbonyls, and 8-oxo-7,8-dihydro-2′-deoxyguanosine. Vázquez-Medina et al. (2011b) suggested that this enhanced antioxidant protection with maturation might also be a way to tackle the development of the diving behaviour, which exposes seals to oxidative challenges (Chap. 6). In another cross-sectional study on hooded seals, Vázquez-Medina et al. (2011a) found that the antioxidant competence of animals increased through the transition from a terrestrial to a semi-aquatic life. They found that levels of glutathione, glutathione peroxidase, peroxyredoxin VI, thioredoxin 1, thioredoxin reductase, and glutaredoxin 1 in the longissimus dorsi muscle increased with maturation (from neonates to adults), likely contributing to the prevention of the accumulation of oxidative damage.

Studies on chickens *Gallus gallus domesticus* showed that the maturation of the antioxidant machinery also occurs in birds. Gaál et al. (1995) found that vitamin A was preferentially removed from the yolk before day 14 of development and accumulated preferentially in the liver and the yolk sac membrane, but not in the brain. The major period of vitamin E transfer occurred later, during the final week of embryogenesis, when it was preferentially accumulated in the liver and the yolk sac membrane (Fig. 7.12). By contrast, levels of reduced glutathione, superoxide dismutase, and glutathione peroxidase did not show notable changes over the embryo development. It might be that the transfer of vitamins from the yolk into the embryo was strictly regulated to meet the specific demands of embryogenesis. For example, the transfer of vitamin E from the yolk into the embryo mainly occurred during a phase of strong increase in the content and degree of unsaturation of lipids in tissues (Noble et al. 1993). One reason for this allocation pattern might lie with the lipophilic properties of vitamin E that would enable it to provide a direct antioxidant protection of unsaturated fatty acids against oxidative damage. Similarly, Surai et al. (1996) found that the concentrations of vitamin E and carotenoids in the yolk decreased between day 15 of development and hatching at day 21, and increased in

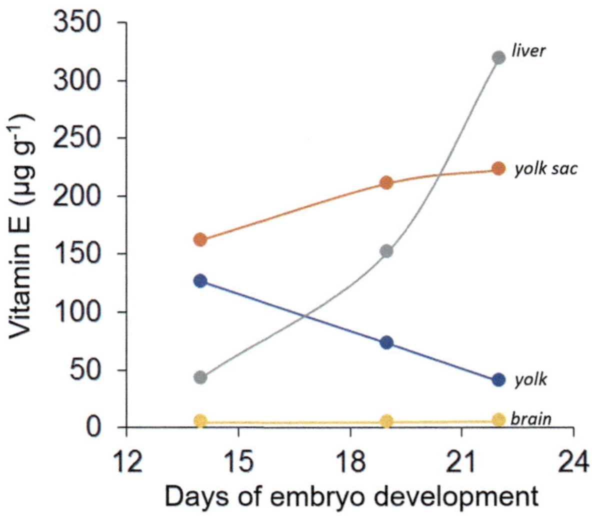


**Fig. 7.12** Change in the concentration of vitamin E in tissues of the chick embryo. Values are shown only as mean per day of development. Data were collected from Gaál et al. (1995)

the yolk sac membrane during the same period. In addition, levels of both vitamin E and carotenoids in the liver increased dramatically between day 18 of embryonic development and day 1 after hatching. By contrast, vitamin E and carotenoids were particularly low and undetectable in the brain, respectively. Further work showed that the liver, brain, yolk sac membrane, kidney, lung, heart, and skeletal muscles displayed distinct development strategies with regard to the acquisition of antioxidant capacity, possibly because the embryo gives priority to the protection of those tissues that are more relevant for its development and survival (Surai 1999, 2002; Surai et al. 1999). Moreover, the role of antioxidant enzymes in the protection against oxidative damage became more important after hatching, a phase characterised by (1) increased metabolic activity, oxygen consumption, and generation of reactive oxygen species, and (2) reduced levels of maternal antioxidants (Surai 1999, 2002; Surai et al. 1999). Similarly, a number of studies on a diverse range of invertebrate and fish species found that, while offspring rely predominantly on maternally derived dietary and other low-molecular weight antioxidants in the initial stages of development, they rapidly develop their own endogenous antioxidant systems during later stages of development (e.g. Rudneva 1999; Fontagné et al. 2008; Wagner et al. 2019). Interestingly, offspring may develop their endogenous antioxidant systems at varying rates across populations (Taylor et al. 2015).

It is not always possible to infer about the role of maturation of the antioxidant machinery in driving the changes of oxidative status during development. This is because changes in the levels of oxidative status markers might also reflect to some degree the response of the offspring to new environmental conditions, such as those that occur after birth, during the transition from a strictly terrestrial to a marine lifestyle, or during the transitions across metamorphic stages, which all require physiological adjustments. For example, human foetal life occurs in a relatively intrauterine hypoxic environment, with a partial pressure of oxygen of 20–25 mmHg and a low presence of reactive oxygen species (Robles et al. 2001; Millán et al. 2018). At birth, the newborn is exposed to a partial pressure of oxygen of about 100 mmHg, and its pulmonary epithelial cells are exposed to pressures of about 140 mmHg in inhaled air (Robles et al. 2001). Consequently, they are at greater risk of oxidative stress simply due to the higher levels of oxygen in the new extrauterine environment. Studies on humans showed that, compared to the condition at birth, newborn have higher levels of hydroperoxides and lower levels of vitamin E and $Q_{10}$ coenzyme (Robles et al. 2001). In birds, it has been found that levels of reactive oxygen metabolites decreased from hatching to fledging in common kestrels *Falco tinnunculus* (Costantini et al. 2006, 2007), but they increased in pigeons *Columba livia* (Costantini 2010) and common cuckoos *Cuculus canorus* (Hargitai et al. 2012). These contrasting results might be due to differences between species in growth and development, or in aspects like diet or brood size. For example, the first days of life are very critical for common kestrels; food is provided by fathers, and mortality rates can be high during this period compared to the final days that precede fledging. By contrast, pigeon chicks are exclusively fed on crop milk by both parents during the first days of life (Levi 1969; Johnston and Janiga 1995). Crop milk contains many nutrients (lipids and proteins) and antioxidants (Bharathi et al. 1997). Therefore, the high nutritional value of crop milk could help hatchlings cope with the pro-oxidant effects of hatching. Cuckoos are brood parasites (i.e. they lay their eggs in the nests of other bird species), and their nestlings evict the host's eggs or hatchlings over the rim of the nest typically within the first 3 days of age. Hence, carry-over effects of eviction behaviour could manifest later in development as higher levels of oxidative damage markers (Hargitai et al. 2012).

We can find other relevant examples of the interaction between organismal oxidative status and environment in those species where the natal and adult environments differ in many respects. For example, Rey et al. (2016) found that the transition from a terrestrial to a marine environment in immature king penguins *Aptenodytes patagonicus* induced a number of functional, cellular, and molecular adjustments to better handle the pro-oxidant conditions faced by marine diving birds. In addition, Rey et al. (2016) suggested that the initial pro-oxidant experience of cold-water immersion might trigger a short-term hormetic response that would favour these physiological adjustments to marine life. Amphibians are renowned for their astonishing capacity to metamorphose in order to move from an aquatic to a terrestrial life. Many studies documented changes in the levels of oxidative damage markers, non-enzymatic and enzymatic antioxidants, or expression of particular antioxidant enzymes during the transition from an aquatic to a terrestrial life (e.g. Amicarelli et al. 2004; Menon and Rozman 2007; Prokić et al. 2019).

We have evidence that some species may also rely on other metabolic mechanisms to mitigate the consequences of oxidative stress for the organism. For example, some species can upregulate the expression of uncoupling proteins in order to keep their mitochondria uncoupled. Dissipation of the proton electrochemical gradient generated by the complexes of the respiratory chain through these uncoupling proteins enables to reduce the generation of reactive oxygen species. However, uncoupled mitochondria are less efficient in the generation of energy that the organism needs to sustain its growth. Salin et al. (2012) found that chronic exposure of common frog *Rana temporaria* tadpoles to a mitochondrial uncoupler (2,4-dinitrophenol) reduced the mitochondrial generation of hydrogen peroxide and the levels of non-enzymatic antioxidants and thiobarbituric acid reactive substances. However, compared to controls, tadpoles that had their mitochondria uncoupled grew more slowly. These tadpoles had also an enhanced metabolic rate, but they did not increase their food consumption, so that they could not compensate for the energy loss elicited by the induced mitochondrial uncoupling.

A final relevant example lies in those animal species that undergo multiples phases of moult during the development period. For example, crustaceans go through a number of larval and immature stages before reaching their adult form, and each of these stages is separated from each other by a moult, in which the hard exoskeleton is separated from the soft parts to enable the organismal growth. The rate of growth of crustaceans is influenced by the environmental conditions, and organismal demands of oxygen dramatically increase while moulting (Fanjul-Moles and Gonsebatt 2012). Eggs and larvae may also enter periods of dormancy/quiescence during which the organism shifts from aerobic to anaerobic metabolism. Upregulation of antioxidant enzymes during sensitive periods of development is therefore required, but this upregulation needs to be well regulated because reactive oxygen species also play a role as signalling molecules in the orchestration of cell growth and development. Many studies documented differences between crustacean species in how the expression of antioxidant enzymes and generation of oxidative damage vary across development until the adult phase (e.g. Arun and Subramanian 1998; Correia et al. 2003; Barata et al. 2005). In addition, some studies found that the developmental trajectory of crustacean larvae is influenced by the interaction between enzymatic and non-enzymatic antioxidants (Dandapat et al. 2003). These changes in the levels of antioxidants and oxidative damage across development were also documented in other taxa of arthropods (e.g. Jovanović-Galović et al. 2004; Klichko et al. 2004; Dmochowska-Ślęzak et al. 2015; Mitić et al. 2023), indicating a highly conserved pattern across these organisms.

## 7.6 Priming for Life: Early Life Oxidative Stress and the Transhormesis Hypothesis

Developmental plasticity is a main type of phenotypic plasticity that is induced by the interaction between genome and the pre- and/or post-natal environmental conditions that organisms experience. Plasticity is not expressed only within a single generation, but it can be transmitted by parents to their offspring. One type of environmentally-induced phenotypic variation transmitted across generations is called transgenerational plasticity. It is increasingly recognised that epigenetic mechanisms would underlie much of this capacity of transmitting acquired characters (what we call Lamarckian inheritance). Epigenetics refers to the non-Mendelian transgenerational transfer of acquired phenotypic traits without modification of gene sequences (Ho and Burggren 2010; Danchin 2022). For example, epigenetic alterations may be induced by oxidative stress early in life and have long-lasting consequences for the organism (e.g. Jaenisch and Bird 2003; Delcuve et al. 2009; Crews et al. 2012; Oleson et al. 2024). These modifications comprise molecular mechanisms, such as DNA methylation, histone modification, or micro-RNA-regulated transcriptional silencing. Although the strongest epigenetic effects emerge when these modifications occur early in life, they are not confined to this temporal window, but they can also occur in adulthood (Vaiserman 2011).

There is growing evidence that epigenetic variation may occur in natural animal populations and may be inherited by future generations (Bossdorf et al. 2008). It is also increasingly recognised that at least some of this epigenetic reorganisation of the phenotype might be due to hormetic mechanisms, e.g. through hormetic priming of the developing individual (e.g. Costantini et al. 2010, 2012; Costantini 2013, 2014, 2022; Costantini and Borremans 2019). Many hormetic responses induced by a wide range of factors, including irradiation, thermal stress, dietary restriction, reactive oxygen species, or dietary antioxidants were characterised by large-scale changes in gene expression (Vaiserman 2008, 2010, 2011). In 1998, Rutherford and Lindquist proposed that the heat shock response could be a molecular mechanism underlying the emergence of epigenetic effects in natural animal populations. They found that the experimental alteration of the *Drosophila* Hsp90 gene functions generated phenotypic variation that affected nearly any adult structure, with specific variants depending on the genetic background (Rutherford and Lindquist 1998). They also found that multiple, previously silent, genetic determinants generated these phenotypic variants, suggesting that there is much silent variation that is capable of driving morphogenetic pathways when its expression is stimulated (Rutherford and Lindquist 1998). It was later discovered that heat shock proteins also affect the expression of genetic variation and developmental stability in plants (Sangster et al. 2008), pointing to a highly conserved pathway in living organisms. The results of these studies provided evidence in support of a molecular mechanism contributing to the organismal adaptability to the environment. Importantly, these studies identified molecular mechanisms (i.e. expression of heat shock proteins) involved in the protection against oxidative stress.

Further work suggested that the long-term consequences of hormesis-induced changes in gene expression are driven by epigenetic memory, and that these effects may carry forward to future generations (e.g. Zalizniak and Nugegoda 2006; Champagne and Curley 2009; Cutler et al. 2009; Scott et al. 2009; Whittle et al. 2009; Drake and Liu 2010; Piiroinen et al. 2013; Kishimoto et al. 2017; Rix and Cutler 2018; Chen et al. 2020). Transgenerational effects also involve protective mechanisms against oxidative stress. For example, embryos of the Antarctic Sea urchin *Sterechinus neumayeri* that were derived from contaminant-experienced mothers were endowed with higher baseline levels of antioxidants from the eggs, which conferred an enhanced capacity to minimise oxidative damage when exposed to additional contaminants (Lister et al. 2015). Recently, Costantini (2022) proposed the *transhormesis hypothesis*, whereby hormetic priming of the parents is transmitted to their offspring so that they are better prepared to withstand future challenges. For example, exposure to low or mild doses of a given abiotic stressor or contaminant may stimulate transcriptional responses and establish long-term somatic molecular memory that promote survival and even reproduction in some study models (Sani et al. 2013).

We can identify at least two relevant weak points of much of this research. First, most studies have been done using artificially selected lines of laboratory animals, so that it is complicated to translate these results to natural animal populations. Second, it is not straightforward to determine whether an effect of parental experience on a phenotypic trait of their offspring is truly a transgenerational effect. This is because when a female (F0 generation) is being exposed to a given stressor during gestation or egg production, its F1 generation and, in mammals even its F2 germline, are exposed to some degree to the same stressful stimulus. Thus, any effects on progeny cannot be unequivocally related only to parental experience. Although there is no consensus yet on the role of hormesis in driving transgenerational effects, we have many studies that documented long-term consequences of hormetic priming during development. For example, exposure of nematodes *Caenorhabditis elegans* early in life to heat stress (35 °C for 2 h), hyperbaric oxygen, or juglone (a chemical that stimulates the generation of reactive oxygen species) increased significantly subsequent resistance of nematodes to the same or to a different challenge, resulting in a longer lifespan (Cypser and Johnson 2002). Similarly, Olsen et al. (2006) found that repeated exposure to heat stress (30 °C for 2 or 4 h) of nematodes throughout life had a larger beneficial effect on longevity than a single exposure early in life. Further studies found that exposure to mild heat stress early in life also extended lifespan in *Drosophila melanogaster* (e.g. Hercus et al. 2003; Le Bourg 2005), and that this hormesis-induced effect on lifespan was associated with a higher expression of the heat shock protein 70 (Hercus et al. 2003). However, there was also a reduction in fecundity (number of eggs), supporting the idea that trade-offs might occur when a trait is hormetically stimulated.

A number of studies also documented hormesis-induced extensions of lifespan in experimental settings where animals were exposed to heat stress only early in life. For example, a single heat shock exposure (37 °C for 1 h at 4 days of age) extended mean longevity in *Drosophila buzzatii* but not in *Drosophila koepferae* (Scannapieco et al. 2007). A detailed demographic analysis showed that there was no effect on longevity in *Drosophila koepferae* because, while the rate of senescence was slowed down (hormetic effect), the rate of mortality was increased. The heat-induced expression of heat shock proteins 70 was similar between species, but *Drosophila koepferae* was more resistant to knock-down by heat stress (Scannapieco et al. 2007). These findings showed that species (also when closely related as in the case of these two *Drosophila* species) can exhibit different patterns of hormesis in response to a same stressor intensity or dose of a certain chemical as shown in Fig. 7.1. Such differences might be due to the different life histories of the two *Drosophila* species (Scannapieco et al. 2007). *Drosophila buzzatii* live longer than *Drosophila koepferae* under both standard laboratory conditions and free-living conditions (Sambucetti et al. 2005; Scannapieco et al. 2007). Moreover, during the first days of life, *Drosophila koepferae* flies invest more massively into reproduction and have higher fecundity than *Drosophila buzzatii* flies (Sambucetti et al. 2005; Scannapieco et al. 2007). Therefore, for *Drosophila koepferae* flies, it might be more advantageous to invest massively in resistance to stressors in early life (when reproduction occurs and natural selection is stronger) than in a long-lasting phenotype. However, investing lots of resources in self-maintenance might be traded-off against investment in reproduction under natural conditions where resources may occur in more limited supply.

We have also evidence for hormetic priming in vertebrates. Probably, the first evidence of hormetic priming comes from studies on poultry. Yahav and McMurtry (2001) found that exposure of male broiler chickens to 36 or 37.5 °C increased their thermal tolerance to an exposure to 35 °C later in life compared to chickens that were exposed to higher and more stressful temperatures early in life. Importantly, Yahav and McMurtry (2001) found that this priming effect was only evident for those chickens that were exposed to heat stress at 3 days of age, while it did not occur in chickens that were exposed at 1, 2, 4, or 5 days of age. These results indicated that, at least in this particular case, there was a very strict sensitive temporal window during which the phenotype was responsive to this thermal stimulus.

A potential problem with this research on poultry is that the selection lines are not replicated, and this implies that we cannot exclude an effect of artificial selection. Experimental support for hormetic priming in birds was, however, also found in other studies. Costantini et al. (2012) tested the hypotheses that (1) hormetic priming can also occur in wild-type phenotypes, and (2) the hormetic priming induced by an early life exposure to mild heat stress involves a greater capability of zebra finches to cope with oxidative stress when faced with high heat stress in adulthood compared to zebra finches that did not experience any heat stress. In addition, it has also been tested whether such a hormetic effect does not occur when the initial exposure involved high heat stress early in life. Zebra finches are small songbirds that occur throughout much of semi-arid and arid zones of Australia, which are characterised by high temperatures and wide temperature fluctuations. Thus, zebra finches have evolved many metabolic mechanisms to survive in these challenging environments, especially relating to tolerance of high ambient temperature and limited water availability (Zann 1996). The timescale over which zebra finches experience high temperatures is also likely to be important in relation to their thermal performance. Zebra finches vary in their thermal experiences, depending on the prevailing weather conditions of the season when and the location where they were born. Thus, we could expect that the extent to which ambient temperatures cause oxidative stress depends on the thermal experiences of the development period. Previous work in zebra finches found that the metabolic rate is minimum when the ambient temperature was around 35 °C and increases when the ambient temperature was below ca. 29.5 °C (lower critical temperature) or was above ca. 40 °C (higher critical temperature) (Calder 1964; Cade et al. 1965). In addition, the body temperature of zebra finches was relatively constant when the ambient temperature is between ca. 10 and 30 °C, while a gradual progression to mild hyperthermia occurred when the ambient temperature increases beyond the 30 °C. Finally, the pulmocutaneous water loss (ratio of water loss to oxygen consumption) increased gradually between 10 and 30 °C, more rapidly between 30 and 40 °C, and sharply above 40 °C (Calder 1964). It was also estimated that the lethal body temperature is around

46 °C (Calder 1964). Costantini et al. (2012) allocated sexually immature zebra finches to three experimental groups, which involved exposure of individuals early in life at control conditions or varying levels of heat stress and then in adulthood either at no stressful conditions or at highly stressful conditions, respectively. The results of this experiment showed that early life exposure to mild heat stress (38 °C, which is close to the upper critical ambient temperature of the thermoneutral zone of the zebra finch) enhanced the capacity of zebra finches to cope with high heat stress (42 °C, which is above the upper limit of the thermoneutral zone, but below the lethal limit in adulthood). Specifically, plasma levels of reactive oxygen metabolites did not increase in these birds after exposure to high heat stress. By contrast, those zebra finches that had either no previous experience of heat stress or experienced high heat stress (42 °C) early in life showed increased plasma levels of reactive oxygen metabolites. In addition, these birds also exhibited a strong decrease in the concentration of total thiols in red blood cells, while the concentration of thiols increased in those birds that experienced mild heat stress in early life, although this effect was evident only in females. Interestingly, zebra finches that experienced mild heat stress in early life, and then were not exposed again to heat stress, also showed a significant decrease in total thiols and a higher mortality rate in the first 3 years of life (Costantini et al. 2012, 2014). This result is particularly relevant because many studies suggested that a disruption of homeostatic mechanisms through overoxidation of thiols might make the cells less capable of efficiently controlling the activity of reactive oxygen species (Jones 2006; Sohal and Orr 2012). It might be that a long-term impairment in the regulation of mechanisms that control the oxidative status is a manifestation of the cost of hormetic priming. Resources required to maintain a molecular "memory" of the hormetic priming may be costly when such stressful events are no longer experienced later in life (environmental mismatch), particularly when these resources are taken away from other organismal functions that impact on survival (Costantini et al. 2014).

In mammals, we have evidence of hormetic priming only for laboratory strains of rodents. For example, Caratero et al. (1998) showed that mice *Mus musculus* exposed to very low doses of gamma radiation in early life lived longer than controls. Damelin et al. (2000) found that mouse cells exposed to low concentrations of mercuric, cadmium, or cupric chloride exhibited significant increases in heat shock proteins 70 and metallothioneins. Gomez-Cabrera et al. (2008) found that moderate exercise stimulated the upregulation of antioxidant enzymes in rats *Rattus rattus*. Finally, Sharma et al. (2019) found that low-dose X-rays radiation exposure stimulated endogenous antioxidant mechanisms and also caused an increase in whole blood lymphocytes and eosinophils in female rats. We have also other potential examples of hormetic-like responses in rodents. For example, while a prolonged separation from the mother may impair a number of physiological and cognitive functions of offspring later in life, brief episodes of separation during the first weeks of life may stimulate a number of mechanisms that lead to reduced stress responsiveness, enhanced spatial working memory, and competitive performance for food in adulthood (e.g. Meaney et al. 1991; Liu et al. 1997; Macrí et al. 2004; Tang et al. 2006; Coutellier 2013). Moreover, postnatal exposure of pups to low levels of stress hormones reduced stress responsiveness and enhanced the cognitive capacities in rats (Catalani et al. 1993) or enhanced the cognitive capacities in mice (Macrì et al. 2009). By contrast, post-natal exposure of pups to high levels of stress hormones increased stress responsiveness and behavioural anxiety in rats (Brummelte et al. 2006; Macrí and Würbel 2007), and reduced the cognitive capacities in mice (Macrì et al. 2009). Although these studies suggest that there might be an hormetic priming of stress response in mammals, it is unknown if this also involves the regulation of cellular oxidative status.

## 7.7 Conclusions and Future Directions

Phenotypic and fitness consequences of early life experiences might be better appreciated if measurements of oxidative damage and antioxidant responses are integrated with classic studies of other physiological metrics, such as hormones. In this chapter, we have seen that mothers have profound effects on the development of the endogenous machinery that regulates the oxidative status and the capability to cope with oxidative stress of their offspring. The extent to which pre-natal conditions created by mothers (e.g. transfer of hormones to their progeny) guide the development trajectory also depends on the conditions that the organism will experience during its post-natal life, pointing to the relevance of the interaction between pre- and post-natal conditions.

We have also seen that reactive oxygen species play a relevant role in the organisation of phenotypic development and are likely to be important to mould the organismal oxidative status. In this context, we have solid evidence to say that oxidative stress plays important roles as both a constraint on and a cost of growth. Moreover, we can expect an interaction between reactive oxygen species and the passive antioxidant protection that mothers pass over to their offspring to guide the maturation of the antioxidant machinery and determine the impact of oxidative stress on growth and development.

Finally, this chapter has shown how exposure to mild stressors during sensitive periods of development can prime through hormesis the organismal capability to cope with stressors in adulthood, and that this acquired resistance or tolerance can be transmitted to the next generations. However, we still have a poor understanding of the ecological and evolutionary significance of hormetic priming, mainly due to the difficulty of experimentally testing hormetic effects in free-living animals.

## References

Allen RG, Venkatraj VS (1992) Oxidants and antioxidants in development and differentiation. J Nutr 122:631–635

Alonso-Alvarez C, Bertrand S, Devevey G, Prost J, Faivre B, Chastel O, Sorci G (2006) An experimental manipulation of life history trajectories and resistance to oxidative stress. Evolution 60:1913–1924

Alonso-Alvarez C, Bertrand S, Faivre B, Sorci G (2007) Increased susceptibility to oxidative damage as a cost of accelerated somatic growth in zebra finches. Funct Ecol 21:873–879

Amicarelli F, Falone S, Cattani F, Alamanou MT, Bonfigli A, Zarivi O, Miranda M, Ragnelli AM, Di Ilio C (2004) Amphibian transition to the oxidant terrestrial environment affects the expression of glutathione S-transferases isoenzymatic pattern. Biochim Biophys Acta 1691:181–192

Apirajkamol NB, James B, Gordon KHJ, Walsh TK, McGaughran A (2020) Oxidative stress delays development and alters gene expression in the agricultural pest moth, *Helicoverpa armigera*. Ecol Evol 10:5680–5693

Armitage JA, Poston L, Taylor PD (2008) Developmental origins of obesity and the metabolic syndrome: the role of maternal obesity. Front Horm Res 36:73–84

Arun S, Subramanian P (1998) Antioxidant enzymes in freshwater prawn *Macrobrachium malcolmsonii* during embryonic and larval development. Comp Biochem Physiol Part B 121:273–277

Atanasova S, Wieland E, Schlumbohm C, Korecka M, Shaw L, von Ahsen N, Fuchs E, Oellerich M, Armstrong V (2009) Prenatal dexamethasone exposure in the common marmoset monkey enhances gene expression of antioxidant enzymes in the aorta of adult offspring. Stress 12:215–224

Aziza AE, Awadin WF, Orma AM (2013) Effect of dietary substitution of cod liver oil by vegetable oils on growth performance, body composition, lipid peroxidation, liver and muscle histopathological state in Nile tilapia (*Oreochromis niloticus*). J Fish Aquacult 4:87–94

Badyaev AV (2011) Origin of the fittest: Link between emergent variation and evolutionary change as a critical question in evolutionary biology. Proc R Soc Lond B 278:1921–1929

Barata C, Navarro JC, Varo I, Riva MC, Arun S, Porte C (2005) Changes in antioxidant enzyme activities, fatty acid composition and lipid peroxidation in Daphnia magna during the aging process. Comp Biochem Physiol Part B 140:81–90

Barker DJ (1998) In utero programming of chronic disease. Clin Sci 95:115–128

Barker DJ, Gluckman PD, Godfrey KM, Harding JE, Owens JA, Robinson JS (1993) Fetal nutrition and cardiovascular disease in adult life. Lancet 341:938–941

Barros MP, Bechara EJ (1998) Bioluminescence as a possible auxiliary oxygen detoxifying mechanism in elaterid larvae. Free Radic Biol Med 24:767–777

Bartho LA, Holland OJ, Moritz KM, Perkins AV, Cuffe JSM (2019) Maternal corticosterone in the mouse alters oxidative stress markers, antioxidant function and mitochondrial content in placentas of female fetuses. J Physiol 597:3053–3067

Bateson P, Barker D, Clutton-Brock T, Deb D, D'Udine B, Foley RA, Gluckman P, Godfrey K, Kirkwood T, Lahr MM, McNamara J, Metcalfe NB, Monaghan P, Spencer HG, Sultan SE (2004) Developmental plasticity and human health. Nature 430:419–421

Bharathi L, Shenoy K, Hegde S (1997) Biochemical differences between crop tissue and crop milk of pigeons (*Columba livia*). Comp Biochem Physiol Part A 116:51–55

Blount JD, Metcalfe NB, Arnold KE, Surai PF, Devevey GL, Monaghan P (2003) Neonatal nutrition, adult antioxidant defences and sexual attractiveness in the zebra finch. Proc R Soc Lond B 270:1691–1696

Bossdorf O, Richards CL, Pigliucci M (2008) Epigenetics for ecologists. Ecol Lett 11:106–115

Bräutigam L, Schütte LD, Godoy JR, Prozorovski T, Gellert M, Hauptmann G, Holmgren A, Lillig CH, Berndt C (2011) Vertebrate-specific glutaredoxin is essential for brain development. Proc Natl Acad Sci USA 108:20532–20537

Brummelte S, Pawluski JL, Galea LA (2006) High post-partum levels of corticosterone given to dams influence postnatal hippocampal cell proliferation and behavior of offspring: a model of post-partum stress and possible depression. Horm Behav 50:370–382

Burraco P, Valdés AE, Orizaola G (2020) Metabolic costs of altered growth trajectories across life transitions in amphibians. J Anim Ecol 89:855–866

Cade TJ, Tobin CA, Gold A (1965) Water economy and metabolism of two estrildine finches. Physiol Zool 38:9–33

Calder WA (1964) Gaseous metabolism and water relations of the zebra finch, *Taeniopygia guttata*. Physiol Zool 37:400–413

Can E, Kurtoğlu İZ, Benzer F, Erişir M, Kocabaş M, Kizak V, Kayim M, Çelik HT (2012) The effects of different dosage of kefir with different durations on growth performances and antioxidant system in the blood and liver tissues of Çoruh trout (*Salmo coruhensis*). Turk J Fisher Aquat Sci 12:277–283

Caratero A, Courtade M, Bonnet L, Planel H, Caratero C (1998) Effect of continuous gamma irradiation at a very low dose on the life span of mice. Gerontology 44:272–276

Carere C, Balthazart J (2007) Sexual versus individual differentiation: the controversial role of avian maternal hormones. Trends Endocrinol Metab 18:73–80

Casagrande S, Costantini D, Dell'Omo G, Tagliavini J, Groothuis TGG (2012a) Differential effects of testosterone metabolites oestradiol and dihydrotestosterone on oxidative stress and carotenoid-dependent colour expression in a bird. Behav Ecol Sociobiol 66:1319–1331

Casagrande S, Costantini D, Groothuis TGG (2012b) Interaction between sexual steroids and immune response in affecting oxidative status of birds. Comp Biochem Physiol Part A 163:296–301

Catalani A, Marinelli M, Scaccianoce S, Nicolai R, Muscolo LA, Porcu A, Korányi L, Piazza PV, Angelucci L (1993) Progeny of mothers drinking corticosterone during lactation has lower stress-induced corticosterone secretion and better cognitive performance. Brain Res 624:209–215

Celorio-Mancera MDLP, Ahn SJ, Vogel H, Heckel DG (2011) Transcriptional responses underlying the hormetic and detrimental effects of the plant secondary metabolite gossypol on the generalist herbivore *Helicoverpa armigera*. BMC Genomics 12:575

Champagne FA, Curley JP (2009) Epigenetic mechanisms mediating the long-term effects of maternal care on development. Neurosci Biobehav Rev 33:593–600

Chen YJ, Liu YJ, Yang HJ, Yuan Y, Liu FJ, Tian LX, Liang GY, Yuan RM (2012) Effect of dietary oxidized fish oil on growth performance, body composition, antioxidant defence mechanism and liver histology of juvenile largemouth bass *Micropterus salmoides*. Aquacult Nutr 18:321–331

Chen YJ, Liu YJ, Tian LX, Niu J, Liang GY, Yang HJ, Yuan Y, Zhang YQ (2013) Effect of dietary vitamin E and selenium supplementation on growth, body composition, and antioxidant defense mechanism in juvenile largemouth bass (*Micropterus salmoides*) fed oxidized fish oil. Fish Physiol Biochem 39:593–604

Chen XD, Seo M, Ebert TA, Ashfaq M, Qin W, Stelinski LL (2020) Hormesis in the brown citrus aphid, *Toxoptera citricida* (Kirkaldy) (Hemiptera: Aphididae) exposed to sublethal doses of imidacloprid. Fla Entomol 103:337–343

Cheron M, Costantini D, Brischoux F (2022) Nicosulfuron, a sulfonylurea herbicide, alters embryonic development and oxidative status of hatchlings at environmental concentrations in an amphibian species. Ecotoxicol Environ Saf 232:113277

Chien YH, Shiau WC (2005) The effects of dietary supplementation of algae and synthetic astaxanthin on body astaxanthin, survival, growth, and low dissolved oxygen stress resistance of kuruma prawn, *Marsupenaeus japonicus* Bate. J Exp Mar Biol Ecol 318:201–211

Cinar M, Yildirim E, Yigit AA, Yalcinkaya I, Duru O, Kisa U, Atmaca N (2014) Effects of dietary supplementation with vitamin C and vitamin E and their combination on growth performance, some biochemical parameters, and oxidative stress induced by copper toxicity in broilers. Biol Trace Elem Res 158:186–196

Coffman JA, Coluccio A, Planchart A, Robertson AJ (2009) Oral-aboral axis specification in the sea urchin embryo III. Role of mitochondrial redox signaling via $H_2O_2$. Dev Biol 330:123–130

Coffman JA, Wessels A, DeSchiffart C, Rydlizky K (2014) Oral-aboral axis specification in the sea urchin embryo, IV: hypoxia radializes embryos by preventing the initial spatialization of nodal activity. Dev Biol 386:302–307

Correia AD, Costa MH, Luis OJ, Livingstone DR (2003) Age-related changes in antioxidant enzyme activities, fatty acid composition and lipid peroxidation in whole body *Gammarus locusta* (Crustacea: Amphipoda). J Exp Mar Biol Ecol 289:83–101

Costantini D (2010) Effects of diet quality on growth pattern, serum oxidative status, and corticosterone in pigeons (*Columba livia*). Can J Zool 88:795–802

Costantini D (2013) Oxidative stress and hormetic responses in the early life of birds. In: Laviola G, Macrì S (eds) Adaptive and maladaptive aspects of developmental stress. Current topics in neurotoxicity 3. Springer, New York, pp 257–273
Costantini D (2014) Does hormesis foster organism resistance to extreme events? Front Ecol Environ 12:209–210
Costantini D (2022) For better or worse: benefits and costs of transgenerational plasticity and the transhormesis hypothesis. In: Costantini D, Marasco V (eds) Development strategies and biodiversity: Darwinian fitness and evolution in the Anthropocene. Springer, Cham, pp 37–49
Costantini D, Borremans B (2019) The linear no-threshold model is less realistic than threshold or hormesis-based models: an evolutionary perspective. Chem Biol Interact 301:26–33
Costantini D, Marasco V (2022) Development strategies and biodiversity: Darwinian fitness and evolution in the Anthropocene. Springer, Cham. pp 316
Costantini D, Casagrande S, De Filippis S, Brambilla G, Fanfani A, Tagliavini J, Dell'Omo G (2006) Correlates of oxidative stress in wild kestrel nestlings (*Falco tinnunculus*). J Comp Physiol B 176:329–337
Costantini D, Fanfani A, Dell'Omo G (2007) Carotenoid availability does not limit the capability of nestling kestrels (*Falco tinnunculus*) to cope with oxidative stress. J Exp Biol 210:1238–1244
Costantini D, Metcalfe NB, Monaghan P (2010) Ecological processes in a hormetic framework. Ecol Lett 13:1435–1447
Costantini D, Monaghan P, Metcalfe N (2012) Early life experience primes resistance to oxidative stress. J Exp Biol 215:2820–2826
Costantini D, Monaghan P, Metcalfe N (2014) Prior hormetic priming is costly under environmental mismatch. Biol Lett 10:20131010
Costantini D, Angeletti D, Strinati C, Trisolino P, Carlini A, Nascetti G, Carere C (2018) Dietary antioxidants, food deprivation and growth affect differently oxidative status of blood and brain in juvenile European seabass (*Dicentrarchus labrax*). Comp Biochem Physiol Part A 216:1–7
Cothran RD, Gervasi SS, Murray C, French BJ, Bradley PW, Urbina J, Blaustein AR, Relyea RA (2015) Carotenoids and amphibians: effects on life history and susceptibility to the infectious pathogen, *Batrachochytrium dendrobatidis*. Conserv Physiol 3:cov005
Coutellier L (2013) Adaptive regulations in developing rodents following neonatal challenges. In: Laviola G, Macrì S (eds) Adaptive and maladaptive aspects of developmental stress. Current topics in neurotoxicity 3. Springer, New York, pp 229–241
Covarrubias L, Hernández-García D, Schnabel D, Salas-Vidal E, Castro-Obregón S (2008) Function of reactive oxygen species during animal development: passive or active? Dev Biol 320:1–11
Crews D, Gillette R, Scarpino SV, Manikkam M, Savenkova MI, Skinner MK (2012) Epigenetic transgenerational inheritance of altered stress responses. Proc Natl Acad Sci USA 109:9143–9148
Curley JP, Branchi I (2013) Ontogeny of stable individual differences: gene-environment and epigenetic mechanisms. In: Carere C, Maestripieri D (eds) Animal personalities. Behavior, physiology, and evolution. University of Chicago Press, Chicago, pp 279–316
Cutler GC, Ramanaidu K, Astatkie T, Isman MB (2009) Green peach aphid, *Myzus persicae* (Hemiptera: Aphididae), reproduction during exposure to sublethal concentrations of imidacloprid and azadirachtin. Pest Manag Sci 65:205–209
Cypser JR, Johnson TE (2002) Multiple stressors in *Caenorhabditis elegans* induce stress hormesis and extended longevity. J Gerontol A 57:B109–B114
Dahlöf L-G, Hård E, Larsson K (1977) Influence of maternal stress on offspring sexual behaviour. Anim Behav 25:958–968
Damelin LH, Vokes S, Whitcutt JM, Damelin SB, Alexander JJ (2000) Hormesis: a stress response in cells exposed to low levels of heavy metals. Hum Exp Toxicol 19:420–430
Danchin E (2022) More than fifty shades of epigenetics for the study of early in life effects in medicine, ecology, and evolution. In: Costantini D, Marasco V (eds) Development strategies and biodiversity: Darwinian fitness and evolution in the Anthropocene. Springer, Cham, pp 3–35

Dandapat J, Chainy GBN, Rao KJ (2003) Improved post-larval production in giant prawn, *Macrobrachium rosenbergii*, through modulation of antioxidant defence system by dietary vitamin-E. Ind J Biotech 2:195–202
Dantzer B, van Kesteren F, Westrick SE, Boutin S, McAdam AG, Lane JE, Gillespie R, Majer A, Haussmann MF, Monaghan P (2020) Maternal glucocorticoids promote offspring growth without inducing oxidative stress or shortening telomeres in wild red squirrels. J Exp Biol 223:jeb212373
de Ayala RM, Martinelli R, Saino N (2006) Vitamin E supplementation enhances growth and condition of nestling barn swallows (*Hirundo rustica*). Behav Ecol Sociobiol 60:619–630
De Block M, Stoks R (2008) Compensatory growth and oxidative stress in a damselfly. Proc R Soc Lond B 275:781–785
De Block M, McPeek MA, Stoks R (2008) Stronger permanent growth in a permanent-pond *Lestes* damselfly relative to temporary-pond *Lestes*. Oikos 117:245–254
Delcuve GP, Rastegar M, Davie JR (2009) Epigenetic control. J Cell Physiol 219:243–250
Díaz-Cruz A, Serret M, Ramírez G, Avila E, Guinzberg R, Piña E (2003) Prophylactic action of lipoic acid on oxidative stress and growth performance in broilers at risk of developing ascites syndrome. Avian Pathol 32:645–653
Djuric Z, Lewis SM, Lu MH, Mayhugh M, Tang N, Hart RW (2009) Effect of varying dietary fat levels on rat growth and oxidative DNA damage. Nutr Cancer 39:214–219
Dmochowska-Ślęzak K, Giejdasz K, Fliszkiewicz M, Żółtowska K (2015) Variations in antioxidant defense during the development of the solitary bee *Osmia bicornis*. Apidologie 46:432–444
Dong GF, Yang YO, Song XM, Yu L, Zhao TT, Huang GL, Hu ZJ, Zhang JL (2013) Comparative effects of dietary supplementation with maggot meal and soybean meal in gibel carp (Carassius auratus gibelio) and darkbarbel catfish (Pelteobagrus vachelli): growth performance and antioxidant responses. Aquacult Nutr 19:543–554
Drake AJ, Liu L (2010) Intergenerational transmission of programmed effects: public health consequences. Trends Endocrinol Metab 21:206–213
Edwards CRW, Benediktsson R, Lindsay RS, Seckl JR (1993) Dysfunction of the placental glucocorticoid barrier: a link between the foetal environment and adult hypertension? Lancet 341:355–357
Fanjul-Moles ML, Gonsebatt ME (2012) Oxidative stress and antioxidant systems in crustacean life cycles. In: Abele D, Vázquez-Medina JP, Zenteno-Savín T (eds) Oxidative stress in aquatic ecosystems. Wiley-Blackwell, Oxford, pp 208–223
Fantel AG (1996) Reactive oxygen species in developmental toxicity: review and hypothesis. Teratology 53:196–217
Filho DW, Torres MA, Zaniboni-Filho E, Pedrosa RC (2005) Effect of different oxygen tensions on weight gain, feed conversion, and antioxidant status in piapara, *Leporinus elongatus* (Valenciennes, 1847). Aquaculture 244:349–357
Foerder CA, Klebanoff SJ, Shapiro BM (1978) Hydrogen peroxide production, chemiluminescence, and the respiratory burst of fertilization: interrelated events in early sea urchin development. Proc Natl Acad Sci USA 75:3183–3187
Fontagné S, Bazin D, Brèque J, Vachot C, Bernarde C, Rouault T, Bergot P (2006) Effects of dietary oxidized lipid and vitamin A on the early development and antioxidant status of Siberian sturgeon (*Acipenser baeri*) larvae. Aquaculture 257:400–411
Fontagné S, Lataillade E, Brèque J, Kaushik S (2008) Lipid peroxidative stress and antioxidant defence status during ontogeny of rainbow trout (*Oncorhynchus mykiss*). Br J Nutr 100:102–111
Gaál T, Mézes M, Noble RC, Dixon J, Speake BK (1995) Development of antioxidant capacity in tissues of the chick embryo. Comp Biochem Physiol Part B 112:711–716
Galvan I, Alonso-Alvarez C (2010) Yolk testosterone shapes the expression of a melanin-based signal in great tits: an antioxidant-mediated mechanism? J Exp Biol 213:3127–3130
Gao J, Koshio S, Ishikawa M, Yokoyama S, Mamauag REP, Han Y (2012) Effects of dietary oxidized fish oil with vitamin E supplementation on growth performance and reduction of lipid peroxidation in tissues and blood of red sea bream *Pagrus major*. Aquacult 356–357:73–79

Gao J, Koshio S, Ishikawa M, Yokoyama S, Nguyen BT, Mamauag RE (2013a) Effect of dietary oxidized fish oil and vitamin C supplementation on growth performance and reduction of oxidative stress in Red Sea Bream *Pagrus major*. Aquacult Nutr 19:35–44

Gao J, Koshio S, Ishikawa M, Yokoyama S, Nose D, Ren T (2013b) Interactive effects of vitamin C and E supplementation on growth, fatty acid composition, and lipid peroxidation of sea cucumber, *Apostichopus japonicus*, fed with dietary oxidized fish oil. J World Aquacult Soc 44:536–546

Gao J, Koshio S, Ishikawa M, Yokoyama S, Mamauag REP (2014a) Interactive effects of vitamin C and E supplementation on growth performance, fatty acid composition and reduction of oxidative stress in juvenile Japanese flounder *Paralichthys olivaceus* fed dietary oxidized fish oil. Aquaculture 422–423:84–90

Gao J, Koshio S, Wang W, Li Y, Huang S, Cao X (2014b) Effects of dietary phospholipid levels on growth performance, fatty acid composition and antioxidant responses of Dojo loach *Misgurnus anguillicaudatus* larvae. Aquaculture 426–427:304–309

Gavrić JP, Despotović SG, Gavrilović BR, Radovanović TB, Petrović TG, Ajduković M, Vučić T, Cvijanović M, Faggio C, Prokić MD (2021) Oxidative stress parameters in goitrogen-exposed crested newt larvae (Triturus spp.): arrested metamorphosis. Int J Environ Res Public Health 18:9653

Geiger S, Le Vaillant M, Lebard T, Reichert S, Stier A, Le Maho Y, Criscuolo F (2012) Catching-up but telomere loss: half-opening the black box of growth and ageing trade-off in wild king penguin chicks. Mol Ecol 21:1500–1510

Gerhart J, Kirschner M (2007) The theory of facilitated variation. Proc Natl Acad Sci USA 104:8582–8589

Gilbert SF (2001) Ecological developmental biology: developmental biology meets the real world. Dev Biol 233:1–12

Gilbert SF, Epel D (2009) Ecological developmental biology: integrating epigenetics, medicine, and evolution. Sinauer Associates, Sunderland, MA

Gomez-Cabrera M-C, Domenech E, Viña J (2008) Moderate exercise is an antioxidant: upregulation of antioxidant genes by training. Free Radic Biol Med 44:126–131

Gomez-Mestre I, Kulkarni S, Buchholz DR (2013) Mechanisms and consequences of developmental acceleration in tadpoles responding to pond drying. PLoS One 8:e84266

Groothuis TGG, Schwabl H (2008) Hormone-mediated maternal effects in birds: mechanisms matter but what do we know of them? Philos Trans R Soc Lond B 363:1647–1661

Groothuis TGG, Müller W, von Engelhardt N, Carere C, Eising C (2005) Maternal hormones as a tool to adjust offspring phenotype in avian species. Neurosci Biobehav Rev 29:329–352

Groothuis TGG, Hsu BY, Kumar N, Tschirren B (2019) Revisiting mechanisms and functions of prenatal hormone-mediated maternal effects using avian species as a model. Philos Trans R Soc Lond B 374:20180115

Guerra C, Zenteno-Savín T, Maeda-Martínez AN, Philipp EER, Abele D (2012) Changes in oxidative stress parameters in relation to age, growth and reproduction in the short-lived catarina scallop *Argopecten ventricosus* reared in its natural environment. Comp Biochem Physiol Part A 162:421–430

Han Y, Ishibashi S, Iglesias-Gonzalez J, Chen Y, Love NR, Amaya E (2018) $Ca^{2+}$-induced mitochondrial ROS regulate the early embryonic cell cycle. Cell Rep 22:218–231

Hanada H, Kashiwagi A, Takehara Y, Kanno T, Yabuki M, Sasaki J, Inoue M, Utsumi K (1997) Do reactive oxygen species underlie the mechanism of apoptosis in the tadpole tail? Free Radic Biol Med 23:294–301

Hargitai R, Costantini D, Moskát C, Bán M, Muriel J, Hauber ME (2012) Variation in plasma oxidative status and testosterone level in relation to egg eviction effort and age of brood parasitic common cuckoo nestlings. Condor 114:782–791

Harrison JF, Cease AJ, VandenBrooks JM, Albert T, Davidowitz G (2013) Caterpillars selected for large body size and short development time are more susceptible to oxygen-related stress. Ecol Evol 3:1305–1316

Hauser J (2013) Developmental consequences of prenatal administration of glucocorticoids in rodents and primates. In: Laviola G, Macrì S (eds) Adaptive and maladaptive aspects of developmental stress. Current topics in neurotoxicity 3. Springer, New York, pp 195–209
Haussmann MF, Longenecker AS, Marchetto NM, Juliano SA, Bowden RM (2012) Embryonic exposure to corticosterone modifies the juvenile stress response, oxidative stress and telomere length. Proc R Soc Lond B 279:1447–1456
Hayward LS, Wingfield JC (2004) Maternal corticosterone is transferred to avian yolk and may alter offspring growth and adult phenotype. Gen Comp Endocrinol 135:365–371
Henriksen R, Rettenbacher S, Groothuis TGG (2011) Prenatal stress in birds: pathways, effects, function and perspectives. Neurosci Biobehav Rev 35:1484–1501
Hercus MJ, Loeschcke V, Rattan SIS (2003) Lifespan extension of *Drosophila melanogaster* through hormesis by repeated mild heat stress. Biogerontology 4:149–156
Hisar O, Yanik T, Kocaman EM, Arslan M, Slukvin A, Goncharova R (2012) Effects of diludine supplementation on growth performance, liver antioxidant enzyme activities and muscular trace elements of rainbow trout (Oncorhynchus mykiss) juveniles at a low water temperature. Aquacult Nutr 18:211–219
Ho DH, Burggren WW (2010) Epigenetics and transgenerational transfer: a physiological perspective. J Exp Biol 213:3–16
Hsu B-Y, Doligez B, Gustafsson L, Ruuskanen S (2019) Transient growth-enhancing effects of elevated maternal thyroid hormones at no apparent oxidative cost during early postnatal period. J Avian Biol 50:e01919
Huang CH, Huang SL (2004) Effect of dietary vitamin E on growth, tissue lipid peroxidation, and liver glutathione level of juvenile hybrid tilapia, *Oreochromis niloticus*×*O. aureus*, fed oxidized oil. Aquaculture 237:381–389
Huang CH, Lin WY (2004) Effects of dietary vitamin E level on growth and tissue lipid peroxidation of soft-shelled turtle, *Pelodiscus sinensis* (Wiegmann). Aquacult Res 35:948–954
Huang CH, Higgs DA, Balfry SK, Devlin RH (2004) Effect of dietary vitamin E level on growth, tissue lipid peroxidation, and erythrocyte fragility of transgenic coho salmon, Oncorhynchus kisutch. Comp Biochem Physiol Part A 139:199–204
Hung SS, Cho CY, Slinger SJ (1981) Effect of oxidized fish oil, DL-α-tocopheryl acetate and ethoxyquin supplementation on the vitamin E nutrition of rainbow trout (*Salmo gairdneri*) fed practical diets. J Nutr 111:648–654
Ijiri D, Nakamura S, Tatsugawa K, Ijiri S, Ohtsuka A (2013) Effects of feeding dried concentrated rice-washing water on growth performance and skeletal muscle lipid peroxidation in broiler chickens. J Poult Sci 50:370–374
Jaenisch R, Bird A (2003) Epigenetic regulation of gene expression, how the genome integrates intrinsic and environmental signals. Nat Genet 33:245–254
Janssens L, Stoks R (2018) Rapid larval development under time stress reduces adult life span through increasing oxidative damage. Funct Ecol 32:1036–1045
Johnson J, Manzo W, Gardner E, Menon J (2013) Reactive oxygen species and anti-oxidant defenses in tail of tadpoles, *Xenopus laevis*. Comp Biochem Physiol Part C 158:101–108
Johnston RF, Janiga M (1995) Feral pigeons. Oxford University Press, New York
Jones DP (2006) Redefining oxidative stress. Antiox Redox Signal 8:1865–1879
Jovanović-Galović A, Blagojević DP, Grubor-Lajsić G, Worland R, Spasić MB (2004) Role of antioxidant defense during different stages of preadult life cycle in European corn borer (*Ostrinia nubilalis*, Hubn.): diapause and metamorphosis. Arch Insect Biochem Physiol 55:79–89
Kader MA, Koshio S, Ishikawa M, Yokoyama S, Bulbul M, Honda Y, Mamauag RE, Laining A (2010) Growth, nutrient utilization, oxidative condition, and element composition of juvenile red sea bream *Pagrus major* fed with fermented soybean meal and scallop by-product blend as fishmeal replacement. Fish Sci 77:119–128
Kato N, Kawai K, Yoshida A (1981) Effect of dietary level of ascorbic acid on the growth, hepatic lipid peroxidation, and serum lipids in guinea pigs fed polychlorinated biphenyls. J Nutr 111:1727–1733

Keleştemur GT, Seven İ (2013) Effects of dietary propolis and vitamin E on growth performance and antioxidant status in juvenile rainbow trout (*Oncorhynchus mykiss*) under different flow rate. Aquacult Res 44:1120–1131

Khempaka S, Pudpila U, Molee W (2013) Effect of dried peppermint (*Mentha cordifolia*) on growth performance, nutrient digestibility, carcass traits, antioxidant properties, and ammonia production in broilers. J Appl Poult Res 22:904–912

Kilgas P, Tilgar V, Külavee R, Saks L, Hõrak P, Mänd R (2010) Antioxidant protection, immune function and growth of nestling great tits *Parus major* in relation to within-brood hierarchy. Comp Biochem Physiol Part B 157:288–293

Kishimoto S, Uno M, Okabe E, Nono M, Nishida E (2017) Environmental stresses induce transgenerationally inheritable survival advantages via germline-to-soma communication in Caenorhabditis elegans. Nat Commun 8:14031

Klichko VI, Radyuk SN, Orr WC (2004) Profiling catalase gene expression in Drosophila melanogaster during development and aging. Arch Insect Biochem Physiol 56:34–50

Koshio S, Ackman RG, Lall SP (1994) Effects of oxidized herring and canola oils in diets on growth, survival, and flavor of Atlantic salmon, *Salmo salar*. J Agric Food Chem 42:1164–1169

Koyama T, Nunes C, Khong H, Suzuki Y (2022) Adaptive meaning of early life experience in species that go through metamorphosis. In: Costantini D, Marasco V (eds) Development strategies and biodiversity: Darwinian fitness and evolution in the Anthropocene. Springer, Cham, pp 51–95

Kütter MT, Monserrat JM, Primel EG, Caldas SS, Tesser MB (2012) Effects of dietary α-lipoic acid on growth, body composition and antioxidant status in the Plata pompano *Trachinotus marginatus* (Pisces, Carangidae). Aquaculture 368–369:29–35

Langley-Evans SC, Sculley DV (2006) The association between birthweight and longevity in the rat is complex and modulated by maternal protein intake during fetal life. FEBS Lett 580:4150–4153

Laviola G, Macrì S (2013) Adaptive and maladaptive aspects of developmental stress. Current topics in neurotoxicity 3. Springer, New York

Le Bourg E (2005) Hormetic protection of *Drosophila melanogaster* middle-aged male flies from heat stress by mildly stressing them at young age. Naturwissenschaften 92:293–296

Lee KJ, Dabrowski K (2003) Interaction between vitamins C and E affects their tissue concentrations, growth, lipid oxidation, and deficiency symptoms in yellow perch (*Perca flavescens*). Brit J Nutr 89:589–596

Lee JS, Cheng H, Damte D, Lee SJ, Kim JC, Rhee MH, Suh JW, Park SC (2013a) Effects of dietary supplementation of *Lactobacillus pentosus* PL11 on the growth performance, immune and antioxidant systems of Japanese eel Anguilla japonica challenged with *Edwardsiella* tarda. Fish Shellfish Immunol 34:756–761

Lee W-S, Monaghan P, Metcalfe NB (2013b) Experimental demonstration of the growth rate-lifespan trade-off. Proc R Soc Lond B 280:20122370

Levi W (1969) The pigeon. Levi, Sumter, SC

Li ZH, Xie S, Wang JX, Sales J, Li P, Chen DQ (2009) Effect of intermittent starvation on growth and some antioxidant indexes of *Macrobrachium nipponense* (De Haan). Aquacult Res 40:526–532

Lindström J (1999) Early development and fitness in birds and mammals. Trends Ecol Evol 14:343–348

Lipar JL, Ketterson ED, Nolan V Jr, Casto JM (1999) Egg yolk layers vary in the concentration of steroid hormones in two avian species. Gen Comp Endocrinol 115:220–227

Lister KN, Lamare MD, Burritt DJ (2015) Pollutant resilience in embryos of the Antarctic sea urchin Sterechinus neumayeri reflects maternal antioxidant status. Aquat Toxicol 161:61–72

Liu D, Diorio J, Tannenbaum B, Caldji C, Francis D, Freedman A, Sharma S, Pearson D, Plotsky PM, Meaney MJ (1997) Maternal care, hippocampal glucocorticoid receptors, and hypothalamic-pituitary-adrenal responses to stress. Science 277:1659–1661

Liu HW, Dong XF, Tong JM, Zhang Q (2010) Alfalfa polysaccharides improve the growth performance and antioxidant status of heat-stressed rabbits. Livest Sci 131:88–93
Lloyd D, Aon MA, Cortassa S (2001) Why homeodynamics, not homeostasis? Sci World J 1:133–145
Love OP, Williams TD (2008a) Plasticity in the adrenocortical response of a free-living vertebrate: the role of pre- and post-natal developmental stress. Horm Behav 54:496–505
Love OP, Williams TD (2008b) The adaptive value of stress-induced phenotypes: effects of maternally derived corticosterone on sex-biased investment, cost of reproduction, and maternal fitness. Am Nat 172:E135–E149
Love NR, Chen Y, Ishibashi S, Kritsiligkou P, Lea R, Koh Y, Gallop JL, Dorey K, Amaya E (2013) Amputation-induced reactive oxygen species are required for successful Xenopus tadpole tail regeneration. Nat Cell Biol 15:222–228
Lu T, Harper AF, Zhao J, Dalloul RA (2014) Effects of a dietary antioxidant blend and vitamin E on growth performance, oxidative status, and meat quality in broiler chickens fed a diet high in oxidants. Poult Sci 93:1649–1657
Luckey TD (1968) Insecticide hormoligosis. J Econ Entomol 61:7–12
Luo Z, Tan XY, Li XD, Yin GJ (2012) Effect of dietary arachidonic acid levels on growth performance, hepatic fatty acid profile, intermediary metabolism and antioxidant responses for juvenile *Synechogobius hasta*. Aquacult Nutr 18:340–348
Ma JJ, Xu ZR, Shao QJ, Xu JZ, Hung SSO, Hu WL, Zhuo LY (2008) Effect of dietary supplemental l-carnitine on growth performance, body composition and antioxidant status in juvenile black sea bream, *Sparus macrocephalus*. Aquacult Nutr 14:464–471
Macrí S, Würbel H (2007) Effects of variation in postnatal maternal environment on maternal behaviour and fear and stress responses in rats. Anim Behav 73:171–184
Macrí S, Mason GJ, Würbel H (2004) Dissociation in the effects of neonatal maternal separations on maternal care and the offspring's HPA and fear responses in rats. Eur J Neurosci 20:1017–1024
Macrì S, Granstrem O, Shumilina M, Antunes Gomes dos Santos FJ, Berry A, Saso L, Laviola G (2009) Resilience and vulnerability are dose-dependently related to neonatal stressors in mice. Horm Behav 56:391–398
Marasco V, Robinson J, Herzyk P, Spencer KA (2012) Pre- and post-natal stress in context: effects on the stress physiology in a precocial bird. J Exp Biol 215:3955–3964
Marasco V, Spencer KA, Robinson J, Herzyk P, Costantini D (2013) Developmental post-natal stress can alter the effects of pre-natal stress on the adult redox balance. Gen Comp Endocrinol 191:239–246
Marasco V, Herzyk P, Robinson J, Spencer KA (2016) Pre- and post-natal stress programming: developmental exposure to glucocorticoids causes long-term brain-region specific changes to transcriptome in the precocial Japanese quail. J Neuroendocrinol 28. https://doi.org/10.1111/jne.12387
Marri V, Richner H (2014) Differential effects of vitamins E and C and carotenoids on growth, resistance to oxidative stress, fledging success and plumage colouration in wild great tits. J Exp Biol 217:1478–1484
Matrková J, Remeš V (2014) Vitamin E improves growth of collared flycatcher *Ficedula albicollis* young: a supplementation experiment. J Avian Biol 45:475–483
McFarland CA, Talent LG, Quinn MJ Jr, Bazar MA, Wilbanks MS, Nisanian M, Gogal RM Jr, Johnson MS, Perkins EJ, Gust KA (2012) Multiple environmental stressors elicit complex interactive effects in the western fence lizard (*Sceloporus occidentalis*). Ecotoxicology 21:2372–2390
Meaney MJ, Mitchell JB, Aitken DH, Bhatnagar S, Bodnoff SR, Iny LJ, Sarrieau A (1991) The effects of neonatal handling on the development of the adrenocortical response to stress: implications for neuropathology and cognitive deficits in later life. Psychoneuroendocrinology 16:85–103

Mendez-Armenta M, Villeda-Hernandez J, Barroso-Moguel R, Nava-Ruiz C, Jimenez-Capdeville ME, Rios C (2003) Brain regional lipid peroxidation and metallothionein levels of developing rats exposed to cadmium and dexamethasone. Toxicol Lett 144:151–157

Meniri M, Hebinger D, Sorlin M, Ramirez M, Kauffmann E, Vallat AJ, Glauser G, Fasel N, Helfenstein F (2020) Morphological and physiological consequences of a dietary restriction during early life in bats. Behav Ecol 31:475–486

Menon J, Rozman R (2007) Oxidative stress, tissue remodeling and regression during amphibian metamorphosis. Comp Biochem Physiol Part C 145:625–631

Metcalfe NB, Monaghan P (2001a) Compensation for a bad start: grow now, pay later? Trends Ecol Evol 16:254–260

Metcalfe NB, Monaghan P (2001b) Growth versus life-span: perspectives from evolutionary ecology. Exp Gerontol 38:935–940

Millán I, Piñero-Ramos JD, Lara I, Parra-Llorca A, Torres-Cuevas I, Vento M (2018) Oxidative stress in the newborn period: useful biomarkers in the clinical setting. Antioxidants 7:193

Mitić BM, Borković-Mitić SS, Vranković JS, Stojanović DZ, Pavlović SZ (2023) Age-related changes in antioxidant defenses of the Mediterranean centipede *Scolopendra cingulata* (Chilopoda). J Comp Physiol B 193:249–260

Monaghan P (2008) Early growth conditions, phenotypic development and environmental change. Philos Trans R Soc B 363:1635–1645

Montoya B, Ancona S, Beamonte-Barrientos R, Martínez-Gómez M (2020) Does vitamin E supplementation enhance growth benefits of breeding helpers at no oxidative costs? Physiol Biochem Zool 93:37–48

Moore MP, Whiteman HH, Martin RA (2019) A mother's legacy: the strength of maternal effects in animal populations. Ecol Lett 22:1620–1628

Mustafa SA, Al-Subiai SN, Davies SJ, Jha AN (2011) Hypoxia-induced oxidative DNA damage links with higher level biological effects including specific growth rate in common carp, *Cyprinus carpio* L. Ecotoxicol 20:1455–1466

Nain S, Ling B, Bandy B, Alcorn J, Wojnarowicz C, Laarveld B, Olkowski AA (2008) The role of oxidative stress in the development of congestive heart failure in a chicken genotype selected for rapid growth. Avian Pathol 37:367–373

Niu J, Li CH, Liu YJ, Tian LX, Chen X, Huang Z, Lin HZ (2012) Dietary values of astaxanthin and canthaxanthin in *Penaeus monodon* in the presence and absence of cholesterol supplementation: effect on growth, nutrient digestibility and tissue carotenoid composition. Br J Nutr 108:80–91

Niu J, Lin HZ, Jiang SG, Chen X, Wu KC, Liu YJ, Wang S, Tian LX (2013) Comparison of effect of chitin, chitosan, chitosan oligosaccharide and N-acetyl-d-glucosamine on growth performance, antioxidant defenses and oxidative stress status of Penaeus monodon. Aquaculture 372–375:1–8

Noble RC, Cocchi M, Bath HM (1993) α-tocopherol absorption and polyunsaturated fatty acid metabolism in the developing chick embryo. Br Poult Sci 34:815–818

Noguera JC, Lores M, Alonso-Álvarez C, Velando A (2011a) Thrifty development: early-life diet restriction reduces oxidative damage during later growth. Funct Ecol 25:1144–1153

Noguera JC, Alonso-Alvarez C, Kim SY, Morales J, Velando A (2011b) Yolk testosterone reduces levels of oxidative damages during postnatal development. Biol Lett 7:93–95

Obata F, Fons CO, Gould AP (2018) Early-life exposure to low-dose oxidants can increase longevity via microbiome remodelling in Drosophila. Nat Commun 9:975

Ohtsuka A, Ohtani T, Horiguchi H, Kojima H, Hayashi K (1998) Vitamin E reduces glucocorticoid-induced growth inhibition and lipid peroxidation in rats. J Nutr Sci Vitaminol 44:237–247

Oleson BJ, Bhattrai J, Zalubas SL, Kravchenko TR, Ji Y, Jiang EL, Lu CC, Madden CR, Coffman JG, Bazopoulou D, Jones JW, Jakob U (2024) Early life changes in histone landscape protect against age-associated amyloid toxicities through HSF-1-dependent regulation of lipid metabolism. Nat Aging 4:48–61

Olsen A, Vantipalli MC, Lighgow GJ (2006) Lifespan extension of *Caenorhabditis elegans* following repeated mild hormetic heat treatments. Biogerontology 7:221–230

Onderci M, Sahin K, Sahin N, Cikim G, Vijaya J, Kucuk O (2005) Effects of dietary combination of chromium and biotin on growth performance, carcass characteristics, and oxidative stress markers in heat-distressed Japanese quail. Biol Trace Elem Res 106:165–176

Özlüer-Hunt A, Berköz M, Özkan F, Yalin S, Erçen Z, Erdoğan E, Gündüz SG (2011) Effect of mannan oligosaccharide on growth, body composition, and antioxidant enzyme activity of tilapia (*Oreochromis niloticus*). Israeli J Aquacult 63:1–8

Parolini M, Romano M, Caprioli M, Rubolini D, Saino N (2015) Vitamin E deficiency in last-laid eggs limits growth of yellow-legged gull chicks. Funct Ecol 29:1070–1077

Parolini M, Romano A, Possenti CD, Caprioli M, Rubolini D, Saino N (2017) Contrasting effects of increased yolk testosterone content on development and oxidative status in gull embryos. J Exp Biol 220:625–633

Parolini M, Possenti CD, Romano A, Caprioli M, Rubolini D, Saino N (2018a) Physiological increase of yolk testosterone level does not affect oxidative status and telomere length in gull hatchlings. PLoS One 13:e0206503

Parolini M, Possenti CD, Karadas F, Colombo G, Romano M, Caprioli M, Dalle-Donne I, Rubolini D, Milzani A, Saino N (2018b) Yolk vitamin E positively affects prenatal growth but not oxidative status in yellow-legged gull embryos. Curr Zool 64:285–292

Peng S, Chen L, Qin JG, Hou J, Yu N, Long Z, Li E, Ye J (2009) Effects of dietary vitamin E supplementation on growth performance, lipid peroxidation and tissue fatty acid composition of black sea bream (Acanthopagrus schlegeli) fed oxidized fish oil. Aquacult Nutr 15:329–337

Pérez-Rodríguez L, Redondo T, Ruiz-Mata R, Camacho C, Moreno-Rueda G, Potti J (2019) Vitamin E supplementation—but not induced oxidative stress—influences telomere dynamics during early development in wild passerines. Front Ecol Evol 7:173

Piiroinen S, Lyytinen A, Lindström L (2013) Stress for invasion success? Temperature stress of preceding generations modifies the response to insecticide stress in an invasive pest insect. Evol Appl 6:313–323

Prokić MD, Gavrić JP, Petrović TG, Despotović SG, Gavrilović BR, Radovanović TB, Krizmanić II, Pavlović SZ (2019) Oxidative stress in *Pelophylax esculentus* complex frogs in the wild during transition from aquatic to terrestrial life. Comp Biochem Physiol Part A 234:98–105

Radhakrishnan S, Bhavan PS, Seenivasan C, Shanthi R, Poongodi R (2013) Influence of medicinal herbs (*Alternanthera sessilis*, *Eclipta alba* and Cissus quadrangularis) on growth and biochemical parameters of the freshwater prawn Macrobrachium rosenbergii. Aquacult Int 22:551–572

Rey B, Dégletagne C, Bodennec J, Monternier PA, Mortz M, Roussel D, Romestaing C, Rouanet JL, Tornos J, Duchamp C (2016) Hormetic response triggers multifaceted anti-oxidant strategies in immature king penguins (*Aptenodytes patagonicus*). Free Radic Biol Med 97:577–587

Ricklefs RE (2006) Embryo development and ageing in birds and mammals. Proc R Soc Lond B 273:2077–2082

Ricklefs RE, Wikelski M (2002) The physiology/life-history nexus. Trends Ecol Evol 17:462–468

Rix RR, Cutler GC (2018) Does multigenerational exposure to hormetic concentrations of imidacloprid precondition aphids for increased insecticide tolerance? Pest Manag Sci 74:314–322

Roark AM, Bjorndal KA, Bolten AB, Leeuwenburgh C (2009) Biochemical indices as correlates of recent growth in juvenile green turtles (*Chelonia mydas*). J Exp Mar Biol Ecol 376:59–67

Robles R, Palomino N, Robles A (2001) Oxidative stress in the neonate. Early Hum Dev 65:S75–S81

Rollo CD (2002) Growth negatively impacts the life span of mammals. Evol Dev 4:55–61

Romano A, Possenti CD, Caprioli M, De Felice B, Rubolini D, Parolini M (2021) Prenatal yolk corticosterone exposure promotes skeletal growth and induces oxidative imbalance in yellow-legged gull embryos. J Exp Biol 224:jeb242943

Romero LM (2004) Physiological stress in ecology: lessons from biomedical research. Trends Ecol Evol 19:249–255

Romero-Haro AA, Alonso-Alvarez C (2015) The level of an intracellular antioxidant during development determines the adult phenotype in a bird species: a potential organizer role for glutathione. Am Nat 185:390–405

Romero-Haro AA, Alonso-Alvarez C (2020) Oxidative stress experienced during early development influences the offspring phenotype. Am Nat 196:704–716

Rudneva II (1999) Antioxidant system of Black Sea animals in early development. Comp Biochem Physiol Part C 122:265–271
Rutherford SL, Lindquist S (1998) Hsp90 as a capacitor for morphological evolution. Nature 396:336–342
Saïdi SA, Azaza MS, Abdelmouleh A, Pelt JV, Kraïem MM, El-Feki A (2010) The use of tuna industry waste in the practical diets of juvenile Nile tilapia (*Oreochromis niloticus*, L.): effect on growth performance, nutrient digestibility and oxidative status. Aquacult Res 41:1875–1886
Salin K, Luquet E, Rey B, Roussel D, Voituron Y (2012) Alteration of mitochondrial efficiency affects oxidative balance, development and growth in frog (*Rana temporaria*) tadpoles. J Exp Biol 215:863–869
Salomons HM (2009) Fighting for fitness: telomeres, oxidative stress and life history trade-offs in a colonial corvid (Chapter 3). Ph.D. Thesis, University of Groningen, The Netherlands
Sambucetti P, Sørensen JG, Loeschcke V, Norry FM (2005) Variation in senescence and associated traits between sympatric cactophilic sibling species of *Drosophila*. Evol Ecol Res 7:915–930
Sangster TA, Salathia N, Uneurraga S, Schellenberg K, Lindquist S, Queitsch C (2008) Hsp90 affects the expression of genetic variation and developmental stability in quantitative traits. Proc Natl Acad Sci USA 105:2963–2968
Sani E, Herzyk P, Perrella G, Colot V, Amtmann A (2013) Hyperosmotic priming of Arabidopsis seedlings establishes a long-term somatic memory accompanied by specific changes of the epigenome. Genome Biol 14:R59
Sapolsky RM, Romero LM, Munck AU (2000) How do glucocorticoids influence stress responses? Integrating permissive, suppressive, stimulatory, and preparative actions. Endocr Rev 21:55–89
Sarraude T, Hsu B-Y, Groothuis TGG, Ruuskanen S (2020) Manipulation of prenatal thyroid hormones does not influence growth or physiology in nestling pied flycatchers. Physiol Biochem Zool 93:255–266
Savary-Auzeloux I, Durand D, Gruffat D, Bauchart D, Ortigues-Marty I (2008) Food restriction and refeeding in lambs influence muscle antioxidant status. Animal 2:738–745
Scannapieco AC, Sørensen JG, Loeschcke V, Norry FM (2007) Heat-induced hormesis in longevity of two sibling *Drosophila* species. Biogerontology 8:315–325
Schwabl H (1993) Yolk is a source of maternal testosterone for developing birds. Proc Natl Acad Sci USA 90:11446–11450
Schwabl H (1996) Maternal testosterone in the avian egg enhances postnatal growth. Comp Biochem Physiol Part A 114:271–276
Scott BR, Belinsky SA, Leng S, Lin Y, Wilder JA, Damiani LA (2009) Radiation-stimulated epigenetic reprogramming of adaptive-response genes in the lung: an evolutionary gift for mounting adaptive protection against lung cancer. Dose Response 7:104–131
Seckl JR (1998) Physiologic programming of the fetus. Clin Perinatol 25:939–964
Sharma S, Singla N, Chadha VD, Dhawan DK (2019) A concept of radiation hormesis: stimulation of antioxidant machinery in rats by low dose ionizing radiation. Hellenic J Nucl Med 22:43–48
Shearer C (1922) On the oxidation processes of the echinoderm egg during fertilisation. Proc R Soc Lond B 93:213–229
Shen WY, Fu LL, Li WF, Zhu YR (2010) Effect of dietary supplementation with Bacillus subtilis on the growth, performance, immune response and antioxidant activities of the shrimp (*Litopenaeus vannamei*). Aquacult Res 41:1691–1698
Sheriff MJ, Love OP (2013) Determining the adaptive potential of maternal stress. Ecol Lett 16:271–280
Simitzis PE, Kalogeraki E, Goliomytis M, Charismiadou MA, Triantaphyllopoulos K, Ayoutanti A, Niforou K, Hager-Theodorides AL, Deligeorgis SG (2012) Impact of stocking density on broiler growth performance, meat characteristics, behavioural components and indicators of physiological and oxidative stress. Br Poult Sci 53:721–730
Sitjà-Bobadilla A, Peña-Llopis S, Gómez-Requeni P, Médale F, Kaushik S, Pérez-Sánchez J (2005) Effect of fish meal replacement by plant protein sources on non-specific defence mechanisms and oxidative stress in gilthead sea bream (*Sparus aurata*). Aquaculture 249:387–400

Smith F, Hu D, Young NM, Lainoff AJ, Jamniczky HA, Maltepe E, Hallgrimsson B, Marcucio RS (2013) The effect of hypoxia on facial shape variation and disease phenotypes in chicken embryos. Dis Model Mech 6:915–924

Smith SM, Nager RG, Costantini D (2016) Meta-analysis indicates that oxidative stress is both a constraint on and a cost of growth. Ecol Evol 6:2833–2842

Sockman KW, Schwabl H (2000) Yolk androgens reduce offspring survival. Proc R Soc Lond B 267:1451–1456

Sohal RS, Orr WC (2012) The redox stress hypothesis of aging. Free Radic Biol Medic 52:539–555

Stearns SC (1992) The evolution of life histories. Oxford University Press, Oxford

Stebbing ARD (2009) Interpreting 'dose-response' curves using homeodynamic data: with an improved explanation for hormesis. Dose Response 7:221–233

Steinberg CEW (2018) Aquatic animal nutrition. Springer, Cham

Stier A, Delestrade A, Zahn S, Arrivé M, Criscuolo F, Massemin-Challet S (2014) Elevation impacts the balance between growth and oxidative stress in coal tits. Oecologia 175:791–800

Stipanovic RD, Williams HJ, Smith LA (1986) Cotton terpenoid inhibition of *Heliothis virescens* development. Am Chem Soc Symp Ser 296:79–94

Stoks R, McPeek MA (2003) Predators and life histories shape *Lestes* damselfly assemblages along a freshwater habitat gradient. Ecology 84:1576–1587

Stoks R, McPeek MA (2006) A tale of two diversifications: reciprocal habitat shifts to fill ecological space along the pond permanence gradient. Am Nat 168:S50–S72

Stoks R, Janssens L, Delnat V, Swaegers J, Tüzün N, Verheyen J (2022) Adaptive and maladaptive consequences of larval stressors for metamorphic and postmetamorphic traits and fitness. In: Costantini D, Marasco V (eds) Development strategies and biodiversity: Darwinian fitness and evolution in the Anthropocene. Springer, Cham, pp 217–265

Sun H, Wang W, Geng L, Chen Y, Yang Z (2013) In situ studies on growth, oxidative stress responses, and gene expression of juvenile bighead carp (*Hypophthalmichthys nobilis*) to eutrophic lake water dominated by cyanobacterial blooms. Chemosphere 93:421–427

Surai P (1999) Tissue-specific changes in the activities of antioxidant enzymes during the development of the chicken embryo. Br Poult Sci 40:397–405

Surai P (2002) Natural antioxidants in avian nutrition and reproduction. Nottingham University Press, Nottingham

Surai PF, Noble RC, Speake BK (1996) Tissue-specific differences in antioxidant distribution and susceptibility to lipid peroxidation during development of the chick embryo. Biochim Biophys Acta 1304:1–10

Surai PF, Speake BK, Noble RC, Sparks NHC (1999) Tissue-specific antioxidant profiles and susceptibility to lipid peroxidation of the newly hatched chick. Biol Trace Elem Res 68:63–78

Szyf M, Weaver IC, Champagne FA, Diorio J, Meaney MJ (2005) Maternal programming of steroid receptor expression and phenotype through DNA methylation in the rat. Front Neuroendocrinol 26:139–162

Tang AC, Akers KG, Reeb BC, Romeo RD, McEwen BS (2006) Programming social, cognitive, and neuroendocrine development by early exposure to novelty. Proc Natl Acad Sci USA 103:15716–15721

Taniguchi N, Ohtsuka A, Hayashi K (1999) Effect of dietary corticosterone and vitamin E on growth and oxidative stress in broiler chickens. Anim Sci J 70:195–200

Tarry-Adkins JL, Martin-Gronert MS, Chen JH, Cripps RL, Ozanne SE (2008) Maternal diet influences DNA damage, aortic telomere length, oxidative stress, and antioxidant defense capacity in rats. FASEB J 22:2037–2044

Taylor JJ, Wilson SM, Sopinka NM, Hinch SG, Patterson DA, Cooke SJ, Willmore WG (2015) Are there intergenerational and population-specific effects of oxidative stress in sockeye salmon (*Oncorhynchus nerka*)? Comp Biochem Physiol Part A 184:97–104

Taylor JJ, Sopinka NM, Wilson SM, Hinch SG, Patterson DA, Cooke SJ, Willmore WG (2016) Examining the relationships between egg cortisol and oxidative stress in developing wild sockeye salmon (*Oncorhynchus nerka*). Comp Biochem Physiol Part A 200:87–93

Thiamhirunsopit K, Phisalaphong C, Boonkird S, Kijparkorn S (2014) Effect of chili meal (*Capsicum frutescens* LINN.) on growth performance, stress index, lipid peroxidation and ileal

nutrient digestibility in broilers reared under high stocking density condition. Anim Feed Sci Technol 192:90–100
Tobler M, Sandell MI (2009) Sex-specific effects of prenatal testosterone on nestling plasma antioxidant capacity in the zebra finch. J Exp Biol 212:89–94
Tobler M, Sandell MI, Chiriac S, Hasselquist D (2013) Effects of prenatal testosterone exposure on antioxidant status and bill color in adult zebra finches. Physiol Biochem Zool 86:333–345
Treidel LA, Whitley BN, Benowitz-Fredericks ZM, Haussmann MF (2013) Prenatal exposure to testosterone impairs oxidative damage repair efficiency in the domestic chicken (*Gallus gallus*). Biol Lett 9:20130684
Tüzün N, De Block M, Stoks R (2020) Live fast, die old: oxidative stress as a potential mediator of an unexpected life-history evolution. Oikos 129:1330–1340
Vaiserman AM (2008) Epigenetic engineering and its possible role in anti-aging intervention. Rejuv Res 11:39–42
Vaiserman AM (2010) Hormesis, adaptive epigenetic reorganization, and implications for human health and longevity. Dose Response 8:16–21
Vaiserman AM (2011) Hormesis and epigentics: is there a link? Ageing Res Rev 10:413–421
Vázquez-Medina JP, Soñanez-Organis JG, Burns JM, Zenteno-Savín T, Ortiz RM (2011a) Antioxidant capacity develops with maturation in the deep-diving hooded seal. J Exp Biol 214:2903–2910
Vázquez-Medina JP, Olguín-Monroy NO, Maldonado PD, Santamaría A, Königsberg M, Elsner R, Hammill MO, Burns JM, Zenteno-Savín T (2011b) Maturation increases superoxide radical production without increasing oxidative damage in the skeletal muscle of hooded seals (*Cystophora cristata*). Can J Zool 89:206–212
Wagner JT, Knapp MJ, Podrabsky JE (2019) Antioxidant capacity and anoxia tolerance in Austrofundulus limnaeus embryos. J Exp Biol 222:jeb204347
Wang X, Kim K, Bai SC (2002) Effects of different dietary levels of L-ascorbyl-2-polyphosphate on growth and tissue vitamin C concentrations in juvenile olive flounder, *Paralichthys olivaceus* (Temminck et Schlegel). Aquacult Res 33:261–267
Wang YJ, Chien YH, Pan CH (2006) Effects of dietary supplementation of carotenoids on survival, growth, pigmentation, and antioxidant capacity of characins, *Hyphessobrycon callistus*. Aquaculture 261:641–648
Wang H, Yang W, Wang Y, Yang Z, Cui Y (2011) The study on the effects of Chinese herbal mixtures on growth, activity of post-ruminal digestive enzymes and serum antioxidant status of beef cattle. Agricult Sci China 10:448–455
Wang JP, Chi F, Kim IH (2012a) Effects of montmorillonite clay on growth performance, nutrient digestibility, vulva size, faecal microflora, and oxidative stress in weaning gilts challenged with zearalenone. Anim Feed Sci Technol 178:158–166
Wang J, Ji HF, Wang SX, Zhang DY, Liu H, Shan DC, Wang YM (2012b) *Lactobacillus plantarum* ZLP001: in vitro assessment of antioxidant capacity and effect on growth performance and antioxidant status in weaning piglets. Asian Austral J Anim Sci 25:1153–1158
Wang J, Zhang D, Sun Y, Wang S, Li P, Gatlin DM, Zhang L (2016) Effect of a dairy-yeast prebiotic (GroBiotic® -A) on growth performance, body composition, antioxidant capacity and immune functions of juvenile starry flounder (*Platichthys stellatus*). Aquacult Res 47:398–408
Warburg O (1908) Beobachtungen über die Oxydationsprozesse im Seeigelei. Hoppe-Seyler's Zeitschrift für physiologische Chemie 57:1–16
Ward IL (1972) Prenatal stress feminizes and demasculinizes the behavior of males. Science 175:82–84
Waterland RA, Garza C (1999) Potential mechanisms of metabolic imprinting that lead to chronic disease. Am J Clin Nutr 69:179–197
Weimerskirch H, Stahl JC, Jouventin P (1992) The breeding biology and population dynamics of king penguins *Aptenodytes patagonica* on the Crozet islands. Ibis 134:107–117
West-Eberhard MJ (2003) Developmental plasticity and evolution. Oxford University Press, Oxford
Whittle CA, Otto SP, Johnston MO, Krochko JE (2009) Adaptive epigenetic memory of ancestral temperature regime in *Arabidopsis thaliana*. Botany 87:650–657

Wingfield JC, Maney DL, Breuner CW, Jacobs JD, Lynn S, Ramenofsky M, Richardson RD (1998) Ecological bases of hormone–behavior interactions: the "emergency life history stage". Am Zool 38:191–206
Xia XJ, Zhou YH, Shi K, Zhou J, Foyer CH, Yu JQ (2015) Interplay between reactive oxygen species and hormones in the control of plant development and stress tolerance. J Exp Bot 66:2839–2856
Xu WJ, Pan LQ (2014) Evaluation of dietary protein level on selected parameters of immune and antioxidant systems, and growth performance of juvenile *Litopenaeus vannamei* reared in zero-water exchange biofloc-based culture tanks. Aquaculture 426–427:181–188
Yahav S, McMurtry JP (2001) Thermotolerance acquisition in broiler chickens by temperature conditioning early in life–the effect of timing and ambient temperature. Poult Sci 80:1662–1666
Yang SP, Wu ZH, Jian JC, Zhang XZ (2010) Effect of marine red yeast *Rhodosporidium paludigenum* on growth and antioxidant competence of *Litopenaeus vannamei*. Aquaculture 309:62–65
Yates FE (1994) Order and complexity in dynamical systems: homeodynamics as a generalized mechanics for biology. Math Comp Model 19:49–74
Yengkokpam S, Debnath D, Pal AK, Sahu NP, Jain KK, Norouzitallab P, Baruah K (2013) Short-term periodic feed deprivation in *Labeo rohita* fingerlings: effect on the activities of digestive, metabolic and anti-oxidative enzymes. Aquaculture 412–413:186–192
Yi D, Gu L, Ding B, Li M, Hou Y, Wang L, Gong J (2012) Effects of dietary silymarin supplementation on growth performance and oxidative status in Carassius auratus gibelio. J Anim Vet Adv 11:3399–3404
Yurtseven S, Çetin M, Şengül T, Sögüt B (2008) Effect of sage extract (*Salvia officinalis*) on growth performance, blood parameters, oxidative stress and DNA damage in partridges. S Afr J Anim Sci 38:145–152
Zalizniak L, Nugegoda D (2006) Effect of sublethal concentrations of chlorpyrifos on three successive generations of *Daphnia carinata*. Ecotoxicol Environ Saf 64:207–214
Zann RA (1996) Zebra finch: a synthesis of field and laboratory studies. Oxford University Press, Oxford
Zera AJ, Harshman LG (2001) The physiology of life history trade-offs in animals. Annu Rev Ecol Syst 32:95–126
Zhang GF, Yang ZB, Wang Y, Yang WR, Jiang SZ, Gai GS (2009) Effects of ginger root (*Zingiber officinale*) processed to different particle sizes on growth performance, antioxidant status, and serum metabolites of broiler chickens. Poult Sci 88:2159–2166
Zhang W, Chen Q, Mai K, Xu W, Wang X, Liufu Z (2010) Effects of dietary α-lipoic acid on the growth and antioxidative responses of juvenile abalone *Haliotis discus* hannai Ino. Aquacult Res 41:781–787
Zhang J, Liu YJ, Tian LX, Yang HJ, Liang GY, Yue YR, Xu DH (2013a) Effects of dietary astaxanthin on growth, antioxidant capacity and gene expression in Pacific white shrimp Litopenaeus vannamei. Aquacult Nutr 19:917–927
Zhang GG, Yang ZB, Wang Y, Yang WR (2013b) Effects of Astragalus membranaceus root processed to different particle sizes on growth performance, antioxidant status, and serum metabolites of broiler chickens. Poult Sci 92:178–183
Zheng ZL, Tan JYW, Liu HY, Zhou XH, Xiang X, Wang KY (2009) Evaluation of oregano essential oil (*Origanum heracleoticum* L.) on growth, antioxidant effect and resistance against *Aeromonas hydrophila* in channel catfish (*Ictalurus punctatus*). Aquaculture 292:214–218
Zheng J, Chen LL, Zhang HH, Hu X, Kong W, Hu D (2012) Resveratrol improves insulin resistance of catch-up growth by increasing mitochondrial complexes and antioxidant function in skeletal muscle. Metab Clin Exp 61:954–965
Zheng L, Ma YE, Gu LY, Yuan D, Shi ML, Guo XY, Zhan XA (2013) Growth performance, antioxidant status, and nonspecific immunity in broilers under different lighting regimens. J Appl Poult Res 22:798–807
Zhou QC, Wang LG, Wang HL, Wang T, Elmada CZ, Xie FJ (2013) Dietary vitamin E could improve growth performance, lipid peroxidation and non-specific immune responses for juvenile cobia (*Rachycentron canadum*). Aquacult Nutr 19:421–429

# Chapter 8
# The Oxidative Costs of a Colourful Life

**Abstract** Scientists have always been fascinated by the extravagant body colourations of animals. Both females and males produce colourations or even achromatic signals that are used in the context of mate choice, but also in social interactions. Young can also produce colourations that advertise their quality to parents or siblings. Moreover, many species rely on body colourations as warning signals (e.g. Batesian and Müllerian mimicry). The expression of these signals is costly, but the underlying physiological mechanisms are not well understood. In this chapter, I have reviewed the evidence about the role of oxidative stress as a possible mediator of the expression of many of these signals, particularly carotenoid- and melanin-based colourations, or the conspicuous body colourations that are used by some species to signal their unprofitability to predators or to warn and ward off potential competitors.

## 8.1 Introduction

Life is plenty of colours, and for animals, colours can be a matter of life or death. How some animals use their extravagant colourations and patterns has attracted the attention of scientists since more than a century. The production of body colourations is a key component of the reproductive investment of males in many sexually reproducing species (Darwin 1871; Andersson 1994). But body colourations are also used to warn and ward off potential predators or competitors. Thus, extravagant male ornaments expressed during reproduction could be sexually selected or also evolve through social competition. Body colourations are not, however, a peculiarity of males. Females and young can also develop colourations that are used in several contexts.

What do these different types of colourations have in common? They are costly to produce and can convey information about the state or condition of the individual. Thus, they are considered biological signals, which are *any act or structure which alters the behaviour of other organisms, which evolved because of that effect, and*

© The Author(s), under exclusive license to Springer Nature Switzerland AG 2024

D. Costantini, *The Role of Organismal Oxidative Stress in the Ecology and Life-History Evolution of Animals*, Fascinating Life Sciences,
https://doi.org/10.1007/978-3-031-65183-0_8

*which is effective because the receiver's response has also evolved* (p. 15 in Maynard Smith and Harper 2003). Animals can also produce multiple signals thanks to their capacity of colouring different body regions with different colours. Why? It is hypothesised that (1) each colouration signals a different quality of the male (*multiple messages hypothesis*) or the overall quality to reduce errors in assessment of single colourations by the receiver (*redundant signal hypothesis*), (2) one signal colouration could be an unreliable indicator of the overall male quality (*unreliable signals hypothesis*), or (3) additional colourations evolve as a way for males to hinder female mate choice by interfering with the propagation and reception of sexual signals of other males, hence females have responded by evolving the ability to glean meaningful information from male signals (*sexual interference hypothesis*) (e.g. Zuk et al. 1992; Møller and Pomiankoski 1993; Johnstone 1996; Candolin 2003; Lozano 2009).

The conspicuousness of colourations varies significantly between males as a function of factors like genetic background, stress level, or parasite burden. In other words, the colourations can convey information about the male quality that the female may use to select a mate that will bring to her and to her offspring some fitness benefits (Fisher 1915; Kirkpatrick and Ryan 1991; Andersson 1994). Thus, male colourations are generally honest, which means that the information they convey is reliable, otherwise the ornaments would be quickly counterselected if males were attempting to cheat (Zahavi 1975; Hamilton and Zuk 1982; Andersson 1994; Zahavi and Zahavi 1997). For example, Pike et al. (2007c) showed that male sticklebacks *Gasterosteus aculeatus* maintained on a low-carotenoid diet allocated more carotenoids to their nuptial colouration in an attempt to maintain sexual attractiveness. Not only were they not chosen by females in mate-choice trials, but their capital investment in sexual colouration was paid in terms of higher oxidative damage and lower survival (Pike et al. 2007c). The results of this study also indicated that the honesty of the signal was guaranteed by the costs of producing it that only males in good condition could sustain (Hamilton and Zuk 1982; Andersson 1994).

There are also alternative hypotheses stating that the production of a signal does not need to be necessarily costly. For example, a signal that is not difficult to produce or maintain can still be honest if cheating run the risk of paying high costs (Weaver et al. 2017). Moreover, honesty of the signal can also be achieved via cheap or cost-free mechanisms if colour production is causally linked to physiological pathways that are costly to maintain, so that they are determinants of individual condition (Weaver et al. 2017). The question then is which endogenous mechanisms link the expression of sexual ornaments to individual condition.

von Schantz et al. (1999) suggested that the costs induced by oxidative stress may be a reliable currency in the trade-off between individual health and condition-dependent ornamental sexual traits, the expression of which in many cases appears to be particularly sensitive to oxidative stress. In this book chapter, I have illustrated how the building up of both carotenoid- and melanin-based colourations is linked to oxidative stress, and how oxidative stress itself may constrain the expression of body colourations. I have also discussed the role of oxidative stress as a modulator of colour expression in females and young, and in contexts beyond mate choice, such as sociality and warning signalling.

## 8.2 Carotenoid-Based Colourations

Carotenoids form the basis of many yellow, orange, and red colourations in vertebrates (Hill and McGraw 2006a, b). Some species use carotenoids to colour their plumage, while others to colour their bare parts, such as the integument of legs, bellies, flanks, feet, and eye rings (Olson and Owens 2005; Hill and McGraw 2006a, b). Carotenoids are lipophilic pigments that animals are unable to synthesise de novo, hence they must rely on dietary sources (Hill and McGraw 2006a, b). Carotenoids occur in limited supply in food. Moreover, the quantity of carotenoids that is acquired from food may be limited further by factors like parasitic infections or competitive interactions with other nutrients that limit their intestinal absorption (e.g. Tyczkowski et al. 1991; Surai 2002). Moreover, a biotransformation cost derived from maintaining a well-adjusted redox machinery could coexist with a cost linked to carotenoid acquisition and allocation (e.g. García-de Blas et al. 2016).

Earlier studies on guppies *Poecilia reticulata* showed that in some, but not all, populations of this small fish species, females preferred to mate with males that had brighter carotenoid-based orange spots on their skin, probably because they were in better condition (Endler 1980, 1983). Rodd et al. (2002) proposed an alternative hypothesis for this mate choice. The origin of the female preference for brighter males would be a pleiotropic effect of a sensory bias for the colour orange, which might have arisen in the context of food detection. This would imply a sensory bias that was positively selected because feeding on orange-coloured food would provide benefits in terms of a larger intake of antioxidants, or mating with brighter males would carry genetic or resource-based benefits for the female provided that the male colouration is condition-dependent (Rodd et al. 2002).

In addition to their limited availability, once absorbed, carotenoids enter a number of physiological pathways (e.g. immune function, antioxidant protection, vision), which make their availability for each single function even more restricted. Thus, the allocation of carotenoids to the production of body colourations would mean that less carotenoids are available for other important functions, giving rise to costs that only high-quality males would be able to afford (Lozano 1994; Møller et al. 2000; Hill and McGraw 2006a, b).

von Schantz et al. (1999) suggested that diverting carotenoids from antioxidant protection to the production of sexual colourations would expose the organism to oxidative stress because less carotenoids are available to cope with reactive oxygen species. In 2004, Hartley and Kennedy (2004) proposed that the antioxidant role of carotenoids should be reconsidered because it was unlikely to be so important as previously thought. Hartley and Kennedy's proposition was later supported by a number of studies, involving correlative, experimental, or meta-analytical analyses (Costantini and Møller 2008; Cohen and McGraw 2009; Simons et al. 2012; Weaver et al. 2018; Koch et al. 2019; Marton et al. 2022), which found that (1) the antioxidant role of carotenoids is generally weak in birds and (2) variation in carotenoid pigmentation among bird species is not associated with variation in blood-based levels of oxidative damage markers (Fig. 8.1). For example, studies on common

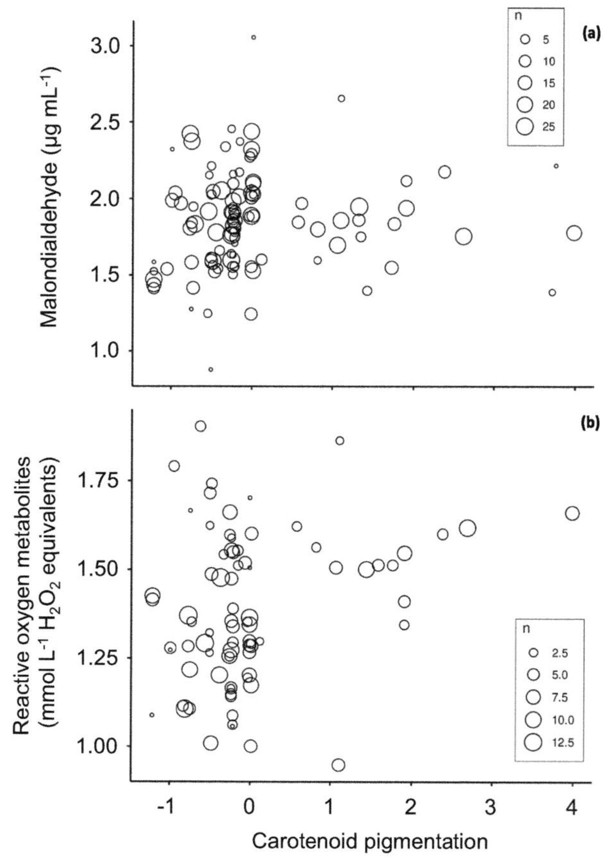


**Fig. 8.1** Comparative models in bird species sampled during the breeding season show that plasma levels of (**a**) malondialdehyde and (**b**) reactive oxygen metabolites are not correlated with carotenoid pigmentation. Reprinted with minor modifications from Marton et al. (2022). © 2022 The Authors. Ecology and Evolution published by John Wiley & Sons Ltd; distributed under the terms of the Creative Commons Attribution License (CC BY)

kestrels *Falco tinnunculus* found that the serum concentration of carotenoids was not correlated with levels of reactive oxygen metabolites nor with the non-enzymatic antioxidant capacity (Costantini et al. 2006), and that supplementation of diet with carotenoids did not affect levels of reactive oxygen metabolites nor the non-enzymatic antioxidant capacity (Costantini et al. 2007b). Further experiments carried out under controlled laboratory conditions also found little support for an important antioxidant function of carotenoids in birds (Fig. 8.2) (e.g. Tomášek et al. 2016; Alonso-Alvarez et al. 2018). The importance of carotenoids as antioxidants has also been questioned in studies on lizards, although some variations in the antioxidant role of carotenoids may occur between different male colour morphs (Olsson et al. 2008, 2009). Contrasting results about the antioxidant role of carotenoids were also found in fish (Wang et al. 2006; Pike et al. 2007c). The conclusions of these studies were also supported by in vitro experiments, which found that carotenoids like lutein or β-carotene have low scavenging activity against reactive oxygen species (Zhang et al. 2014).

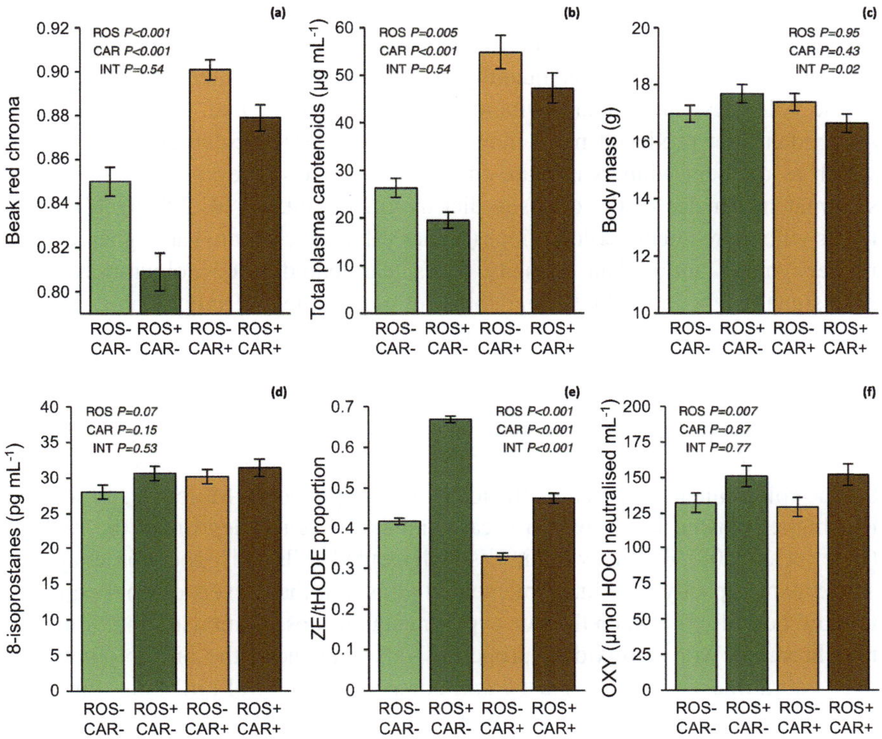


**Fig. 8.2** Effects of an oxidative challenge (diquat dibromide) and carotenoid intake on beak colouration, circulating carotenoids, body mass, and blood-based markers of oxidative status in adult male zebra finches *Taeniopygia guttata*. Birds were exposed to either low (ROS−) or high oxidative load (ROS+) and to either low (CAR−) or high carotenoid intake (CAR+). (**a**, **b**) Both beak red chroma (higher values indicate more saturated red colour) and total plasma carotenoids were reduced by oxidative challenge and enhanced by high carotenoid intake. (**c**) Neither treatment factor had a significant effect on body mass, though their interaction resulted in body mass in ROS + CAR + group that was lower than would be expected if both effects were additive. (**d**) Neither treatment factor had a significant effect on lipid peroxidation (8-isoprostanes). (**e**, **f**) The oxidative challenge elicited a marked increase in two markers of non-enzymatic antioxidant capacity (ZE/tHODE and OXY), while the carotenoid supplementation counteracted the effect of oxidative challenge on ZE/tHODE ratio. Values are shown as mean and standard error. *P*-values are shown for the main effects of oxidative challenge (ROS) and carotenoid intake (CAR), as well as their interaction (INT). Reprinted with minor modifications from Tomášek et al. (2016). © 2016, The Author(s); distributed under the terms of the Creative Commons Attribution License (CC BY)

Hartley and Kennedy (2004) proposed alternative hypotheses to explain why carotenoid-based colourations are widespread in animals and are able to convey information about the condition of the bearer. Since carotenoids are prone to oxidation and consequent bleaching, they might signal the availability of other, more potent and limiting antioxidant resources, so that they are not directly involved in physiological trade-offs (named *protection hypothesis* by Pérez et al. 2008). The protection hypothesis received support by studies on fish and birds that documented an increased expression of carotenoid-based colourations following the supplementation of diet with non-carotenoid antioxidants. Pike et al. (2007a) fed male sticklebacks a diet containing a fixed quantity of carotenoids and either low or high levels of the colourless antioxidants vitamins C and E. Males that were on the high-antioxidant diet produced more intensely, but not larger, coloured body regions, which made them winners in mate-choice trials because they were preferred over size-matched males of the opposite diet treatment (Pike et al. 2007a). Moreover, males with more intense colourations (higher values of chroma) had higher levels of non-enzymatic antioxidant capacity. Thus, carotenoid-based colourations could have signalled the individual's availability of non-carotenoid antioxidants, enabling females to make adaptive mate-choice decisions (Pike et al. 2007a). Similarly, Pérez et al. (2008) modified the availability of the non-pigmentary antioxidant vitamin E in the diet of free-living yellow-legged gulls *Larus michahellis* before egg laying by means of a supplementation experiment. During the incubation period, males that received vitamin E supplements had larger red spots in the bill than control birds, but the colour intensity was not affected by the supplements. Moreover, these males had higher plasma levels of those carotenoids that gulls deposit in the red spot (Pérez et al. 2008). Finally, vitamin E-supplemented gulls had higher non-enzymatic antioxidant capacity than unsupplemented gulls, but their levels of lipid peroxidation did not differ between the two experimental groups. Administration of melatonin (hormone with antioxidant properties) also enhanced the bill colouration of male zebra finches *Taeniopygia guttata* compared to control birds, and this effect was stronger when melatonin was supplemented in combination with carotenoids (Bertrand et al. 2006). Although melatonin may directly react with reactive oxygen species, it can also regulate the activity of various antioxidant enzymes and physiological functions, which might indirectly impact on the metabolism of carotenoids (Costantini 2010).

We have also evidence against the protection hypothesis. For example, while the supplementation of diet with carotenoids enhanced feather chroma of male greenfinches *Carduelis chloris* housed in captivity, the supplementation with vitamin E did not influence the colouration of the plumage (Karu et al. 2008). It might be that the high concentrations of vitamin E in the food might have masked any effect of the supplementation because birds were already plenty of vitamin E (Karu et al. 2008). However, even supplementation studies carried out on free-living birds did not find any evidence in support of the protection hypothesis (Marri and Richner 2014). Moreover, a meta-analysis showed that the redness of plumage or bare parts was not associated with any markers of oxidative status in birds (Simons et al. 2012), raising doubt on the role of carotenoid-based colourations as signals of the male resistance against oxidative stress.

The *sparing hypothesis* is similar to the protection hypothesis, but it provides different mechanistic explanations (Svensson and Wong 2011; see also von Schantz et al. 1999). According to this hypothesis, animals could sacrifice certain antioxidants in order to spare other antioxidants, without compromising the homeostasis of the oxidative status (von Schantz et al. 1999). Thus, animals could save carotenoids, relying on other antioxidants for their protection against oxidative stress and vice versa, depending on their relative availabilities (von Schantz et al. 1999). This means that an increase in the redness of the colouration does not reflect a better protection of carotenoids against oxidation. Rather it would indicate that the organism has plenty of other antioxidants that can be used to replace carotenoids in the antioxidant protection, so that carotenoids can be invested in the expression of the colourations. In order to determine whether carotenoids are protected or spared, we need to explore in depth the molecular mechanisms that would regulate the metabolic fate of carotenoids. Is there an active control that determines the protection or the sparing of carotenoids? Or is it an (almost) entirely passive process that is dependent on the relative concentrations of the different antioxidants? In other words, are carotenoids protected only when other antioxidants occur at much higher concentrations?

If carotenoids are wholly or partly the resource being advertised, then they are probably signalling other functions, such as their contribution to cell signalling, gene activation, regulation of immune function, tissue repair, or embryonic development (named *alternative function hypothesis* by Metcalfe and Alonso-Alvarez 2010). In particular, the role of carotenoids as stimulants of the immune function has received considerable research attention (Lozano 1994; Hill 1999), which also provided mechanistic explanations for the Hamilton and Zuk's *parasite model* (Hamilton and Zuk 1982) and for the Folstad and Karter's *immunocompetence handicap hypothesis* (Folstad and Karter 1992; Peters 2007). It is not entirely clear how carotenoids can boost the immune function. They might protect immune cells against oxidative damage generated by the oxidative burst (see Chap. 2), maintain the redox environment of immune cells, or affect tissue repair mechanisms and gene regulation through the action of retinoids, which are derived from carotenoids (de la Fuente 2002; Chew and Park 2004). The alternative function hypothesis received experimental and meta-analytical support. For example, male zebra finches supplemented with carotenoids produced stronger immune responses than control birds towards (1) the antigen phytohaemagglutinin, which induces a non-specific response of T lymphocytes and macrophages (Blount et al. 2003; McGraw and Ardia 2003), and (2) sheep red blood cells, which stimulate a humoral response to T-dependent antigens (McGraw and Ardia 2003). Moreover, the zebra finches that were carotenoid-supplemented were more likely to be chosen in a mate-choice experiment than control birds (Blount et al. 2003). The stimulation of the immune function also caused a decrease of bill colour in captive male blackbirds *Turdus merula*, suggesting that the activation of the immune response triggers the mobilisation of stored carotenoids towards those sites of the organism where they are needed to sustain the immune response (Faivre et al. 2003). A subsequent meta-analysis of the experimental studies on the link between colourations and immune function revealed that it was only that intensity of the response to a phytohaemagglutinin injection that was significantly and positively associated with the degree of redness of the colouration and the levels of carotenoids in blood (Simons et al. 2012). However, the degree of redness of the colouration was not associated with the intensity of experimentally stimulated antibody responses, the parasite load, or white blood cell counts (Simons et al. 2012). Interestingly, the role of carotenoids as enhancers of the immune function was similar between bird species that express or do not express carotenoid-based colourations, indicating that positive selection of this function was not probably reinforced by its connection to sexual traits. Inflammatory response, however, generates large quantities of reactive oxygen species, implying that carotenoids could be rapidly bleached through oxidation. Production of body colouration could be impaired as a consequence. The question then is how the organism regulates this interplay among carotenoids, immune response, and reactive oxygen species in order to clear pathogens off without compromising the expression of the colourations.

A fourth hypothesis proposes that carotenoids might have detrimental effects on cell membranes, potentially acting as pro-oxidants (*toxicity hypothesis*), when occurring in the body at high concentrations (Zahavi and Zahavi 1997; El Agamey et al. 2004; Costantini et al. 2007a), or under certain circumstances (Simons 2013; Beamonte-Barrientos et al. 2014). For example, high but still physiologically relevant concentrations of metabolites of β-carotene can enhance damage to DNA induced by neutrophils (van Helden et al. 2009). These β-carotene metabolites can inhibit the myeloperoxidase (mammalian enzyme involved in the oxidative burst), reducing the conversion of hydrogen peroxide as a consequence. The accumulation of hydrogen peroxide would enhance the generation of the more reactive hydroxyl radicals through the Fenton reaction, which in turn cause damage to DNA (van Helden et al. 2009). Further work also showed that carotenoids may cause oxidative stress in mitochondria that led to cytochrome c release, proteolytic activation of caspase 3 and PARP1, and stimulation of the apoptotic pathway (Lobo et al. 2012).

One reason for the toxic effects that carotenoids may have at high concentrations might lie in the delicate equilibrium between concentrations of the different types of dietary antioxidants. Carotenoids, vitamin C, and vitamin E can guarantee a synergistic protection against oxidative stress provided that there is a balance in their relative concentrations (e.g. Palozza 1998; Jomova and Valko 2013; see also Chap. 1). Some studies showed that an increase in the concentration of carotenoids, but not of the two vitamins C and E, may cause the formation of carotenoid radicals or adducts to a level that prevents the ascorbate/tocopherol pool from effectively protecting against oxidative stress, leading to pro-oxidant effects as a result (Young and Lowe 2001). Moreover, carotenoids tend to aggregate or crystallise out of solution when they occur at high concentrations (Gruszecki 1999). These aggregates were directly observed in cell membranes, where they might dramatically affect their properties, such as the fluidity and permeability to substances.

The hypothesis that carotenoid-based colourations would convey the capability to deal with the toxic effects of carotenoids is intriguing, but there is no formal investigation yet of this hypothesis. Some studies on birds suggested that carotenoids might have pro-oxidant effects (e.g. Costantini et al. 2007a; Huggins et al. 2010; Leclaire et al. 2015). For example, an experiment on adult common kestrels tested the effects of a prolonged carotenoid supplementation on serum oxidative status and carotenoid-based colourations (Costantini et al. 2007a). The common kestrel is a raptor species that displays sexual dimorphism in the carotenoid-based colourations of bare parts (Casagrande et al. 2006). Compared to controls, supplemented birds had an increase in the levels of carotenoids and reactive oxygen metabolites in serum, an increase in the red and yellow colorimetric components of tarsi colouration, and a decrease in body mass, respectively (Costantini et al. 2007a). Hence, the increase in colouration intensity was associated to some degree with an increase in the generation of oxidative damage. The conclusions of this study are, however, limited to a captive context, where the dietary regime and the physiological status of birds might not accurately reflect those of free-living birds. Moreover, there is no evidence yet of pro-oxidant activity of carotenoids in free-living birds, and the correlations between skin colouration metrics and oxidative damage markers in free-living kestrels vary in strength and direction (Casagrande et al. 2011). Levels of serum carotenoids in captive control common kestrels (95% confidence interval: 34.96–46.48 μg $mL^{-1}$) were within the range of carotenoid levels recorded in free-living adult common kestrels during the courtship phase (95% confidence interval: 27.81–46.80 μg $mL^{-1}$; Casagrande et al. 2006). By contrast, levels of serum carotenoids were about twofold higher in supplemented than in free-living common kestrels (95% confidence interval, e.g. 2 weeks after the start of the supplementation, 56.56–82.14 μg $mL^{-1}$). However, these carotenoid concentrations did not appear to be unnatural because they may be up to 60–80 μg $mL^{-1}$ in free-living common kestrels (Casagrande et al. 2007, 2011). Although there is no clear demonstration of how common kestrel females use the information content of bare part colourations to select their mates, we have evidence that males with more conspicuous colourations are those that deliver more prey to the nest and have larger home ranges with a higher availability of prey (Casagrande et al. 2006). Hence, the colourations of bare parts might function as an indicator of the hunting capacity of male common kestrels. Being good hunters is a very important quality in the common kestrel because males provide food to their mates during the incubation, and to both their mates and offspring during the first couple of weeks after hatching of eggs (Costantini and Dell'Omo 2020).

Evidence of toxic effects of carotenoids was also found in granivorous birds that express a colourful plumage (Fig. 8.3). Huggins et al. (2010) assessed whether the accumulation of high levels of carotenoids in feathers during moult is physiologically costly. Male American goldfinches *Carduelis tristis*, a small passerine bird that accumulates high levels of carotenoids in feathers, were maintained on diets either rich or poor in carotenoids during the moult period. After carotenoid supplementation, birds on the high-carotenoid diet had greater values of yellow chroma in feathers compared to birds on the low-carotenoid diet (Huggins et al. 2010). Although the two diet regimes produced similar short-term effects on the physiology of birds, in the long-term birds that had a higher intake of carotenoids had no measurable effect on aspartate amino transferase (indicator of hepatocellular damage), but they had increased levels of creatine kinase (indicator of skeletal muscle breakdown), and a decreased performance in the vertical flight (indicator of skeletal muscle integrity) (Huggins et al. 2010).

Other authors suggested that the metabolic effects of carotenoids might vary among species as a function of the type of carotenoid, the capacity to transform the ingested carotenoid in a different carotenoid, or the intensity of sexual selection.

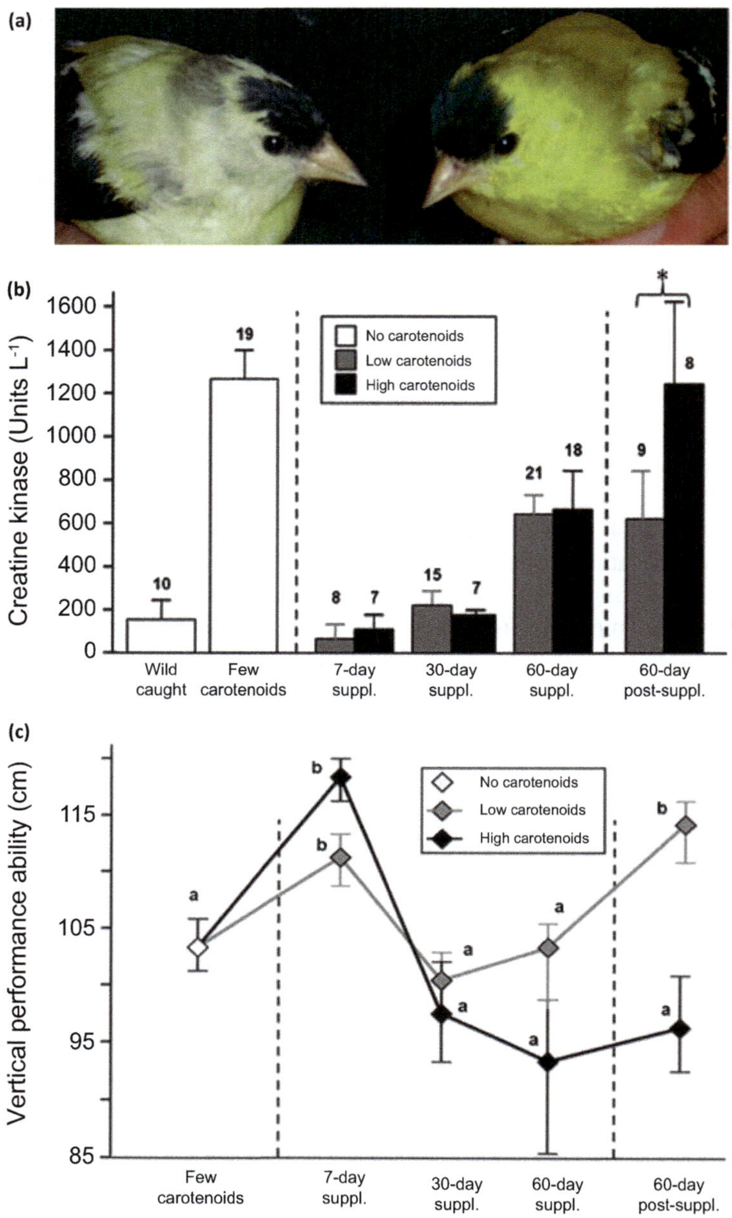


**Fig. 8.3** Toxic effects of high intakes of carotenoids in songbirds. (**a**) Carotenoid-based yellow feather colouration in male American goldfinches *Carduelis tristis* resulting from low- (left) and high- (right) carotenoid supplementation during moult. (**b**) Plasma creatine kinase (CK) assessed in wild-caught birds, after 10–30 days in captivity on a diet containing few carotenoids and at four experimental time points during and after carotenoid supplementation. Significant differences are indicated by an asterisk and sample sizes are included above the bars. (**c**) Vertical flight performance assessed at similar experimental time points to those describe in (**b**). Significant differences are represented by different lowercase letters. Values are shown as mean and standard error. Reprinted with minor modifications from Huggins et al. (2010) with permission of Springer Nature

Johnson and Hill (2013) proposed that, in those species that modify the ingested carotenoids before being allocated to the production of colourations, the oxidation of carotenoids occurs in the inner mitochondria membrane where they would interact with the electron transport chain and oxidative phosphorylation (*inner mitochondria membrane carotenoid oxidation hypothesis*). In so doing, the expression of colourations would be linked to some extent to the mitochondrial function (Hill et al. 2019; Cantarero et al. 2020). Johnson and Hill (2013) argued that ubiquinone and highly oxidised ornamental carotenoids share a stereochemically-conserved binding region, suggesting that these two molecules may have shared similar pathways in the past. Hence, carotenoids and ubiquinones may cooperate as redox participants in anti-radical reactions or independently in helping to maintain the membrane or supra-complex stabilisation during times of high-energy demand. Johnson and Hill (2013) further proposed that the hepatic inner mitochondrial membrane in birds provides a sufficient and sustainable oxidative environment with potentially compatible enzymes necessary for the sequential oxidation of dietary carotenoids to more highly oxidised forms.

There is evidence showing that carotenoids preferentially concentrate in the outer and inner membranes of mitochondria in many cell types, but these results are strongly dependent on species and tissue sampled (Johnson and Hill 2013). For example, in avian hepatocytes and lymphocytes, carotenoids are highly concentrated within their mitochondria following supplementation (Mayne and Parker 1986). The inner mitochondria membrane carotenoid oxidation hypothesis is based on the potential role that redox reactions of carotenoids may play into mitochondria. It implies that the metabolic transformation of carotenoids (e.g. from reduced to oxidised forms) plays a biochemical role in the protection against oxidative stress. In order to demonstrate this role, it is essential to show that oxidative non-enzymatic chemical modifications (damage) to a target molecule occur. Costs of biotransformation of carotenoids might also be related to the consumption of energy and substrates needed for the reaction to work. Costs might also arise from the carotenoid-induced impairment of mitochondrial functionality and consequent stimulation of antioxidant mechanisms (Amengual et al. 2011; Lobo et al. 2012). However, species might differ in their tolerance of carotenoids that occur in mitochondria (Johnson and Hill 2013). We know very little about how these mechanisms work and whether they are peculiar of some species or are highly conserved across animal taxa. The relevance of carotenoids for mitochondrial function is also recognised in human health and disease (Lu et al. 2020), but this research is still in its infancy.

Although evidence in favour of a link between carotenoid-based colourations and organismal oxidative status is generally weak, many studies found stimulatory effects of carotenoid supplementation on traits as diverse as escape performance (Blount and Matheson 2006; Silla et al. 2016), parental care (Pike et al. 2007b), sperm quality (Helfenstein et al. 2010), song (Van Hout et al. 2011), or immune function (Simons et al. 2012; Edwards et al. 2017). Hence, either the antioxidant action of carotenoids is so specific that the available methods to estimate the organismal oxidative status did not enable to accurately quantify it, or we should look at other physiological functions that carotenoids might have. There are also other explanations. For example, there may be a loss of integration between the expression of the colouration and other body traits (Badyaev et al. 2001; Badyaev 2004). This scenario would imply that the mechanisms regulating the expression of the colouration would work as a module, which means that they work almost independently from other physiological mechanisms, such as those that regulate the oxidative status. One possible consequence of this loss in integration might be a limitation of the costs associated with the production of the colouration because metabolism of carotenoids would be weakly associated with the regulation of oxidative status (Badyaev 2004). This would mean that a greater expression of carotenoid-based colourations would not cause a corresponding increase in oxidative damage. It might also be that the carotenoid-based colourations are the byproduct of distinct and partially developmentally independent components, such as type and quality of carotenoids, or size and symmetry of the body trait (Badyaev et al. 2001). As a consequence, the various components that build up a colouration might be under different selective pressures (Hill 1992, 1993; Badyaev et al. 2001).

Another important implication of the complexity of carotenoid-based colourations is that costs afforded for the production of their components probably differ between individuals and environmental circumstances (Badyaev et al. 2001). For example, while some carotenoids are deposited unchanged through a metabolically cheap route, other carotenoids are metabolically converted in other forms through costly metabolic routes (Hill 1996; Stradi et al. 1997; Lopes et al. 2016; Weaver et al. 2018; Toomey et al. 2022). Although biotransformation of carotenoids is genetically controlled (e.g. Lopes et al. 2016; Mundy et al. 2016; Weaver et al. 2018; Toomey et al. 2022), the metabolic costs associated with their transformation might be relevant for the overall energetic budget and might vary across individuals and contexts.

## 8.3 Melanin-Based Colourations

Melanins are pigments that animals synthesise from the aromatic amino acids phenylalanine and tyrosine (Hearing 1993). Melanins occur in two chemically distinct forms. Eumelanin is a brown-black polymer of dihydroxyindole carboxylic acids and their reduced forms. Pheomelanin is a cysteine-containing red-brown polymer of benzothiazine units. Melanin-based colourations are under control of the pro-opiomelanocortin gene, which is responsible for the synthesis of melanocyte-stimulating hormones and the adrenocorticotropic hormone (Lin and Fisher 2007). Importantly, the expression of melanin-based colourations appears to be condition-dependent and influenced by environmental conditions (Griffith et al. 2006).

Melanocyte-stimulating hormones and the adrenocorticotropic hormone regulate the activity of the enzyme tyrosinase, which favours the shift from eumelanogenesis to pheomelanogenesis when the activity of the enzyme is high (Ozeki et al. 1997; Benathan et al. 1999). Melanogenesis is also regulated by the availability of thiol groups, such as free cysteine and cysteine-containing peptides (Ozeki et al. 1997; Benathan et al. 1999). The reaction between dopaquinone (metabolite of L-DOPA, which is derived from the amino acid tyrosine) and thiol groups generates pheomelanin (Fig. 8.4). Otherwise, dopaquinone undergoes a cyclisation that leads to the synthesis of eumelanin (Fig. 8.4) (García-Borrón and Olivares Sánchez 2011). Given that thiol compounds like glutathione play an important role in the regulation of cellular oxidative status and protection against oxidative damage, it has been hypothesised that the production of melanin-based colourations would generate oxidative stress that is affordable only by high-quality individuals that can, therefore, display high-quality signals compared to low-quality individuals (e.g. Galván and Alonso-Alvarez 2008; Galván and Solano 2009; Grunst et al. 2014; Galván et al. 2015; Arai et al. 2017).

The property of melanin-based colourations to signal the individual condition would be favoured by the differentiation in colouration between eumelanic (black and grey) and pheomelanic (yellowish, reddish, chestnut, and brown) traits (Toral et al. 2008). This is because the production of pheomelanic traits needs glutathione. Hence, melanin-based colourations might signal the capability to cope with oxidative stress because using glutathione in melanogenesis would indicate that less glutathione is available for coping with oxidative stress. Thus, the individual should have sufficient alternative antioxidant resources to compensate for the decrease in glutathione availability (Galván and Alonso-Alvarez 2008; Galván and Solano

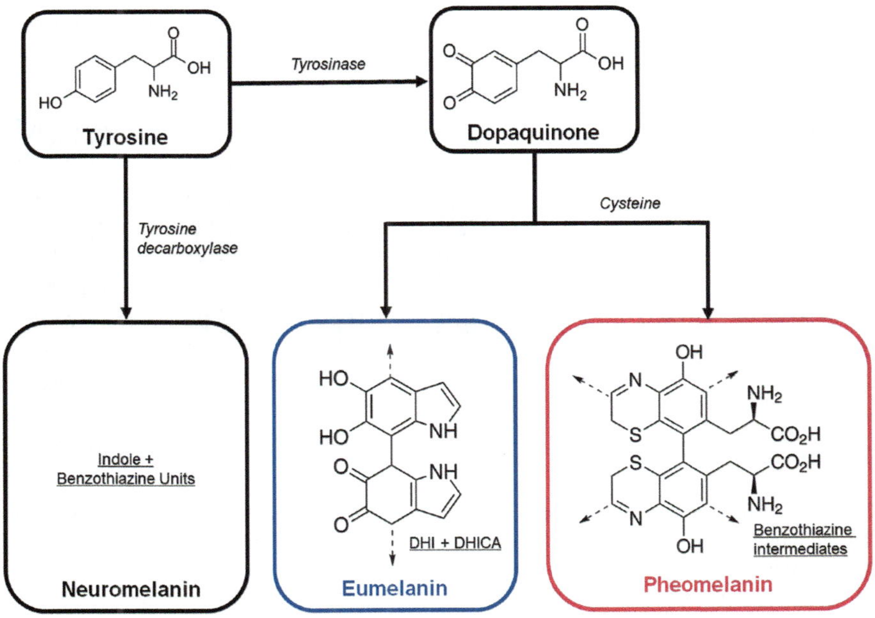


**Fig. 8.4** Biosynthetic pathways of three different types of natural melanins in birds. Reprinted from Jeon et al. (2021). © 2021 The Author(s); distributed under the terms of the Creative Commons Attribution License (CC BY)

2009). This hypothesis received partial support from correlative and experimental studies in birds and mammals.

Experimental inhibition of the γ-glutamyl cysteine synthetase (rate-limiting enzyme in glutathione synthesis) in free-living great tits *Parus major* through the subcutaneous injection of DL-buthionine-*S*,*R*-sulfoximine (BSO) reduced, as expected, the concentration of glutathione in red blood cells, and it increased the size of the eumelanin-based black breast stripe and the plasma non-enzymatic antioxidant capacity (Fig. 8.5) (Galván and Alonso-Alvarez 2008). Administration of diquat (herbicide that increases the generation of reactive oxygen species) in captive red-legged partridges *Alectoris rufa* increased the size of eumelanin-based colourations and reduced that of pheomelanic-based colourations, respectively (Galván and Alonso-Alvarez 2009). Diquat also decreased the concentration of glutathione in red blood cells (Galván and Alonso-Alvarez 2009), the concentration of carotenoids in plasma (Alonso-Alvarez and Galvan 2011), and, unexpectedly, the levels of thiobarbituric acid reactive substances in red blood cells (Galván and Alonso-Alvarez 2009).

In another experiment on a granivorous bird species, the administration of BSO to captive wild-caught greenfinches increased the blackness of the tips of tail feathers, decreased the concentration of glutathione in red blood cells, and increased the plasma concentration of malondialdehyde, an end-product of lipid peroxidation (Hõrak et al. 2010). In another study, male common yellowthroats *Geothlypis trichas* with larger black masks were more resistant against oxidative stress, as measured by an in vitro assay of the resistance of erythrocytes to haemolysis induced by free radicals (Henschen et al. 2016). Furthermore, males with larger masks also tended to have lower levels of glutathione, which was predicted because glutathione inhibits eumelanin production (Henschen et al. 2016). Conversely to these studies,

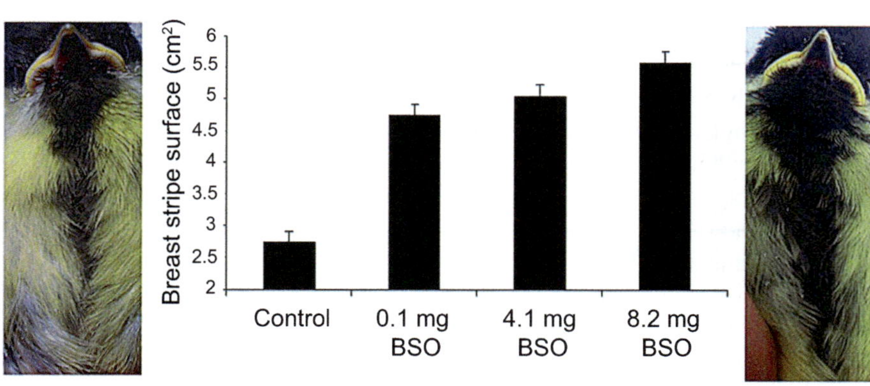


**Fig. 8.5** Nestling great tits *Parus major* that were treated with different doses of an inhibitor (DL-buthionine-S,R-sulfoximine, BSO) of glutathione synthesis throughout their development period produced larger breast stripes compared to control nestlings, indicating that the experimental reduction of glutathione synthesis affected the expression of the melanin-based colouration. Pictures on left and right sides of the figure illustrate a control nestling and a nestling that received the highest BSO dose, respectively. Values are shown as least squared mean and standard error. Reprinted with minor modifications from Galván and Alonso-Alvarez (2008). © 2008 Galvan et al.; distributed under the terms of the Creative Commons Attribution License (CC BY)

Stewart and Westneat (2013) did not find evidence that the bib size in house sparrows *Passer domesticus* was related to glutathione levels. Rather, the bib size was negatively related to dietary calcium. Interestingly, studies on chicks *Gallus gallus domesticus* showed that the duodenal calcium absorption may be reduced by oxidative stress (Rivoira et al. 2012). Similarly, pheomelanin-based colourations were not significantly related to two plasma markers of oxidative status in blue petrels *Halobaena caerulea* (Leclaire et al. 2019). In other studies on zebra finches, the experimental inhibition of the synthesis of glutathione reduced the intensity of pheomelanic colourations, but, unexpectedly, it did not influence levels of thiobarbituric acid reactive substances (Rodríguez-Martínez and Galván 2019), and the exposure to the pro-oxidant diquat unexpectedly increased the intensity of pheomelanic colourations, but did not affect levels of thiobarbituric acid reactive substances (Rodríguez-Martínez and Galván 2020). Finally, a comparative study, involving 1387 individuals of 104 European bird species sampled during the breeding season did not support the role of glutathione in driving variation in melanin synthesis across species and found weak evidence for a link between blood oxidative status and melanin-based colourations (Fig. 8.6) (Marton et al. 2022).

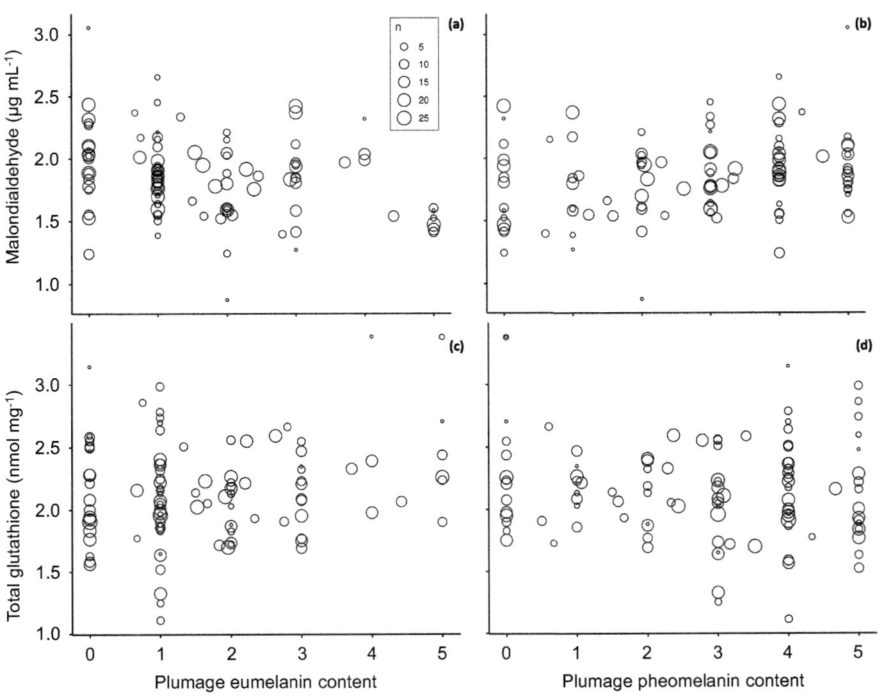


**Fig. 8.6** Comparative models in birds show that plasma levels of malondialdehyde are not correlated with (**a**) eumelanin nor with (**b**) pheomelanin scores. Similarly, concentration of glutathione in erythrocytes is not correlated with (**c**) eumelanin nor with (**d**) pheomelanin scores. The sample included European bird species sampled during the breeding season. Reprinted with minor modifications from Marton et al. (2022). © 2022 The Authors. Ecology and Evolution published by John Wiley & Sons Ltd; distributed under the terms of the Creative Commons Attribution License (CC BY)

Other studies suggested that the existence of a physiological mechanism uncoupling cysteine and pheomelanin in low-quality males makes them fairly insensitive to variation in cysteine availability (Galván and Alonso-Alvarez 2017). In other words, these males would be unable of increasing or decreasing the expression of a given signal in response to oxidative stress and related availability of cysteine, indicating that the occurrence of oxidative costs is not necessarily needed for the evolution of signal honesty (Galván and Alonso-Alvarez 2017). Thus, the inconsistencies between studies could be due to some extent to the lack of control for individual quality in statistical models.

The link between oxidative stress and melanogenesis does not occur only in birds (examples in mammalian species: Benedetto et al. 1981; Arck et al. 2006; Wood et al. 2009; Galván et al. 2012). For example, wild boars *Sus scrofa* with the highest levels of pheomelanin in their hair had lower levels of glutathione in their muscles, probably because glutathione was used for pheomelanin synthesis (Galván and Solano 2009), and higher levels of thiobarbituric acid reactive substances in muscle (Galván et al. 2012).

Beyond the use of glutathione in pheomelanin synthesis, it is often forgotten that another important aspect of melanin lies with its capability to extract metals from the environment and chelate them to coordination sites within the polymer framework. For example, $Fe^{3+}$ and $Cu^{2+}$ can cause damage to biological systems through Fenton reactions, and thus when melanin sequesters iron and copper, it prevents their reduction and subsequent generation of oxidative stress (Pilas et al. 1988; Zaręba et al. 1995). Moreover, absorption of solar radiation by melanins induces a transformation to an excited state, which can generate reactive oxygen species (Chiarelli-Neto et al. 2011). Given these additional properties of melanins, it will be important to refine our mechanistic predictions of how melanins interact with the oxidative status in order to better clarify the information that melanin-based colourations can convey.

## 8.4 The Role of Testosterone: Pro-oxidant, Antioxidant, or What?

Identification of proximate mechanisms that are linked to the expression of sexually dimorphic colourations is crucial for our understanding of how such traits are regulated, which selective pressures may affect their function and evolution, and which costs make these sexual signals honest. In 1992, Folstad and Karter proposed that the depression of the immune function might be one cost that would make the sexual signal reliable (*immunocompetence handicap hypothesis*). This immune depression would be caused by the high production of testosterone that occurs in males during the breeding season. This hypothesis is based on at least two lines of evidence. First, testosterone can regulate the expression of body traits as diverse as skeletal muscle, reproductive organs, or body colourations (Wingfield et al. 2001). Second, testosterone generally weakens the immune function, making the organism more vulnerable to pathogens (Grossman 1985). Therefore, only high-quality males would be able to display intense sexual colourations and, at the same time, to cope effectively with pathogens. The immunocompetence handicap hypothesis has, however, received equivocal support. A meta-analysis involving experimental studies that tested the effects of testosterone on immunity or ectoparasite load found a significant suppressive effect of testosterone on immunity, but this effect disappeared when multiple studies on the same species were controlled for in the statistical analyses (Roberts et al. 2004). This meta-analysis also showed that there was generally no effect of testosterone on direct measures of immunity, and that testosterone increased abundance of ectoparasites, particularly in reptiles (Roberts et al. 2004). The results of this meta-analysis reinforced the criticisms that were raised on the immunosuppressive activity of testosterone. Alternatively, Braude et al. (1999) proposed the *immunoredistribution hypothesis*, which predicts that when a male is under stress, immune cells are redeployed to target tissues that need protection. This hypothesis would explain why some studies found a decrease of immune cells in the blood. Implicit in this hypothesis is that testosterone is not necessarily immunosuppressive, which would not support the theoretical underpinnings of the immunocompetence handicap hypothesis. Studies on humans also gave more support to the hypothesis that testosterone is immunomodulatory rather than immunosuppressive (e.g. Trumble et al. 2016; Nowak et al. 2018).

The question then is how we can reconcile the immunoredistribution hypothesis with the handicap model. Kurtz et al. (2000) suggested that the redistribution of immune cells between tissues would give rise to trade-offs in the allocation of the different components of the immune system, when more tissues compete for the same components (Kurtz et al. 2000). We do not have evidence, however, of the extent to which these trade-offs are costly. An alternative hypothesis was proposed by Poiani et al. (2000). They suggested that testosterone might have either immunosuppressive or immunoenhancing effects (*integrated immunocompetence hypothesis*), a possibility that would be supported by the hormetic responses that were documented for some immune cells in relation to other hormonal stimuli.

Oxidative stress is another potential consequence of the metabolic action of testosterone (von Schantz et al. 1999). The *oxidation handicap hypothesis* states that, if testosterone would increase the allocation of pigments to body colourations, these pigments would be taken away from antioxidant protection, favouring the generation of oxidative damage (Alonso-Alvarez et al. 2007). However, there is also controversial evidence in favour of this hypothesis. Male zebra finches that received subcutaneous implants filled with a selective antagonist of the androgen receptor (flutamide), in order to prevent the effects of testosterone, had lower levels of blood antioxidants than control birds (Alonso-Alvarez et al. 2007). However, antioxidant levels of birds that received implants of testosterone did not differ from those of control birds (Alonso-Alvarez et al. 2007). Moreover, the hormonal treatment did not affect the carotenoid-based bill colouration. In another study, male captive red-legged partridges *Alectoris rufa*, whose testosterone levels were experimentally increased, showed a steeper shading of the periorbital ring colouration during reproduction (Alonso-Alvarez et al. 2008). These birds had also higher levels of circulating carotenoids compared to control birds, which was explained by a reallocation of carotenoids from the periorbital ring to the antioxidant protection in order to withstand the pro-oxidant action of testosterone. This explanation would be supported by the fact that testosterone also induced a transient elevation of the levels of glutathione in red blood cells (Alonso-Alvarez et al. 2008). This transient increase of glutathione might suggest that testosterone induced an hormetic response in testosterone-treated partridges, which would explain why control birds and testosterone-treated birds had similar levels of thiobarbituric acid reactive substances. However, this study only measured the reduced form of the glutathione, which provides an uncomplete picture of the impact of an oxidative stressor on the regulation of the oxidative status (Isaksson et al. 2005). Taff and Freeman-Gallant (2014) tested the oxidation handicap hypothesis experimentally by administering testosterone and control implants to free-living common yellowthroat warblers *Geothlypis trichas* and quantifying testosterone and oxidative status markers before and over a month later after implantation. Upon recapture, they found that testosterone-implanted males had higher levels of plasma non-enzymatic antioxidant capacity than control males, but oxidative damage to DNA of testosterone-implanted birds did not differ from control birds. Thus, these results did not support the key prediction of the oxidation handicap hypothesis.

Other studies found more convincing support of the oxidation handicap hypothesis. Mougeot et al. (2009) allocated free-living male red grouses *Lagopus lagopus scoticus* that were free of parasites to four different experimental groups, including control birds, testosterone-implanted birds, birds infected with larvae of the parasite *Trichostrongylus tenuis*, and birds that received both a subcutaneous implant of testosterone and the parasite infection, respectively. Compared to control birds, testosterone-implanted birds had an increase of the comb area (a secondary sexual trait in red grouses), but also increased levels of malondialdehyde, a finding consistent with a handicap mechanism. The parasite challenge also increased plasma levels of lipid peroxidation, especially when combined with the testosterone implant and reduced the comb area in birds that were not given a testosterone implant, respectively (Mougeot et al. 2009).

One explanation for the inconsistencies among studies might lie in the capacity of testosterone of stimulating antioxidant mechanisms or of reducing metabolic rate under certain circumstances (reviewed in Costantini 2010). Two studies carried out on captive birds suggested that another reason for the inconsistencies among studies could lie in the metabolites of testosterone (17β-estradiol and 5α-dihydrotestosterone) and in their effects on the organismal oxidative status (Casagrande et al. 2012a, b). Testosterone can be metabolised into either estradiol by the enzyme aromatase or dihydrotestosterone by the enzyme 5α-reductase, but there is no transformation of the secondary metabolites into each other (Pfaff et al. 2002; Ball and Balthazart 2008). Casagrande et al. (2012a) showed that estradiol administration by implants increased levels of reactive oxygen metabolites and reduced non-enzymatic antioxidant capacity of plasma in adult common kestrels, but it did not affect circulating carotenoids nor the expression of carotenoid-based colourations. By contrast, dihydrotestosterone, which binds to the same receptor of testosterone, did not affect the oxidative status of birds nor their circulating carotenoids, but increased the redness of the colouration. Finally, neither changes in reactive oxygen metabolites nor in antioxidant levels in plasma were correlated with those in body colourations and plasma levels of carotenoids. Thus, the impact of testosterone implants on oxidative status might depend on whether testosterone is converted into estradiol or dihydrotestosterone. For example, conversion of testosterone into estradiol by aromatase can affect several physiological and behavioural patterns in male birds (e.g. Balthazart and Ball 1998; Soma et al. 1999; Rosvall et al. 2012). Importantly, the occurrence and rate of aromatase activity also varies between species (Fusani et al. 2001, 2003), and/or time of year and sexual context (Hutchison et al. 1986; Sharp et al. 1986; Foidart et al. 1998; Soma et al. 1999). This context-dependent variation of aromatase activity might also contribute to generate inconsistencies among studies in the effects of testosterone implants on the oxidative status (Casagrande et al. 2012a).

In another experiment, Casagrande et al. (2012b) analysed the effects of testosterone and of its metabolites on the oxidative status of diamond doves *Geopelia cuneata* that were also concomitantly challenged with an immune challenge. In so doing, the authors could assess whether any effects of hormones on oxidative status depend on current individual state or condition. Diamond doves have a red-orange orbital ring, whose redness and size are increased by testosterone and dihydrotestosterone (Casagrande et al. 2012b), but are decreased by immunological challenges through a reduction in androgen levels (Casagrande et al. 2012b). Immunisation of diamond doves with sheep red blood cells increased levels of reactive oxygen metabolites in all experimental groups, regardless of the hormonal treatment, and decreased the plasma non-enzymatic antioxidant capacity in all groups, except for those birds that were given subcutaneous implants of testosterone. The results of this experiment did not give support to the oxidation handicap hypothesis because neither testosterone nor dihydrotestosterone affected the oxidative status of birds, even under challenging conditions. Rather, these results suggested that testosterone had a buffering effect on the plasma non-enzymatic antioxidant capacity of immune-challenged birds. It might be that there was a hormonal stimulation of antioxidant mobilisation from storage tissues towards those sites that needed protection against reactive oxygen species. However, the interpretation of these hormonal treatments is not straightforward for a number of reasons. For example, both oestrogens and androgens can induce hormetic responses or even change following hormetic patterns (Calabrese 2001a, b). Moreover, the action of sexual hormones varies between tissues, which might make blood-based markers of oxidative status poorly representative of what happens in those tissues where colourations are expressed.

## 8.5 When Life Is Not So Colourful: The Oxidative Costs of a Life in Black and White

In many species, males did not evolve any colourations to attract females. Rather, they rely on the size of achromatic secondary sexual traits. For example, males of collared flycatchers *Ficedula albicollis* and pied flycatchers *Ficedula hypoleuca* have a depigmented area on the forehead that works as a sexual signal (e.g. Potti and

Montalvo 1991; Saetre et al. 1997; Török et al. 2003; Qvarnström et al. 2006). Some studies found that the forehead patch size was correlated with several blood-based oxidative status markers in both flycatcher species (Markó et al. 2011; Moreno et al. 2011; López-Arrabé et al. 2014), suggesting that this depigmented area of the body might convey information about the oxidative status of the bearers. Pied flycatcher males with a larger forehead patch size had lower levels of oxidative damage and higher levels of non-enzymatic antioxidant capacity during the nestling period, respectively (Moreno et al. 2011). Similarly, collared flycatcher males with a larger white forehead patch had higher levels of plasma non-enzymatic antioxidant capacity (Markó et al. 2011). In addition, collared flycatcher females with higher plasma non-enzymatic antioxidant capacity mated with males having larger forehead patches, and smaller collared flycatcher females mated with males suffering more oxidative stress (Markó et al. 2011). Similarly, pied flycatcher females paired with attractive males were more successful in reproduction and had reduced levels of oxidative damage (Moreno et al. 2013). In conclusion, the results of these studies suggest that (1) high-quality males of both flycatcher species are capable of investing in the expression of sexual signals, without paying important oxidative costs, (2) assortative mating for the oxidative stress resistance, where higher quality females mate with males that are more resistant against oxidative stress, and (3) females may suffer less oxidative damage when they mate with high-quality males.

## 8.6 Body Colourations and Egg Pigmentation in Sexual Signalling of Females

The role of female colourations in the contexts of sexual selection and sociality has received comparatively less attention than that of male colourations. In recent decades, there has been growing interest in elucidating the adaptive significance of body colourations that females display during courtship, around the time of ovulation, or beyond during the rearing of offspring (e.g. Cunningham and Birkhead 1998; Amundsen 2000; Kraaijeveld et al. 2007; Rooker and Gavrilets 2018; Lüdtke and Foerster 2019). This research also involved studies on the role of eggshell pigmentation as a potential post-mating signal of females in some oviparous species (Moreno and Osorno 2003; see the *extended phenotype hypothesis*, Dawkins 1982).

Production and maintenance of pigmented or non-pigmented areas of the body demand resources that are costly to find and use (Searcy and Nowicki 2005). Moreover, females may produce badges of status (Dawkins and Krebs 1978), whose size or conspicuousness may be associated with the hormonal status, aggressiveness, and status of dominance (Senar et al. 1993; González et al. 2001). To test if social stress imposes costs on signalling, Moreno et al. (2013) painted white patches on females that did not have a natural patch and covered natural patches of females with a white paint, respectively. They found that female pied flycatchers use the white forehead patch to signal their status to other females during aggressive interactions to acquire a breeding site. Moreover, the addition of a badge to females without one caused an increase in plasma levels of malondialdehyde, but did not affect reproduction. These results suggest that the maintenance of this badge of status might induce physiological costs that only some females can afford without sacrificing their investment into reproduction.

Blas et al. (2013) proposed that the carotenoid-based colourations of bare parts may play a significant signalling role in wider competitive contexts (e.g. social interactions) than those enforced by sexual selection. They found that the physiological regulation of signal expression differs between breeders and floaters (i.e. non-breeding individuals) in the black kite *Milvus migrans*, which is a highly social bird of prey. Blas et al. (2013) proposed two non-mutually exclusive hypotheses to explain these results: the lower colouration displayed by floaters reflects underlying physiological limitations and mediates the access to breeding resources through social competition (*constraint hypothesis*); young floaters have evolved mechanisms to restrain colour expression, so that they signal their poor competition capabilities, avoiding physiological and social costs (*restraint hypothesis*; see also Williams 1966; Curio 1983). Carotenoid-based colourations also appear to convey information about the antioxidant status of the female. For example, both colouration of bare parts and plasma non-enzymatic antioxidant capacity of female common kestrels increased from the mating to the chick-rearing period (Casagrande et al. 2011). It is, however, difficult to explain this temporal covariation between body colourations and antioxidant levels in the context of the sexual selection theory since it is likely that the signalling function of body colourations is less important in common kestrels during the chick-rearing period than the mating period. In common kestrels, males provide food to the female during the incubation period, and to both the female and the chicks during the first couple of weeks from hatching. Thus, we might speculate that males adjust their foraging efforts upon the information they gather from the colourations of their mates.

In those oviparous species that produce coloured eggs, it has also been hypothesised that the eggshell pigmentation might work as a potential post-mating signal that enables females to convey information about their quality to the partners, which would adjust their investment in reproduction as a consequence (Moreno and Osorno 2003). In particular, Moreno and Osorno based their hypothesis on the role of blue–green colourations of eggs as signals of the antioxidant capacity of the female. The green-blue colouration of eggs that occurs in many bird species is based on the deposition of the biliverdin pigment into the eggshell (Kennedy and Vevers 1973; Miksik et al. 1996). Biliverdin is derived from the oxidation of bilirubin, a bile pigment derived from the haem catabolism. Biliverdin is deposited in the eggshell through the shell gland (Solomon 1997). In the organism, the enzyme biliverdin reductase reduces biliverdin back to bilirubin (Jansen and Daiber 2012). Antioxidant properties of biliverdin are, however, unclear (Kaur et al. 2003; Butler et al. 2023).

In pied flycatchers, Morales et al. (2008) found that females with higher levels of plasma non-enzymatic antioxidant capacity laid eggs with smaller values of blue-green chroma of the eggshell. Interestingly, this significant covariation between eggshell colouration and antioxidant levels was only evident in those females that experienced an intense physical effort before the start of egg laying. In female canaries, Hargitai et al. (2016a) found that blue-green eggshell colouration was not related to the female oxidative status during or after the laying period. However, eggs with lower yolk antioxidant concentration had higher average eggshell brown chroma.

We might expect that the deposition of biliverdin in the eggshell is constrained by the availability of other antioxidants if colourless antioxidants protect biliverdin from oxidation. For example, in blue-footed boobies *Sula nebouxii*, Morales et al. (2011) found that the eggshell colouration declined with the laying order of the egg, and those females that were carotenoid-supplemented after they laid their first egg, laid a second egg with an enhanced colouration compared to control females.

We can identify at least two main weak points of the hypothesised role of eggshell colouration as a signal of female quality. First, an effective engagement of eggshell colouration in sexual signalling has not been convincingly demonstrated (i.e. the capacity of males to discriminate clutches by colour; Reynolds et al. 2009). Second, although the antioxidant properties of biliverdin were described in some mammalian models (Kaur et al. 2003), they were not in avian species (Reynolds et al. 2009; Butler et al. 2023). There is not in-depth investigation of the biochemical and metabolic roles of biliverdin in the regulation of oxidative status. Finally, for those species that breed inside cavities, it is unclear how males may perceive the supposed signal of eggshell pigmentation.

As with biliverdin, many bird species deposit the red-brown protoporphyrin pigment in the eggshell, either as a background colour or as spotting (With 1973). Protoporphyrin-based pigmentation provides structural overcompensation for shell thinning caused, for example, by a deficit of dietary calcium (Gosler et al. 2011). Moreover, there is biochemical evidence for both pro- and antioxidant properties of protoporphyrin (e.g. Williams et al. 1994; Afonso et al. 1999; Shan et al. 2000). Thus, is protoporphyrin-based eggshell colouration related to the oxidative status of females? Giordano et al. (2015) documented a positive relationship between eggshell pigment distribution and both maternal and yolk antioxidant protection, suggesting that eggshell pigmentation might be a cue of female (and offspring) quality in great tits (Fig. 8.7). However, spot size did not correlate with plasma reactive

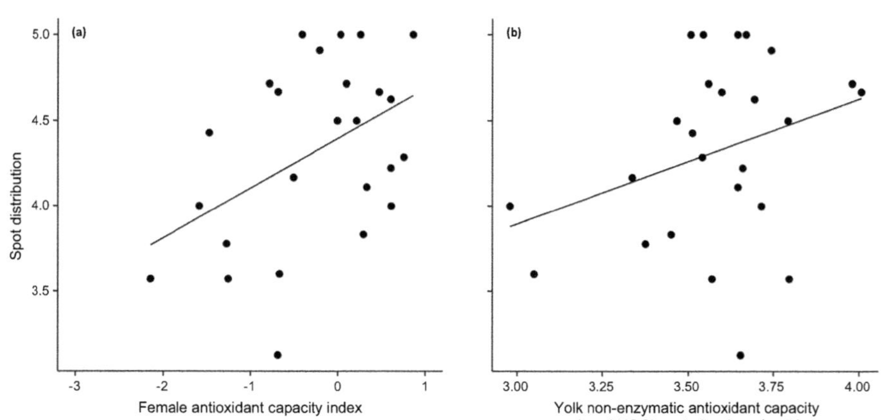


**Fig. 8.7** Relationships between eggshell pigmentation distribution and (**a**) female antioxidant capacity index (including both enzymatic and non-enzymatic antioxidants) and (**b**) yolk non-enzymatic antioxidant capacity (log-transformed) in great tits *Parus major*. Reprinted with minor modifications from Giordano et al. (2015) with permission of Springer Nature

oxygen metabolites nor with plasma non-enzymatic antioxidant capacity of females. In a similar study on great tits, Hargitai et al. (2016b) found that eggshell pigmentation patterns (e.g. spot intensity, size, and distribution) were not correlated with plasma reactive oxygen metabolites nor with plasma non-enzymatic antioxidant capacity in female great tits. Moreover, they found that eggs with darker spots contained lower concentrations of non-enzymatic antioxidants in the yolk. In conclusion, we do not have convincing evidence about the potential of eggshell colouration to signal the oxidative status of the female.

## 8.7 Beyond Sex: Social Interactions, Offspring Condition, and Warning Signals

Beyond the context of mate choice, body colourations may mediate social interactions. There is also potential for the colouration to affect the physiological status of the bearer if the colouration alters interactions with other individuals (e.g. acquisition of dominance status, competition for resources) that can in turn affect the physiology of the individual. Vitousek et al. (2013) provided support for the presence of a social mechanism that links colourations of the plumage with the organismal oxidative status during trait advertisement, long after the completion of signal development. They found that naturally darker barn swallows *Hirundo rustica* had lower levels of plasma oxidative damage. Importantly, females whose ventral plumage was made experimentally darker rapidly decreased oxidative damage to a concentration similar to that of naturally darker individuals (Vitousek et al. 2013). Although we do not know the endogenous mechanisms underlying the link between colourations and social interactions, the pattern of physiological changes observed in female barn swallows could result from darker plumage conferring a social advantage (e.g. darker females receiving fewer challenges from conspecifics during the early reproductive period; Vitousek et al. 2013). It might also be that males paired with those females that were manipulated to display darker ventral plumage invested more into reproduction, reducing the oxidative cost of reproduction in females as a consequence. Implicit in this second explanation is a mechanistic link with the individual reproductive strategies.

Interactions between family members are some of the most widely studied social behaviours in animals (Morales and Velando 2013). Although we can think that interests of family members are the same, interests are actually not identical and give rise to three main forms of evolutionary conflicts: those between parents and offspring, those among siblings (which all compete for the same resources), and those between parents (Trivers 1974; O'Connor 1978; Lessells 1999). The hypothesis of parent–offspring conflicts, in particular, proposes that a conflict arises over the amount of parental care, where the offspring tries to elicit more care than would be optimal for the parents. One way for offspring to elicit parents to provide more care is by showing signals that convey honestly their quality. Whether colourations of nestlings are honest signals of their quality, as they are for adults, remains poorly understood. That parents assess the colourations of their offspring was first demonstrated by Lyon et al. (1994). They altered the appearance of half of the chicks in each brood of coots *Fulica americana* by trimming the orange tips off their body feathers. Parent coots fed ornamented chicks preferentially over non-ornamented chicks, resulting in higher growth rates and greater survival for ornamented chicks compared to non-ornamented chicks. Lyon et al. (1994) also showed that the experimental manipulation of chick plumage did not influence their begging behaviour. Parental favouritism for offspring with brighter colourations was later described in various other bird species (e.g. Götmark and Ahlström 1997; Saino et al. 2000; Tschirren et al. 2005).

Parental favouritism for brighter offspring raises the question of what colourations of offspring would convey to their parents. For example, we have evidence that carotenoid-based colourations of nestlings are associated with fitness-related traits, such as body mass, stress hormones, or immune function (e.g. Saino et al. 2000, 2003; Casagrande et al. 2007; Loiseau et al. 2008; Dugas and McGraw 2011; Sternalski et al. 2012; Martínez-Padilla et al. 2013). As with sexual ornaments, colourations of nestlings can also be correlated with oxidative status markers. For example, Galván et al. (2010) found differences in antioxidant levels between melanic- and light-morphs of nestlings of the two birds of prey, the booted eagle *Hieraaetus pennatus* and the Eleonora's falcon *Falco eleonorae* (Galván et al. 2010). In another study, Roulin et al. (2011) found that the degrees of black pigmentation and of reddish pigmentation in nestlings of the barn owl *Tyto alba* were positively and negatively correlated with the resistance of red blood cells to haemolysis induced by an in vitro exposure to free radicals, respectively.

In those species that express carotenoid-based colourations in their bare parts, Casagrande et al. (2007) showed that the body levels of carotenoids and the expression of their colourations did not change synchronously. Carotenoids are absorbed in the gut and pass relatively rapidly into the bloodstream. Casagrande et al. (2007) found that an increase in the serum levels of carotenoids was already detectable at least 5 days after the start of a carotenoid supplementation protocol. By contrast, it took around 11 days since the start of the carotenoid supplementation before they observed a significant change in the colouration of tarsi skin. By contrast, 11 days after the end of the supplementation, the serum levels of carotenoids returned to those recorded before the start of the experiment, probably because they were rapidly stored in other tissues (e.g. skin) or consumed in metabolic reactions (Casagrande et al. 2007). Thus, there seems to be an endogenous constraint that limits the capability of the body colouration to rapidly change as a consequence of any change in the nestling condition. This constraint should not prevent the skin colouration from signalling the nestling condition in the long-term, which is relevant for those species with prolonged parental care.

Finally, body colourations play important roles as warning signals. As an antipredator strategy, many animal species have evolved conspicuous body colourations that are used to signal their unprofitability to predators (e.g. Batesian and Müllerian mimicry). Given this antipredator role, it has been assumed for years that changes in the colouration should be negligible in order to facilitate predator learning and avoidance. However, we now know that even warning colourations vary between individuals and are significantly influenced by the development conditions (Mappes et al. 2005; Stevens and Ruxton 2012). Aposematic species are often warningly coloured early in life, but their chemical defences may be weak or absent because they have yet to be acquired through the diet (e.g. alkaloids acquired from dietary arthropods) or synthesised endogenously (e.g. Daly et al. 1994; Nishida 2002; Saporito et al. 2009). Hence, investment in the production of body colourations during growth might expose individuals with poor chemical defences to a higher risk of predation. A strategy to avoid this state of being conspicuous, yet poorly defended against predators, could be to invest lots of resources in growth rather than in the production of body colourations (Flores et al. 2013). However, there may be costs associated with this strategy, such as oxidative stress.

Poison dart frogs (family *Dendrobatidae*) are renowned for their toxicity and brightly coloured bodies. Flores et al. (2013) reared larvae of green and black poison frogs *Dendrobates auratus* on a relatively low or a high food supply and tested the hypotheses that individuals having access to a larger amount of resources should grow larger while reducing their investment in warning signals at metamorphic completion, and oxidative stress might affect resource-allocation trade-offs. They found that, when food was abundant, frogs grew larger and invested less in signalling, resulting in less conspicuous as a consequence. Signal luminance and levels of oxidative stress were positively correlated in high-food froglets, but were negatively correlated in low-food froglets, suggesting a resource-allocation trade-off when the food availability was relatively low (Flores et al. 2013). Hence, oxidative stress might have constrained the investment in growth and signalling of resource-limited froglets.

In another experiment, Blount et al. (2023) raised monarch butterflies *Danaus plexippus* on different species of their milkweed host plants that vary in quantities of cardenolides to test whether the sequestration of cardenolides as a secondary defence is associated with costs in the form of oxidative damage and reduced antioxidant defences, which in turn can affect the capacity to produce aposematic displays. They found that levels of malondialdehyde were higher in butterflies with higher body levels of cardenolides (Fig. 8.8). They also found that males with high levels of malondialdehyde became less conspicuous with increased sequestration of cardenolides, whereas those with low levels of malondialdehyde became more

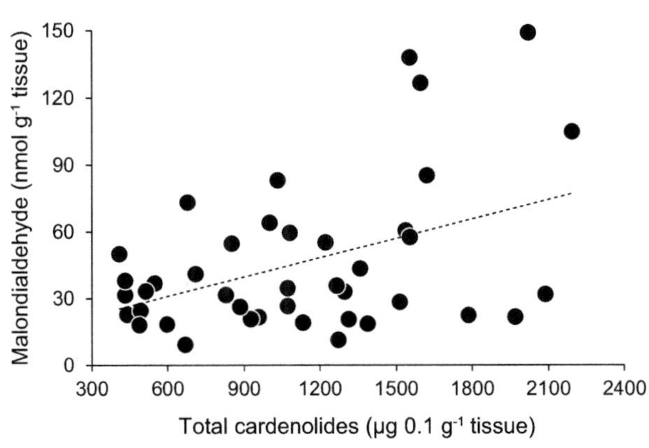


**Fig. 8.8** Relationship between total cardenolide concentration and malondialdehyde concentration in monarch butterflies *Danaus plexippus*. Reproduced using data collected from Blount et al. (2023)—Rowland, Hannah, 2022, "Data on oxidative state markers and colouration of monarch butterflies", https://doi.org/10.17617/3.YC1BR4, Edmond, V1

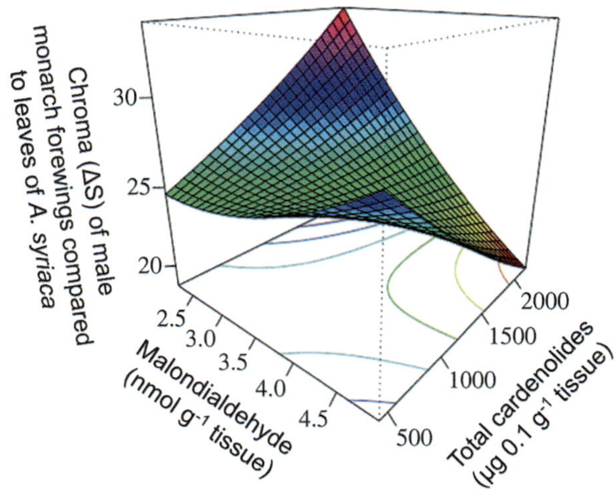


**Fig. 8.9** Three-dimensional surface plot for the response surface predicting the relationship between the conspicuousness of the male monarch's forewing, oxidative damage measured as malondialdehyde, and total cardenolide concentration. Males with high levels of malondialdehyde become less conspicuous with increased sequestration of cardenolides, whereas those with low levels of malondialdehyde become more conspicuous with increased levels of cardenolides. Reprinted with minor modifications from Blount et al. (2023). © 2023 The Authors; published by the Royal Society under the terms of the Creative Commons Attribution License

conspicuous with increased levels of cardenolides (Fig. 8.9). The results of this study suggested that oxidative status might be a key physiological mechanism that links warning colourations to sequestration costs, and that the contribution of pigments to antioxidant activity for maintaining redox homeostasis is important.

## 8.8 Conclusions and Future Directions

The expression of colourations and achromatic traits impose physiological costs on the organism that only high-quality individuals are able to afford. Much work focussed on the costs associated with the expression of body colourations that are relevant for mate choice. However, we have also evidence that it is costly to express body colourations that are used in sociality or as warning signals.

We have weak evidence in support of a general role of oxidative stress as a cost of and constraint on the expression of colourations. Rather, the relevance of oxidative stress seems to be circumstantial. We can identify some fundamental limitations on inferences imposed by study design in this research field. One main limitation is that we have an elusive understanding of the physiological roles and metabolic fate of pigments across the wide range of species that display body colourations. Another main limitation is that in many studies, the investigators seek to link markers of current organismal oxidative status to metrics of body colourations, even if the colouration was developed days to months earlier. This asynchrony in the data collection makes complicated the interpretation of results, because the condition of the individual at the time of sampling might not reflect the condition of the individual at the time the colouration was produced.

Other factors might also be relevant to explain discrepancies between studies. The strength of the relationship between body colouration and oxidative status might differ among species, depending on how strong the selection pressure on the choosy sex is for the accuracy of its evaluation of mate's qualities. For example, this relationship may be stronger in short-lived species, where there is obligate paternal care or when one mate invests weakly in rearing the offspring, which comes at a cost for the other mate. This second scenario is well exemplified by studies on zebra finches, which found that a lower investment of a pair member comes at a cost for the survival of the mate (Monaghan et al. 2012). For the opposite reasons, selection for the capability to judge the quality of potential mates might be relaxed in long-lived species, where there are better chances of fixing an initial mistake in the mate choice.

Finally, individual age seems to be an important determinant of the reliability of the sexual signal (Torres and Velando 2007; Alonso-Alvarez et al. 2009). Older individuals should show a stronger decline of the colouration, particularly when they are being exposed to a stressor, compared to younger individuals. This is particularly relevant because oxidative status also changes with age. The free radical and oxidative stress hypotheses of ageing (see Chap. 10) state that age-associated functional losses are owing to the accumulation of oxidative damage to lipids, DNA, and proteins by reactive oxygen species. Thus, it might be that the significance of oxidative stress for the honesty of the signal varies with age.

## References

Afonso S, Vanore G, Batlle A (1999) Protoporphyrin IX and oxidative stress. Free Radic Res 31:161–170

Alonso-Alvarez C, Galvan I (2011) Free radical exposure creates paler carotenoid-based ornaments: a possible interaction in the expression of black and red traits. PLoS One 6:e19403

Alonso-Alvarez C, Bertrand S, Faivre B, Chastel O, Sorci G (2007) Testosterone and oxidative stress: the oxidation handicap hypothesis. Proc R Soc Lond B 274:819–825

Alonso-Alvarez C, Pérez-Rodríguez L, Mateo R, Chastel O, Viñuela J (2008) The oxidation handicap hypothesis and the carotenoid allocation trade-off. J Evol Biol 21:1789–1797

Alonso-Alvarez C, Pérez-Rodríguez L, Garcia JT, Viñuela J (2009) Testosterone-mediated trade-offs in the old age: a new approach to the immunocompetence handicap and carotenoid-based sexual signalling. Proc R Soc Lond B 276:2093–2101

Alonso-Alvarez C, García-de Blas E, Mateo R (2018) Dietary canthaxanthin reduces xanthophyll uptake and red coloration in adult red-legged partridges. J Exp Biol 221:jeb185074

Amengual J, Lobo GP, Golczak M, Li HN, Klimova T, Hoppel CL, Wyss A, Palczewski K, von Lintig J (2011) A mitochondrial enzyme degrades carotenoids and protects against oxidative stress. FASEB J 25:948–959

Amundsen T (2000) Why are female birds ornamented? Trends Ecol Evol 15:149–155

Andersson M (1994) Sexual selection. Princeton University Press, Princeton, NJ

Arai E, Hasegawa M, Makino T, Hagino A, Sakai Y, Ohtsuki H, Wakamatsu K, Kawata M (2017) Physiological conditions and genetic controls of phaeomelanin pigmentation in nestling barn swallows. Behav Ecol 28:706–716

Arck PC, Overall R, Spatz K, Liezman C, Handjiski B, Klapp BF, Birch-Machin MA, Peters EMJ (2006) Towards a "free radical theory of graying:" melanocyte apoptosis in the aging human hair follicle is an indicator of oxidative stress induced tissue damage. FASEB J 20:1567–1569

Badyaev AV (2004) Developmental perspective on the evolution of sexual ornaments. Evol Ecol Res 6:1–17

Badyaev AV, Hill GE, Dunn PO, Glen JC (2001) Plumage color as a composite trait: developmental and functional integration of sexual ornamentation. Am Nat 158:221–235

Ball GF, Balthazart J (2008) Individual variation and the endocrine regulation of behaviour and physiology in birds: a cellular/molecular perspective. Philos Trans R Soc Lond B 363:1699–1710

Balthazart J, Ball GF (1998) New insights into the regulation and function of brain estrogen synthase (aromatase). Trends Neurosci 21:243–249

Beamonte-Barrientos R, Velando A, Torres R (2014) Age-dependent effects of carotenoids on sexual ornaments and reproductive performance of a long-lived seabird. Behav Ecol Sociobiol 68:115–126

Benathan M, Virador V, Furumura M, Kobayashi N, Panizzon RG, Hearing VJ (1999) Co-regulation of melanin precursors and tyrosinase in human pigment cells: roles of cysteine and glutathione. Cell Mol Biol 45:981–990

Benedetto JP, Ortonne JP, Voulot C, Khatchadourian C, Prota G, Thivolet J (1981) Role of thiol compounds in mammalian melanin pigmentation. I. Reduced and oxidized glutathione. J Invest Dermatol 77:402–405

Bertrand S, Faivre B, Sorci G (2006) Do carotenoid-based sexual traits signal the availability of non-pigmentary antioxidants? J Exp Biol 209:4414–4419

Blas J, Cabezas S, Figuerola J, López L, Tanferna A, Hiraldo F, Sergio F, Negro JJ (2013) Carotenoids and skin coloration in a social raptor. J Raptor Res 47:174–184

Blount JD, Matheson SM (2006) Effects of carotenoid supply on escape flight responses in zebra finches, *Taeniopygia guttata*. Anim Behav 72:595–601

Blount JD, Metcalfe NB, Birkhead TR, Surai PF (2003) Carotenoid modulation of immune function and sexual attractiveness in zebra finches. Science 300:125–127

Blount JD, Rowland HM, Mitchell C, Speed MP, Ruxton GD, Endler JA, Brower LP (2023) The price of defence: toxins, visual signals and oxidative state in anaposematic butterfly. Proc R Soc Lond B 290:20222068

Braude S, Tang-Martinez Z, Taylor GT (1999) Stress, testosterone, and the immunoredistribution hypothesis. Behav Ecol 10:345–350

Butler MW, Cullen ZE, Garti CM, Howard DE, Corpus BA, McNish BA, Hines JK (2023) Physiologically relevant levels of biliverdin do not significantly oppose oxidative damage in plasma in vitro. Physiol Biochem Zool 96:294–303

Calabrese EJ (2001a) Estrogen and related compounds: biphasic dose responses. Crit Rev Toxicol 31:503–515

Calabrese EJ (2001b) Androgens: biphasic dose responses. Crit Rev Toxicol 31:517–522

Candolin U (2003) The use of multiple cues in mate choice. Biol Rev 78:575–595

Cantarero A, Mateo R, Camarero PR, Alonso D, Fernandez-Eslava B, Alonso-Alvarez C (2020) Testing the shared-pathway hypothesis in the carotenoid-based coloration of red crossbills. Evolution 74:2348–2364

Casagrande S, Csermely D, Pini E, Bertacche V, Tagliavini J (2006) Skin carotenoid concentration correlates with male hunting skill and territory quality in the kestrel (*Falco tinnunculus*). J Avian Biol 37:190–196

Casagrande S, Costantini D, Fanfani A, Tagliavini J, Dell'Omo G (2007) Patterns of serum carotenoid accumulation and skin color variation in nestling kestrels in relation to breeding conditions and different terms of carotenoid supplementation. J Comp Physiol B 177:237–245

Casagrande S, Dell'Omo G, Costantini D, Tagliavini J, Groothuis T (2011) Variation of a carotenoid-based trait in relation to oxidative stress and endocrine status during the breeding season in the Eurasian kestrel: a multi-factorial study. Comp Biochem Physiol Part A 160:16–26

Casagrande S, Costantini D, Dell'Omo G, Tagliavini J, Groothuis TGG (2012a) Differential effects of testosterone metabolites oestradiol and dihydrotestosterone on oxidative stress and carotenoid-dependent colour expression in a bird. Behav Ecol Sociobiol 66:1319–1331

Casagrande S, Costantini D, Groothuis TGG (2012b) Interaction between sexual steroids and immune response in affecting oxidative status of birds. Comp Biochem Physiol Part A 163:296–301

Chew BP, Park JS (2004) Carotenoid action on the immune response. J Nutr 134:257S–261S

Chiarelli-Neto O, Pavani C, Ferreira AS, Uchoa AF, Severino D, Baptista MS (2011) Generation and suppression of singlet oxygen in hair by photosensitization of melanin. Free Radic Biol Med 51:1195–1202

Cohen AA, McGraw KJ (2009) No simple measures for antioxidant status in birds: complexity in inter- and intraspecific correlations among circulating antioxidant types. Funct Ecol 23:310–320

Costantini D (2010) Redox physiology in animal function: the struggle of living in an oxidant environment. Curr Zool 56:687–702

Costantini D, Dell'Omo G (2020) The kestrel: ecology, behaviour and conservation of an open-land predator. Cambridge University Press, Cambridge, UK

Costantini D, Møller AP (2008) Carotenoids are minor antioxidants for birds. Funct Ecol 22:367–370

Costantini D, Casagrande S, De Filippis S, Brambilla G, Fanfani A, Tagliavini J, Dell'Omo G (2006) Correlates of oxidative stress in wild kestrel nestlings (*Falco tinnunculus*). J Comp Physiol B 176:329–337

Costantini D, Coluzza C, Fanfani A, Dell'Omo G (2007a) Effects of carotenoid supplementation on colour expression, oxidative stress and body mass in rehabilitated captive adult kestrels (*Falco tinnunculus*). J Comp Physiol B 177:723–731

Costantini D, Fanfani A, Dell'Omo G (2007b) Carotenoid availability does not limit the capability of nestling kestrels (*Falco tinnunculus*) to cope with oxidative stress. J Exp Biol 210:1238–1244

Cunningham EJA, Birkhead TR (1998) Sex roles and sexual selection. Anim Behav 56:1311–1321

Curio E (1983) Why do young birds reproduce less well? Ibis 125:400–404

Daly JW, Secunda SI, Garrafo HM, Spande TF, Wisnieski A, Cover JFJ (1994) An uptake system for dietary alkaloids in poison frogs (Dendrobatidae). Toxicon 32:657–663

Darwin C (1871) The descent of man, and selection in relation to sex. John Murray, London

Dawkins R (1982) The extended phenotype. Oxford University Press, Oxford

Dawkins R, Krebs JR (1978) Animal signals: information or manipulation? In: Krebs JR, Davies NB (eds) Behavioural ecology. Blackwell, Oxford, pp 282–309

de la Fuente M (2002) Effects of antioxidants on immune system ageing. Eur J Clin Nutr 56:S5–S8

Dugas MB, McGraw KJ (2011) Proximate correlates of carotenoid-based mouth coloration in nestling house sparrows. Condor 113:691–700

Edwards CL, Byrne PG, Harlow P, Silla AJ (2017) Dietary carotenoid supplementation enhances the cutaneous bacterial communities of the critically endangered Southern corroboree frog (Pseudophryne corroboree). Microb Ecol 73:435–444

El Agamey A, Lowe GM, McGarvey DJ, Mortensen A, Phillip DM, Truscott TG, Young AJ (2004) Carotenoid radical chemistry and antioxidant/pro-oxidant properties. Arch Biochem Biophys 430:37–48

Endler JA (1980) Natural-selection on color patterns in *Poecilia reticulata*. Evolution 34:76–91

Endler JA (1983) Natural and sexual selection on color patterns in poeciliid fishes. Environ Biol Fish 9:173–190

Faivre B, Grégoire A, Préault M, Cézilly F, Sorci G (2003) Immune activation rapidly mirrored in a secondary sexual trait. Science 300:103

Fisher RA (1915) The evolution of sexual preference. Eugen Rev 7:184–192

Flores EE, Stevens M, Moore AJ, Blount JD (2013) Diet, development and the optimization of warning signals in post-metamorphic green and black poison frogs. Funct Ecol 27:816–829

Foidart A, Silverin B, Baillien M, Harada N, Balthazart J (1998) Neuroanatomical distribution and variations across the reproductive cycle of aromatase activity and aromatase-immunoreactive cells in the pied flycatcher (*Ficedula hypoleuca*). Horm Behav 33:180–196

Folstad I, Karter AJ (1992) Parasites, bright males, and the immunocompetence handicap. Am Nat 139:603–622

Fusani L, Hutchison JB, Gahr M (2001) Testosterone regulates the activity and expression of aromatase in the canary neostriatum. J Neurobiol 49:1–8

Fusani L, Metzdorf R, Hutchison JB, Gahr M (2003) Aromatase inhibition affects testosterone-induced masculinization of song and the neural song system in female canaries. J Neurobiol 54:370–379

Galván I, Alonso-Alvarez C (2008) An intracellular antioxidant determines the expression of a melanin-based signal in a bird. PLoS One 3:e3335

Galván I, Alonso-Alvarez C (2009) The expression of melanin-based plumage is separately modulated by exogenous oxidative stress and a melanocortin. Proc R Soc Lond B 276:3089–3097

Galván I, Alonso-Alvarez C (2017) Individual quality via sensitivity to cysteine availability in a melanin-based honest signaling system. J Exp Biol 220:2825–2833

Galván I, Solano F (2009) The evolution of eu- and pheomelanic traits may respond to an economy of pigments related to environmental oxidative stress. Pigment Cell Melan Res 22:339–342

Galván I, Gangoso L, Grande JM, Negro JJ, Rodríguez A, Figuerola J, Alonso-Alvarez C (2010) Antioxidant machinery differs between melanic and light nestlings of two polymorphic raptors. PLoS One 5:e13369

Galván I, Alonso-Alvarez C, Negro JJ (2012) Relationships between hair melanization, glutathione levels, and senescence in wild boars. Physiol Biochem Zool 85:332–347

Galván I, Wakamatsu K, Camarero PR, Mateo R, Alonso-Alvarez C (2015) Low-quality birds do not display high-quality signals: the cysteine-pheomelanin mechanism of honesty. Evolution 69:26–38

García-Borrón JC, Olivares Sánchez MC (2011) Biosynthesis of melanins. In: Borovanský J, Riley PA (eds) Melanins and melanosomes: biosynthesis, biogenesis, physiological, and pathological functions. Wiley-Black-well, Weinheim, pp 87–116

García-de Blas E, Mateo R, Alonso-Alvarez C (2016) Specific carotenoid pigments in the diet and a bit of oxidative stress in the recipe for producing red carotenoid-based signals. PeerJ 4:e2237

Giordano M, Costantini D, Pick JL, Tschirren B (2015) Female oxidative status, egg antioxidant protection and eggshell pigmentation: a supplemental feeding experiment in great tits. Behav Ecol Sociobiol 69:777–785

González G, Sorci G, Smith LC, de Lope F (2001) Testosterone and sexual signalling in male house sparrows (*Passer domesticus*). Behav Ecol Sociobiol 50:557–562

Gosler AG, Connor OR, Bonser RHC (2011) Protoporphyrin and eggshell strength: preliminary findings from a passerine bird. Avian Biol Res 4:214–223

Götmark F, Ahlström M (1997) Parental preference for red mouth of chicks in a songbird. Proc R Soc Lond B 264:959–962

Griffith SC, Parker TH, Olson VA (2006) Melanin- versus carotenoid-based sexual signals: is the difference really so black and red? Anim Behav 71:749–763

Grossman CJ (1985) Interactions between the gonadal steroids and the immune system. Science 227:257–261

Grunst AS, Salgado-Ortiz J, Rotenberry JT, Grunst ML (2014) Phaeomelanin- and carotenoid-based pigmentation reflect oxidative status in two populations of the yellow warbler (*Setophaga petechia*). Behav Ecol Sociobiol 68:669–680

Gruszecki WI (1999) Carotenoids in membranes. In: Frank HA, Young AJ, Britton G, Cogdell RJ (eds) The photochemistry of carotenoids. Kluwer, Dordrecht, pp 363–379

Hamilton WD, Zuk M (1982) Heritable true fitness and bright birds: a role for parasites? Science 218:384–387

Hargitai R, Boross N, Nyiri Z, Eke Z (2016a) Biliverdin- and protoporphyrin-based eggshell pigmentation in relation to antioxidant supplementation, female characteristics and egg traits in the canary (*Serinus canaria*). Behav Ecol Sociobiol 70:2093–2110

Hargitai R, Nagy G, Herenyi M, Nyiri Z, Laczi M, Hegyi G, Eke Z, Torok J (2016b) Darker eggshell spotting indicates lower yolk antioxidant level and poorer female quality in the Eurasian Great Tit (*Parus major*). Auk 133:131–146
Hartley RC, Kennedy MW (2004) Are carotenoids a red herring in sexual display? Trends Ecol Evol 19:353–354
Hearing VJ (1993) Unraveling the melanocyte. Am J Hum Genet 52:1–7
Helfenstein F, Losdat S, Møller AP, Blount JD, Richner H (2010) Sperm of colourful males are better protected against oxidative stress. Ecol Lett 13:213–222
Henschen AE, Whittingham LA, Dunn PO (2016) Oxidative stress is related to both melanin- and carotenoid-based ornaments in the common yellowthroat. Funct Ecol 30:749–758
Hill GE (1992) The proximate basis of variation in carotenoid pigmentation in male house finches. Auk 109:1–12
Hill GE (1993) Geographic variation in the carotenoid plumage pigmentation of male house finches (*Carpodacus mexicanus*). Biol J Linn Soc 49:63–86
Hill GE (1996) Redness as a measure of the production cost of ornamental coloration. Ethol Ecol Evol 8:157–175
Hill GE (1999) Is there an immunological cost to carotenoid-based ornamental coloration? Am Nat 154:589–595
Hill GE, McGraw KJ (2006a) Bird coloration. Volume I. Mechanisms and measurements. Harvard University Press, Cambridge, MA
Hill GE, McGraw KJ (2006b) Bird coloration. Volume II. Function and evolution. Harvard University Press, Cambridge, MA
Hill GE, Hood WR, Ge Z, Grinter R, Greening C, Johnson JD, Park NR, Taylor HA, Andreasen VA, Powers MJ, Justyn NM, Parry HA, Kavazis AN, Zhang Y (2019) Plumage redness signals mitochondrial function in the house finch. Proc R Soc Lond B 286:20191354
Hõrak P, Sild E, Soomets U, Sepp T, Kilk K (2010) Oxidative stress and information content of black and yellow plumage coloration: an experiment with greenfinches. J Exp Biol 213:2225–2233
Huggins KA, Navara KJ, Mendonça MT, Hill GE (2010) Detrimental effects of carotenoid pigments: the dark side of bright coloration. Naturwissenschaften 97:637–644
Hutchison JB, Steimer T, Jaggard P (1986) Effects of photoperiod on formation of oestradiol-17b in the dove brain. J Endocrinol 109:371–377
Isaksson C, Örnborg J, Stephensen E, Andersson S (2005) Plasma glutathione and carotenoid coloration as potential biomarkers of environmental stress in great tits. EcoHealth 2:138–146
Jansen T, Daiber A (2012) Direct antioxidant properties of bilirubin and biliverdin. Is there a role for biliverdin reductase? Front Pharmacol 3:1–10
Jeon DJ, Paik S, Ji S, Yeo JS (2021) Melanin-based structural coloration of birds and its biomimetic applications. Appl Microsc 51:14
Johnson JD, Hill GE (2013) Is carotenoid ornamentation linked to the inner mitochondria membrane potential? A hypothesis for the maintenance of signal honesty. Biochimie 95:436–444
Johnstone RA (1996) Multiple displays in animal communication: 'backup signals' and 'multiple messages'. Philos Trans R Soc B 351:329–338
Jomova K, Valko M (2013) Health protective effects of carotenoids and their interactions with other biological antioxidants. Eur J Med Chem 70:102–110
Karu U, Saks L, Hõrak P (2008) Carotenoid-based plumage coloration is not affected by vitamin E supplementation in male greenfinches. Ecol Res 23:931–935
Kaur H, Hughes MN, Green CJ, Naughton P, Foresti R, Motterlini R (2003) Interaction of bilirubin and biliverdin with reactive nitrogen species. FEBS Lett 543:113–119
Kennedy GY, Vevers HG (1973) Eggshell pigments of Araucano Fowl. Comp Biochem Physiol Part B 44:11–25
Kirkpatrick M, Ryan MJ (1991) The evolution of mating preferences and the paradox of lek. Nature 350:33–38
Koch RE, Staley M, Kavazis AN, Hasselquist D, Toomey MB, Hill GE (2019) Testing the resource trade-off hypothesis for carotenoid-based signal honesty using genetic variants of the domestic canary. J Exp Biol 222:jeb188102

Kraaijeveld K, Kraaijeveld-Smit FJL, Komdeur J (2007) The evolution of mutual ornamentation. Anim Behav 74:657–677
Kurtz J, Reinhold K, Engqvist L (2000) Immunosuppression under stress: necessary for condition-dependent signalling? Trends Ecol Evol 15:418–419
Leclaire S, Bourret V, Blanchard P, de Franceschi C, Merkling T, Hatch SA, Danchin E (2015) Carotenoids increase immunity and sex specifically affect color and redox homeostasis in a monochromatic seabird. Behav Ecol Sociobiol 69:1097–1111
Leclaire S, Perret S, Galván I, Bonadonna F (2019) Pheomelanin-based coloration is related to individual quality and oxidative stress in blue petrels. Evol Ecol 33:873–887
Lessells CM (1999) Sexual conflict. In: Keller L (ed) Levels of selection in evolution. Princeton University Press, Princeton, NJ, pp 75–99
Lin JY, Fisher DE (2007) Melanocyte biology and skin pigmentation. Nature 445:843–850
Lobo GP, Isken A, Hoff S, Babino D, von Lintig J (2012) BCDO2 acts as a carotenoid scavenger and gatekeeper for the mitochondrial apoptotic pathway. Development 139:2966–2977
Loiseau C, Fellous S, Haussy C, Chastel O, Sorci G (2008) Condition-dependent effects of corticosterone on a carotenoid-based begging signal in house sparrows. Horm Behav 53:266–273
Lopes RJ, Johnson JD, Toomey MB, Ferreira MS, Araujo PM, Melo-Ferreira J, Andersson L, Hill GE, Corbo JC, Carneiro M (2016) Genetic basis for red coloration in birds. Curr Biol 26:1427–1434
López-Arrabé J, Cantarero A, Pérez-Rodríguez L, Palma A, Moreno J (2014) Plumage ornaments and reproductive investment in relation to oxidative status in the Pied Flycatcher *Ficedula hypoleuca* iberiae. Can J Zool 92:1019–1027
Lozano GA (1994) Carotenoids, parasites, and sexual selection. Oikos 70:309–311
Lozano GA (2009) Multiple cues in mate selection: the sexual interference hypothesis. BioSci Hypotheses 2:37–42
Lu P, Wong SY, Wu L, Lin D (2020) Carotenoid metabolism in mitochondrial function. Food Qual Saf 4:115–122
Lüdtke DU, Foerster K (2019) A female color ornament honestly signals fecundity. Front Ecol Evol 7:432
Lyon BE, Eadie JM, Hamilton LD (1994) Parental choice selects for ornamental plumage in American coot chicks. Nature 371:240–243
Mappes J, Marples NM, Endler JA (2005) The complex business of survival by aposematism. Trends Ecol Evol 20:598–603
Markó G, Costantini D, Michl G, Török J (2011) Oxidative damage and plasma antioxidant capacity in relation to body size, age, male sexual traits and female reproductive performance in the collared flycatcher (*Ficedula albicollis*). J Comp Physiol B 181:73–81
Marri V, Richner H (2014) Differential effects of vitamins E and C and carotenoids on growth, resistance to oxidative stress, fledging success and plumage colouration in wild great tits. J Exp Biol 217:1478–1484
Martínez-Padilla J, Mougeot F, García JT, Arroyo B, Bortolotti GR (2013) Feather corticosterone levels and carotenoid-based coloration in common buzzard (*Buteo buteo*) nestlings. J Raptor Res 47:161–173
Marton A, Vágási CI, Vincze O, Bókony V, Pap PL, Pătraş L, Pénzes J, Bărbos L, Fülöp A, Osváth G, Ducatez S, Giraudeau M (2022) Oxidative physiology is weakly associated with pigmentation in birds. Ecol Evol 12:e9177
Maynard Smith J, Harper D (2003) Animal signals. Oxford University Press, Oxford
Mayne ST, Parker RS (1986) Subcellular distribution of dietary beta-carotene in chick liver. Lipids 21:164–169
McGraw KJ, Ardia DR (2003) Carotenoids, immunocompetence, and the information content of sexual colors: an experimental test. Am Nat 162:704–712
Metcalfe NB, Alonso-Alvarez C (2010) Oxidative stress as a life-history constraint: the role of reactive oxygen species in shaping phenotypes from conception to death. Funct Ecol 24:984–996
Miksik I, Holan V, Deyl Z (1996) Avian eggshell pigments and their variability. Comp Biochem Physiol Part B 113:607–612

Møller AP, Pomiankoski A (1993) Why have birds got multiple sexual ornaments? Behav Ecol Sociobiol 32:167–176

Møller AP, Biard C, Blount JD, Houston DC, Ninni P, Saino N, Surai PF (2000) Carotenoid-dependent signals: indicators of foraging efficiency, immunocompetence or detoxification ability? Avian Poult Biol Rev 11:137–159

Monaghan P, Heidinger BJ, D'Alba L, Evans NP, Spencer KA (2012) For better or worse: reduced adult lifespan following early-life stress is transmitted to breeding partners. Proc R Soc Lond B 279:709–714

Morales J, Velando A (2013) Signals in family conflicts. Anim Behav 86:11–16

Morales J, Velando A, Moreno J (2008) Pigment allocation to eggs decreases plasma antioxidants in a songbird. Behav Ecol Sociobiol 63:227–233

Morales J, Velando A, Torres R (2011) Biliverdin-based egg coloration is enhanced by carotenoid supplementation. Behav Ecol Sociobiol 65:197–203

Moreno J, Osorno JL (2003) Avian egg colour and sexual selection: does eggshell pigmentation reflect female condition and genetic quality? Ecol Lett 6:803–806

Moreno J, Velando A, Ruiz-de-Castañeda R, Cantarero A, González-Braojos S, Redondo A (2011) Plasma antioxidant capacity and oxidative damage in relation to male plumage ornamental traits in a montane Iberian pied flycatcher *Ficedula hypoleuca* population. Acta Ornithol 46:65–70

Moreno J, Velando A, Ruiz-de-Castañeda R, González-Braojos S, Cantarero A (2013) Oxidative damage in relation to a female plumage badge: evidence for signalling costs. Acta Ethol 16:65–75

Mougeot F, Martínez-Padilla J, Webster LMI, Blount JD, Pérez-Rodríguez L, Piertney SB (2009) Honest sexual signalling mediated by parasite and testosterone effects on oxidative balance. Proc R Soc Lond B 276:1093–1100

Mundy NI, Stapley J, Bennison C, Tucker R, Twyman H, Kim KW, Burke T, Birkhead TR, Andersson S, Slate J (2016) Red carotenoid coloration in the zebra finch is controlled by a cytochrome P450 gene cluster. Curr Biol 26:1435–1440

Nishida R (2002) Sequestration of defensive substances from plants by Lepidoptera. Annu Rev Entomol 47:57–92

Nowak J, Pawłowski B, Borkowska B, Augustyniak D, Drulis-Kawa Z (2018) No evidence for the immunocompetence handicap hypothesis in male humans. Sci Rep 8:7392

O'Connor RJ (1978) Brood reduction in birds: selection for fratricide, infanticide and suicide? Anim Behav 26:79–96

Olson VA, Owens IPF (2005) Interspecific variation in the use of carotenoid-based coloration in birds: diet, life history and phylogeny. J Evol Biol 18:1534–1546

Olsson M, Wilson M, Isaksson C, Uller T, Mott B (2008) Carotenoid intake does not mediate a relationship between reactive oxygen species and bright coloration: an experimental test in a lizard. J Exp Biol 211:1257–1261

Olsson M, Wilson M, Isaksson C, Uller T (2009) Polymorphic ROS scavenging revealed by CCCP in a lizard. Naturwissenschaften 96:845–849

Ozeki H, Ito S, Wakamatsu K, Ishiguro I (1997) Chemical characterization of pheomelanogenesis starting from dihydroxyphenylalanine or tyrosine and cysteine. Effects of tyrosinase and cysteine concentrations and reaction time. Biochim Biophys Acta 1336:539–548

Palozza P (1998) Prooxidant actions of carotenoids in biological systems. Nutr Rev 56:257–265

Pérez C, Lores M, Velando A (2008) Availability of nonpigmentary antioxidant affects red coloration in gulls. Behav Ecol 19:967–973

Peters A (2007) Testosterone and carotenoids: an integrated view of trade-offs between immunity and sexual signalling. Bioessays 29:427–430

Pfaff DW, Arnold AP, Etgen AM, Fahrbach SE, Rubin RT (2002) Hormones, brain and behavior. Academic Press, San Diego

Pike TW, Blount JD, Lindström J, Metcalfe NB (2007a) Availability of non-carotenoid antioxidants affects the expression of a carotenoid-based sexual ornament. Biol Lett 3:353–356

Pike TW, Blount JD, Lindstrom J, Metcalfe NB (2007b) Dietary carotenoid availability influences a male's ability to provide parental care. Behav Ecol 18:1100–1105

Pike TW, Blount JD, Bjerkeng B, Lindström J, Metcalfe NB (2007c) Carotenoids, oxidative stress and female mating preference for longer lived males. Proc R Soc Lond B 274:1591–1596

Pilas B, Sarna T, Kalyanaraman B, Swartz HM (1988) The effect of melanin on iron associated decomposition of hydrogen peroxide. Free Radic Biol Med 4:285–293

Poiani A, Goldsmith AR, Evans MR (2000) Ectoparasites of house sparrows (*Passer domesticus*): an experimental test of the immunocompetence handicap hypothesis and a new model. Behav Ecol Sociobiol 47:230–242

Potti J, Montalvo S (1991) Male arrival and female mate choice in pied flycatchers *Ficedula hypoleuca* in Central Spain. Ornis Scand 22:45–54

Qvarnström A, Brommer JE, Gustafsson L (2006) Testing the genetics underlying the co-evolution of mate choice and ornament in the wild. Nature 441:84–86

Reynolds SJ, Martin GR, Cassey P (2009) Is sexual selection blurring the functional significance of eggshell coloration hypotheses? Anim Behav 78:209–215

Rivoira MA, Marchionatti AM, Centeno VA, Díaz de Barboza GE, Peralta López ME, Tolosa de Talamoni NG (2012) Sodium deoxycholate inhibits chick duodenal calcium absorption through oxidative stress and apoptosis. Comp Biochem Physiol Part A 162:397–405

Roberts ML, Buchanan KL, Evans MR (2004) Testing the immunocompetence handicap hypothesis: a review of the evidence. Anim Behav 68:227–239

Rodd FH, Hughes KA, Grether GF, Baril CT (2002) A possible non-sexual origin of mate preference: are male guppies mimicking fruit? Proc R Soc Lond B 269:475–481

Rodríguez-Martínez S, Galván I (2019) Exposure to a competitive social environment activates an epigenetic mechanism that limits pheomelanin synthesis in zebra finches. Mol Ecol 28:3698–3708

Rodríguez-Martínez S, Galván I (2020) A source of exogenous oxidative stress improves oxidative status and favors pheomelanin synthesis in zebra finches. Comp Biochem Physiol Part C 228:108667

Rooker K, Gavrilets S (2018) On the evolution of visual female sexual signalling. Proc R Soc Lond B 285:20172875

Rosvall KA, Bergeon Burns CM, Barske J, Goodson JL, Schlinger BA, Sengelaub DR, Ketterson ED (2012) Neural sensitivity to sex steroids predicts individual differences in aggression: implications for behavioural evolution. Proc R Soc Lond B 279:3547–3555

Roulin A, Almasi B, Meichtry-Stier KS, Jenni L (2011) Eumelanin- and pheomelanin-based colour advertise resistance to oxidative stress in opposite ways. J Evol Biol 24:2241–2247

Saetre GP, Moum T, Bures S, Kral M, Adamjan M, Moreno J (1997) A sexually selected character displacement in flycatchers reinforces premating isolation. Nature 387:589–592

Saino N, Ninni P, Calza S, Martinelli R, De Bernardi F, Møller AP (2000) Better red than dead: carotenoid-based mouth colouration reveals infection in barn swallow nestlings. Proc R Soc Lond B 267:57–61

Saino N, Ambrosini R, Martinelli R, Ninni P, Møller AP (2003) Gape coloration reliably reflects immunocompetence of barn swallow (*Hirundo rustica*) nestlings. Behav Ecol 14:16–22

Saporito RA, Spande TF, Garraffo HM, Donnelly MA (2009) Arthropod alkaloids in poison frogs: a review of the "dietary hypothesis". Heterocycles 79:277–297

Searcy WA, Nowicki S (2005) The evolution of animal communication. Princeton University Press, Princeton, NJ

Senar JC, Camerino M, Copete JL, Metcalfe NB (1993) Variation in black bib of the Eurasian siskin (*Carduelis spinus*) and its role as a reliable badge of dominance. Auk 110:924–927

Shan Y, Pepe J, Lu TH, Elbirt KK, Lambrecht RW, Bonkovsky HL (2000) Induction of the heme oxygenase-1 gene by metalloporphyrins. Arch Biochem Biophys 380:219–227

Sharp PJ, Armstrong DG, Moss R (1986) Changes in aromatase activity in the neuroendocrine tissues of red grouse (*Lagopus lagopus scoticus*) in relation to the development of long-day refractoriness. J Endocrinol 108:129–135

Silla AJ, McInerney EP, Byrne PG (2016) Dietary carotenoid supplementation improves the escape performance of the southern corroboree frog. Anim Behav 112:213–220

Simons MJP (2013) Sexual coloration and aging. Ph.D. Thesis, University of Groningen, Groningen, The Netherlands

Simons MJP, Cohen AA, Verhulst S (2012) What does carotenoid-dependent coloration tell? Plasma carotenoid level signals immunocompetence and oxidative stress state in birds–a meta-analysis. PLoS One 7:e43088

Solomon SE (1997) Egg and eggshell quality. Iowa State University Press, Ames, IA

Soma KK, Bindra RK, Gee J, Wingfield JC, Schlinger BA (1999) Androgen-metabolizing enzymes show region-specific changes across the breeding season in the brain of a wild songbird. J Neurobiol 41:176–188

Sternalski A, Mougeot F, Pérez-Rodriguez L, Bretagnolle V (2012) Carotenoid-based coloration, condition, and immune responsiveness in the nestlings of a sexually dimorphic bird of prey. Physiol Biochem Zool 85:364–375

Stevens M, Ruxton GD (2012) Linking the evolution and form of warning coloration in nature. Proc R Soc Lond B 279:417–426

Stewart IRK, Westneat DF (2013) Dietary calcium, but not a glutathione inhibitor, affects bib size in juvenile male house sparrows. Condor 115:921–930

Stradi R, Celentano G, Boles M, Mercato F (1997) Carotenoids in bird plumage: the pattern in a series of red-pigmented Carduelinae. Comp Biochem Physiol Part B 117:85–91

Surai P (2002) Natural antioxidants in avian nutrition and reproduction. Nottingham University Press, Nottingham

Svensson PA, Wong BBM (2011) Carotenoid-based signals in behavioural ecology: a review. Behaviour 148:131–189

Taff CC, Freeman-Gallant CR (2014) An experimental test of the testosterone mediated oxidation handicap hypothesis in a wild bird. Horm Behav 66:276–282

Tomášek O, Gabrielová B, Kačer P, Maršík P, Svobodová J, Syslová K, Vinkler M, Albrecht T (2016) Opposing effects of oxidative challenge and carotenoids on antioxidant status and condition-dependent sexual signalling. Sci Rep 6:23546

Toomey MB, Marques CI, Araújo PM, Huang D, Zhong S, Liu Y, Schreiner GD, Myers CA, Pereira P, Afonso S, Andrade P, Gazda MA, Lopes RJ, Viegas I, Koch RE, Haynes ME, Smith DJ, Ogawa Y, Murphy D, Kopec RE, Parichy DM, Carneiro M, Corbo JC (2022) A mechanism for red coloration in vertebrates. Curr Biol 32:4201–4214

Toral GM, Figuerola J, Negro JJ (2008) Multiple ways to become red: pigment identification in red feathers using spectrometry. Comp Biochem Physiol Part B 150:147–152

Török J, Hegyi G, Garamszegi LZ (2003) Depigmented wing patch size is condition dependent indicator of viability in male collared flycatchers. Behav Ecol 14:382–388

Torres R, Velando A (2007) Male reproductive senescence: the price of immune-induced oxidative damage on sexual attractiveness in the blue-footed boobies. J Anim Ecol 76:1161–1168

Trivers R (1974) Parent-offspring conflict. Am Zool 14:249–264

Trumble BC, Blackwell AD, Stieglitz J, Thompson ME, Suarez IM, Kaplan H, Gurven M (2016) Associations between male testosterone and immune function in a pathogenically stressed forager-horticultural population. Am J Phys Anthropol 161:494–505

Tschirren B, Fitze PS, Richner H (2005) Carotenoid-based nestling colouration and parental favouritism in the great tit. Oecologia 143:477–482

Tyczkowski JK, Schaeffer JL, Hamilton PB (1991) Measurement of malabsorption of carotenoids in chickens with pale-bird syndrome. Poult Sci 70:2275–2279

van Helden YG, Keijer J, Knaapen AM, Heil SG, Briedé JJ, van Schooten FJ, Godschalk RW (2009) Beta-carotene metabolites enhance inflammation-induced oxidative DNA damage in lung epithelial cells. Free Radic Biol Med 46:299–304

Van Hout AJ-M, Eens M, Pinxten R (2011) Carotenoid supplementation positively affects the expression of a non-visual sexual signal. PLoS One 6:e16326

Vitousek MN, Stewart RA, Safran RJ (2013) Female plumage colour influences seasonal oxidative damage and testosterone profiles in a songbird. Biol Lett 9:20130539

von Schantz T, Bensch S, Grahn M, Hasselquist D, Wittzell H (1999) Good genes, oxidative stress and condition-dependent sexual signals. Proc R Soc Lond B 266:1–12

Wang Y-J, Chien Y-H, Pan C-H (2006) Effects of dietary supplementation of carotenoids on survival, growth, pigmentation, and antioxidant capacity of characins, *Hyphessobrycon callistus*. Aquacult 261:641–648

Weaver RJ, Koch RE, Hill GE (2017) What maintains signal honesty in animal colour displays used in mate choice? Philos Trans R Soc Lond B 372:20160343
Weaver RJ, Santos ESA, Tucker AM, Wilson AE, Hill GE (2018) Carotenoid metabolism strengthens the link between feather coloration and individual quality. Nat Commun 2018:73
Williams CG (1966) Natural selection, the costs of reproduction, and a refinement of Lack's principle. Am Nat 100:687–690
Williams M, Krootjes BB, van Steveninck J, van der Zee J (1994) The pro- and antioxidant properties of protoporphyrin IX. Biochim Biophys Acta 1211:310–316
Wingfield JC, Lynn SE, Soma KK (2001) Avoiding the "costs" of testosterone: ecological bases of hormone-behavior interactions. Brain Behav Evol 57:239–251
With TK (1973) Porphyrins in egg shells. Biochem J 137:597–598
Wood JM, Decker H, Hartmann H, Chavan B, Rokos H, Spencer JD, Hasse S, Thornton MJ, Shalbaf M, Paus R, Schallreuter KU (2009) Senile hair graying: H2O2-mediated oxidative stress affects human hair color by blunting methionine sulfoxide repair. FASEB J 23:2065–2075
Young AJ, Lowe GM (2001) Antioxidant and prooxidant properties of carotenoids. Arch Biochem Biop 385:20–27
Zahavi A (1975) Mate selection: a selection for a handicap. J Theor Biol 53:205–214
Zahavi A, Zahavi A (1997) The handicap principle: a missing piece of Darwin's puzzle. Oxford University Press, Oxford
Zaręba M, Bober A, Korytowski W, Zecca L, Sarna T (1995) The effect of a synthetic neuromelanin on yield of free hydroxyl radicals generated in model systems. Biochim Biophys Acta 1271:343–348
Zhang J, Hou X, Ahmad H, Zhang H, Zhang L, Wang T (2014) Assessment of free radicals scavenging activity of seven natural pigments and protective effects in AAPH-challenged chicken erythrocytes. Food Chem 145:57–65
Zuk M, Ligon JD, Thornhill R (1992) Effects of experimental manipulations of male secondary sex characters on mate preference in red jungle fowl. Anim Behav 44:999–1006

# Chapter 9
# Oxidative Stress and Reproductive Trade-Offs: From Courtship to Parental Care

**Abstract** There has been considerable work to test the role of organismal oxidative stress as a mediator of reproductive trade-offs. Intense research focussed on whether parental effort in offspring rearing results in oxidative stress. Many other studies carried out comparisons between non-reproducing and reproducing individuals in a narrow range of taxa. By contrast, we have a very limited number of studies that tested whether pre-breeding oxidative stress constraints reproductive decisions and quality of parental care (and reproductive success as a consequence). In this chapter, I have examined how the need to manage oxidative stress has possibly influenced the evolution of reproductive strategies, from courtship to parental care. I have also illustrated (1) the impact of oxidative stress on fertility, (2) differences between sexes in the regulation of oxidative status homeostasis, and (3) oxidative benefits and costs of cooperative breeding. Finally, I have highlighted a number of weaknesses in experimental designs that have been used to test the oxidative cost of reproduction hypothesis, calling for novel and more realistic studies.

## 9.1 Introduction

Reproduction is a fundamental but demanding phase of animals' lives. These demands at their most basic physiological level involve the consumption of energy and nutrients that parents need to acquire to successfully generate and, in some species, rear offspring (e.g. Drent and Daan 1980; Masman et al. 1986; Hammond 1997; Angilletta and Sears 2000; Nilsson 2002; McGrosky and Pontzer 2023). In sexually reproducing species, animals also invest a considerable amount of resources in producing attractive ornaments and displays (see Chap. 8) or fighting to obtain mates, which also contribute to determine the cost of reproduction.

The cost of reproduction is a central paradigm of life-history theory (Bell 1980; Stearns 1992; Roff 2002). This paradigm is based on the premise that organisms face relevant trade-offs during reproduction. It predicts that a large investment of resources (e.g. nutrients, energy, and time) in current reproduction results in less

© The Author(s), under exclusive license to Springer Nature Switzerland AG 2024

D. Costantini, *The Role of Organismal Oxidative Stress in the Ecology and Life-History Evolution of Animals*, Fascinating Life Sciences,
https://doi.org/10.1007/978-3-031-65183-0_9

resources available for maintenance of organismal homeostasis, which would reduce probabilities of survival or future investment into reproduction (Williams 1966; Stearns 1992). It is also predicted that pre-reproductive conditions that elicit stress or reduce organismal energetic stores might constrain the reproductive decisions (when to reproduce and how many offspring to generate) and success. For example, experimental enlargements or reductions of the brood size of common kestrels *Falco tinnunculus* increased or decreased the energy expenditure of parents, respectively (Deerenberg et al. 1995). More than half of the parents that raised two extra nestlings (brood size enlargement) died before the end of the first winter compared to 29% of those parents that raised control or reduced broods (Daan et al. 1996). Costs of reproduction can also emerge later in life. For example, common guillemots *Uria aalge* that raised more chicks at an early stage of their lives had shorter lifespans and produced later-in-life offspring with reduced probabilities of survival, compared to guillemots that raised fewer chicks at an early stage of their lives (Reed et al. 2008).

We have learned over time that a number of factors can influence the trade-off between reproduction and survival, such as pace of life, food availability, age, or sex (e.g. Snell and King 1977; Attisano et al. 2012; Santos and Nakagawa 2012; Rughetti et al. 2015). For example, there are species that reproduce once during their lifetime (semelparous), putting all available energy into a single reproductive event with death of the individual after reproduction (e.g. mayflies and Pacific salmon), or reproduce multiple times throughout their lifetime (iteroparous), so that they adjust their reproductive effort in relation to their condition and environmental circumstances (Stearns 1992; Roff 2002; Stephens et al. 2009).

The question then is how costs of reproduction are actually incurred, since the majority of studies have focussed on the ultimate outcomes rather than on the proximate mechanisms. At a deeper level, we have a limited understanding of the mechanisms underlying such trade-offs. It has been suggested that oxidative stress might be one key cellular mechanism that underlies reproductive trade-offs (*oxidative cost of reproduction hypothesis*; e.g. Costantini 2008; Metcalfe and Alonso-Alvarez 2010; Metcalfe and Monaghan 2013). This hypothesis is based on the prediction that a large investment into reproduction would increase oxidative damage and/or perturb antioxidant protection, resulting in faster somatic deterioration, impairment of organismal functions, and reduced investment into future reproduction or shorter life expectancy since resources allocated to reproduction are no longer available for self-maintenance.

In this chapter, I have examined how the need to manage oxidative stress has likely contributed to generate diversity in reproductive strategies, from the range of mating tactics to parental care. I have discussed the relevance of oxidative stress as a constraint on and cost of reproductive effort, and the possible role of hormesis as an analytical framework in the interpretation of the link between reproduction and oxidative stress. I have also examined the role of oxidative stress in the contexts of fertility, sex differentiation, and evolution of cooperative breeding. Finally, I have illustrated a number of weaknesses in experimental designs that have been used to test the oxidative cost of reproduction hypothesis, calling for novel and more realistic studies.

## 9.2 In the Beginning: The Costs of Courtship and Social Hierarchy

Courtship display is a special set of behaviours that males of some species have evolved as part of their strategy to attract a mate. Courtship behaviours can involve special calls, postures, movements, the expression of extravagant ornaments and bright colourations (see Chap. 8), and intense fighting between rivals. Thus, there are a number of potential routes through which performing courtship displays is costly for the individual, such as increased metabolic rate and energy expenditure, reduced food intake, injuries and tissue inflammation, and development of ornaments that take resources away from endogenous functions that maintain organismal homeostasis. All of these routes can potentially pave the way for oxidative stress.

There are many relevant examples in nature, where the physical effort of courtship may be important. Manakins are renowned for their elaborate courtship. Manakins perform some of the most physically demanding courtship dances in the animal kingdom, including acrobatic routines and mechanically produced sounds (Fusani et al. 2007; Barske et al. 2011). Display performance might be an integrative indicator of male quality, including sensorimotor coordination, physiological condition, and development history (Fusani et al. 2007; Barske et al. 2011). Manakins offer an exciting model to address questions related to the role of oxidative stress as a mediator of the links between skeletal muscle contraction, heart activity, courtship, and reproductive fitness. For example, oxidative stress might affect the muscle performance and the motor skills of the male, and dance-induced oxidative stress might carry over affecting sperm quality and fertility. A comparison of oxidative status markers between females and males of the red-headed manakin *Ceratopipra rubrocapilla* and of the white-crowned manakin *Pseudopipra pipra* found that males had higher levels of lipid peroxidation than females, supporting the hypothesis that the special courtship display of manakins might expose males to high oxidative stress (Gomes et al. 2019).

There are many other examples of courtship activities that require substantial investments of resources, such as the vocalisations in birds or in frogs, or the leg-waving behaviour in spiders (Byers et al. 2010). Acoustic signals are condition-dependent in a variety of invertebrates and vertebrates, and advertise to conspecifics the social or mating status of the singer (e.g. Hunt et al. 2004; Ball et al. 2006; Humfeld 2013). Avian song is a renowned acoustic trait that may convey several attributes of individual quality, including dominance and territory ownership (e.g. Duffy and Ball 2002; Gil and Gahr 2002; MacDougall-Shackleton et al. 2009; Kelm-Nelson et al. 2011; Ritschard and Brumm 2012; DeVries et al. 2016). Several studies identified a number of factors that can constrain the expression of song behaviour, including experimental manipulations of immunological reactions (Garamszegi et al. 2004; Casagrande et al. 2015; Costantini et al. 2015a; York et al. 2016) or of testosterone levels (De Ridder et al. 2000; Van Hout et al. 2012). Correlative studies also suggested that the expression of song rate might be linked to cellular oxidative status (Baldo et al. 2015; Casagrande et al. 2015, 2016;

Costantini et al. 2015a). Experimental induction of oxidative stress enabled to reveal an effect of oxidative stress on song production, particularly in subordinate individuals (Messina et al. 2017), providing the first experimental evidence that oxidative stress might be causally linked to the expression of condition-dependent acoustic signals (Fig. 9.1). Specifically, Messina et al. (2017) experimentally inhibited the synthesis of the key cellular antioxidant glutathione of wild-caught male common starlings *Sturnus vulgaris* that were housed in outdoor aviaries. As a consequence of the inhibition of glutathione synthesis, the levels of plasma oxidative damage increased and the song rate decreased, respectively. These effects were, however, evident only in subordinate individuals, pointing to a role of individual quality or condition as expected by the honest signalling hypothesis. To date, there is no experimental evidence that manipulation of song rate increases oxidative stress.

Costs of courtship can also come through strong fights between males. Fights to maintain a territory and defend or acquire mates are very common in animals, from beetles and lizards to lekking birds and mammals. Oxidative stress may arise from the intense physical effort required for fighting, and the associated reduced food intake and activation of immune and inflammatory processes caused by injuries. Another source of oxidative stress might lie in the increased levels of androgens that boost metabolic rate, aggressiveness, and tolerance of injuries. The quantification of oxidative stress in these species may inform behavioural ecologists about the proximate mechanisms underlying the evolution of mating strategies.

Finally, we have also evidence that oxidative stress may impair chemical communication. Relying on a superoxide dismutase knockout mouse, Garratt et al. (2014) found that oxidative stress may directly impair investment in morphological (preputial glands) and molecular (major urinary proteins) components of olfactory signalling essential for mate attraction. Males of the yellow mealworm beetle *Tenebrio molitor* use pheromones to attract mating partners, and the level of pheromone production is condition-dependent in this particular species. Ruiz-Guzmán et al. (2021) found that females preferred healthy males to those that were experiencing oxidative stress experimentally induced by paraquat. Although males are the main target of research on the costs of courtship, we have many examples of species where females compete with each other to acquire males (e.g. Gwynne 1981; Berglund et al. 1986; Alcock 2009). However, we have a limited understanding of the role of oxidative stress as a mediator of courtship strategies of females.

The results of these studies also suggest that in some species the social hierarchy might have a strong influence on the costs of courtship and mating. There are many benefits of being dominant, such as priority of access to partners and food. However, there can also be severe costs to attaining and maintaining high social status. Subordinate individuals might be unable of tolerating the energetic or other metabolic costs that the maintenance of a dominant position may impose. On the other hand, avoidance of these costs might counterbalance the benefits of having more mating opportunities if subordinates live longer than dominants and have future access to mates, or fertility of dominant individuals is reduced in the long-term because of these physiological costs. We have a growing number of studies that are documenting the nature of these costs.

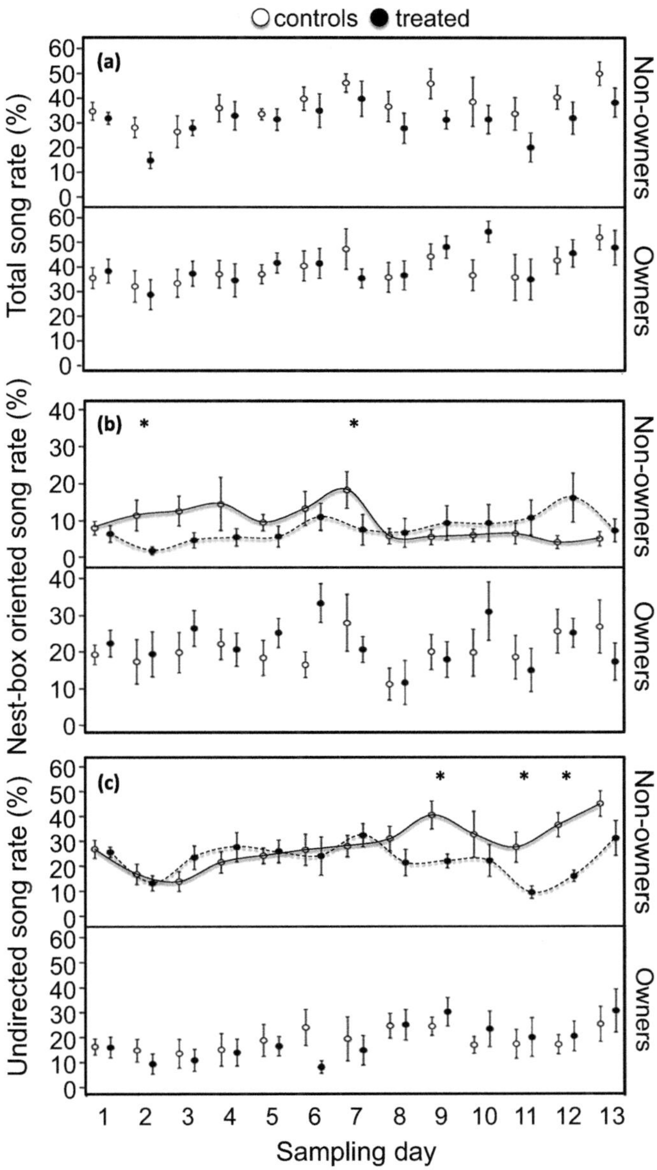


**Fig. 9.1** Temporal trends of (**a**) total song rate, (**b**) nest box-oriented song rate, and (**c**) undirected song rate throughout the experiment in relation to treatment group (C, control; T, treated) and nest-box ownership (N, non-owner; Y, owner) in male European starlings *Sturnus vulgaris*. Pre-treatment (i.e. sampling day 1) values of each song metric are expressed as the average of the values recorded over the 10 days before the start of the experiment. Asterisks indicate significant differences between C and T groups by sampling day. Values are shown as mean and standard error. Reprinted with minor modifications from Messina et al. (2017) with permission of John Wiley and Sons

Subordinate baboons can form friendships with females and other males in order to have some chances of mating and to avoid fights with rivals, respectively (e.g. Palombit et al. 1997; Alberts et al. 2003). Similarly, in many other animal species, weaker males do better by switching to alternative mating tactics rather than trying to compete with rivals. For example, satellite males or sneaker males patrol around until they come across females that are not being guarded by the dominant males. They take this opportunity to mate avoiding the attack of the dominant. At a proximate level, this carpe diem strategy might pay in terms of reduced physiological costs because a subordinate male would not experience increased production of high testosterone, intense physical effort, and inflammation. For example, in polymorphic species, males differ in many ways, including morphology, colouration, life-history, or behaviour. In ruffs *Philomachus pugnax*, we can recognise at least three male morphs. Two of these morphs, called independents and satellites, have elaborate courtship behaviours and plumages; the third morph looks like a female, and is called faeder or female mimic (Lozano et al. 2013). During the breeding season, independents establish lek mating courts and defend them against other independents. Satellites do not fight but are pecked and chased by independents if they fail to behave submissively; by contrast, faeders are not attacked by independents (Lozano et al. 2013). If maintenance of the independent status is costly, we might expect that they would suffer more oxidative stress than the other morphs. Lozano et al. (2013) found that both the innate and cell-mediated immunity decreased from independent to females, supporting a risk-of-injury explanation. By contrast, male morphs had similar plasma levels of oxidative damage and antioxidants (Lozano et al. 2013). By contrast, in the polymorphic Australian-painted dragon *Ctenophorus pictus*, males of different morphs (head colour) differed in their oxidative status. For example, they produced different amounts of the superoxide free radical (Friesen et al. 2017). Similarly, in the polymorphic Gouldian finch *Erythrura gouldiae*, the oxidative status of birds differed between morphs (head colour) when they were experimentally exposed to different energetically demanding conditions, with individuals of the dominant morph being less affected than individuals of the subordinate morph (Fragueira et al. 2019).

Costs of acquiring or maintaining a social rank may vary across seasons. Outside the mating season, high-ranking free-ranging male mandrills *Mandrillus sphinx* had the lowest levels of reactive oxygen metabolites in plasma (marker of oxidative damage), whilst during the mating season, they were the only males to experience increased oxidative damage (Beaulieu et al. 2014). By contrast, oxidative damage increased during the mating season in all the females, irrespective of their social rank (Beaulieu et al. 2014). Similarly, outside the mating season, high-ranking free-ranging male rhesus macaques *Macaca mulatta* had lower plasma levels of thiobarbituric acid reactive substances than low-ranking males (Georgiev et al. 2015), and the experimental reduction of glutathione synthesis did not affect plasma levels of reactive oxygen metabolites in dominant common starlings, whilst it increased oxidative damage in subordinate common starlings (Messina et al. 2017). To some extent, seasonal hormonal changes might explain this variation in oxidative status. High testosterone was associated with lower oxidative damage in both dominants and subordinates of the cichlid fish *Astatotilapia burtoni*; by contrast, high levels of cortisol were associated with lower oxidative damage in dominants and higher oxidative damage in subordinates, respectively (Culbert et al. 2023).

Other studies found evidence for vulnerability of dominant individuals to oxidative stress. Male cheetahs *Acinonyx jubatus* exhibit two distinct spatial tactics: they can defend a small territory (territorial individuals) or roam over large areas (floaters) encompassing several territories of territory holders. Costantini et al. (2017) found that territory holders had significantly higher serum levels of protein carbonyls and lower levels of superoxide dismutase in erythrocytes than floaters, indicating that dominant cheetahs might experience higher oxidative stress than floaters (Fig. 9.2). In fish, Border et al. (2019) found that dominant males of the cichlid fish *Astatotilapia burtoni* experienced higher concentrations of circulating reactive oxygen metabolites without notably increasing their antioxidant defences. In particular,

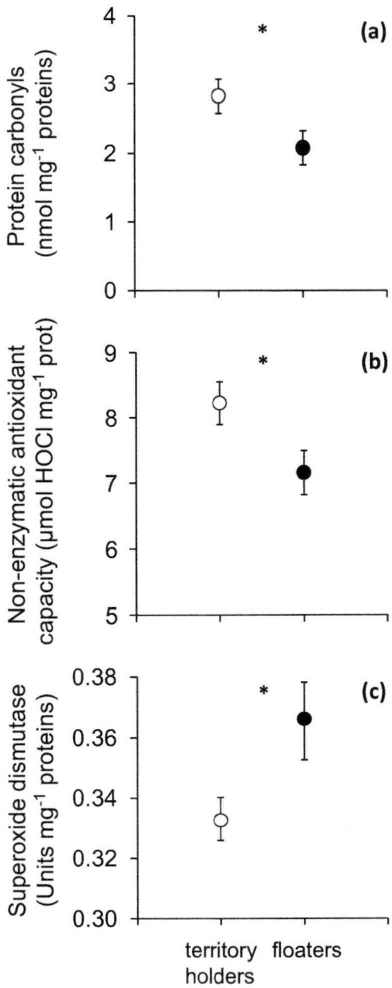


**Fig. 9.2** Variation in oxidative status markers between dominant (territory holders) and floater (non-territorial) male cheetahs *Acinonyx jubatus*. Dominant cheetahs had (**a**) higher serum protein carbonyls, (**b**) higher serum non-enzymatic micro-molecular antioxidant capacity, and (**c**) lower levels of superoxide dismutase than floater males. Asterisks indicate significant differences between dominants and floaters. Values are shown as mean and standard error. Reprinted with minor modifications from Costantini et al. (2017). © The Author 2017, published by Oxford University Press and the Society for Experimental Biology; distributed under the terms of the Creative Commons Attribution License (CC BY)

yellow but not blue males showed signs of increased oxidative damage with increasing agonistic effort. Fialkowski et al. (2021) also found that, after removing the dominant male, males of the cichlid fish *Astatotilapia burtoni* that ascended from subordinate to dominant status experienced changes in their antioxidant and oxidative damage levels. Conversely to males, dominant females did not have higher levels of oxidative damage compared to same-sex subordinates, suggesting that associations between rank and oxidative stress were also probably driven by investment in other reproductive traits beyond aggressiveness expressed in agonistic interactions (Funnell et al. 2022).

Oxidative costs of dominance were not documented only in vertebrates. In insects, for example the visual exposure to a mating competitor promoted a higher calling effort in male field crickets *Gryllus bimaculatus*, and this effort resulted in a decline in the haemolymph antioxidant content (Noguera 2018). The haemolymph antioxidant content was also negatively covaried with the antioxidant content of the sperm, but interestingly, only when the males were exposed to a competitor. These results suggested that when mating competition is high, males might prioritise antioxidant protection of sperm against somatic maintenance. Moreover, these males that, when exposed to a mating competitor, lost a significant proportion of body antioxidant defences and body mass over time, had also a reduced lifespan (Noguera 2018).

The hierarchical position is not only an issue for males. Females can also experience oxidative costs of being dominant. Silva et al. (2018) found that in females, but not in males, of the sociable weaver *Philetairus socius*, dominance was associated with higher plasma levels of reactive oxygen metabolites than in same-sex subordinates, suggesting that the physiological cost of dominance is underpinned by oxidative stress in a sex-specific manner. This sex difference in oxidative damage was associated with sex-specific differences in antioxidant defences, with males (the dominant sex) showing higher antioxidant levels than females, independently of their social rank.

Finally, we have evidence that the hierarchical position of an individual may affect the oxidative status of the partner. For example, early in the rearing period, female Western jackdaws *Coloeus monedula* mated to subdominant males had higher oxidative damage than females mated to dominant males (Salomons 2009). This difference disappeared later during the period of chick rearing. One reason for this result might lie with the male quality. Early in the breeding season, the female is largely dependent on the partner for food provisioning. Thus, subordinate individuals might not be particularly good at providing food to their mates, deteriorating their physiological condition as a result.

## 9.3 (In)Fertility

It is increasingly recognised that oxidative stress is harmful to sperm and can reduce male fertility as a result (Surai et al. 1997; Sikka 2001). In humans and domesticated animals, oxidative stress to sperm can cause decreased motility and viability, an inability to fuse with the oocyte, and decreased embryo survival (e.g. Fujihara and Howarth 1978; Wishart 1984; Aitken et al. 1989; de Lamirande and Gagnon 1992, 1993; Baumber et al. 2000; Bilodeau et al. 2002). Moreover, oxidative stress may also reduce the capacity of testes to differentiate normal spermatozoa, which would ultimately result in reduced levels of sperm with good fertilising ability (Wu et al. 1973; Aitken and Roman 2008). Studies on free-living birds and mammals corroborate the results obtained from studies on captive animals. Oxidative stress contributed to reduce the quality of sperm in animals living under various highly demanding conditions, such as heavy metal exposure (Reglero et al. 2009), parental care (Helfenstein et al. 2010), or radioactive contamination (Bonisoli-Alquati et al. 2011).

The results of these studies point to oxidative stress as a potential mechanism that affects post-copulatory sexual selection, provided that females can filter sperm that, owing to its oxidative stress, is poor at fertilisation or at building a viable zygote. It is, therefore, possible that oxidative stress of sperm has played a relevant role in a number of evolutionary processes, including the evolution of polyandry, sexual conflict over mating, and sperm competition dynamics (e.g. Siva-Jothy 2000; Arenas-Ríos et al. 2005; Dean et al. 2007; Pizzari et al. 2008; Rojas Mora et al. 2017; Friesen et al. 2020). The jungle fowl *Gallus gallus* is a valuable species to test these hypotheses. Females often mate multiply to force the sperms of different males into competition, and the capacity of males to mate and fertilise successfully declines with age (Dean et al. 2010), possibly because of the accumulation of oxidative damage (Velando et al. 2011; Noguera et al. 2012). Similarly, male houbara bustards *Chlamydotis undulata* invest large amounts of resources in sexual displays and, probably due to their large investment, they experience a rapid deterioration of spermatogenic function (Preston et al. 2011). Investment in the protection of sperm against oxidative damage is therefore essential for the male in order to maintain healthy sperm that will have higher probabilities of winning in the competition for egg fertilisation. To this end, some studies documented that antioxidant levels are particularly high in the testes and in the seminal fluid (e.g. Surai et al. 1997; Sikka 2001; Weirich et al. 2002; Poiani 2006; Lahnsteiner and Mansour 2010; Simmons et al. 2018). For example, carotenoids are present in the ejaculates of birds (Rowe and McGraw 2008) and insects (Heller et al. 2000), hence they might have a direct protective role on sperm. Several studies found that the supplementation of diet with carotenoids or other dietary antioxidants increased sperm quality and/or fertilisation (Gupta and Kumar 2002; Ahmadi et al. 2006; Locatello et al. 2006; Helfenstein et al. 2010; Pike et al. 2010; Tomášek et al. 2017). Supplementation of diet with other antioxidants had also positive effects on sperm. For example, intramuscular injection of dexamethasone (to mimic a stress condition) for a week reduced significantly the ejaculate volume and increased levels of thiobarbituric acid reactive substances in the seminal plasma of dogs (Hatamoto et al. 2006). Oral supplementation of vitamin E prevented some of the negative effects that stress had on sperm: it increased sperm motility, vigour and concentration, and decreased the percentage of major sperm defects and the levels of thiobarbituric acid reactive substances, respectively (Hatamoto et al. 2006). However, there are also studies that found beneficial effects of dietary antioxidants on sperm that were not related to their presumed capacity to protect against lipid peroxidation (Helfenstein et al. 2010), or did not even find any beneficial effects of dietary antioxidants on sperm quality (Sullivan et al. 2014; Rahman et al. 2015; Keogh et al. 2018). Other studies suggested that antioxidant enzymes, like superoxide dismutase, glutathione peroxidase, and catalase, may play a very important role in the regulation of oxidative status during testes involution–recrudescence, epididymal maturation, and prolonged storage of spermatozoa (Arenas-Ríos et al. 2005, 2007).

One reason for these discrepancies between studies might lie in the dual role of reactive oxygen species in sperm physiology. Although reactive oxygen species can impair sperm functions through their pro-oxidant effects, they can also stimulate several functions, such as the ability to achieve and sustain hyperactivation of sperm, and the promotion of capacitation (de Lamirande and Gagnon 1993; Zini et al. 1995; Godeas et al. 1997; de Lamirande et al. 1997). Mild oxidative conditions promote or are even required for binding of spermatozoa to the zona pellucida of the egg and acrosome reaction (de Lamirande et al. 1997). Although these studies were performed in vitro with human sperm, they support the hypothesis that reactive oxygen species can also have positive effects on fertilisation (Costantini et al. 2010b).

Investment in sperm protection varies as a function of male condition, level of competition, or reproductive stage. For example, in the bluegill sunfish *Lepomis macrochirus*, sneaker males, that dart in and out of nests in order to ejaculate between the spawning pair and larger cuckolders can fertilise more eggs than guard males during sperm competition, and are superior to satellite males in sperm competition (Fu et al. 2001). In side-blotched lizards *Uta stansburiana*, reproductive morphs differed in both hormonal and immune traits (Mills et al. 2008), which both can affect the oxidative status. Studies on bats showed that sneaker males produced sperm of higher quality than harem males, and that this higher quality might be due to some extent to lower oxidation processes (Fig. 9.3) (Wesseling et al. 2016; Fasel et al. 2017). Rojas Mora et al. (2017) also found that dominant male house sparrows *Passer domesticus* produced more oxidised and less viable ejaculates than subordinate male house sparrows, indicating a possible fertility penalty for dominant males (Fig. 9.4). However, this fertility cost was not evident when dominant males were compared to males being at the bottom end of the hierarchy (subordinate 3 in Fig. 9.4), probably because of the very low quality of this particular group of subordinate males. This variation between males in the link between oxidative stress and fertility was also associated with body colourations (e.g. Peters et al. 2004; Locatello et al. 2006; Pitcher et al. 2007; Helfenstein et al. 2010; Velando et al. 2011). For example, Locatello et al. (2006) found that male guppies *Poecilia reticulata* with relatively large areas of orange pigmentation had significantly faster and more viable sperm than their less ornamented counterparts, suggesting a possible link between dietary carotenoid intake and sperm quality. Velando et al. (2011) found that old male blue-footed boobies *Sula nebouxii* with a duller foot colour (sexual signal) had higher DNA damage in sperm, whilst foot colour and damage were not associated with middle-age males.

The high vulnerability of sperm to oxidative stress is potentially relevant because it may prevent eggs from being fertilised by more than one sperm (polyspermy; Boldt et al. 1981; Coburn et al. 1981), and mediate patterns of sperm competition (Costantini

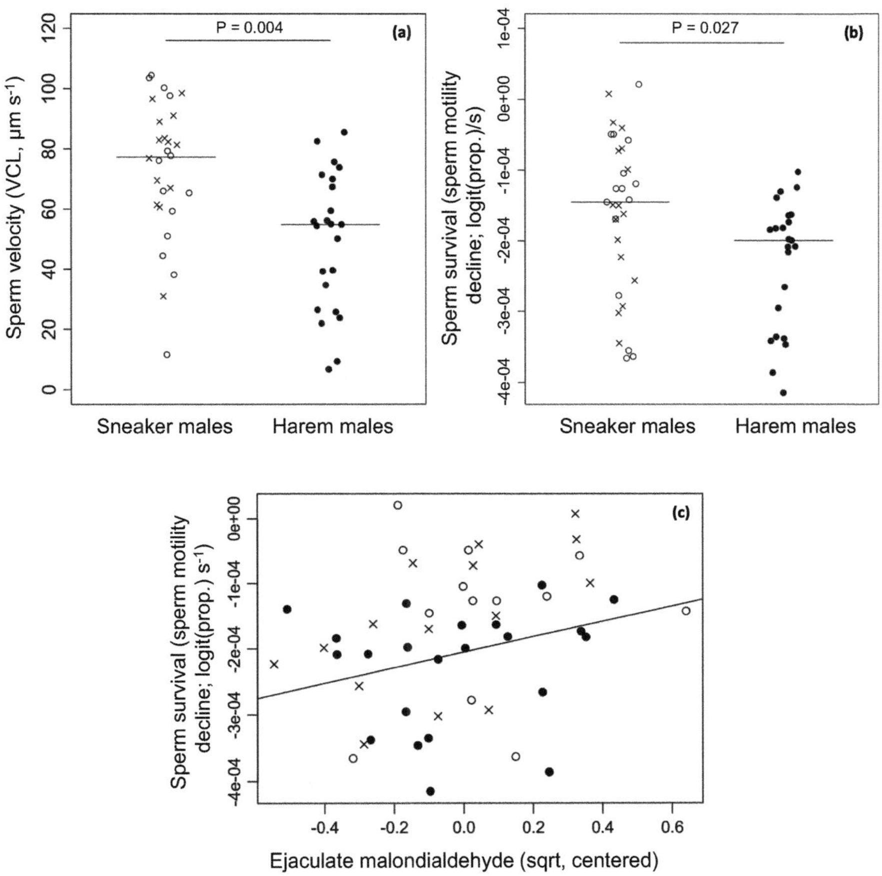


**Fig. 9.3** Sperm quality and oxidative stress in Seba's short-tailed bats *Carollia perspicillata*. (**a**) Sperm velocity (VCL, curvilinear velocity) and (**b**) sperm survival of harem (filled circles) and sneaker males (empty circles for bachelors and crosses for peripheral males); the bars show the medians. (**c**) Correlation between ejaculate malondialdehyde and sperm survival for harem (filled circles) and sneaker males (empty circles for bachelors and crosses for peripheral males). Reprinted with minor modifications from Fasel et al. (2017) with permission of Springer Nature

et al. 2010b). In those species that have evolved sperm competition, males that are better at defending sperm against oxidative stress might gain a greater share of paternity. As a result, sexual selection may target those cellular mechanisms that enable sperm to avoid oxidative stress (Costantini et al. 2010b). Although these hypotheses remain largely unexplored, a comparative study found that the protamine 2 gene (protamines are sperm nuclear proteins responsible for maintaining DNA in a compacted form in chromosomes) is under positive selection in those mouse species with the highest inferred levels of sperm competition (Martin-Coello et al. 2009). In addition, variation in sperm competition intensity between species was strongly associated with high divergence in protamine 2 promoters that, in turn, were associated with sperm swimming speed, a major determinant of fertilisation success under sperm

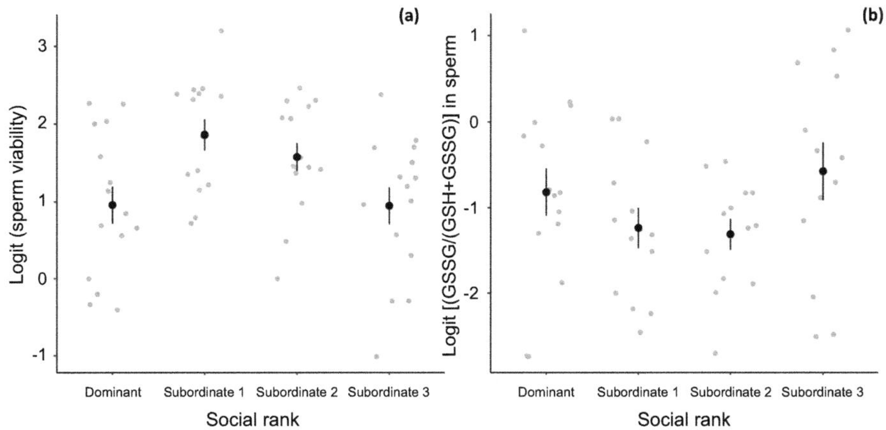


**Fig. 9.4** Proportion of (**a**) motile sperm and (**b**) oxidised over total glutathione [GSSG/(GSH+GSSG)] in sperm of male house sparrows *Passer domesticus* with a different social status. Males lower in the social ladder were referred to as subordinate-1, subordinate-2, and subordinate-3, subordinate-3 males being at the bottom end of the hierarchy. Ejaculates with higher motility had lower proportions of GSSG, whilst the greater the proportion of GSSG (e.g. the more oxidatively stressed the sperm cells) the less motile sperm was. Proportions are logit-transformed. Values are shown as mean and standard error. Reprinted with minor modifications from Rojas Mora et al. (2017)

competition. The results of this study suggested that sexual promiscuity might have shaped selective pressures that favoured the evolution of these molecular differences given their links with changes in both sperm design and sperm performance (Martin-Coello et al. 2009). Avoiding egg fertilisation from sperm suffering high oxidative stress can also bring benefits for the progeny. Lane et al. (2014) tested experimentally whether a chemically induced increase of reactive oxygen species in sperm impaired embryo, pregnancy, and offspring health in laboratory mice. They found that sperm treated with hydrogen peroxide delayed development of subsequent embryos, reduced implantation rates, and made shorter the length of the embryo or foetus from the top of its head to bottom of the body (Fig. 9.5). They also found that female, but not male, offspring from treated sperm were smaller, became glucose intolerant, and accumulated increased levels of adipose tissue compared to control female offspring (Fig. 9.5). Thus, paternal health cues were able to program the health of the next generation through oxidative stress carried out by sperm.

Females of many vertebrate species have the capacity to store sperm within their reproductive tracts for prolonged periods of time (Birkhead and Møller 1993). Queens of the narrow-headed ant *Formica exsecta* can maintain viable populations of stored sperm for periods as long as 30 years (Pamilo 1991). Female western diamond-backed rattlesnakes *Crotalus atrox* can store sperm for approximately 6 years (Levine et al. 2021). Stored sperm might show reduced capacity to fertilise an egg because semen quality deteriorates with time (e.g. Siva-Jothy 2000; Tarín et al. 2000; Wagner et al. 2004). Thus, how can sperm remain viable during this very long storage? The allocation of antioxidants to sperm by males and to spermatheca by

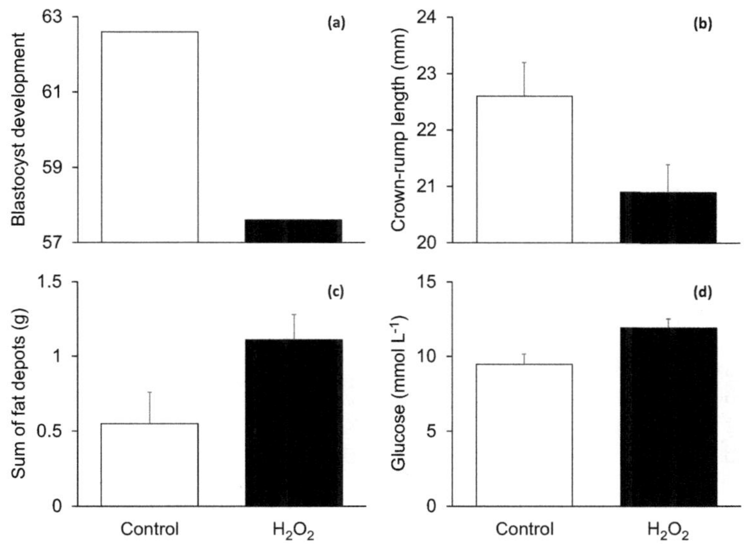


**Fig. 9.5** Effects of mouse sperm treatment with 1500 mmol $L^{-1}$ $H_2O_2$ on (**a**) development of blastocysts (96 h of culture), (**b**) crown-rump length at day 18 of gestation, (**c**) fat depots at 17 weeks of age, and (**d**) glucose concentration at 17 weeks of age. Data were collected from Lane et al. (2014)

females might contribute to maintaining this long-term viability (Collins et al. 2004; Costantini et al. 2010b). Much more work will be needed to elucidate whether there is selection on the mechanisms of antioxidant allocation to sperm protection in both females and males.

We know comparatively much less about the ecological relevance of the impact of oxidative stress on the quality of oocytes. Research on humans showed that oxidative stress has dramatic impacts on female fertility (e.g. Ruder et al. 2009; Agarwal et al. 2012; Hardy et al. 2021). As with sperm, however, reactive oxygen species can also play important roles in egg production. Many studies on mammals showed that reactive oxygen species are generated throughout the pre-implantation period, and are required for crucial embryonic events, such as pronuclear formation, first cleavage, and cell proliferation (e.g. Deluao et al. 2022). For example, the addition of glutathione to mouse embryo cultures reduced their levels of reactive oxygen species and increased the percentage of blastocysts (Gardiner and Reed 1994). Similarly, Tamura et al. (2008) found that the concentration of 8-OHdG (a marker of oxidative DNA damage) was significantly greater in the follicular fluid with a high rate of degenerative oocytes and reduced pregnancy. They also found that the supplementation of diet with antioxidants could reduce levels of 8-OHdG, improving the quality of oocytes as a result. In non-mammalian organisms, we know very little about the impact of oxidative stress on the production of eggs. For example, the experimental suppression of glutathione synthesis, with consequent increase of oxidative damage, reduced the number of eggs laid by canaries *Serinus canaria* (Fig. 9.6) (Costantini et al. 2016). Thus, females need to balance the benefits of

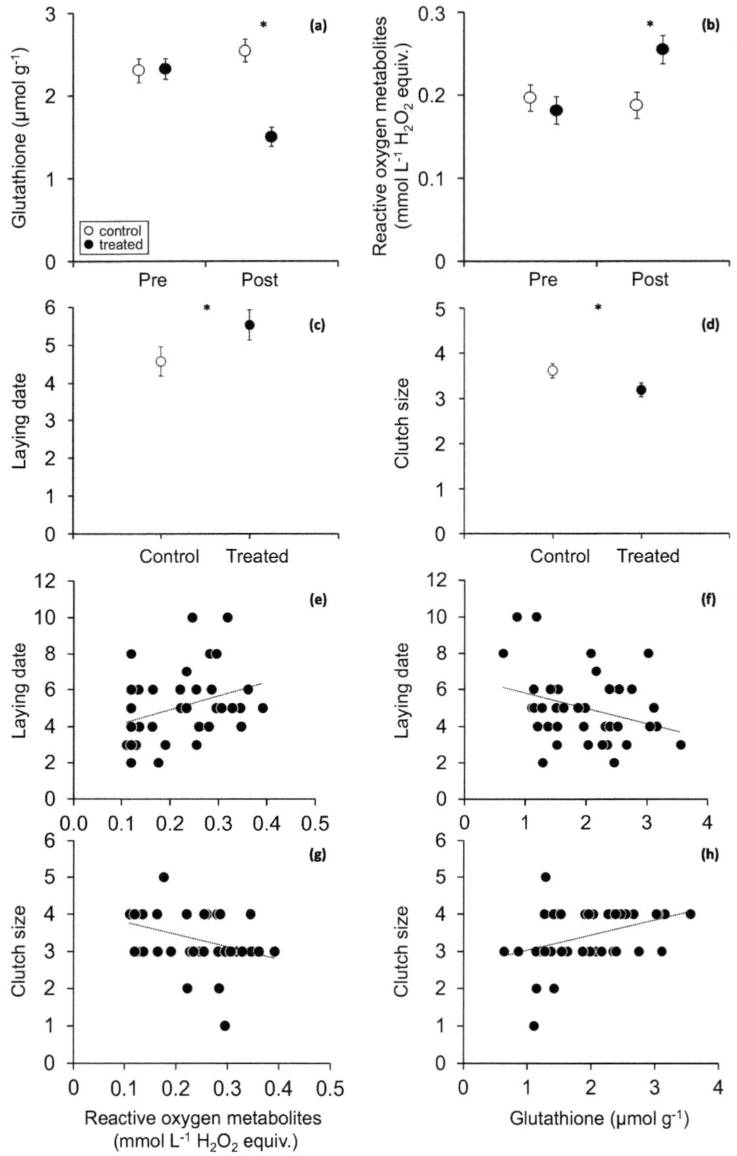


**Fig. 9.6** Pre- and post-manipulation levels of (**a**) reduced glutathione and (**b**) plasma reactive oxygen metabolites (marker of oxidative damage) of female canaries *Serinus canaria* in relation to treatment (block of glutathione synthesis). (**c**) Laying date and (**d**) clutch size in relation to treatment. Asterisks indicate significant differences between control and treated birds. Values are shown as mean and standard error. Panels (**e**) and (**f**) show that females with higher post-treatment levels of reactive oxygen metabolites and lower post-treatment levels of glutathione delayed significantly egg laying, respectively. Panels (**g**) and (**h**) show that females with higher post-treatment levels of reactive oxygen metabolites and lower post-treatment levels of glutathione laid smaller clutch sizes, respectively. Laying date is expressed as days after the second bleeding (i.e. day 0). Panels (**a**) to (**d**) reprinted with minor modifications from Costantini et al. (2016) with permission of John Wiley and Sons

generating small amounts of reactive oxygen species with the potential risks that their antioxidant defences are overwhelmed. The endogenous regulation of this balance is likely one important way through which females modulate their investment in offspring generation. Experimental testing of the role of reactive oxygen species in mediating reproductive decisions of females (e.g. how many offspring to generate and investment in egg quality) is an important area for future research.

## 9.4 Maternal Effects

Production of eggs is particularly costly for females (e.g. Wang et al. 2001; Travers et al. 2010; Webb et al. 2019; Lin et al. 2022). Females deposit a number of molecules, including antioxidants, into the eggs, depriving themselves of important tools for the maintenance of body homeostasis (Fig. 9.7). For example, studies on female lesser black-backed gulls *Larus fuscus* that were experimentally stimulated to lay extra eggs by egg removal had a significant reduction of energy reserves and capacity of chick rearing, laid lower-quality eggs, and had lower return rates to the breeding site the next year, and those that did return were less likely to produce a clutch compared to control birds (Monaghan et al. 1998; Nager et al. 2000, 2001). Egg removal experiments coupled with food supplementation, which was needed to avoid any resource limitation, also showed that females laying many extra eggs suffered higher levels of oxidative damage and lower levels of plasma antioxidants (Travers et al. 2010). Similarly, female Australian painted dragons *Ctenophorus pictus* that produced more clutches over time suffered more DNA damage than females that produced fewer clutches (Olsson et al. 2012).

A number of studies quantified the correlations between markers of oxidative status of females and clutch size or egg volume. For example, female alpine swifts *Apus melba* with higher blood antioxidant capacity laid larger clutches, and their eggs were more likely to hatch than those laid by females with lower blood antioxidants (Bize et al. 2008). By contrast, female great tits *Parus major* with higher plasma non-enzymatic antioxidant capacity during incubation laid smaller eggs and clutch sizes (Costantini et al. 2010a). As with great tits, female common starlings with higher plasma non-enzymatic antioxidant capacity during incubation laid smaller eggs, but they also laid larger clutch sizes and had lower plasma levels of reactive oxygen metabolites (Costantini et al. 2010a). Finally, female collared flycatchers *Ficedula albicollis* that produced heavier eggs had relatively lower levels of plasma reactive oxygen metabolites and higher levels of plasma non-enzymatic antioxidant capacity than females that produced lighter eggs, respectively (Markó et al. 2011). In other oviparous taxa, we have also contrasted results about the oxidative costs of egg production. For example, Wang et al. (2001) found that an experimental stimulation of the production of eggs in female fruit flies *Drosophila melanogaster* reduced their resistance to experimentally induced oxidative stress, resulting in higher mortality compared to control females. Stahlschmidt et al. (2013) found that female Children's pythons *Antaresia childreni* that laid heavier clutches

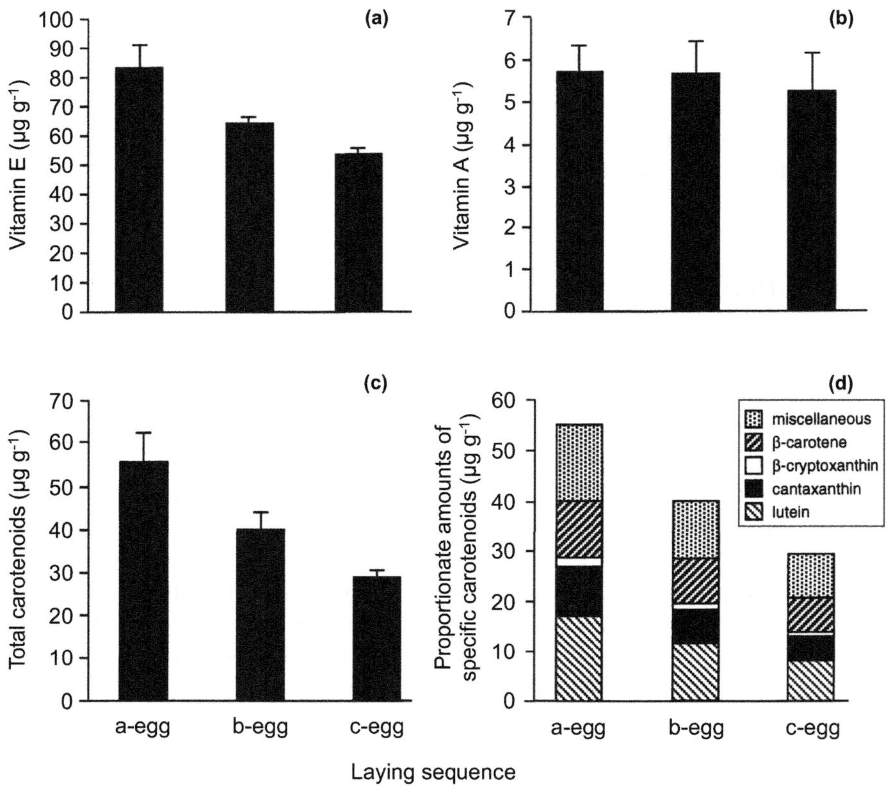


**Fig. 9.7** Variation in the concentrations of (**a**) vitamin E, (**b**) vitamin A, (**c**) total carotenoids, and (**d**) specific carotenoids in yolk across the egg-laying sequence in lesser black-backed gulls *Larus fuscus* (a-egg, first laid egg; b-egg, middle laid egg; c-egg, last laid egg). Values are shown as mean and standard error. Reprinted with minor modifications from Royle et al. (1999) with permission of John Wiley and Sons

had lower plasma levels of thiobarbituric acid reactive substances during early brooding. In addition, the percentage of muscle lost (a proxy for maternal protein allocation) from the onset of reproduction to early brooding was positively correlated with the level of thiobarbituric acid reactive substances accrued during early brooding. Webb et al. (2019) found that female Allen Cays rock iguanas *Cyclura cychlura inornata* sampled during the early (vitellogenic) reproductive season had higher reactive oxygen metabolites than during later phases of reproduction or than non-reproducing females. Finally, studies on Roatán spiny-tailed iguanas *Ctenosaura oedirhina* and Colorado checkered whiptails *Aspidoscelis neotesselatus* found that females with larger eggs and clutches had higher reactive oxygen metabolites in their plasma, which could indicate that reproducing females were facing with oxidative stress (French et al. 2021; Caracalas et al. 2021). Overall, these correlative studies do not enable us to infer with reasonable support whether egg production generates oxidative stress or oxidative status reflects the quality or state of the females and their potential investment into reproduction.

Assuming that egg production generally exposes females to oxidative stress, which might be the proximate mechanisms that give rise to these oxidative costs? A number of studies on birds found that metabolic rate may be 20–30% higher in egg-laying than in control females, and an increased metabolism might result in higher generation of reactive oxygen species in mitochondria (e.g. Carey 1996; Nilsson and Raberg 2001). The deposition of antioxidants (e.g. vitamins C and E and carotenoids) in eggs is also relevant because it could reduce the antioxidants available for the female. For example, female zebra finches *Taeniopygia guttata* that laid many eggs had a decrease in blood levels of non-enzymatic antioxidants compared to females that also laid many eggs but that were supplemented with carotenoids (Bertrand et al. 2006). The deposition of antioxidants into the eggs is also influenced by the female's own antioxidant requirements. Supplementation of the maternal diet with antioxidants, such as vitamin E or carotenoids, can increase their concentration in the egg, and the antioxidant capacity of yolk as a consequence (e.g. Royle et al. 1999, 2003; Blount et al. 2002a, b; Surai 2002; McGraw et al. 2005). The results of these studies give support to the *resource limitation hypothesis*, which states that the quantity of antioxidants deposited in the eggs depends on the antioxidant content of food. It is, therefore, predicted a significant strong positive correlation between antioxidant levels of the female and antioxidant contents of food. However, several studies also found support for the *active regulation hypothesis*, which states that females can regulate to some degree the investment of antioxidants in the eggs (e.g. Royle et al. 1999, 2003; Blount et al. 2002a, b; Surai 2002; Costantini 2010a). Accordingly, it is expected (1) either a weak positive or a negative correlation between antioxidant levels of the female and antioxidant content of food, or (2) similar antioxidant levels between first- and last-laid eggs. It might be that the importance of the passive and active routes of antioxidant deposition in the eggs is contingent on the circumstances under which the formation of eggs takes place, the metabolic pathways through which the different types of antioxidants are deposited in the yolk and in the albumen, or the chemical properties of substances (e.g. hydrophilic or lipophilic). For example, a study on lesser black-backed gulls found that eggs did not differ in the concentration of vitamin A in the yolk in relation to their laying order, whilst concentrations of vitamin E and carotenoids decreased considerably with laying order (Fig. 9.7) (Royle et al. 1999). By contrast, a study conducted on Audouin's gulls *Larus audouinii* found little difference in the antioxidant deposition in the yolk over the laying sequence (García-Tarrasón et al. 2014). Similarly, a study on captive pigeons *Columba livia* found (1) a negative correlation between the yolk levels of hydrophilic and lipophilic antioxidants and (2) that females that laid eggs with higher yolk levels of non-enzymatic hydrophilic antioxidant capacity had lower serum levels of non-enzymatic antioxidant capacity, and led eggs with a lower content of antioxidants in albumen (Costantini 2010a). The complexity of these correlations suggests that there might be trade-offs between different types of antioxidants in the way they are deposited into the eggs (see also Blount et al. 2002a for a specific example on the deposition of various types of carotenoids in the egg). If these trade-offs are owing to an active regulation of antioxidant deposition or to competitive interactions (e.g. for transport proteins or other uptake mechanisms) between molecules that affect their transfer from the mother into the eggs is an open question.

Much research has been dedicated to understanding the patterns of antioxidant deposition in eggs. Comparatively less attention has been given to antioxidant allocation of mothers to offspring through colostrum and milk (e.g. Kumagai and Chaipan 2004; Hu et al. 2011; Abuelo et al. 2014). For example, Kumagai and Chaipan (2004) reported that the concentration of α-tocopherol in the colostrum of multiparous cows was significantly higher than that in the colostrum of primiparous cows from 2 months before expected calving to 3 months of lactation. As a consequence, plasma concentrations of α-tocopherol in calves borne from multiparous cows were significantly higher than those in calves born from primiparous cows (Kumagai and Chaipan 2004). However, milk can also bring pro-oxidant molecules (Hu et al. 2011; Rizzo et al. 2013; Abuelo et al. 2014). For example, Rizzo et al. (2013) showed that lipid peroxides were constantly present in cow milk until 16 days after delivery. Moreover, the concentration of lipid peroxides in milk was positively correlated with levels of both vitamins A and E. Domesticated animals, like cows, were selected for extreme reproductive output, so that they might not be considered representative of free-living mammals. However, the results of these studies suggest that it is very relevant to analyse whether mammalian mothers regulate the content of antioxidants and pro-oxidants in their milk, respectively.

## 9.5 Oxidative Stress as a Cost of and a Constraint on Reproduction

It is hypothesised that a greater investment in reproduction, be it in terms of egg production or parental care, would result in faster somatic deterioration due to oxidative stress (*oxidative cost of reproduction hypothesis*, Costantini 2008). In particular, iteroparous species are expected to be exposed to these oxidative costs given their multiple reproductive cycles over the course of their lifetime. It is also hypothesised that pre-breeding oxidative stress would constrain reproductive investment and success (*oxidative constraint hypothesis of reproduction*) through effects of macromolecular oxidation on important functions, such as fertility and quality of parental care. In the following paragraphs, I have illustrated the research that has addressed the links between oxidative stress and parental care, and how pre-breeding oxidative stress turns out to be relevant for different reproductive strategies due to its role as a constraint.

Parental care of offspring is a stage of reproduction that can potentially result in oxidative stress. However, we lack clear experimental demonstrations of the link between oxidative stress and reproductive effort (e.g. Metcalfe and Monaghan 2013; Blount et al. 2016; Costantini et al. 2014a; Costantini 2016). Limitations in experimental designs may explain inconsistencies between studies (Costantini

2016; Alonso-Alvarez et al. 2017). We have many studies that did not properly manipulate the reproductive effort or are correlative, meaning that oxidative stress would be difficult to detect if individuals manage their reproductive effort so as to minimise damage or the size of costs vary amongst individuals of different quality. It is also often complicated to determine whether any increases in oxidative damage are a consequence of increased effort in parental care or reflect changes in the hormonal status that occur in animals during the reproductive season. Finally, which should be the real control in studies looking at reproductive costs? Some authors suggested that non-reproducing individuals might be ideal controls to test the oxidative cost of reproduction hypothesis. If reproduction causes oxidative stress, we should expect lower levels of oxidative damage in non-reproducing individuals. However, the comparison between non-reproducing and reproducing individuals does not appear to be suitable to assess the costs of parental care. There might be a number of caveats in the use of non-reproducing individuals as controls that, if not being carefully considered, might lead to erroneous interpretations and conclusions. For example, social isolation induced both a rise in plasma corticosterone and a reduced vocal activity in captive zebra finches (Perez et al. 2012). Similarly, chronic social isolation resulted in depression-like behaviour associated with increased oxidative damage and accelerated telomere shortening in female prairie voles *Microtus ochrogaster* (Stevenson et al. 2019). We should recognise that natural selection does not generally favour individuals that do not reproduce. Hence, we should also consider individuals that choose their own level of reproduction as appropriate controls in experimental investigations of the oxidative cost of reproduction hypothesis.

What evidence do we have about the link between parental effort and oxidative stress? Some studies on captive and free-living birds found that an increase in workload may decrease levels of antioxidants in parents. For example, male zebra finches that, subsequently to an enlargement of their brood size, experienced an increase of workload for rearing their offspring did not consume more energy, but had lower levels of superoxide dismutase and glutathione peroxidase in their pectoral muscle than male zebra finches that experienced a decrease of workload (Wiersma et al. 2004). By contrast, female zebra finches that experienced an increase in workload for rearing their offspring increased their energy expenditure but did not change their levels of superoxide dismutase and glutathione peroxidase in their pectoral muscle (Wiersma et al. 2004). In a similar experiment on zebra finches, males that raised six nestlings had a reduction of blood levels of non-enzymatic antioxidants compared to non-breeding males, or to males that raised only two nestlings (Alonso-Alvarez et al. 2004). By contrast, breeding females had lower blood levels of non-enzymatic antioxidants than non-breeding females, although this pattern was mainly due to an increase in antioxidant levels in non-breeding females (Alonso-Alvarez et al. 2004). Similar results were also found on free-living birds. Studies on great tits found that parents rearing experimentally enlarged broods had lower levels of blood non-enzymatic antioxidants than control parents (Losdat et al. 2011; Christie et al. 2012).

These studies focussed on the impact of parental effort on antioxidant defences. It is, however, unclear if these decreased levels of antioxidants were associated with increased levels of oxidative damage. Experimental increase of parental workload by means of brood size enlargement did not affect the plasma concentration of 8-OHdG, reactive oxygen metabolites, and non-enzymatic antioxidants in captive zebra finches (Reichert et al. 2014), and the plasma concentration of reactive oxygen metabolites and non-enzymatic antioxidant capacity in free-living great tits (Wegmann et al. 2015), suggesting that, under certain circumstances, adults might favour maintenance of cellular homeostasis over current reproduction. By contrast, another experiment on zebra finches found that birds raising an experimentally enlarged brood had higher levels of 8-OHdG in red blood cells in comparison to birds rearing a reduced brood (Noguera 2017). A correlative study on northern gannets *Morus bassanus* found higher levels of oxidative damage during the breeding stage of chick rearing than during the incubation (Pelletier et al. 2023).

The results of these studies suggest that the oxidative cost of reproduction is circumstantial and might differ between females and males (see also Sect. 9.6). For example, a member of the pair may increase its breeding effort in order to compensate for a lower breeding effort of its mate who had previously stressful experiences (Spencer et al. 2010), and this increased effort may reduce its life expectancy (Monaghan et al. 2012). Females and males may also strongly differ in their investment in reproduction. For example, Casagrande et al. (2011) found that plasma levels of reactive oxygen metabolites did not differ between female and male common kestrels during the mating season, and were significantly higher in male than in female common kestrels during the rearing season, probably because in this bird species males are responsible for providing most of the food to the family. Studies on other bird species found that, compared to males, rearing of nestlings was more antioxidant demanding for female barn swallows *Hirundo rustica* (Costantini et al. 2014b; Pap et al. 2018), female tawny owls *Strix aluco* (Emaresi et al. 2016), and female Adélie penguins *Pygoscelis adeliae* (Colominas-Ciuró et al. 2017).

In mammals, females invest substantially in parental care compared to males (Clutton-Brock 1991). As a consequence, females might be more exposed to oxidative stress than males (Speakman 2008). As with birds, there is little direct experimental evidence that reproductive effort elevates oxidative stress. For example, Nussey et al. (2009) found that the reproductive effort of female Soay sheep *Ovis aries* was not correlated to their plasma levels of malondialdehyde. Garratt et al. (2011) found that levels of malondialdehyde and oxidation of protein thiols in liver were lower in reproducing female house mice *Mus musculus* than in non-reproducing females. Moreover, the concentration of malondialdehyde in liver of female mice that did not reproduce was higher than that of females at peak of lactation, or that had just weaned their litter. However, when looking at the variation in oxidative status markers between reproducing females, Garratt et al. (2011) observed that levels of protein thiols in liver and gastrocnemius muscle were lower in those lactating females that had the heaviest litters. Bergeron et al. (2011) and Stier et al. (2012) also found that free-living eastern chipmunks *Tamias striatus* and captive mice with larger litter sizes had higher levels of plasma oxidative damage markers, respectively. By contrast, Fletcher et al. (2013) found that levels of protein carbonyls were higher during lactation than in non-breeding periods in free-living female American red squirrels *Tamiasciurus hudsonicus*. Plumel et al. (2014) found that female mice (C57BL/6J) rearing experimentally enlarged litters had higher plasma levels of reactive oxygen metabolites than female mice rearing experimentally reduced litters. Similarly, Armstrong et al. (2023) found higher levels of lipid peroxidation in lactating-fasting than in foraging female grey seals *Halichoerus grypus*.

The results of these studies do not enable us to make general conclusions about the oxidative cost of reproduction. Moreover, inferences are further complicated by the fact that results may vary between different tissues and markers analysed. Yang et al. (2013) compared a number of oxidative status markers between lactating and non-reproducing female Mongolian gerbils *Meriones unguiculatus*. They found that lactating females had lower serum levels of superoxide dismutase and glutathione peroxidase, and higher serum levels of protein carbonyls than non-reproducing females. However, lactating females had also lower hepatic levels of thiobarbituric acid reactive substances and protein carbonyls, and higher hepatic levels of superoxide dismutase and non-enzymatic antioxidant capacity than non-reproducing gerbils (Yang et al. 2013). Similarly, reproducing female Damaraland mole-rats *Fukomys damarensis* exhibited significantly lower levels of protein carbonyls in heart, kidney, liver, skeletal muscle, and plasma, and lower levels of malondialdehyde in heart, kidney, and liver than non-reproducing female Damaraland mole-rats (Schmidt et al. 2014). Similar differences between tissues and markers were also obtained using wild-caught bank voles *Myodes glareolus* (Ołdakowski and Taylor 2018). One potential explanation for these results might lie in the capacity of reproducing individuals to strategically allocate protection to those tissues that have a major impact on fitness, meaning that they sacrifice protection of those tissues that are less relevant during reproduction (e.g. Yang et al. 2013; Wilson et al. 2014; Zhao et al. 2015). This differential antioxidant protection of tissues might also be important to protect offspring from maternal oxidative stress (e.g. Costantini et al. 2014a; Blount et al. 2016; Costantini et al. 2017). Thus, we might expect oxidative damage to decline during the gestation-lactation period to limit any carry-over effects from mothers to their offspring (Fig. 9.8). However, we lack mechanistic evidence in support of this hypothesis.

The results of these correlative studies are in line with those of studies that experimentally manipulated the reproductive effort of females. For example, Garratt et al. (2013) found that a bunch of oxidative status markers analysed in the heart and gastrocnemius muscle were not affected by reproductive effort, and hepatic levels of protein carbonyls of females with larger litters were lower than those of females with smaller litters or that did not reproduce. In a similar study, Xu et al. (2014) compared markers of oxidative status between females with natural large or small litters and between females with experimentally enlarged or reduced litters in Brandt's voles *Lasiopodomys brandtii*, respectively. Levels of liver thiobarbituric acid reactive substances and superoxide dismutase were significantly higher in voles with more pups than in those with fewer pups. Litter size did not predict levels of protein carbonyls in liver and serum, and thiobarbituric acid reactive substances and superoxide dismutase in serum. By contrast, the litter size manipulation did not affect any of the markers analysed. Despite this result, combined data for all voles

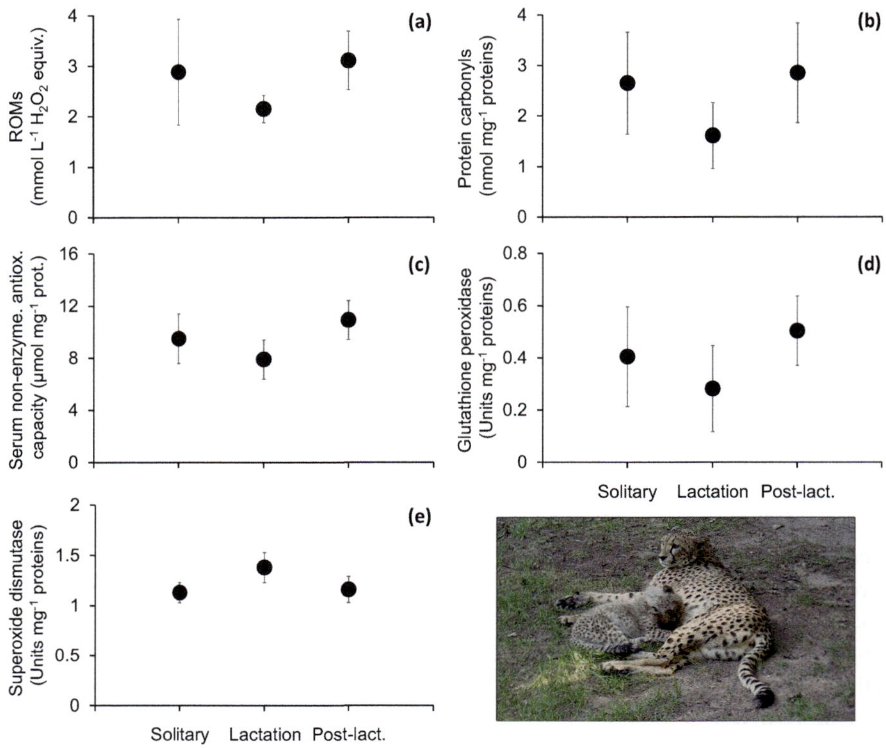


**Fig. 9.8** Comparison of levels of (**a**) serum reactive oxygen metabolites, (**b**) serum protein carbonyls, (**c**) serum non-enzymatic antioxidant capacity, (**d**) glutathione peroxidase in whole blood, and (**e**) superoxide dismutase in whole blood amongst solitary female cheetahs *Acinonyx jubatus*, lactating cheetahs, and post-lactation cheetahs. Values are shown as mean and standard deviation. Data were collected from Costantini et al. (2017)

with natural litters showed that litter size at day 18 of lactation was positively correlated with levels of liver thiobarbituric acid reactive substances and superoxide dismutase (Xu et al. 2014).

In summary, the results of these studies provided evidence that oxidative damage does not necessarily increase with lactating effort in rodents and that the inferred effects of rearing effort on oxidative status are critically dependent on the markers used and the tissues in which they are measured. However, mice and voles were maintained under optimal diet conditions, and this diet regime might not be appropriate to reflect conditions that occur under natural conditions. Although dietary restriction may induce confounding effects (e.g. due to malnutrition), a solution for this problem might lie in the increase of the effort required to obtain food rather than in limiting the abundance of food available (Metcalfe and Monaghan 2013). However, we should not lose sight of the fact that either shortage of food or of specific nutrients can occur in the real world. It is therefore ecologically relevant to also test the oxidative cost of reproduction hypothesis through the manipulation of the food composition and availability in a way that reflects conditions that the species can experience in the wild, or the quantification of oxidative stress in animals that experienced prolonged fasting during reproduction, such as elephant seals (Sharick et al. 2015).

Another explanation for this complex relationship between reproductive effort and oxidative status might lie with the strategy that females use to offset particularly costly periods of reproduction. For example, King et al. (2013) found that female mice with a greater reproductive load had no consistent changes in oxidative status markers compared to females that had a smaller reproductive load. These two groups of females differed, however, in food consumption, reproductive scheduling, and mean offspring mass. Importantly, females with a very high reproductive load delayed blastocyst implantation of their second litter, potentially mitigating the costs of energetically demanding reproductive periods. Thus, simple comparisons of oxidative status markers between reproducing and non-reproducing individuals, or individuals rearing different numbers of offspring might be a too simplistic way of looking at oxidative costs of reproduction.

We can identify several other explanations for these discrepancies between studies. Research on birds and mammals indicated that the nutritional resources available during the rearing period are particularly important. For example, pigeons that lost more body mass during reproduction had a higher increase in serum levels of reactive oxygen metabolites when maintained on a low-quality diet (Costantini 2010b). By contrast, there was no increase in oxidative damage in pigeons that were on a diet of better quality (Costantini 2010b). Fletcher et al. (2013) found that female American red squirrels that were food-supplemented had increased antioxidant protection and reduced plasma levels of protein carbonyls during lactation. Early life nutritional conditions might also have a long-term influence on the effects of reproductive effort on parental oxidative status. For example, female zebra finches that experienced either a constant or a fluctuating diet composition early in life showed similar reductions in their levels of glutathione peroxidase in red blood cells when raising a reduced brood (Noguera 2017). By contrast, when raising an enlarged brood, female zebra finches that experienced a fluctuating diet composition early in life experienced a significantly greater decline of glutathione peroxidase compared to birds that were on a constant diet during development (Noguera 2017). In contrast to females, levels of glutathione peroxidase in males were not affected by early life nutrition.

Individual coping style might be another important factor that affects the way animals respond to an experimentally increased breeding effort. Nicolaus et al. (2015) showed that female great tits with a reactive coping style performed best at their natural brood size, whereas female great tits with a proactive coping style performed best when raising an enlarged brood, respectively. This study did not quantify the oxidative status of birds. However, we know that proactive and reactive individuals differ in oxidative status and stress reactivity (Chap. 4). Thus, it is reasonable to conclude that the role of coping style in mediating the oxidative costs of reproduction needs to be considered.

Another relevant explanation for the discrepancies between studies might lie with the conditions experienced early in life because reduced resistance to oxidative stress during reproduction seems to be induced by pre-natal exposure to oxidative stress (Zimmer and Spencer 2015; Romero-Haro et al. 2016). Age at which non-reproducing and reproducing animals are being compared (Heemann et al. 2017), or brood sex ratio are also relevant (Merkling et al. 2017). Finally, environmental conditions experienced during the breeding stage also play an important role. Dupoué et al. (2020) found a context-dependent oxidative cost of reproduction in common lizards *Zootoca vivipara* that was concomitant with a conflict in water demand from offspring. Jacobs et al. (2021) also suggested that the oxidative cost of reproduction identified in highveld mole-rats *Cryptomys hottentotus pretoriae* might be affected by abiotic conditions that animals experience during the breeding season.

In contrast to the oxidative cost of reproduction hypothesis, comparatively much less research has been done to test whether pre-breeding oxidative status affects future reproductive decisions (e.g. when and how many eggs to lay) and reproductive success (*oxidative constraint hypothesis of reproduction*). Correlative studies found that: female mice with lower pre-reproduction plasma levels of reactive oxygen metabolites produced larger litters at birth compared to females with higher plasma levels of reactive oxygen metabolites (Stier et al. 2012); female great tits with higher levels of the antioxidant enzymes glutathione peroxidase and catalase in red blood cells had higher fledging success (Berglund et al. 2014); female common starlings with higher plasma levels of reactive oxygen metabolites during incubation had a lower fledging success compared to females with lower plasma levels of reactive oxygen metabolites (Costantini 2014); Scopoli's shearwaters *Calonectris diomedea* with higher levels of plasma reactive oxygen metabolites during the breeding season raised a lower number of chicks in the next years (Costantini and Dell'Omo 2015); non-breeder male wandering albatrosses *Diomedea exulans* with higher levels of thiobarbituric acid reactive substances at time t had reduced future breeding probabilities (Costantini et al. 2015b); female brown boobies *Sula leucogaster* with (1) higher generation of reactive oxygen species during courtship laid smaller first eggs and had chicks with lower rates of body mass gain and (2) higher plasma levels of thiobarbituric acid reactive substances during courtship spent less time attending their chicks (Montoya et al. 2016); and blue tits with higher plasma concentrations of malondialdehyde produced offspring that were structurally smaller and had reduced fledging success (Plummer et al. 2018). However, other correlative studies did not support a role of oxidative stress as a constraint on reproduction. For example, levels of antioxidants in red blood cells of female pied flycatchers *Ficedula hypoleuca* did not predict fledging success (Berglund et al. 2014), and plasma levels of reactive oxygen metabolites and of non-enzymatic antioxidants prior to hatching did not predict the amount of nest guarding during the nestling stage in superb starlings *Lamprotornis superbus* (Guindre-Parker and Rubenstein 2018a). Experimental manipulations of oxidative status of females prior to reproduction are clearly very important to elucidate whether and under which conditions oxidative stress can constrain reproduction. Michalkova et al. (2014) found that suppression of superoxide dismutase (gene knockdown) in tsetse flies decreased fecundity in subsequent reproductive cycles in young mothers, and nearly abolished fecundity in older females and impaired their longevity. Generation of oxidative stress in virgin females through exogenous treatment with hydrogen peroxide at times corresponding to pregnancy intervals reduced survival, which was exacerbated by superoxide dismutase knockdown. In order to induce a short-term increase of oxidative damage in birds, Costantini et al. (2016) reduced experimentally the body concentration of the antioxidant glutathione in female canaries before the start of reproduction. Afterwards, they randomly mated each female with a non-related sexual partner, and monitored their reproductive activity. Those females whose oxidative stress level was transiently increased delayed the start of egg laying and laid significantly smaller clutches compared to those females whose oxidative status was not experimentally modified (Fig. 9.6). However, the reproductive success was similar between control and stressed females. If oxidative stress constrains reproduction, we might infer that the observed lower levels of oxidative damage in reproducing than in non-reproducing females are due to some degree to the stimulatory effect of reproduction on the expression of those mechanisms that reduce oxidative damage in certain tissues, or those types of damage that are more detrimental for fitness. This stimulatory effect of reproduction could be beneficial because it would protect reproducing mothers, gametes, and developing offspring against oxidative stress (Fig. 9.8) (Costantini et al. 2014a; Blount et al. 2016; Sudyka et al. 2016; Costantini et al. 2017; Tasaki et al. 2017; Meniri et al. 2022).

The conclusions of these studies, which are based on whole organisms, are supported by more mechanistic approaches, involving experiments on cell cultures or molecular and histological analyses. As illustrated in previous paragraphs, high levels of oxidative stress can reduce the probability of oocytes to mature or to be fertilised (e.g. Knapen et al. 1999; Agarwal and Allamaneni 2004), degenerate yolk nutrients (e.g. Grune et al. 2001), or determine defective embryogenesis and pregnancy loss (e.g. Gupta et al. 2007; Al-Gubory et al. 2010). Quantification of oxidative status markers in primary fibroblast cells isolated from virgin females, reproductively experienced females, and male Sprague-Dawley rats showed that reproductively experienced females had significantly lower rates of lipid peroxidation compared to virgin females and males (Winward et al. 2018). Maternal levels of oxidative damage may also constrain the expression of reproductive traits in offspring, indicating that oxidative costs of reproduction may carry over and express in the next generations (Romero-Haro et al. 2022). However, in some organisms a burst of reactive oxygen species generation occurs soon after egg fertilisation, and this burst is required to develop a coat that protects the embryo (Shapiro 1991). In humans, both male and female germ lines rely on oxidative processes to achieve a wide range of biological functions from capacitation and chromatin cross linking in the spermatozoa to the maturation and ovulation of oocytes, pre-implantation embryo development, implantation, or foetal development to term (Aitken et al. 2022). Thus, natural selection has also likely moulded the reproductive physiology to optimise a compromise between the beneficial and detrimental effects of reactive oxygen species on both mothers and offspring.

In conclusion, we do not have yet convincing evidence in favour of a role of oxidative stress as a general cost of and constraint on reproduction. This casts doubt on the universal role of oxidative stress as a mediator of reproductive strategies. However, trade-offs are central to life-history theory, and an appraisal of current literature shows that experimental animals are not always forced into trade-offs between self-maintenance and reproduction. For example, using *Bicyclus anynana* butterflies, Beaulieu et al. (2015) showed that a trade-off between reproduction and self-maintenance occurs only under challenging conditions, and this is indeed a necessary condition for reproduction to alter oxidative status.

One argument against the oxidative cost of reproduction hypothesis lies with the incongruencies of results between tissues analysed. If reproduction is costly, it should be expected a general increase of oxidative damage across tissues. Evidence shows that this is not always the case. This general expectation seems to be misguided. Tissues are not always metabolically active to the same extent, meaning that it is probably too simplistic to expect an increase in oxidative damage or depletion of antioxidants across all tissues. For example, male house mice that defended a territory tended to have higher oxidative damage in the gastrocnemius muscle, but lower lipid peroxidation in serum and liver (Garratt et al. 2012). Mice that defended a territory might have experienced an increase in muscle effort needed to sustain an intense physical activity. This increase in physical effort may have induced localised oxidative damage in muscle (Garratt et al. 2012). Whether localised oxidative damage has fitness consequences is an open question. Moreover, we should also clarify if the sampling design is appropriate to compare markers across tissues. It might be that the rate at which markers respond to an increased reproductive effort differs amongst tissues, meaning that we need to have multiple measurements over time of the same marker analysed in a given tissue. For those tissues whose sampling requires the sacrifice of the animal, we need to have many animals that are sampled at different points in time.

## 9.6 Sex Differences in Oxidative Status

In Sect. 9.5, we have seen that sex differences may occur in both antioxidant machinery and oxidative damage across diverse species, and that these differences can be generated by a number of factors, including reproductive stage or nutrition. Sexually antagonistic selection has promoted the evolution of different trait optima in sexes for many traits, for example over the mode and frequency of mating (Parker 2006) or the amount of parental care provided to offspring (Alonso-Alvarez and Velando 2012; Balshine 2012). Physiological and epidemiological evidence also shows that sex is an important biological variable in shaping immune and endocrine functions (Norris and Lopez 2010; Wilkinson et al. 2022). A meta-analysis including 732 effect size estimates from

100 articles and 82 vertebrate species also identified relevant differences between sexes in oxidative status (Costantini 2018). This meta-analysis documented larger sexual differences in oxidative status in fish and reptiles, in those species that provide parental care, and in oviparous species. Moreover, the meta-analysis identified a larger vulnerability of females to oxidative stress, particularly in those species that do not provide any parental care, suggesting a strong cost of egg production (Costantini 2018). It might be that selection for higher resistance to oxidative stress has contributed to favour evolution of viviparity, for example because given the strong intimacy between mother and foetus in viviparous species, being more resistant to oxidative stress would protect the foetus from the pathological consequences associated with the accumulation of oxidative damage during embryogenesis (e.g. Blount et al. 2016; Vitikainen et al. 2016). Elucidating the role of oxidative stress in the evolution of viviparity is an important topic that requires further research.

Although egg production is generally costly for females, the meta-analysis also suggested that male competition could also be particularly costly in terms of oxidative stress (Costantini 2018). It is generally hypothesised that sexual conflicts originate from the difference in gamete size between female and male gametes. If sperm are tiny and not produced in vastly greater numbers than eggs, at any mating the gametic investment of females would be larger than that of males (Kokko and Jennions 2008). Although large variation exists, it is expected that the investment of resources in gametes is much higher for females than for males (Hayward and Gillooly, 2011). However, it is often forgotten that investment in gametes is only one component of reproductive costs. In many species, males invest massively in mate search, mate competition, courtship, or territoriality. All of these behavioural and non-behavioural traits require resources (e.g. energy, nutrients, and time), so that they can contribute to the overall costs that sexes pay for reproduction. Hermaphrodites have both sexes tied together in the same body, offering the advantage of estimating the reproductive cost whilst minimising confounding factors associated with sexes being expressed in different bodies. In addition, simultaneous hermaphrodites are able to vary their relative investment into the two sexual functions. Relying on a simultaneously hermaphroditic polychaete worm species that can invest either relatively more in the female function or more in the male function, Lorenzi et al. (2023) found that less female-biased hermaphrodites (i.e. those that invest relatively more in the male function) produced fewer eggs but, unexpectedly, had lower levels of antioxidant protection than more female-biased hermaphrodites (i.e. those that invest relatively more in the female function), which produced more eggs (Fig. 9.9) (Lorenzi et al. 2023). Male-biased hermaphrodites compete for mating as males (hermaphrodites in pairs do not), suggesting that, in the short-term, male competition over egg fertilisation might be costlier than egg production in terms of maintenance of oxidative status homeostasis.

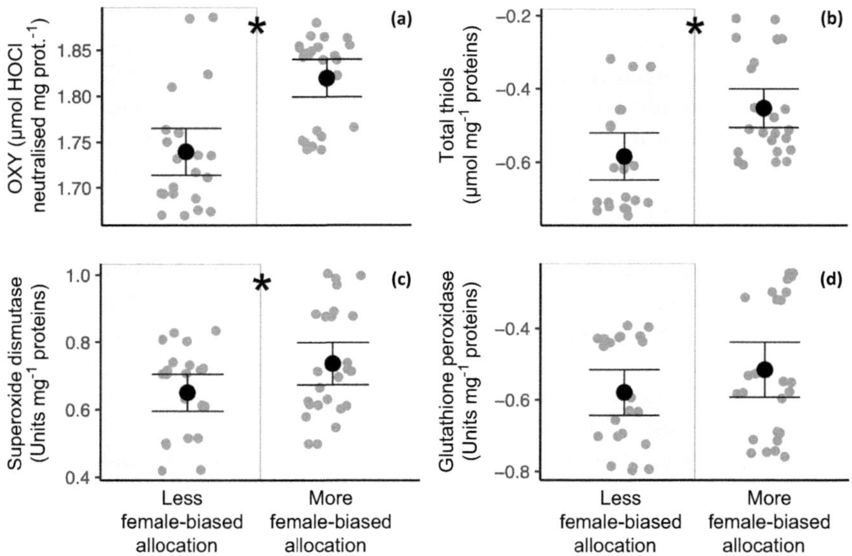


**Fig. 9.9** Polychaete worms *Ophryotrocha diadema* (two sexual functions in the same body) in pairs (more female-biased allocation) experienced less oxidative stress than those in groups (less female-biased allocation, more competition amongst makes), as suggested by their higher levels of (**a**) non-enzymatic antioxidant capacity (OXY), (**b**) total thiols, and (**c**) superoxide dismutase. By contrast, (**d**) glutathione peroxidase levels were similar between worms in pairs and in groups, although the trend was consistent with the other markers. Asterisks indicate significant differences between experimental groups. Values are shown as estimated marginal mean and 95% confidence interval (full black circles), and raw predicted values from the model (grey circles). Reprinted with minor modifications from Lorenzi et al. (2023) with permission of the Oxford University Press

## 9.7 Oxidative Benefits and Costs of Being Cooperative

Cooperative breeding is a social system in which offspring receive care not only from their parents but also from additional members of the group, often called helpers (Brown 1987; Stacey and Koenig 1990; Cornwallis 2018). Cooperative breeding can be costly for helpers, for example in terms of delayed sexual maturation or high energetic demands. However, several studies also documented benefits for helpers, such as territory inheritance or inclusive fitness. The magnitude of benefits and costs is also influenced by the characteristics of the group an individual lives in, such as size and phenotypic composition.

Some studies investigated the role of oxidative stress as a proximate mechanism underlying the evolution of cooperative breeding. van de Crommenacker et al. (2011) found that, before breeding, subordinate female Seychelles warblers *Acrocephalus sechellensis* that subsequently refrained to help (non-helpers) had higher plasma levels of reactive oxygen metabolites than helper or dominant females, respectively. Female dominants and helpers had higher levels of plasma non-enzymatic antioxidant capacity than female non-helpers during nest-care. In males, plasma levels of reactive oxygen metabolites were higher in dominants than in helpers during nest guarding. By contrast, plasma levels of reactive oxygen metabolites were lower in dominants than in non-helpers during nestling feeding. These correlative results suggested that oxidative stress might be one proximate mechanism underlying the expression of cooperative behaviour during the breeding season.

Subsequent studies have, however, found contrasting results. For example, the social status did not appear to influence the deposition of carotenoids in eggs in the sociable weaver *Philetairus socius*. Paquet et al. (2013) found that the number of helpers affected the deposition of the hormones testosterone, androstenedione, and corticosterone in eggs by dominant females, but did not affect the deposition of carotenoids. In white-browed sparrow weavers *Plocepasser mahali*, Cram et al. (2015a) found that, after breeding, females had elevated plasma levels of malondialdehyde, and dominant females (the only birds to lay and incubate eggs, and the primary nestling provisioners) experienced differential declines in antioxidant protection (Cram et al. 2015a). Whilst males also showed reduced antioxidant capacity after breeding, this decline was not dependent on rank, and it was not associated with elevated plasma levels of malondialdehyde (Cram et al. 2015a). Individuals may also decrease their helping behaviour to minimise the associated oxidative costs, suggesting that oxidative stress can influence helping decisions and underlie a trade-off between cooperation and self-maintenance (Covas et al. 2022). Guindre-Parker and Rubenstein (2018b) found that reactive oxygen metabolites increased proportionally to workload in non-cooperatively breeding greater blue-eared glossy starlings *Lamprotornis chalybaeus*, but not in cooperatively breeding superb starlings *Lamprotornis superbus*. Mendonça et al. (2020) found experimental evidence in captive Damaraland mole-rats *Fukomys damarensis* that oxidative stress might represent a key physiological cost of energetically demanding cooperative behaviours.

The consequences of reproducing or helping may also depend on the size of the social group. For example, Cram et al. (2015b) found that white-browed sparrow weavers that raised offspring experienced an associated cost (elevated oxidative damage and reduced body mass) only if they were in small groups containing fewer or no helpers, suggesting that the cooperative contributions of helpers may offset the costs of reproduction for all of the group members. Similarly, Lardy et al. (2016) found a marked decrease in both plasma thiobarbituric acid reactive substances and plasma superoxide dismutase in Alpine marmots *Marmota marmota* from larger groups, highlighting a potential effect of group size on the oxidative status of individuals (Fig. 9.10).

Other factors can also affect the oxidative benefits and costs of being dominant or helpers in cooperative breeders. In animal societies, the competition for and the acquisition of a dominant rank can be very intense. In many cooperatively breeding mammals, a single female has a dominant status that prevents subordinate females from breeding (Solomon and French 1997). For example, dominant female meerkats *Suricata suricatta* that experienced a higher level of competition for the dominant position had also lower breeding success and higher rates of reproductive senescence. Furthermore, females that were more frequently evicted as subordinates from the group had lower breeding success when they obtained a dominant position (Sharp and

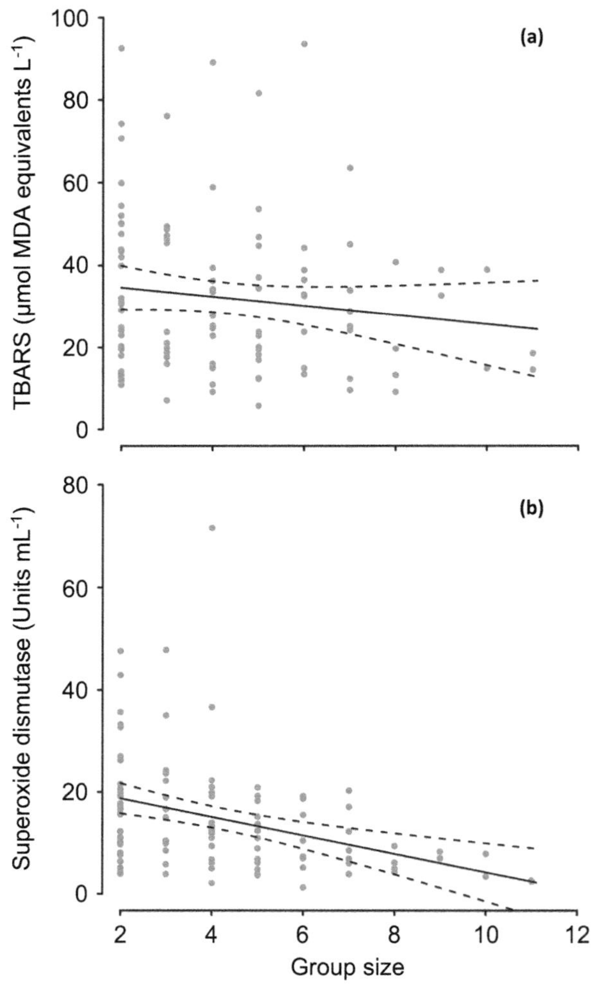


**Fig. 9.10** Effect of group size on (**a**) plasma thiobarbituric acid reactive substances levels (TBARS) and (**b**) plasma superoxide dismutase activity in free-living dominant Alpine marmots *Marmota marmota*. The lines represent the model predictions (plain) and its associated standard errors (dotted), and dots represent raw data. Reprinted with minor modifications from Lardy et al. (2016) with permission of the Oxford University Press

Clutton-Brock 2011). One of the consequences of this social stress was increased secretion of glucocorticoids. For example, subordinate meerkat females have double levels of faecal stress hormones when evicted than when they lived within their group, and during this time they either failed to conceive or aborted if pregnant (Young et al. 2006). Both short- and long-term elevation of glucocorticoids may affect the cellular oxidative status (Costantini et al. 2011; Birnie-Gauvin et al. 2017; Vágási et al. 2018; Majer et al. 2019; Stier et al. 2019; Josserand et al. 2020), providing evidence for different resistance of species to acute or chronic stress. In turn, oxidative stress can promote cellular senescence, and reduce fertility or embryo viability.

Finally, the role of group phenotypic composition in driving the benefits and costs of cooperative breeding is currently unexplored. The physiological or behavioural composition of a group can potentially affect competition and acquisition of resources. It could also determine how advantageous is to be cooperative in function of the frequencies at which different styles of coping with social stress occur in the population.

## 9.8 Conclusions and Future Directions

There has been considerable interest in testing whether oxidative stress is a universal physiological cost of and constraint on the expression of different reproductive strategies and the resulting reproductive success. Evidence in favour of this link between oxidative stress and reproduction is, however, conflicting. We can identify a number of reasons that could explain the inconsistencies amongst studies.

A major limitation of our understanding of the role of oxidative stress as a mediator of reproductive strategies probably lies in the narrow taxonomic range of the species studied so far. Although we have some clues to suggest that oxidative stress might have played a role in the evolution of sexual reproduction (Nedelcu and Michod 2003; Nedelcu et al. 2004; Nedelcu 2005) or viviparity (Costantini 2018), we need more work on these topics. Moreover, we need to know more about how oxidative costs of reproduction vary between mating systems (e.g. semelparous vs. iteroparous; monogamous vs. polygamous), sexual and asexual systems (e.g. hermaphroditism, gonochorism, parthenogenesis), or pre- and post-copulatory investments. We definitely need to broaden our perspectives if we aim at clarifying the significance of oxidative stress for reproductive trade-offs in different animal groups.

A second relevant reason might lie with the many limitations of current experimental designs, some of which were illustrated in this chapter. First, it is often assumed that non-breeding individuals are appropriate controls to test the oxidative cost of reproduction hypothesis. However, non-breeding individuals might be in poor condition, stressed by social isolation (e.g. in experiments carried out in captivity), or reproductively quiescent (e.g. due to reproductive suppression in social species). Second, when we experimentally increase the parental effort, parents may respond by decreasing their investment if they give priority to self-maintenance, or the increased physical effort is such that it can stimulate the physiological performance of parents through hormetic mechanisms. We have a bunch of studies that documented hormesis-like relationships between number of offspring and female longevity or stressor intensity and reproductive outcomes (e.g. Parkhurst et al. 1981; Lund et al. 1990; Morse and Zareh 1991; Sclar et al. 1998; Giesy et al. 2000; Cardoso et al. 2002; Calabrese and Baldwin 2003; Calabrese and Blain 2005; Wang et al. 2005; McArdle et al. 2006; Cutler et al. 2009; Heinz et al. 2010, 2012; Schreck 2010; Yu et al. 2010; Kuningas et al. 2011; Zanuncio et al. 2011; López-Martínez and Hahn 2012; Maklakov et al. 2012; Miller et al. 2013; Qu et al. 2015). Thus, we cannot neglect the role of hormesis in reproduction. Third, the long-term effects of oxidative stress on parents or their offspring are generally neglected, so that it is difficult to estimate the impact on lifetime reproductive success or the transgenerational effects.

Another important factor that could influence the results is the individual age, which is rarely considered. This is unfortunate given the number of studies that found a strong impact of age on individual performance, condition of offspring (Lansing effect), and parental resistance against oxidative stress (e.g. Lansing 1947; Lindén and Møller 1989; Fox and Dingle 1994; Priest et al. 2002; Giron and Casas 2003; Jamieson and Santer 2003; Martin 2004; Tarín et al. 2005; Alonso-Alvarez et al. 2010; Bouwhuis et al. 2010; Rodríguez-Graña et al. 2010; Santos and Nakagawa 2012).

Experimental tests of the oxidative cost of reproduction hypothesis mostly focused on parental effort in offspring rearing, but many other aspects contribute to the reproductive effort. We know comparatively much less about the oxidative costs associated with incubation behaviour (e.g. Kulaszewicz et al. 2018), which is also one aspect of parental care, egg production (e.g. Wang et al. 2001; Webb et al. 2019; Lin et al. 2022), or male–male competition (e.g. Lorenzi et al. 2023). Moreover, there is very limited knowledge about the indirect costs of reproduction, for example in the form of male-induced oxidative damage to the female oviduct owing to ejaculate composition or copulation (e.g. Romero-Haro et al. 2023).

The extrinsic mortality rate influences the trade-off between reproduction and lifespan or the duration of reproductive and non-reproductive lifespans (e.g. Reznick et al. 2006). Robert and Bronikowski (2010) found that, compared to the short-lived phenotype, offspring of the western terrestrial garter snake *Thamnophis elegans* from the long-lived ecotype had lower levels of mitochondrial generation of reactive oxygen species, and higher efficiency in repair of oxidative DNA damage. Comparative and experimental work on this topic is an important area for future research.

Observations published in the biomedical literature provide overwhelming evidence that a change in reactive oxygen species can contribute to hormesis, and this might be relevant for testing the oxidative cost of reproduction hypothesis (Costantini 2014; Hood et al. 2018). For example, the main endocrine regulators of reproductive activity (oestrogens and androgens) can induce biphasic dose–response relationships in animal/human models, and across a broad range of cell types and endpoints or even show hormesis-like responses to a number of stimuli (Calabrese 2001a, b).

Finally, the role of oxidative stress as a constraint on reproductive decisions and reproductive success has been poorly investigated so far. We need experiments that can manipulate the oxidative status of individuals before the transition from the non-reproductive to the reproductive phase, and at different phases of the reproductive activity. We also need to elucidate the molecular mechanisms through which oxidative stress can constrain reproduction.

## References

Abuelo A, Pérez-Santos M, Hernández J, Castillo C (2014) Effect of colostrum redox balance on the oxidative status of calves during the first 3 months of life and the relationship with passive immune acquisition. Vet J 199:295–299

Agarwal A, Allamaneni SS (2004) Role of free radicals in female reproductive diseases and assisted reproduction. Reprod Biomed Online 9:338–347

Agarwal A, Aponte-Mellado A, Premkumar BJ, Shaman A, Gupta S (2012) The effects of oxidative stress on female reproduction: a review. Reprod Biol Endocrinol 10:49

Ahmadi MR, Bazyar AA, Safi S, Ytrestøyl T, Bjerkeng B (2006) Effects of dietary astaxanthin supplementation on reproductive characteristics of rainbow trout (*Oncorhynchus mykiss*). J Appl Ichthyol 22:388–394

Aitken RJ, Roman SD (2008) Antioxidant systems and oxidative stress in the testes. Oxid Medic Cell Long 1:15–24

Aitken RJ, Clarkson JS, Fishel S (1989) Generation of reactive oxygen species, lipid peroxidation, and human sperm function. Biol Reprod 40:183–197

Aitken RJ, Bromfield EG, Gibb Z (2022) The impact of oxidative stress on reproduction: a focus on gametogenesis and fertilization. Reproduction 164:F79–F94

Alberts SC, Watts H, Altmann J (2003) Queuing and queue-jumping: long-term patterns of reproductive skew in male savannah baboons, *Papio cynocephalus*. Anim Behav 65:821–840

Alcock J (2009) Animal behavior, 9th edn. Sinauer Associates, Sunderland, MA

Al-Gubory KH, Fowler PA, Garrel C (2010) The roles of cellular reactive oxygen species, oxidative stress and antioxidants in pregnancy outcomes. Int J Biochem Cell Biol 42:1634–1650

Alonso-Alvarez C, Velando A (2012) Benefits and costs of parental care. In: Royle NJ, Smiseth PT, Kölliker M (eds) The evolution of parental care. Oxford University Press, Oxford, pp 40–61

Alonso-Alvarez C, Bertrand S, Devevey G, Prost J, Faivre B, Sorci G (2004) Increased susceptibility to oxidative stress as a proximate cost of reproduction. Ecol Lett 7:363–368

Alonso-Alvarez C, Pérez-Rodríguez L, García JT, Viñuela J, Mateo R (2010) Age and breeding effort as sources of individual variability in oxidative stress markers in a bird species. Physiol Biochem Zool 83:110–118

Alonso-Alvarez C, Canelo T, Romero-Haro AA (2017) The oxidative cost of reproduction: theoretical questions and alternative mechanisms. BioScience 67:258–270

Angilletta MJ Jr, Sears MW (2000) The metabolic cost of reproduction in an oviparous lizard. Funct Ecol 14:39–45

Arenas-Ríos E, León-Galván MA, Mercado PE, Rosado A (2005) Superoxide dismutase, catalase, and glutathione peroxidase during epididymal maturation and prolonged storage of spermatozoa in the Mexican big-eared bat (*Corynorhinus mexicanus*). Can J Zool 83:1556–1565

Arenas-Ríos E, León-Galván MA, Mercado PE, López-Wilchis R, Cervantes DL, Rosado A (2007) Superoxide dismutase, catalase, and glutathione peroxidase in the testis of the Mexican big-eared bat (Corynorhinus mexicanus) during its annual reproductive cycle. Comp Biochem Physiol Part A 148:150–158

Armstrong HC, Russell DJF, Moss SEW, Pomeroy P, Bennett KA (2023) Fitness correlates of blubber oxidative stress and cellular defences in grey seals (Halichoerus grypus): support for the life-history-oxidative stress theory from an animal model of simultaneous lactation and fasting. Cell Stress Chaperones 28:551–566

Attisano A, Moore AJ, Moore PJ (2012) Reproduction-longevity trade-offs reflect diet, not adaptation. J Evol Biol 25:873–880

Baldo S, Mennill DJ, Guindre-Parker S, Gilchrist HG, Love OP (2015) The oxidative cost of acoustic signals: examining steroid versus aerobic activity hypotheses in a wild bird. Ethology 121:1081–1090

Ball GF, Sockman KW, Duffy DL, Gentner TQ (2006) A neuroethological approach to song behavior and perception in European starlings: interrelationships among testosterone, neuroanatomy, immediate early gene expression, and immune function. Adv Study Behav 36:59–121

Balshine S (2012) Patterns of parental care in vertebrates. In: Royle NJ, Smiseth PT, Kölliker M (eds) The evolution of parental care. Oxford University Press, Oxford, pp 62–80

Barske J, Schlinger BA, Wikelski M, Fusani L (2011) Female choice for male motor skills. Proc R Soc Lond B 278:3523–3528

Baumber J, Ball BA, Gravance CG, Medina V, Davies-Morel MCG (2000) The effect of reactive oxygen species on equine sperm motility, viability, acrosomal integrity, mitochondrial membrane potential, and membrane lipid peroxidation. J Androl 21:895–902

Beaulieu M, Mboumba S, Willaume E, Kappeler PM, Charpentier MJ (2014) The oxidative cost of unstable social dominance. J Exp Biol 217:2629–2632

Beaulieu M, Geiger RE, Reim E, Zielke L, Fischer K (2015) Reproduction alters oxidative status when it is traded-off against longevity. Evolution 69:1786–1796

Bell G (1980) The costs of reproduction and their consequences. Am Nat 116:45–77

Bergeron P, Careau V, Humphries MM, Reale D, Speakman JR, Garant D (2011) The energetic and oxidative costs of reproduction in a free-ranging rodent. Funct Ecol 761(25):1063–1071

Berglund A, Rosenqvist G, Svensson I (1986) Mate choice, fecundity and sexual dimorphism in two pipefish species (Syngnathidae). Behav Ecol Sociobiol 19:301–307

Berglund AMM, Rainio MJ, Kanerva M, Nikinmaa M, Eeva T (2014) Antioxidant status in relation to age, condition, reproductive performance and pollution in three passerine species. J Avian Biol 45:235–246

Bertrand S, Alonso-Alvarez C, Devevey G, Faivre B, Prost J, Sorci G (2006) Carotenoids modulate the trade-off between egg production and resistance to oxidative stress in zebra finches. Oecologia 147:576–584

Bilodeau J-F, Blanchette S, Cormier N, Sirard M-A (2002) Reactive oxygen species-mediated loss of bovine sperm motility in egg yolk Tris extender: protection by pyruvate, metal chelators and bovine liver or oviductal fluid catalase. Theriogenology 57:1105–1122

Birkhead TR, Møller AP (1993) Sexual selection and the temporal separation of reproductive events: sperm storage data from reptiles, birds and mammals. Biol J Linn Soc 50:295–311

Birnie-Gauvin K, Peiman KS, Larsen MH, Aarestrup K, Willmore WG, Cooke SJ (2017) Short-term and long-term effects of transient exogenous cortisol manipulation on oxidative stress in juvenile brown trout. J Exp Biol 220:1693–1700

Bize P, Devevey G, Monaghan P, Doligez B, Christe P (2008) Fecundity and survival in relation to resistance to oxidative stress in a free-living bird. Ecology 89:2584–2593

Blount JD, Surai PF, Houston DC, Møller AP (2002a) Patterns of yolk enrichment with dietary carotenoids in gulls: the roles of pigment acquisition and utilization. Funct Ecol 16:445–453

Blount JD, Surai PF, Nager RG, Houston DC, Møller AP, Trewby ML, Kennedy MW (2002b) Carotenoids and egg quality in the lesser black-backed gull *Larus fuscus*: a supplemental feeding study of maternal effects. Proc R Soc Lond B 269:29–36

Blount JD, Vitikainen EI, Stott I, Cant MA (2016) Oxidative shielding and the cost of reproduction. Biol Rev 91:483–497

Boldt J, Schuel H, Schuel R, Dandeckar PV, Troll W (1981) Reaction of sperm with egg-derived hydrogen peroxide helps prevent polyspermy during fertilization in the sea urchin. Gamete Res 4:365–377

Bonisoli-Alquati A, Møller AP, Rudolfsen G, Saino N, Caprioli M, Ostermiller S, Mousseau TA (2011) The effects of radiation on sperm swimming behavior depend on plasma oxidative status in the barn swallow (*Hirundo rustica*). Comp Biochem Physiol Part A 159:105–112

Border SE, DeOliveira GM, Janeski HM, Piefke TJ, Brown TJ, Dijkstra PD (2019) Social rank, color morph, and social network metrics predict oxidative stress in a cichlid fish. Behav Ecol 30:490–499

Bouwhuis S, Charmantier A, Verhulst S, Sheldon BC (2010) Trans-generational effects on ageing in a wild bird population. J Evol Biol 23:636–642

Brown JL (1987) Helping and communal breeding in birds. Princeton University Press, Princeton, NJ

Byers J, Hebets E, Podos J (2010) Female mate choice based upon male motor performance. Anim Behav 79:771–778

Calabrese EJ (2001a) Estrogen and related compounds: biphasic dose responses. Crit Rev Toxicol 31:503–515

Calabrese EJ (2001b) Androgens: biphasic dose responses. Crit Rev Toxicol 31:517–522

Calabrese EJ, Baldwin LA (2003) The hormetic dose response model is more common than the threshold model in toxicology. Toxicol Sci 71:414–428

Calabrese EJ, Blain R (2005) The occurrence of hormetic dose responses in the toxicological literature, the hormesis database: an overview. Toxicol Appl Pharmacol 202:289–301

Caracalas HE, French SS, Hudson SB, Kluever BM, Webb AC, Eifler D, Lehmicke AJ, Aubry LM (2021) Reproductive trade-offs in the Colorado checkered whiptail lizard (Aspidoscelis neotesselatus): an examination of the relationship between clutch and follicle size. Evol Ecol 35:779–794

Cardoso VV, Ferreira MP, Montagner JM, Fernandez CG, Moreira JC, Oliveira AK (2002) The effects of constant and alternating temperatures on the reproductive potential, life span, and life expectancy of *Anastrepha fraterculus* (Wiedemann) (Diptera: Tephritidae). Braz J Biol 62:775–786

Carey C (1996) Avian energetics and nutritional ecology. Chapman and Hall, New York

Casagrande S, Dell'Omo G, Costantini D, Tagliavini J, Groothuis T (2011) Variation of a carotenoid-based trait in relation to oxidative stress and endocrine status during the breeding season in the Eurasian kestrel: a multi-factorial study. Comp Biochem Physiol Part A 160:16–26

Casagrande S, Pinxten R, Zaid E, Eens M (2015) Birds receiving extra carotenoids keep singing during the sickness phase induced by inflammation. Behav Ecol Sociobiol 69:1029–1037

Casagrande S, Pinxten R, Eens M (2016) Honest signaling and oxidative stress: the special case of avian acoustic communication. Front Ecol Evol 4:52

Christie P, Glaizot O, Strepparava N, Devevey G, Fumagalli L (2012) Twofold cost of reproduction: an increase in parental effort leads to higher malarial parasitaemia and to a decrease in resistance to oxidative stress. Proc R Soc Lond B 279:1142–1149

Clutton-Brock TH (1991) The evolution of parental care. Princeton University Press, Princeton, NJ

Coburn M, Schuel H, Troll W (1981) A hydrogen peroxide block to polyspermy in the sea urchin *Arbacia punctulata*. Dev Biol 84:235–238

Collins AM, Williams V, Evans JD (2004) Sperm storage and antioxidative enzyme expression in the honey bee, *Apis mellifera*. Ins Mol Biol 13:141–146

Colominas-Ciuró R, Santos M, Coria N, Barbosa A (2017) Reproductive effort affects oxidative status and stress in an Antarctic penguin species: An experimental study. PLoS One 12:e0177124

Cornwallis CK (2018) Cooperative breeding and the evolutionary coexistence of helper and non-helper strategies. Proc Natl Acad Sci USA 115:1684–1686

Costantini D (2008) Oxidative stress in ecology and evolution: lessons from avian studies. Ecol Lett 11:1238–1251

Costantini D (2010a) Complex trade-offs in the pigeon (*Columba livia*): egg antioxidant capacity and female serum oxidative status in relation to diet quality. J Comp Physiol B 180:731–739

Costantini D (2010b) Effects of diet quality on serum oxidative status and body mass in male and female pigeons during reproduction. Comp Biochem Physiol Part A 156:294–299

Costantini D (2014) Oxidative stress and hormesis in evolutionary ecology and physiology: a marriage between mechanistic and evolutionary approaches. Springer, Berlin. pp 348

Costantini D (2016) Commentary: oxidative stress as a cost of reproduction: beyond the simplistic trade-off model. Front Ecol Evol 4:10

Costantini D (2018) Meta-analysis reveals that reproductive strategies are associated with sexual differences in oxidative balance across vertebrates. Curr Zool 64:1–11

Costantini D, Dell'Omo G (2015) Oxidative stress predicts long-term resight probability and reproductive success in Scopoli's shearwater (Calonectris diomedea). Conserv Physiol 3:cov024

Costantini D, Carello L, Fanfani A (2010a) Relationships among oxidative status, breeding conditions and life-history traits in free-living great tits *Parus major* and common starlings *Sturnus vulgaris*. Ibis 152:793–802

Costantini D, Rowe M, Butler MW, McGraw KJ (2010b) From molecules to living systems: historical and contemporary issues in oxidative stress and antioxidant ecology. Funct Ecol 24:950–959

Costantini D, Marasco V, Møller AP (2011) A meta-analysis of glucocorticoids as modulators of oxidative stress in vertebrates. J Comp Physiol B 181:447–456

Costantini D, Casasole G, Eens M (2014a) Does reproduction protect against oxidative stress? J Exp Biol 217:4237–4243

Costantini D, Bonisoli-Alquati A, Rubolini D, Caprioli M, Ambrosini R, Romano M, Saino N (2014b) Nestling rearing is antioxidant demanding in female barn swallows (Hirundo rustica). Naturwissenschaften 101:541–548

Costantini D, Casagrande S, Casasole G, AbdElgawad H, Asard H, Pinxten R, Eens M (2015a) Immunization reduces vocal communication but does not increase oxidative stress in a songbird species. Behav Ecol Sociobiol 69:829–839

Costantini D, Goutte A, Barbraud C, Faivre B, Sorci G, Weimerskirch H, Delord K, Chastel O (2015b) Demographic responses to oxidative stress and inflammation in the wandering albatross (Diomedea exulans). PLoS One 10:e0133967

Costantini D, Casasole G, AbdElgawad H, Asard H, Eens M (2016) Experimental evidence that oxidative stress influences reproductive decisions. Funct Ecol 30:1169–1174

Costantini D, Wachter B, Melzheimer J, Czirják GÁ (2017) Socioecological and environmental predictors of physiological stress markers in a threatened feline species. Conservation Physiology, 5:cox069

Covas R, Lardy S, Silva LR, Rey B, Ferreira AC, Theron F, Tognetti A, Faivre B, Doutrelant C (2022) The oxidative cost of helping and its minimization in a cooperative breeder. Behav Ecol 33:504–517

Cram DL, Blount JD, Young AJ (2015a) Oxidative status and social dominance in a wild cooperative breeder. Funct Ecol 29:229–238

Cram DL, Blount JD, Young AJ (2015b) The oxidative costs of reproduction are group-size dependent in a wild cooperative breeder. Proc R Soc Lond B 282:20152031

Culbert BM, Border SE, Fialkowski RJ, Bolitho I, Dijkstra PD (2023) Social status influences relationships between hormones and oxidative stress in a cichlid fish. Horm Behav 152:105365

Cutler GC, Ramanaidu K, Astatkie T, Isman MB (2009) Green peach aphid, *Myzus persicae* (Hemiptera: Aphididae), reproduction during exposure to sublethal concentrations of imidacloprid and azadirachtin. Pest Manag Sci 65:205–209

Daan S, Deerenberg C, Dijkstra C (1996) Increased daily work precipitates natural death in the kestrel. J Anim Ecol 65:539–544

Dean R, Bonsall MB, Pizzari T (2007) Aging and sexual conflict. Science 316:383–384

Dean R, Cornwallis CK, Løvlie H, Worley K, Richardson DS, Pizzari T (2010) Male reproductive senescence causes potential for sexual conflict over mating. Curr Biol 20:1192–1196

Deerenberg C, Pen I, Dijkstra C, Arkies BJ, Visser GH, Daan S (1995) Parental energy expenditure in relation to manipulated brood size in the European kestrel, *Falco tinnunculus*. Zoology 99:38–47

de Lamirande E, Gagnon C (1992) Reactive oxygen species and human spermatozoa. I. Effect on the motility of intact spermatozoa and on sperm axonemes. J Androl 13:368–378

de Lamirande E, Gagnon C (1993) Human sperm hyperactivation and capacitation as parts of an oxidative process. Free Radic Biol Medic 14:157–166

de Lamirande E, Jiang H, Zini A, Kodama H, Gagnon C (1997) Reactive oxygen species and sperm physiology. Rev Reprod 2:48–54

Deluao JC, Winstanley Y, Robker RL, Pacella-Ince L, Gonzalez MB, McPherson NO (2022) Reactive oxygen species in the mammalian pre-implantation embryo. Reproduction 164:F95–F108

De Ridder E, Pinxten R, Eens M (2000) Experimental evidence of a testosterone-induced shift from paternal to mating behaviour in a facultatively polygynous songbird. Behav Ecol Sociobiol 49:24–30

DeVries MS, Cordes MA, Rodriguez JD, Stevenson SA, Riters LV (2016) Neural endocannabinoid CB1 receptor expression, social status, and behavior in male European starlings. Brain Res 1644:240–248

Drent RH, Daan S (1980) The prudent parent: energetic adjustments in avian breeding. Ardea 68:225–252

Duffy DL, Ball GF (2002) Song predicts immunocompetence in male European starlings (Sturnus vulgaris). Proc R Soc Lond B 269:847–852

Dupoué A, Blaimont P, Rozen-Rechels D, Richard M, Meylan S, Clobert J, Miles DB, Martin R, Decencière B, Agostini S, Le Galliard J-F (2020) Water availability and temperature induce changes in oxidative status during pregnancy in a viviparous lizard. Funct Ecol 34:475–485

Emaresi G, Henry I, Gonzalez E, Roulin A, Bize P (2016) Sex- and melanism-specific variations in the oxidative status of adult tawny owls in response to manipulated reproductive effort. J Exp Biol 219:73–79

Fasel NJ, Wesseling C, Fernandez AA, Vallat A, Glauser G, Helfenstein F, Heinz Richner H (2017) Alternative reproductive tactics, sperm mobility and oxidative stress in Carollia perspicillata (Seba's short-tailed bat). Behav Ecol Sociobiol 71:11

Fialkowski R, Aufdemberge P, Wright V, Dijkstra P (2021) Radical change: temporal patterns of oxidative stress during social ascent in a dominance hierarchy. Behav Ecol Sociobiol 75:43

Fletcher QE, Selman C, Boutin S, McAdam AG, Woods SB, Seo AY, Leeuwenburgh C, Speakman JR, Humphries MM (2013) Oxidative damage increases with reproductive energy expenditure and is reduced by food-supplementation. Evolution 67:1527–1536

Fox CW, Dingle H (1994) Dietary mediation of maternal age effects on offspring performance in a seed beetle (Coleoptera: Bruchidae). Funct Ecol 8:600–606

Fragueira R, Verhulst S, Beaulieu M (2019) Morph- and sex-specific effects of challenging conditions on maintenance parameters in the Gouldian finch. J Exp Biol 222:jeb196030

French SS, Virgin EE, Ki KC, Maryon DF, Goode AB, Pasachnik SA (2021) Reproductive stage and clutch size incur energetic and oxidative costs in an endangered iguana, Ctenosaura oedirhina. J Herpetol 55:192–200

Friesen CR, Wilson MR, Rollings N, Sudyka J, Whittington CM, Giraudeau M, Olsson M (2017) Conditional handicaps in exuberant lizards: bright color in aggressive males is correlated with high levels of free radicals. Front Ecol Evol 5:1–10

Friesen CR, Noble DWA, Olsson M (2020) The role of oxidative stress in postcopulatory selection. Philos Trans R Soc Lond B 375:20200065

Fu P, Neff BD, Gross MR (2001) Tactic-specific success in sperm competition. Proc R Soc Lond B 268:1105–1112

Fujihara N, Howarth B (1978) Lipid peroxidation in fowl spermatozoa. Poult Sci 57:1766–1768

Funnell TR, Fialkowski RJ, Dijkstra PD (2022) Social dominance does not increase oxidative stress in a female dominance hierarchy of an African cichlid fish. Ethology 128:15–25

Fusani L, Giordano M, Day L, Schlinger B (2007) High-speed video analysis reveals individual variability in the courtship displays of male golden-collared manakins. Ethology 113:964–972

Garamszegi LZ, Møller AP, Torok J, Michl G, Peczely P, Richard M (2004) Immune challenge mediates vocal communication in a passerine bird: an experiment. Behav Ecol 15:148–157

García-Tarrasón M, Sanpera C, Jover L, Costantini D (2014) Levels of antioxidants in breeding female Audouin's gulls and their deposition in eggs across different environments. J Exp Mar Biol Ecol 453:116–122

Gardiner CS, Reed DJ (1994) Status of glutathione during oxidant-induced oxidative stress in the preimplantation mouse embryo. Biol Reprod 51:1307–1314

Garratt M, Vasilaki A, Stockley P, McArdle F, Jackson M, Hurst JL (2011) Is oxidative stress a physiological cost of reproduction? An experimental test in house mice. Proc R Soc Lond B 278:1098–1106

Garratt M, McArdle F, Stockley P, Vasilaki A, Beynon RJ, Jackson MJ, Hurst JL (2012) Tissue-dependent changes in oxidative damage with male reproductive effort in house mice. Funct Ecol 26:423–433

Garratt M, Pichaud N, King EDA, Brooks RC (2013) Physiological adaptations to reproduction I. Experimentally increasing litter size enhances aspects of antioxidant defence but does not cause oxidative damage in mice. J Exp Biol 216:2879–2888

Garratt M, Pichaud N, Glaros EN, Kee AJ, Brooks RC (2014) Superoxide dismutase deficiency impairs olfactory sexual signaling and alters bioenergetic function in mice. Proc Natl Acad Sci USA 111:8119–8124

Georgiev AV, Thompson ME, Mandalaywala TM, Maestripieri D (2015) Oxidative stress as an indicator of the costs of reproduction among free-ranging rhesus macaques. J Exp Biol 218:1981–1985

Giesy JP, Pierens SL, Snyder EM, Miles-Richardson S, Kramer VJ, Snyder SA, Nichols KM, Villeneuve DA (2000) Effects of 4-nonylphenol on fecundity and biomarkers of estrogenicity in fathead minnows (*Pimephales promelas*). Environm Toxicol Chem 19:1368–1377

Gil D, Gahr M (2002) The honesty of bird song: multiple constraints for multiple traits. Trends Ecol Evol 17:133–140
Giron D, Casas J (2003) Mothers reduce egg provisioning with age. Ecol Lett 6:273–277
Godeas C, Tramer F, Micali F, Soranzo MR, Sandri G, Panfili E (1997) Distribution and possible novel role of phospholipid hydroperoxide glutathione peroxidase in rat epididymal spermatozoa. Biol Reprod 57:1502–1508
Gomes AS, Vieira JLF, da Silva JMC (2019) Sexual differences in oxidative stress in two species of Neotropical manakins (Pipridae). J Ornithol 160:1151–1157
Grune T, Kramer K, Hoppe PP, Siems W (2001) Enrichment of eggs with n-3 polyunsaturated fatty acids: effects of vitamin E supplementation. Lipids 36:833–838
Guindre-Parker S, Rubenstein DR (2018a) No short-term physiological costs of offspring care in a cooperatively breeding bird. J Exp Biol 221:jeb186569
Guindre-Parker S, Rubenstein DR (2018b) The oxidative costs of parental care in cooperative and pair-breeding African starlings. Oecologia 188:53–63
Gupta NP, Kumar R (2002) Lycopene therapy in idiopathic male infertility-a preliminary report. Int Urol Nephrol 34:369–372
Gupta S, Agarwal A, Banerjee J, Alvarez JG (2007) The role of oxidative stress in spontaneous abortion and recurrent pregnancy loss: a systematic review. Obstet Gynecolog Surv 62:335–347
Gwynne DT (1981) Sexual difference theory: Mormon crickets show role reversal in mate choice. Science 213:779–780
Hammond KA (1997) Adaptation of the maternal intestine during lactation. J Mammary Gland Biol Neoplasia 2:243–252
Hardy MLM, Day ML, Morris MB (2021) Redox regulation and oxidative stress in mammalian oocytes and embryos developed in vivo and in vitro. Int J Environ Res Publ Health 18:11374
Hatamoto LK, Sobrinho CAB, Nichi M, Barnabe VH, Barnabe RC, Cortada CNM (2006) Effects of dexamethasone treatment (to mimic stress) and vitamin E oral supplementation on the spermiogram and on seminal plasma spontaneous lipid peroxidation and antioxidant enzyme activities in dogs. Theriogenology 66:1610–1614
Hayward A, Gillooly JF (2011) The cost of sex: quantifying energetic investment in gamete production by males and females. PLoS One 6:e16557
Heemann FM, da Silva AC, Salomon TB, Putti JS, Engers VK, Hackenhaar FS, Benfato MS (2017) Redox changes in the brains of reproductive female rats during aging. Exp Gerontol 87:8–15
Heinz GH, Hoffman DJ, Klimstra JD, Stebbins KR (2010) Enhanced reproduction in mallards fed a low level of methylmercury: an apparent case of hormesis. Environ Toxicol Chem 29:650–653
Heinz GH, Hoffman DJ, Klimstra JD, Stebbins KR, Kondrad SL, Erwin CA (2012) Hormesis associated with a low dose of methylmercury injected into mallard eggs. Arch Environ Contam Toxicol 62:141–144
Helfenstein F, Losdat S, Møller AP, Blount JD, Richner H (2010) Sperm of colourful males are better protected against oxidative stress. Ecol Lett 13:213–222
Heller K-G, Fleischmann P, Lutz-Röder A (2000) Carotenoids in the spermatophores of bushcrickets (Orthoptera: Ephippigerinae). Proc R Soc Lond B 267:1905–1908
Hood WR, Zhang Y, Mowry AV, Hyatt HW, Kavazis AN (2018) Life history trade-offs within the context of mitochondrial hormesis. Integr Comp Biol 58:567–577
Hu H, Wang M, Zhan XA, Li X, Zhao RQ (2011) Effect of different selenium sources on productive performance, serum and milk Se concentrations, and antioxidant status of sows. Biol Trace Elem Res 142:471–480
Humfeld SC (2013) Condition-dependent signaling and adoption of mating tactics in an amphibian with energetic displays. Behav Ecol 24:859–870
Hunt J, Brooks R, Jennions MD, Smith MJ, Bentsen CL, Bussière LF (2004) High-quality male field crickets invest heavily in sexual display but die young. Nature 432:1024–1027
Jacobs PJ, Hart DW, Suess T, Janse van Vuuren AK, Bennett NC (2021) The cost of reproduction in a cooperatively breeding mammal: consequences of seasonal variation in rainfall, reproduction, and reproductive suppression. Front Physiol 12:780490

Jamieson CD, Santer B (2003) Maternal aging in the univoltine freshwater copepod *Cyclops kolensis*: variation in egg sizes, egg development times, and naupliar development times. Hydrobiologia 510:75–81

Josserand R, Haussy C, Agostini S, Decencière B, Le Galliard JF, Meylan S (2020) Chronic elevation of glucorticoids late in life generates long lasting changes in physiological state without a life history switch. Gen Comp Endocrinol 285:113288

Kelm-Nelson CA, Forbes-Lorman RM, Auger CJ, Riters LV (2011) Mu-opioid receptor densities are depleted in regions implicated in agonistic and sexual behavior in male European starlings (Sturnus vulgaris) defending nest sites and courting females. Behav Brain Res 219:15–22

Keogh LM, Byrne PG, Silla AJ (2018) Effect of long-term dietary beta-carotene supplementation on sperm concentration and motility in an endangered amphibian. Anim Reprod Sci 195:259–265

King EDA, Garratt M, Brooks R (2013) Manipulating reproductive effort leads to changes in female reproductive scheduling but not oxidative stress. Ecol Evol 3:4161–4171

Knapen MF, Zusterzeel PL, Peters WH, Steegers EA (1999) Glutathione and glutathione-related enzymes in reproduction. A review. Eur J Obstet Gynecol Reprod Biol 82:171–184

Kokko H, Jennions MD (2008) Parental investment, sexual selection and sex ratios. J Evol Biol 21:919–948

Kulaszewicz I, Wojczulanis-Jakubas K, Jakubas D (2018) Breeding phased dependent oxidative balance in a small High Arctic seabird, the little auk. J Avian Biol 49:e01702

Kumagai H, Chaipan Y (2004) Changes of vitamin E status of periparturient dairy cows and newborn calves. Anim Sci J 75:541–547

Kuningas M, Altmäe S, Uitterlinden AG, Hofman A (2011) The relationship between fertility and lifespan in humans. Age 33:615–622

Lahnsteiner F, Mansour N (2010) A comparative study on antioxidant systems in semen of species of the Percidae, Salmonidae, Cyprinidae, and Lotidae for improving semen storage techniques. Aquaculture 307:130–140

Lane M, McPherson NO, Fullston T, Spillane M, Sandeman L, Kang WX, Zander-Fox DL (2014) Oxidative stress in mouse sperm impairs embryo development, fetal growth and alters adiposity and glucose regulation in female offspring. PLoS One 9:e100832

Lansing AI (1947) A transmissible, cumulative, and reversible factor in aging. J Gerontol 2:228–239

Lardy S, Rey B, Salin K, Voituron Y, Cohas A (2016) Beneficial effects of group size on oxidative balance in a wild cooperative breeder. Behav Ecol 27:1820–1825

Levine BA, Schuett GW, Booth W (2021) Exceptional long-term sperm storage by a female vertebrate. PLoS One 16:e0252049

Lin Y, Patterson A, Jimenez AG, Elliott K (2022) Altered oxidative status as a cost of reproduction in a seabird with high reproductive costs. Physiol Biochem Zool 95:35–53

Lindén M, Møller AP (1989) Cost of reproduction and covariation of life history traits in birds. Trends Ecol Evol 4:367–371

Locatello L, Rasotto MB, Evans JP, Pilastro A (2006) Colourful male guppies produce faster and more viable sperm. J Evol Biol 19:1595–1602

López-Martínez G, Hahn DA (2012) Short-term anoxic conditioning hormesis boosts antioxidant defenses, lowers oxidative damage following irradiation and enhances male sexual performance in the Caribbean fruit fly, *Anastrepha suspensa*. J Exp Biol 215:2150–2161

Lorenzi MC, Robles-Guerrero FG, Costantini D (2023) The oxidative cost of competing for egg fertilisation exceeds the cost of egg production. Evolution 77:199–209

Losdat S, Helfenstein F, Gaude B, Richner H (2011) Reproductive effort transiently reduces antioxidant capacity in a wild bird. Behav Ecol 22:1218–1226

Lozano GA, Lank DB, Addison B (2013) Immune and oxidative stress trade-offs in four classes of ruffs (*Philomachus pugnax*) with different reproductive strategies. Can J Zool 91:212–218

Lund E, Arnesen E, Borgan J-K (1990) Pattern of childbearing and mortality in married women-a national prospective study from Norway. J Epidemiol Community Health 44:237–240

MacDougall-Shackleton SA, Dindia L, Newman AEM, Potvin D, Stewart KA, MacDougall-Shackleton EA (2009) Stress, song and survival in sparrows. Biol Lett 5:746–748
Majer AD, Fasanello VJ, Tindle K, Frenz BJ, Ziur AD, Fischer CP, Fletcher KL, Seecof OM, Gronsky S, Vassallo BG, Reed WL, Paitz RT, Stier A, Haussmann MF (2019) Is there an oxidative cost of acute stress? Characterization, implication of glucocorticoids and modulation by prior stress experience. Proc R Soc Lond B 286:20191698
Maklakov AA, Immler S, Løvlie H, Flis I, Friberg U (2012) The effect of sexual harassment on lethal mutation rate in female *Drosophila melanogaster*. Proc R Soc Lond B 280:20121874
Markó G, Costantini D, Michl G, Török J (2011) Oxidative damage and plasma antioxidant capacity in relation to body size, age, male sexual traits and female reproductive performance in the collared flycatcher (*Ficedula albicollis*). J Comp Physiol B 181:73–81
Martin TE (2004) Avian life-history evolution has an eminent past: does it have a bright future? Auk 121:289–301
Martin-Coello J, Dopazo H, Arbiza L, Ausió J, Roldan ERS, Gomendio M (2009) Sexual selection drives weak positive selection in protamine genes and high promoter divergence, enhancing sperm competitiveness. Proc R Soc Lond B 276:2427–2436
Masman D, Gordijn M, Daan S, Dijkstra C (1986) Ecological energetics of the kestrel: field estimates of energy intake throughout the year. Ardea 74:24–39
McArdle PF, Pollin TI, O'Connell JR, Sorkin JD, Agarwala R, Schaffer AA, Streeten EA, King TM, Shuldiner AR, Mitchell BD (2006) Does having children extend life span? A genealogical study of parity and longevity in the Amish. J Gerontol Biol Sci Med Sci 61:190–195
McGraw KJ, Adkins-Regan E, Parker RS (2005) Maternally derived carotenoid pigments affect offspring survival, sex ratio, and sexual attractiveness in a colorful songbird. Naturwissenschaften 92:375–380
McGrosky A, Pontzer H (2023) The fire of evolution: energy expenditure and ecology in primates and other endotherms. J Exp Biol 226:jeb245272
Mendonça R, Vullioud P, Katlein N, Vallat A, Glauser G, Bennett NC, Helfenstein F (2020) Oxidative costs of cooperation in cooperatively breeding Damaraland mole-rats. Proc R Soc Lond B 287:20201023
Meniri M, Evans E, Thompson FJ, Marshall HH, Nichols HJ, Lewis G, Holt L, Davey E, Mitchell C, Johnstone RA, Cant MA, Blount JD (2022) Untangling the oxidative cost of reproduction: An analysis in wild banded mongooses. Ecol Evol 12:e8644
Merkling T, Blanchard P, Chastel O, Glauser G, Vallat-Michel A, Hatch SA, Danchin E, Helfenstein F (2017) Reproductive effort and oxidative stress: effects of offspring sex and number on the physiological state of a long-lived bird. Funct Ecol 31:1201–1209
Messina S, Eens M, Casasole G, Ablelgawad H, Asard H, Pinxten R, Costantini D (2017) Experimental inhibition of a key cellular antioxidant affects vocal communication. Funct Ecol 31:1101–1110
Metcalfe NB, Alonso-Alvarez C (2010) Oxidative stress as a life-history constraint: the role of reactive oxygen species in shaping phenotypes from conception to death. Funct Ecol 24:984–996
Metcalfe NB, Monaghan P (2013) Does reproduction cause oxidative stress? An open question. Trends Ecol Evol 28:347–350
Michalkova V, Benoit JB, Attardo GM, Medlock J, Aksoy S (2014) Amelioration of reproduction-associated oxidative stress in a viviparous insect is critical to prevent reproductive senescence. PLoS One 9:e87554
Miller GM, Watson SA, McCormick MI, Munday PL (2013) Increased $CO_2$ stimulates reproduction in a coral reef fish. Global Change Biol 19:3037–3045
Mills SC, Hazard L, Lancaster L, Mappes T, Miles D, Oksanen TA, Sinervo B (2008) Gonadotropin hormone modulation of testosterone, immune function, performance, and behavioral trade-offs among male morphs of the lizard *Uta stansburiana*. Am Nat 171:339–357
Monaghan P, Nager RG, Houston DC (1998) The price of eggs: increased investment in egg production reduces offspring rearing capacity of parents. Proc R Soc Lond B 265:1731–1735

Monaghan P, Heidinger BJ, D'Alba L, Evans NP, Spencer KA (2012) For better or worse: reduced adult lifespan following early-life stress is transmitted to breeding partners. Proc R Soc Lond B 279:709–714

Montoya B, Valverde M, Rojas E, Torres R (2016) Oxidative stress during courtship affects male and female reproductive effort differentially in a wild bird with biparental care. J Exp Biol 219:3915–3926

Morse JG, Zareh N (1991) Pesticide-induced hormoligosis of citrus thrips (Thysanoptera: Thripidae) fecundity. J Econ Entomol 84:1169–1174

Nager R, Monaghan P, Houston DC (2000) Within clutch trade-off between the number and quality of eggs: experimental study in gulls. Ecology 81:1339–1350

Nager R, Monaghan P, Houston DC (2001) The cost of egg production: increased egg production reduces future fitness in gulls. J Avian Biol 32:159–166

Nedelcu AM (2005) Sex as a response to oxidative stress: stress genes co-opted for sex. Proc R Soc Lond B 272:1935–1940

Nedelcu AM, Michod RE (2003) Sex as a response to oxidative stress: the effect of antioxidants on sexual induction in a facultatively sexual lineage. Biol Lett 270:136–139

Nedelcu AM, Marcu O, Michod RE (2004) Sex as a response to oxidative stress: a twofold increase in cellular reactive oxygen species activates sex genes. Proc R Soc Lond B 271:1591–1596

Nicolaus M, Mathot KJ, Araya-Ajoy YG, Mutzel A, Wijmenga JJ, Kempenaers B, Dingemanse NJ (2015) Does coping style predict optimization? An experimental test in a wild passerine bird. Proc R Soc Lond B 282:20142405

Nilsson J-Å (2002) Metabolic consequences of hard work. Proc R Soc Lond B 269:1735–1739

Nilsson J-Å, Raberg L (2001) The resting metabolic cost of egg laying and nestling feeding in great tits. Oecologia 128:187–192

Noguera JC (2017) Interacting effects of early dietary conditions and reproductive effort on the oxidative costs of Reproduction. PeerJ 5:e3094

Noguera JC (2018) Crickets increase sexual signalling and sperm protection but live shorter in the presence of rivals. J Evol Biol 32:49–57

Noguera JC, Dean R, Isaksson C, Velando A, Pizzari T (2012) Age-specific oxidative status and the expression of pre- and postcopulatory sexually selected traits in male red junglefowl, *Gallus gallus*. Ecol Evol 2:2155–2167

Norris D, Lopez K (2010) Hormones and reproduction of vertebrates. Academic, London

Nussey DH, Pemberton JM, Pilkington JG, Blount JD (2009) Life history correlates of oxidative damage in a free-living mammal population. Funct Ecol 23:809–817

Ołdakowski Ł, Taylor JRE (2018) Oxidative damage and antioxidant defense are assay and tissue-dependent both in captive and wild-caught bank voles (Myodes glareolus) before and after reproduction. Ecol Evol 8:7543–7552

Olsson M, Healey M, Perrin C, Wilson M, Tobler M (2012) Sex-specific SOD levels and DNA damage in painted dragon lizards (*Ctenophorus pictus*). Oecologia 170:917–924

Palombit RA, Seyfarth RM, Cheney DL (1997) The adaptive value of 'friendships' to female baboons: experimental and observational evidence. Anim Behav 54:599–614

Pamilo P (1991) Life span of queens in the ant *Formica exsecta*. Ins Soc 38:111–119

Pap PL, Vincze O, Fülöp A, Székely-Béres O, Pătraş L, Pénzes J, Vágási CI (2018) Oxidative physiology of reproduction in a passerine bird: a field experiment. Behav Ecol Sociobiol 72:18

Paquet M, Covas R, Chastel O, Parenteau C, Doutrelant C (2013) Maternal effects in relation to helper presence in the cooperatively breeding sociable weaver. PLoS One 8:e59336

Parker GA (2006) Sexual conflict over mating and fertilization: an overview. Philos Trans R Soc Lond B 361:235–259

Parkhurst BR, Bradshaw AS, Forte JL, Wright GP (1981) The chronic toxicity to *Daphnia magna* of acridine, a representative azaarene present in synthetic fossil fuel products and wastewaters. Environ Pollut 24:21–30

Pelletier D, Blier P, Vézina F, Guillemette M (2023) Good times bad times—unfavorable breeding conditions, more than divorce, lead to increased parental effort and reduced physiological condition of northern gannets. Front Ecol Evol 11:1108293

Perez EC, Elie JE, Soulage CO, Soula HA, Mathevon N, Vignal C (2012) The acoustic expression of stress in a songbird: does corticosterone drive isolation-induced modifications of zebra finch calls? Horm Behav 61:573–581

Peters A, Denk AG, Delhey K, Kempenaers B (2004) Carotenoid-based bill colour as an indicator of immunocompetence and sperm performance in male mallards. J Evol Biol 17:1111–1120

Pike TW, Blount JD, Lindström J, Metcalfe NB (2010) Dietary carotenoid availability, sexual signaling and functional fertility in sticklebacks. Biol Lett 6:191–193

Pitcher TE, Rodd FH, Rowe L (2007) Sexual colouration and sperm traits in guppies. J Fish Biol 70:165–177

Pizzari T, Dean R, Pacey A, Moore H, Bonsall MB (2008) The evolutionary ecology of pre- and post-meiotic sperm senescence. Trends Ecol Evol 23:131–140

Plumel MI, Stier A, Thiersé D, van Dorsselaer A, Criscuolo F, Bertile F (2014) Litter size manipulation in laboratory mice: an example of how proteomic analysis can uncover new mechanisms underlying the cost of reproduction. Front Zool 11:41

Plummer KE, Bearhop S, Leech DI, Chamberlain DE, Blount JD (2018) Effects of winter food provisioning on the phenotypes of breeding blue tits. Ecol Evol 8:5059–5068

Poiani A (2006) Complexity of seminal fluid: a review. Behav Ecol Sociobiol 60:289–310

Preston BT, Saint-Jalme M, Hingrat Y, Lacroix F, Sorci G (2011) Sexually extravagant males age more rapidly. Ecol Lett 14:1017–1024

Priest NK, Mackowiak B, Promislow DEL (2002) The role of parental age effects on the evolution of aging. Evolution 56:927–935

Qu Y, Xiao D, Li J, Chen Z, Biondi A, Desneux N, Gao X, Song D (2015) Sublethal and hormesis effects of imidacloprid on the soybean aphid Aphis glycines. Ecotoxicology 24:479–487

Rahman MM, Gasparini C, Turchini GM, Evans JP (2015) Testing the interactive effects of carotenoids and polyunsaturated fatty acids on ejaculate traits in the guppy Poecilia reticulata (Pisces: Poeciliidae). J Fish Biol 86:1638–1643

Reed TE, Kruuk LEB, Wanless S, Frederiksen M, Cunningham EJA, Harris MP (2008) Reproductive senescence in a long-lived seabird: rates of decline in late life performance are associated with varying costs of early reproduction. Am Nat 171:89–101

Reglero MM, Taggart MA, Castellanos P, Mateo R (2009) Reduced sperm quality in relation to oxidative stress in red deer from a lead mining area. Environ Pollut 157:2209–2215

Reichert S, Stier A, Zahn S, Arrivé M, Bize P, Massemin S, Criscuolo F (2014) Increased brood size leads to persistent eroded telomeres. Front Ecol Evol 2:1–11

Reznick D, Bryant M, Holmes D (2006) The evolution of senescence and post-reproductive lifespan in guppies (*Poecilia reticulata*). PLoS Biol 4:e7

Ritschard M, Brumm H (2012) Zebra finch song reflects current food availability. Evol Ecol 26:801–812

Rizzo A, Ceci E, Pantaleo M, Mutinati M, Spedicato M, Minoia G, Sciorsci RL (2013) Evaluation of blood and milk oxidative status during early postpartum of dairy cows. Animal 7:118–123

Robert KA, Bronikowski AM (2010) Evolution of senescence in nature: physiological evolution in populations of garter snake with divergent life histories. Am Nat 175:147–159

Rodríguez-Graña L, Calliari D, Tiselius P, Hansen BW, Sköld HN (2010) Gender-specific ageing and non-Mendelian inheritance of oxidative damage in marine copepods. Mar Ecol Prog Ser 401:1–13

Roff DA (2002) Life history evolution. Sinauer Associates, Sunderland, MA

Rojas Mora A, Meniri M, Gning O, Glauser G, Vallat A, Helfenstein F (2017) Antioxidant allocation modulates sperm quality across changing social environments. PLoS One 12:e0176385

Romero-Haro AÁ, Sorci G, Alonso-Alvarez C (2016) The oxidative cost of reproduction depends on early development oxidative stress and sex in a bird species. Proc R Soc Lond B 283:20160842

Romero-Haro AÁ, Pérez-Rodríguez L, Tschirren B (2022) Intergenerational costs of oxidative stress: reduced fitness in daughters of mothers that experienced high levels of oxidative damage during reproduction. Physiol Biochem Zool 95:1–14

Romero-Haro AÁ, Pérez-Rodríguez L, Tschirren B (2023) Increased male induced harm in response to female-limited selection: interactive effects between intra- and interlocus sexual conflict? Proc R Soc Lond B 290:20230140
Rowe M, McGraw KJ (2008) Carotenoids in the seminal fluid of wild birds: interspecific variation in fairy-wrens. Condor 110:694–700
Royle NJ, Surai PF, McCartney RJ, Speake BK (1999) Parental investment and egg yolk lipid composition in gulls. Funct Ecol 13:298–306
Royle NJ, Surai PF, Hartley IR (2003) The effect of variation in dietary intake on maternal deposition of antioxidants in zebra finch eggs. Funct Ecol 17:472–481
Ruder EH, Hartman TJ, Goldman MB (2009) Impact of oxidative stress on female fertility. Curr Opin Obstet Gynecol 21:219–222
Rughetti M, Dematteis A, Meneguz PG, Festa-Bianchet M (2015) Age-specific reproductive success and cost in female Alpine ibex. Oecologia 178:197–205
Ruiz-Guzmán G, Cordero-Molina S, Krams I, Contreras-Garduño J (2021) Interactions between oxidative stress and attractiveness to mates and individual mate choice in the beetle *Tenebrio molitor*. Ethology 2:109–116
Salomons HM (2009) Fighting for fitness: telomeres, oxidative stress and life history trade-offs in a colonial corvid. Ph.D. Thesis, University of Groningen, Groningen, The Netherlands
Santos ESA, Nakagawa S (2012) The costs of parental care: a meta-analysis of the trade-off between parental effort and survival in birds. J Evol Biol 25:1911–1917
Schmidt CM, Blount JD, Bennett NC (2014) Reproduction is associated with a tissue-dependent reduction of oxidative stress in eusocial female Damaraland mole-rats (Fukomys damarensis). PLoS One 9:e103286
Schreck CB (2010) Stress and fish reproduction: the roles of allostasis and hormesis. Gen Comp Endocrinol 165:549–556
Sclar DC, Gerace D, Cranshaw WS (1998) Observation of population increases and injury by spider mites (Acari: Tetranychidae) on ornamental plants treated with imidacloprid. J Econ Entomol 91:250–255
Shapiro BM (1991) The control of oxidant stress at fertilization. Science 252:533–536
Sharick JT, Vazquez-Medina JP, Ortiz RM, Crocker DE (2015) Oxidative stress is a potential cost of breeding in male and female northern elephant seals. Funct Ecol 29:367–376
Sharp SP, Clutton-Brock TH (2011) Competition, breeding success and ageing rates in female meerkats. J Evol Biol 24:1756–1762
Sikka SC (2001) Relative impact of oxidative stress on male reproductive function. Curr Med Chem 8:851–862
Silva LR, Lardy S, Ferreira AC, Rey B, Doutrelant C, Covas R (2018) Females pay the oxidative cost of dominance in a highly social bird. Anim Behav 144:135–146
Simmons LW, Lovegrove M, Lymbery SJ (2018) Dietary antioxidants, but not courtship effort, affect oxidative balance in the testes and muscles of crickets. J Exp Biol 221:jeb184770
Siva-Jothy MT (2000) The young sperm gambit. Ecol Lett 3:172–174
Snell TW, King CE (1977) Lifespan and fecundity patterns in rotifers: the cost of reproduction. Evolution 31:882–890
Solomon NG, French JA (1997) Cooperative breeding in mammals. Cambridge University Press, Cambridge
Speakman JR (2008) The physiological costs of reproduction in small mammals. Philos Trans R Soc B 363:375–398
Spencer KA, Heidinger BJ, D'Alba LB, Evans N, Monaghan P (2010) Then versus now: effect of developmental and current environmental conditions on incubation effort in birds. Behav Ecol 21:999–1004
Stacey PB, Koenig WD (1990) Cooperative breeding in birds: long-term studies of ecology and behavior. Cambridge University Press, Cambridge

Stahlschmidt ZR, Lourdais O, Lorioux S, Butler MW, Davis JR, Salin K, Voituron Y, DeNardo DL (2013) Morphological and physiological changes during reproduction and their relationships to reproductive performance in a capital breeder. Physiol Biochem Zool 86:398–409

Stearns SC (1992) The evolution of life histories. Oxford University Press, Oxford

Stephens PA, Boyd IL, McNamara JM, Houston AI (2009) Capital breeding and income breeding: their meaning, measurement, and worth. Ecology 90:2057–2067

Stevenson JR, McMahon EK, Boner W, Haussmann MF (2019) Oxytocin administration prevents cellular aging caused by social isolation. Psychoneuroendocrin 103:52–60

Stier A, Reichert S, Massemin S, Bize P, Criscuolo F (2012) Constraint and cost of oxidative stress on reproduction: correlative evidence in laboratory mice and review of the literature. Front Zool 9:37

Stier A, Schull Q, Bize P, Lefol E, Haussmann M, Roussel D, Robin JP, Viblanc VA (2019) Oxidative stress and mitochondrial responses to stress exposure suggest that king penguins are naturally equipped to resist stress. Sci Rep 9:8545

Sudyka J, Casasole G, Rutkowska J, Cichoń M (2016) Elevated reproduction does not affect telomere dynamics and oxidative stress. Behav Ecol Sociobiol 70:2223–2233

Sullivan M, Brown AC, Clotfelter ED (2014) Dietary carotenoids do not improve motility or antioxidant capacity in cichlid fish sperm. Fish Physiol Biochem 40:1399–1405

Surai PF (2002) Natural antioxidants in avian nutrition and reproduction. Nottingham University Press, Nottingham

Surai PF, Kutz E, Wishart GJ, Noble RC, Speake BK (1997) The relationship between the dietary provision of α-tocopherol and the concentration of this vitamin in the semen of chicken: effects on lipid composition and susceptibility to peroxidation. J Reprod Fertil 110:47–51

Tamura H, Takasaki A, Miwa I, Taniguchi K, Maekawa R, Asada H, Taketani T, Matsuoka A, Yamagata Y, Shimamura K, Morioka H, Ishikawa H, Reiter RJ, Sugino N (2008) Oxidative stress impairs oocyte quality and melatonin protects oocytes from free radical damage and improves fertilization rate. J Pineal Res 44:280–287

Tarín JJ, Pérez-Albalá S, Cano A (2000) Consequences on offspring of abnormal function in ageing gametes. Hum Reprod Update 6:532–549

Tarín JJ, Gomez-Piquer V, Raussell F, Navarro S, Hermenegildo C, Cano A (2005) Delayed motherhood decreases life expectancy of mouse offspring. Biol Reprod 72:1336–1343

Tasaki E, Kobayashi K, Matsuura K, Iuchi Y (2017) An efficient antioxidant system in a long-lived termite queen. PLoS One 12:e0167412

Tomášek O, Albrechtová J, Němcová M, Opatová P, Albrecht T (2017) Trade-off between carotenoid-based sexual ornamentation and sperm resistance to oxidative challenge. Proc R Soc Lond B 284:20162444

Travers M, Clinchy M, Zanette L, Boonstra R, Williams TD (2010) Indirect predator effects on clutch size and the cost of egg production. Ecol Lett 13:980–988

Vágási CI, Pătraş L, Pap PL, Vincze O, Mureşan C, Németh J, Lendvai ÁZ (2018) Experimental increase in baseline corticosterone level reduces oxidative damage and enhances innate immune response. PLoS One 13:e0192701

van de Crommenacker J, Komdeur J, Richardson DS (2011) Assessing the cost of helping: the roles of body condition and oxidative balance in the Seychelles warbler (*Acrocephalus sechellensis*). PLoS One 6:e26423

Van Hout AJM, Pinxten R, Darras VM, Eens M (2012) Testosterone increases repertoire size in an open-ended learner: an experimental study using adult male European starlings (Sturnus vulgaris). Horm Behav 62:563–568

Velando A, Noguera JC, Drummond H, Torres R (2011) Senescent males carry premutagenic lesions in sperm. J Evol Biol 24:693–697

Vitikainen EIK, Cant MA, Sanderson JL, Mitchell C, Nichols HJ, Marshall HH, Thompson FJ, Gilchrist JS, Hodge SJ, Johnstone RA, Blount JD (2016) Evidence of oxidative shielding of offspring in a wild mammal. Front Ecol Evol 4:58

Wagner RH, Helfenstein F, Danchin E (2004) Female choice of young sperm in a genetically monogamous bird. Proc R Soc Lond B 271:S134–S137

Wang Y, Salmon AB, Harshman LG (2001) A cost of reproduction: oxidative stress susceptibility is associated with increased egg production in *Drosophila melanogaster*. Exp Geront 36:1349–1359

Wang AH, Wu JC, Yu YS, Liu JL, Yu JF, Wang MY (2005) Selective insecticide induced stimulation on fecundity and biochemical changes in *Tryporyza incertulas* (Lepidoptera: Pyralidae). J Econ Entomol 4:1143–1149

Webb AC, Iverson JB, Knapp CR, DeNardo DF, French SS (2019) Energetic investment associated with vitellogenesis induces an oxidative cost of reproduction. J Anim Ecol 88:461–472

Wegmann M, Voegeli B, Richner H (2015) Physiological responses to increased brood size and ectoparasite infestation: Adult great tits favour self-maintenance. Physiol Behav 141:127–134

Weirich GF, Collins AM, Williams VP (2002) Antioxidant enzymes in the honey bee, *Apis mellifera*. Apidologie 33:3–14

Wesseling C, Fasel N, Richner H, Helfenstein F (2016) Modification of sperm quality after sexual abstinence in Seba's short-tailed bat, Carollia perspicillata. J Exp Biol 219:1363–1368

Wiersma P, Selman C, Speakman JR, Verhulst S (2004) Birds sacrifice oxidative protection for reproduction. Biol Lett 271:360–363

Wilkinson NM, Chen HC, Lechner MG, Su MA (2022) Sex differences in immunity. Annu Rev Immunol 40:75–94

Williams GC (1966) Natural selection, the costs of reproduction, and a refinement of Lack's principle. Am Nat 100:687–690

Wilson SM, Taylor JJ, Mackie TA, Patterson DA, Cooke SJ, Willmore WG (2014) Oxidative stress in Pacific salmon (Oncorhynchus spp.) during spawning migration. Physiol Biochem Zool 87:346–352

Winward JD, Ragan CM, Jimenez AG (2018) Cellular metabolic rates and oxidative stress profiles in primary fibroblast cells isolated from virgin females, reproductively experienced females, and male Sprague-Dawley rats. Physiol Rep 16:e13909

Wishart GJ (1984) Effects of lipid peroxide formation in fowl semen on sperm motility, ATP content and fertilizing ability. J Reprod Fertil 71:113–118

Wu SH, Oldfield JE, Whanger PD, Weswig PH (1973) Effect of selenium, vitamin E, and antioxidants on testicular function in rats. Biol Reprod 8:625–629

Xu Y-C, Yang D-B, Speakman JR, Wang D-H (2014) Oxidative stress in response to natural and experimentally elevated reproductive effort is tissue dependent. Funct Ecol 28(2):402–410

Yang D-B, Xu Y-C, Wang D-H, Speakman JR (2013) Effects of reproduction on immunosuppression and oxidative damage, and hence support or otherwise for their roles as mechanisms underpinning life history trade-offs, are tissue and assay dependent. J Exp Biol 216:4242–4250

York JE, Radford AN, Groothuis TG, Young AJ (2016) Dominant male song performance reflects current immune state in a cooperatively breeding songbird. Ecol Evol 6:1008–1015

Young AJ, Carlson AA, Monfort SL, Russell AF, Bennett NC, Clutton-Brock T (2006) Stress and the suppression of subordinate reproduction in cooperatively breeding meerkats. Proc Natl Acad Sci USA 103:12005–12010

Yu Y, Shen G, Zhu H, Lu Y (2010) Imidacloprid-induced hormesis on the fecundity and juvenile hormone levels of the green peach aphid *Myzus persicae* (Sulzer). Pest Biochem Physiol 98:238–242

Zanuncio JC, Jusselino-Filho P, Ribeiro RC, Zanuncio TV, Ramalho Fde S, Serrão JE (2011) Hormetic responses of a stinkbug predator to sublethal doses of pyrethroid. Bull Environ Contam Toxicol 87:608–614

Zhao XY, Zhang JY, Cao J, Zhao ZJ (2015) Oxidative damage does not occur in striped hamsters raising natural and experimentally increased litter size. PLoS One 10:e0141604

Zimmer C, Spencer KA (2015) Reduced resistance to oxidative stress during reproduction as a cost of early-life stress. Comp Biochem Physiol Part A 183:9–13

Zini A, de Lamirande E, Gagnon C (1995) Low levels of nitric oxide promote human sperm capacitation in vitro. J Androl 16:424–431

# Chapter 10
# The Long and the Short of It: How Oxidative Stress Drives Cellular Senescence, Ageing, and Lifespan of Animals

**Abstract** The process of lifetime progressive deterioration in cellular function and organismal performance is what we usually call ageing. Cellular senescence is a form of stable cell cycle arrest that represents a hallmark of ageing, and reactive oxygen species are important triggers of cellular senescence. We have much empirical evidence that ageing occurs in free-living animals, and evolves in connection with life-history strategies. We have a limited understanding of the molecular mechanisms that would drive their ageing. We have evidence that longer-lived species generally have (i) lower generation of reactive oxygen species or oxidative damage, (ii) higher capacity to repair damage, (iii) lower concentrations of endogenous antioxidants, or (iv) cells made of molecules that are more resistant to oxidative damage than shorter-lived species. However, there are a growing number of studies providing contradictory data, raising doubt on the generalisability of the impact of reactive oxygen species on ageing. In this chapter, I have initially presented a general overview of the historical phases that led to the free radical or oxidative stress hypotheses of ageing. Then, I have (i) reviewed the evidence in (dis)favour of the roles of reactive oxygen species and oxidative stress as molecular drivers of cellular senescence and ageing, (ii) discussed under which conditions oxidative stress can be a promoter of longevity, (iii) illustrated how cross-sectional and longitudinal studies of captive or free-living animals do not always support the general hypothesis that oxidative damage and antioxidants should accumulate and decline with age, respectively, and (iv) presented the conflicting results on the link between oxidative status markers and survival probability. In so doing, I critically appraise our current theoretical and methodological approaches, indicating the need for alternative testable hypotheses, and the need to move beyond the oversimplistic and unsupported predictions '*high oxidative damage at time x is linked to lower survival probability*' and '*two individuals of the same age showing, e.g. similar levels of a given marker of oxidative damage at time t, should have a similar probability of survival per unit of time*'. Regulation of oxidative status homeostasis is a dynamic, continuous, and complex process that needs to be better integrated into the research on cellular senescence and ageing of free-living organisms.

© The Author(s), under exclusive license to Springer Nature Switzerland AG 2024

D. Costantini, *The Role of Organismal Oxidative Stress in the Ecology and Life-History Evolution of Animals*, Fascinating Life Sciences,
https://doi.org/10.1007/978-3-031-65183-0_10

## 10.1 Introduction

Bodies of living organisms wear over time, but they do so at widely different rates. Some species live for hundreds of years; the lifetime of some other species is limited to only a few hours (e.g. Kirkwood et al. 2005; de Magalhães and Costa 2009; Tacutu et al. 2013; Jones et al. 2014; see also https://genomics.senescence.info/species/index.html). Conspecific individuals can also widely differ in longevity. Why do animals have such different lifespans? Are there any defined molecular mechanisms underlying this astonishing variation in lifespan? What are the factors that limit how long we can live for? Most animals gradually degenerate after reaching sexual maturity. This process of progressive deterioration in cellular function and organismal performance is what we call ageing. A hallmark of ageing is cellular senescence. It refers to a state of stable cell cycle arrest in which proliferating cells become resistant to growth-promoting stimuli, typically in response to DNA damage, such as that caused by reactive oxygen species (van Deursen 2014; Parry and Narita 2016; Mylonas and O'Loghlen 2022).

Leonard Hayflick was the first to describe cellular senescence. Hayflick observed that human foetal fibroblasts eventually stopped dividing, but remained viable and metabolically active after prolonged time in culture (Hayflick and Moorhead 1961; Hayflick 1965; van Deursen 2014; Parry and Narita 2016; Mylonas and O'Loghlen 2022). Cellular senescence plays an important role in development because it participates in the regulation of several relevant mechanisms, such as tissue remodelling and the elimination of unwanted cells as the tissues develop (Mylonas and O'Loghlen 2022). In particular, cellular senescence is one of the reasons organisms age, and accumulation of senescent cells promotes accelerated biological ageing and age-related diseases. The accumulation of senescent cells may disrupt functionality of tissues, and disrupt the regulation of stem cells limiting the regenerative potential of adult stem cells as a consequence (Sharpless and DePinho 2007; van Deursen 2014; Parry and Narita 2016; Mylonas and O'Loghlen 2022).

The scientific community assumed for decades that all free-living animals do not age because they die due to predation or pathogens before showing signs of ageing. Research of the last 20 years or so showed that this assumption is incorrect. We have now much empirical evidence that ageing indeed occurs in many free-living animals, and evolves in connection with environmental circumstances (e.g. extrinsic mortality risk owing to predation), and life-history strategies (Medawar 1952; Williams 1957; Kirkwood 1977; Kirkwood and Rose 1991; Austad 1993; Blanco and Sherman 2005; Descamps et al. 2006; Møller 2006; Munshi-South and Wilkinson 2006; de Magalhães et al. 2007; Jones et al. 2008; Monaghan et al. 2008; Nussey et al. 2008, 2013; Reed et al. 2008; Bouwhuis et al. 2010; Peterson et al. 2010; Ricklefs 2010; Wasser and Sherman 2010; Shefferson et al. 2017). Moreover, ageing-associated diseases can also be observed (Fig. 10.1). However, we have a limited understanding of the molecular mechanisms that would drive ageing and underlie the wide variation in ageing patterns and rates across animals.

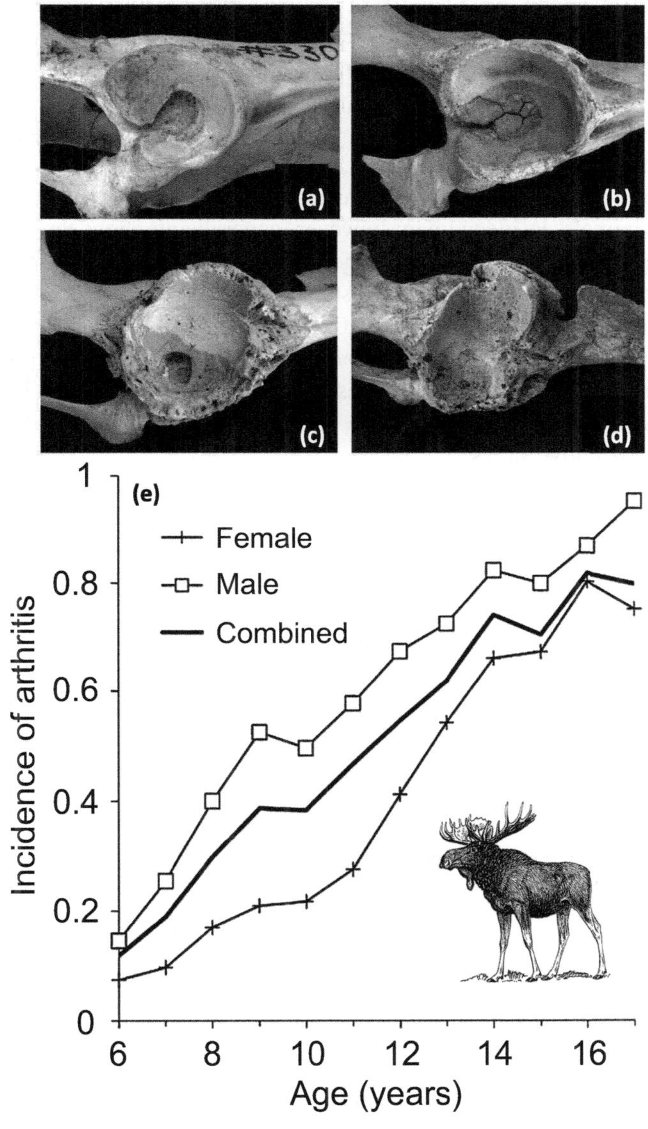


**Fig. 10.1** Right pelvic (coxofemoral) joints of moose *Alces alces* illustrate the progressive bony deterioration associated with osteoarthritis, a degenerative disease that worsens with age. (**a**) Normal joint capsule with open acetabular fossa, which contains the ligament of the head of the femur, (**b**) yields to closure of the acetabular fossa, cartilage deterioration, and dorsal eburnation (bone-on-bone articulation), (**c**) then dorsal displacement of joint articulation from biomechanical weight-bearing forces and, finally, and (**d**) continued displacement as a prelude to dislocation. (**e**) Incidence of osteoarthritis in relation to sex and age of moose. This study provided evidence for a link between early life experiences and the emergence of a degenerative disease during ageing in moose: individuals that experienced poorer nutritional conditions early in life were more likely to develop osteoarthritis, and had also reduced life expectancies. Reprinted with minimal alternations from Peterson and collaborators (2010) with permission of John Wiley and Sons

Research on ageing has been classically performed using a few laboratory animal models or observational data collected from human populations. This research has provided a wealth of information about the possible molecular controllers of ageing. However, our capability to extrapolate across the tree of life is unclear because there are molecular pathways of ageing that do not appear to be conserved across animal taxa. In recent decades, long-term field studies and experimental investigations on a number of animal species has provided widespread evidence that free-living animals do indeed age; however, generalising about the importance of mechanisms that act as drivers of ageing is difficult owing to the enormous variation in life styles (e.g. Møller et al. 2005; Costantini 2008; Monaghan et al. 2008; Nussey et al. 2008, 2013; Bouwhuis et al. 2010; Martin and Festa-Bianchet 2011; Cohen 2018; Kramer et al. 2021). The nature of the mechanisms that determine this variation in ageing patterns and rates or lifespan is therefore unresolved. One potential conserved driver of lifespan lies in the regulation of oxidative status homeostasis and cellular oxidative stress.

In this chapter, I have initially presented a general integrated evolutionary view of ageing. Then, I have reviewed the evidence in (dis)favour of the roles of reactive oxygen species and oxidative stress as molecular generators of the variation in lifespan amongst animals. I have also discussed under which conditions oxidative stress can be a promoter of longevity. Finally, I have (i) discussed how cross-sectional and longitudinal studies of captive or free-living animals do not always support the general hypothesis that oxidative damage and antioxidants should accumulate and decline with age, respectively, and (ii) presented the conflicting results on the link between oxidative status markers and survival probability.

## 10.2 An Integrative Physiological View of Theories of Ageing

Over a century ago, Max Rubner (1908) observed that larger mammals live longer and have slower metabolisms than smaller mammals. Raymond Pearl (1928) expanded and further elaborated Rubner's observations in the book *The Rate of Living*. Pearl suggested a possible causal relationship between having a slow metabolism and a long lifespan (*rate of living theory of ageing*). Pearl observed that fruit flies lived shorter at higher temperatures, and that raising of ambient temperature dramatically increased their metabolic rate. In the 1950s, Gershman and collaborators (1954) and Harman (1956) provided mechanistic support for the rate of living theory of ageing. They suggested that the generation of free radicals might be the mechanism linking metabolic rate to longevity (*free radical hypothesis of ageing*, Harman 1956) because a given amount of oxygen consumed by cells would leak out of the mitochondria in the form of superoxide free radicals (see Chap. 1). In other words, species with fast metabolisms would generate higher amounts of free radicals than species with slower metabolisms, and these free radicals would cause damage to cells, eventually leading to cellular senescence and death (Harman 1956). Ironically, whilst the free radical hypothesis of ageing gained popularity, the rate of living theory lost credibility with time because correlative and experimental investigations later showed that (i) an increase in metabolic rate does not necessarily result in decreased lifespan, (ii) the generation of free radicals is mildly linked to metabolic rate, (iii) birds generally have higher metabolic rates and body temperatures than mammals of similar body size, but live substantially longer, (iv) bats live much longer than other mammalian taxa with similar metabolic rates, and (v) some supposed correlations between longevity and metabolic rate were likely inflated because analyses did not take into account potential confounding effects of body mass and phylogeny (e.g. Storer 1967; Beckman and Ames 1998; Weinert and Timiras 2003; Speakman et al. 2004; Speakman 2005a,b; de Magalhães et al. 2007; Hulbert et al. 2007; Furness and Speakman 2008). For example, studies on microchiropteran bats (Filho et al. 2007), colubrid snakes (Robert et al. 2007), or zebra finches *Taeniopygia guttata* (Moe et al. 2009) did not find any relationship between mortality and basal metabolic rate. Experimental increase in energy expenditure did not shorten lifespan of field voles *Microtus agrestis* (Selman et al. 2008) or laboratory mice (Vaanholt et al. 2009, 2010). Although these studies suggested that metabolic rate is probably not the main controller of longevity, we have also several studies showing that it is certainly part of the picture (e.g. Ruggiero et al. 2008; Cohen et al. 2008; Weirsma et al. 2012; Scholer et al. 2019; Zhang et al. 2021).

Over time, the free radical theory of ageing has been refined several times, such as the *mitochondrial theory of ageing* (Harman 1972; Miquel et al. 1980) and the *oxidative stress theory of ageing* (Yu and Yang 1996; Sohal et al. 2002). These novel theories made the point that the free radical-induced oxidative damage, especially to mitochondria, would be the main driver of the cellular senescence rate and ageing (Beckman and Ames 1998; Finkel and Holbrook 2000). Assumptions of these theories were, however, often criticised because they did not receive any support from empirical evidence. The rise, fall, and resurrection of these theories have actually characterised research on ageing over the last decades. Why is that? Do we have evidence that regulation of cellular oxidative status is relevant for cellular senescence and ageing? Studies on transgenic animals lacking specific antioxidant enzymes, or having the expression of these enzymes modified, supported their key role in health, but provided conflicting results in terms of longevity outcomes (e.g. Van Remmen et al. 2003; Pérez et al. 2009a; Lei et al. 2016). Several of these studies also supported the expected important role of these enzymes in protection against oxidative damage. For example, *Drosophila* fruit flies genetically modified to overexpress superoxide dismutase 1 and catalase lived up to a third longer than control fruit flies (Orr and Sohal 1994). Importantly, the stronger impact on lifespan was recorded when both enzymes were overexpressed in the same individual, because this conferred greater resistance of DNA and proteins against the pro-oxidant effects of ionising radiation (Orr and Sohal 1994). However, other studies did not find any effects of co-overexpression of both superoxide dismutase 1 and catalase on the lifespan of fruit flies (Sun and Tower 1999). Exposure to a pro-oxidant molecule (diquat) of mice lacking superoxide dismutase 1 (SOD1 knockout mice, *Sod1*$^{-/-}$) caused a significant increase in oxidative damage levels within an hour of treatment, causing lethality (Han et al. 2008). By contrast, it increased levels of oxidative damage within a few hours in wild-type mice without causing any lethality (Han et al. 2008). Similarly, mice lacking superoxide dismutase 2 (SOD2 knockout mice, *Sod2*$^{-/-}$) lived for a few weeks, and had increased susceptibility to oxidative mitochondrial injury in the central nervous system neurons, cardiac myocytes, and other metabolically active tissues after postnatal exposure to ambient oxygen concentrations (Lebovitz et al. 1996). Similarly, other experiments that involved SOD2 knockout mice found that they lived for a few weeks and experienced high levels of oxidative damage to DNA (Melov et al. 1998, 1999). Administration of exogenous antioxidants attenuated the negative effects of not having SOD2 suppressed, further strengthening the importance of oxidative damage as a promoter of senescence in these mice (Melov et al. 1998). Studies on other antioxidant genes found similar results. For example, Gpx1 knockout mice (*Gpx1*$^{-/-}$) were highly sensitive to the oxidant paraquat, suffering high mortality (de Haan et al. 1998). Paraquat induced the upregulation of Gpx1 in normal cells, reinforcing a role for Gpx1 in protection against paraquat toxicity (de Haan et al. 1998). Moreover, cortical neurons from *Gpx1*$^{-/-}$ mice were more susceptible to hydrogen peroxide; 30% of neurons from *Gpx1*$^{-/-}$ mice died when exposed to hydrogen peroxide, whereas the wild-type controls were not affected by hydrogen peroxide (Fig. 10.2) (de Haan et al. 1998). Similarly, *Gpx1*$^{-/-}$ mice experimentally exposed to diquat experienced a rapid increase in oxidative damage and higher mortality compared to wild-type mice (Han et al. 2008). Transgenic mice that overexpressed the antioxidant enzyme catalase lived longer, had a delayed development of cardiac pathology and cataract development, and experienced reduced levels of oxidative damage, generation of hydrogen peroxide, and development of mitochondrial deletions (Schriner et al. 2005).

In summary, these studies showed that antioxidant enzymes and oxidative damage have important functional consequences for the organism. However, a critical evaluation of these studies carried out on transgenic models suggested that they have a stronger impact on health than on ageing (Salmon et al. 2010). Some experimental complications were also due to the difficulty of separating the effects of transgene expression on development from its effects on adult lifespan, and the necessity of having identical genetic backgrounds for both control and transgenic animals (Golden et al. 2002). In addition, the interpretation of studies on knockout organisms is not straightforward because knockout of a gene may have a number of downstream consequences. For example, Pérez-Estrada and collaborators (2019) showed that lipid metabolism rather than oxidative stress was the determinant of lifespan of mice lacking the antioxidant enzyme catalase (*Cat*$^{-/-}$). Thus, many other mechanisms, that are relevant for ageing, can be affected in animals with knockout of antioxidant genes, and this has to be addressed in interaction with oxidative stress.

A number of studies suggested that the lipid composition of the cell membrane is an important driver of cellular senescence and ageing because it affects the vulnerability of cells to oxidative stress. In particular, the *homeoviscous-longevity adaptation hypothesis of ageing* (Pamplona et al. 2002) and the *membrane-pacemaker hypothesis of ageing* (Hulbert 2005) state that long-lived species have fatty acid profiles that make their cell membranes more resistant to oxidative damage than short-lived species. Fatty acids actually differ substantially in their vulnerability to free radical-induced oxidation, with polyunsaturated fatty acids being

**Fig. 10.2** Cell viability of Gpx1 wild-type and knockout neuronal cultures after challenge with hydrogen peroxide ($H_2O_2$). Cultured neurons from Gpx1 wild-type and knockout mice were exposed to (**a**) 65 mmol $L^{-1}$ $H_2O_2$, (**b**) 80 mmol $L^{-1}$ $H_2O_2$, or (**c**) 100 mmol $L^{-1}$ $H_2O_2$ on day 8 of culture. Cell viability was determined 24 h after $H_2O_2$ treatment. Asterisks indicate significant differences between wild-type (control) and knockout mice. Values are shown as mean and standard error. Reprinted with minor modifications from de Haan and collaborators (1998). © 1998 ASBMB. Currently published by Elsevier Inc.; originally published by American Society for Biochemistry and Molecular Biology; distributed under the terms of the Creative Commons Attribution License (CC BY)

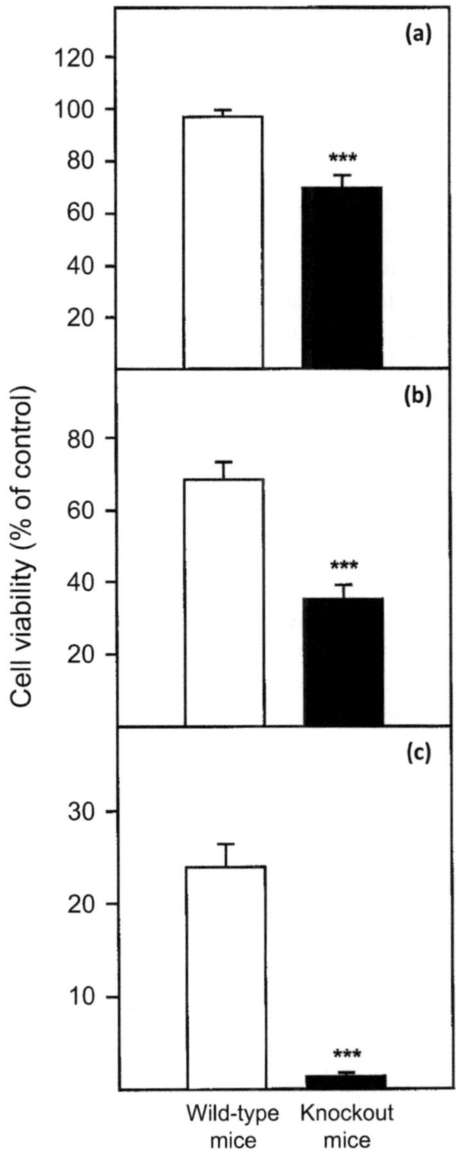


generally the most vulnerable (Chap. 1). Therefore, the susceptibility of cell membranes to peroxidation can be determined from their composition in fatty acids (Holman 1954). The peroxidation index is used as an estimate of cell membrane susceptibility to oxidation; higher values of this index indicate that the cell membrane has a higher content of those fatty acids that are most susceptible to oxidation. Comparative studies indicated that cell membranes of birds are richer in fatty acids and less susceptible to peroxidation than cell membranes of mammals (Hulbert et al. 2002a,b), and the peroxidation index is negatively correlated with the maximum lifespan in birds and mammals (Hulbert et al. 2007; Hulbert 2008). Longer-lived bird species have also more monounsaturated and less polyunsaturated fatty acids in liver than shorter-lived species (Brand et al. 2003; Gálvan et al. 2015). Similarly, longer-lived fish of the genus *Nothobranchius* have more saturated membranes and, therefore, a lower susceptibility to oxidative damage (de Costa et al. 2020), and long-lived marine bivalves have mitochondrial membranes less sensitive to in vitro-initiated peroxidation compared with medium or short-lived marine bivalves (Istomina et al. 2023).

Studies on the long-lived naked mole-rat *Heterocephalus glaber* called into question oxidative stress-related theories of ageing. Naked mole-rats live about 5 times longer than predicted given their body mass (Buffenstein 2005), and although their basal metabolic rate is 30% lower than expected (O'Connor et al. 2002), this does not account for their high life expectancy. Despite its extraordinary longevity, mitochondria of naked mole-rats generate more reactive oxygen species than mitochondria of mice. Moreover, the activity of glutathione peroxidase was almost undetectable in naked mole-rat tissues (Andziak et al. 2005) and, at an early age (7% of maximum lifespan), tissues of naked mole-rats had higher levels of oxidative damage to lipids, proteins, and DNA than those of mice (Andziak et al. 2006). The exceptional longevity of naked mole-rats might be due to their low peroxidation index: they have proportionately large amounts of peroxidation-resistant monounsaturated fatty acids, but low amounts of peroxidation-sensitive polyunsaturated fatty acids, particularly the docosahexaenoic acid (Hulbert et al. 2006, 2007). Moreover, naked mole-rats have proteins that are particularly resistant to oxidative damage or have better protein homeostasis (Fig. 10.3) (Pérez et al. 2009b; De Waal et al. 2013). However, both explanations for the exceptional longevity of naked mole-rats are undermined by the fact that naked mole-rats have high levels of lipid peroxidation and protein damage.

In other taxa, the link between peroxidation index and longevity appears stronger. For example, humans and echidnas live about 5 times longer and have membrane peroxidation indices much lower than expected given their body size (Pamplona et al. 1996; Hulbert et al. 2008). In birds, *Procellariiformes* (petrels and albatrosses) are exceptionally long-lived animals, whereas *Galliformes* (fowl) have short lives. The quantification of the myocardial composition in phospholipids showed that *Procellariiformes* had a lower peroxidation index than *Galliformes*, mainly due to the proportionately higher amounts of monounsaturated fatty acids and lower amounts of polyunsaturated (particularly n-6) fatty acids in seabirds compared to the fowl (Buttemer et al. 2008). Interestingly, *Procellariiformes* had a 36% lower average peroxidation index than *Galliformes*, and this difference was exactly the magnitude expected given the 4.5-fold greater average longevity of petrels compared to the fowl (Buttemer et al. 2008). However, other comparative studies found little support for a direct link between high levels of polyunsaturated fatty acids in cell membranes and reduced longevity (Valencak and Ruf 2007; Kumar et al. 2021).

The fatty acid desaturation pathway (placement of double bonds in a fatty acid molecule) and the deacylation-reacylation cycle (introduction of polyunsaturated fatty acids into cell membranes) are two important mechanisms that contribute to generate

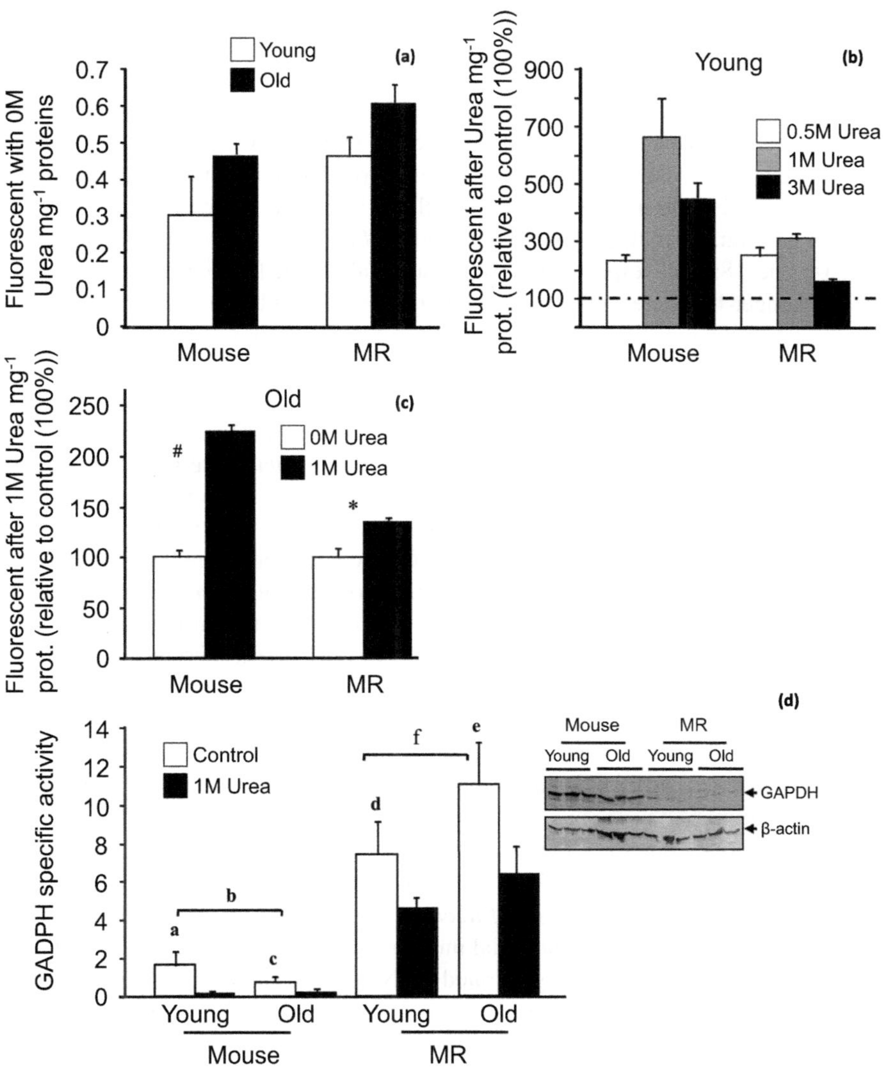


**Fig. 10.3** Proteins from both young and old naked mole-rats are extremely resistant to protein unfolding, whereas those from mice are markedly more susceptible. (**a**) Basal protein unfolding is unchanged with age in mice and naked mole-rats. (**b**) Young mice have markedly higher levels of protein unfolding after urea treatment than do young naked mole-rats. (**c**) BisANS (fluorescent probe, the intensity of its fluorescence is indicative of the changes in protein conformation) incorporation in response to 1 M urea is abrogated with age, although old mice show less resistance to unfolding than do old naked mole-rats. (**d**) GAPDH (glyceraldehyde 3-phosphate dehydrogenase) activity declines less with urea treatment in naked mole-rats than it does in mice regardless of age, indicating that naked mole-rats are better able to maintain protein structure and function than mice. A representative Western blot of GAPDH expression is shown. Letters a–e represent significant differences amongst comparative datasets, whilst f (representing age-related differences in untreated GAPDH activity in naked mole-rats) is not significant. Values are shown as mean and standard error. Reprinted with minor modifications from Pérez and collaborators (2009b)

variation between species in the fatty acid profile of tissues (Pamplona et al. 2002). Estimates of delta-5 and -6 desaturase activities indicated that they are lower in longer-lived than in shorter-lived species (Pamplona and Barja 2003; Pamplona 2008). Accordingly, docosahexaenoic (22:6n-3) and arachidonic (20:4n-6) fatty acids decreased and linoleic (18:2n-6) and α-linolenic (18,3n-3) fatty acids increased from shorter- to longer-lived species, respectively. This pattern was probably due to the role of desaturases as rate-limiting enzymes of the n-3 and n-6 pathways that synthesise the polyunsaturated fatty acids 20:4n-6 and 22:6n-3 from their dietary precursors 18:2n-6 and 18:3n-3, respectively (Pamplona et al. 2002; Pamplona and Costantini 2011). Longer-lived species had also a low degree of membrane fatty acid desaturation based on the redistribution amongst types of polyunsaturated fatty acids without any changes in the total content of polyunsaturated fatty acids, average chain length, and phospholipid distribution. Hence, animal species might rely on the desaturation of fatty acids in order to decrease the sensitivity of cell membranes to lipid peroxidation and lipoxidation-derived damage without strongly influencing the fluidity/microviscosity, a fundamental property of cell membranes, because it affects the proper function of receptors, ion pumps, and transport of metabolites (Pamplona et al. 2002). This possible molecular strategy would be favoured because membrane fluidity increases strongly with the introduction of the first double bond, whilst it increases comparatively less with the introduction of a second double bond; the introduction of additional double bonds has a weak effect on membrane fluidity as a consequence (Brenner 1984). The susceptibility of fatty acids to oxidative damage increases with the increase in double bonds. Hence, substitution of fatty acids with four or six double bonds with fatty acids that have only two or three double bonds would enable animals to increase their resistance to oxidative damage, whilst maintaining adequate membrane fluidity for cell function.

Other authors suggested that the expression of mitochondrial uncoupling proteins is responsible for the weak relationship between metabolic rate and lifespan (*uncoupling to survive hypothesis of ageing*; Brand 2000; Speakman et al. 2004). Uncoupling proteins are mitochondrial inner membrane proteins that are capable of dissipating the proton gradient generated by NADH-powered pumping of protons from the mitochondrial matrix into the mitochondrial intermembrane space. In so doing, they are also capable of reducing the mitochondrial generation of reactive oxygen species. During the transition of mitochondria from state 4 (resting) to state 3 (respiratory active, producing ATP), it has been observed a lower leak of reactive oxygen species at the respiratory chain (Loschen et al. 1971; Barja et al. 1994; Herrero and Barja 1997a; Barja 1998). Uncoupling proteins might be responsible for the decreased generation and leak of reactive oxygen species (e.g. Talbot et al. 2004; Rey et al. 2010). Thus, uncoupling proteins potentially represent a biochemical response to mitigate the generation of free radical-induced oxidative damage (Chap. 1).

Mice lacking superoxide dismutase 2 usually do not survive more than a few weeks owing to the generation of high levels of oxidative damage in mitochondria (Lebovitz et al. 1996). Andrews and Horvath (2009) observed that SOD2 deficiency was lethal, but they also noted that the expression of the uncoupling protein 2

(UCP2) influenced the time of death. SOD2-deficient mice survived only a week if they lacked UCP2 expression (*UCP2*$^{-/-}$), two weeks if they were heterozygous for UCP2 (*UCP2*$^{+/-}$), and three weeks if they were homozygous for UCP2 (*UCP2*$^{+/+}$), respectively. Thus, on one side, mitochondrial oxidative damage can affect longevity and, on the other side, uncoupling proteins are capable of mitigating the potentially deleterious effects of mitochondrial reactive oxygen species.

The question then is whether uncoupling proteins contribute to generate variation between conspecific individuals or species in lifespan. Speakman and collaborators (2004) found that mice in the upper quartile of metabolic intensity had greater resting oxygen consumption, longer lives, and higher expression of the uncoupling protein-3 than mice in the lowest metabolic intensity quartile. Thus, individuals with high metabolism were more uncoupled, and this uncoupling state of mitochondria might have driven their longer lives despite the higher energy expenditure, as predicted by the uncoupling-to-survive hypothesis of ageing. Other studies that transgenically overexpressed the uncoupling protein-1 in murine skeletal muscle also observed an extension of lifespan (Keipert et al. 2011), but this longer life was not associated with a reduction of oxidative damage (Keipert et al. 2013). However, comparative studies provided indirect evidence that variation between species in lifespan is weakly accounted for by variation in uncoupling proteins. The mitochondrial membranes of smaller species of birds and mammals are more uncoupled than those of larger species since the non-phosphorylating respiration rates decrease more strongly than the phosphorylating respiration rates as body mass increases, yet larger species usually live longer (Buttemer et al. 2010; see also Porter and Brand 1993; Brand et al. 2003). Lambert and collaborators (2007) also found that a low generation of reactive oxygen species in mitochondria enhanced longevity in vertebrates. Enhanced longevity may be causally associated with low generation of reactive oxygen species in mitochondria. If we assume a similar expression of uncoupling proteins, other mechanisms might be more relevant as drivers of variation in lifespan. Kern and collaborators (2009) found that the experimental increase of oxidative stress by cadmium exposure or hypoxia-hyperoxic-reoxygenation did not cause the upregulation of the uncoupling protein-5 in eastern oysters *Crassostrea virginica*, indicating that uncoupling proteins are not relevant mitigators of oxidative stress in bivalves. Finally, Stier and collaborators (2014) found that experimentally induced uncoupling of mitochondria did not reduce oxidative damage nor lengthen life in zebra finches.

Reactive oxygen species might drive senescence rate through their action on the length of telomeres, regions of repetitive nucleotide sequences (TTAGGG) that work as protective caps of chromosomes. The *cell senescence-telomere hypothesis of ageing*, as coined by Weinert and Timiras (2003), likely found its roots in 1965, when Hayflick (1965) proposed that the replicative capacity of cells is limited, and this would be one reason for cellular senescence. In the early 1990s, Harley and collaborators (1990) showed that telomeres shorten during ageing of human fibroblasts, indicating that the shrinking of telomeres might be an important mechanism controlling replicative senescence. Telomeres shorten with every mitotic cycle; when telomeres reach a critical length, they induce a permanent arrest in the cell cycle (Hornsby 2003; Haussmann and Marchetto 2010). The rate of telomere shortening is, however, not constant, rather it varies during the lifetime (Heidinger et al. 2012), and accelerates when the organism is exposed to stressful events (Epel et al. 2004; Hau et al. 2015).

Over the last two decades, there has been increasing interest in understanding the ecological and evolutionary relevance of telomere length and dynamics (Nakagawa et al. 2004; Monaghan and Haussmann 2006; Haussmann and Marchetto 2010; Monaghan 2010; Simons 2015; Angelier et al. 2018; Young 2018; Tobler et al. 2022). We have learned that telomere dynamics varies as a function of lifespan (Haussmann et al. 2005; Bize et al. 2009; Salomons et al. 2009; Heidinger et al. 2012; Angelier et al. 2013; Tricola et al. 2018), lifetime reproductive success (Pauliny et al. 2006), physiological stress (Epel et al. 2004; Kotrschal et al. 2007; Hau et al. 2015), growth rate (Jennings et al. 1999; Hall et al. 2004; Heidinger et al. 2012), population (Canestrelli et al. 2021), or infection (Ilmonen et al. 2008). Thus, there is good support for a link between telomere dynamics and lifestyle, and for variation in how the telomere system operates in different kinds of animals. Indirectly, this evidence provides support for the role of reactive oxygen species as controllers of cellular senescence because we have evidence that telomeres are highly sensitive to oxidation (von Zglinicki 2002; Houben et al. 2008; Jurk et al. 2014; Reichert and Stier 2017; Armstrong and Boonekamp 2023). There is, however, much that we do not know about the links between telomere biology and oxidative stress.

Most of the hypotheses of ageing that we have seen so far do not make specific predictions about the impact of different types of molecular damage on cellular senescence and ageing. Moreover, some hypotheses postulate that accrual of such molecular damage over time would be responsible for the loss in physiological performance; however, there are many indications that this damage alone is insufficient to satisfactorily explain the age-associated functional losses. Sohal and Orr (2012) proposed that we should rather look at how the cell redox state changes over time. Their *redox stress hypothesis of ageing* states that ageing-associated functional losses are primarily caused by a progressive shift in the redox state of the cells, which leads to the overoxidation of redox-sensitive protein thiols and the consequent disruption of the redox-regulated signalling mechanisms (Sohal and Orr 2012). The oxidation of protein thiols would cause dysfunction of basic cellular activities, and the cellular redox state would become increasingly more pro-oxidising in the latter part of life (e.g. increases in oxidised glutathione or protein mixed disulphides). Thus, the loss in reducing power (thiol groups) would be the main mechanism underlying ageing. This hypothesis is based on the definition of oxidative stress as a pro-oxidant shift in the thiol redox state and the resulting dysfunction of redox-sensitive proteins (Jones 2006). Sohal and Orr (2012) also raised other criticisms concerning the link between oxidative damage and ageing. For example, the steady-state levels of modified macromolecules are a snap-shot of the dynamic balance between the processes of degradation, repair, and biosynthesis; thus, they do not provide accurate quantifications of the cumulative damage (Sohal and Orr 2012). It is also unclear whether steady-state concentrations of oxidised macromolecules increase with age determining significant decreases in the net amount of the unmodified molecules (Sohal and Orr 2012). Studies on oxidative damage to proteins also showed that a type of damage, called carbonylation, does not affect the functionality of all the proteins (Yarian et al. 2005). Thus, are reactive oxygen species and antioxidants important for longevity? And what evidence do we have about their links with lifespan or survival in free-living animals?

## 10.3 Reactive Oxygen Species, Antioxidants, and Longevity: From Ephemeral Lifespans to Supercentenarians

We have much evidence converging on the conclusion that longer-lived species have generally lower generation of reactive oxygen species or oxidative damage, higher capacity to repair damage, lower concentrations of endogenous antioxidants, or cell membranes made of molecules that are more resistant to oxidative damage than shorter-lived species (e.g. Lopez-Torres et al. 1993; Orr and Sohal 1994; Perez-Campo et al. 1994; Herrero and Barja 1997b, 1998; Brunet-Rossinni 2004; Chaudhuri et al. 2006; Philipp et al. 2006; Sanz et al. 2006; Hulbert et al. 2007; Lambert et al. 2007; Sasaki et al. 2008; Abele et al. 2009; Begum et al. 2009; Pamplona and Costantini 2011; Montgomery et al. 2011, 2012; Barja 2013; Delhaye et al. 2016; Xia and Møller 2018; Martínez-Lendech et al. 2019; Vágási et al. 2019; Domínguez-de-Barros et al. 2023). We have also experimental evidence that increased intake of antioxidants can sometimes slow down age-related declines in physical performance or cognition (Fig. 10.4) (e.g. Cherubini et al. 2005; Pike et al. 2010; Vyas et al. 2024).

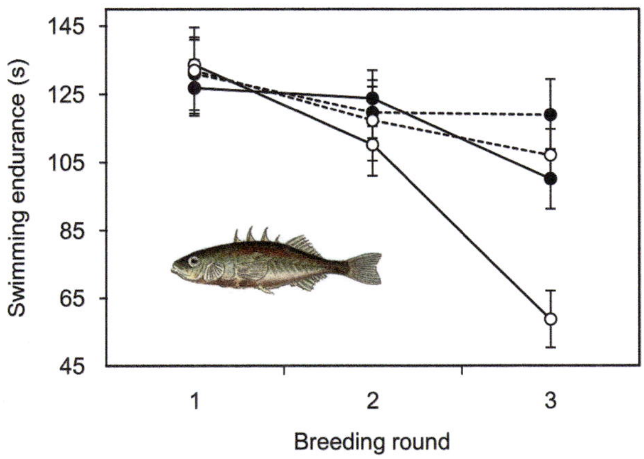


**Fig. 10.4** Change in sustained swimming endurance (a measure of locomotor senescence) over three successive breeding rounds for males on high- (black data points) and low- (white data points) carotenoid diets in the high (solid lines) and low (dashed lines) reproduction groups. Values are shown as mean and standard error. Reprinted with minor modifications from Pike and collaborators (2010) with permission of Oxford University Press

In the 1990s, Barja's group showed that mitochondria of pigeons consumed much more oxygen than those of rats, but also generated smaller levels of reactive oxygen species, and converted a greater proportion of oxygen into water (Barja et al. 1994; Herrero and Barja 1997a,b, 1998; Barja 1998). Using data collected from a large number of free-living species, Xia and Møller (2018) and Vágási and collaborators (2019) showed that relatively long-lived bird species had high levels of plasma non-enzymatic antioxidant capacity and low plasma levels of malondialdehyde, respectively (Fig. 10.5). The bivalve *Artica islandica*, which can live over 500 years of age (Butler et al. 2013), had reduced levels of protein carbonyls in gills compared with *Mercenaria mercenaria* (Ungvari et al. 2011), generated significantly less hydrogen peroxide, and had higher capacity to remove it than the two short-lived species *Mya arenaria* (maximum longevity of 28 years) and *Spisula solidissima* (maximum longevity of 37 years) under various conditions of mitochondrial respiration (Fig. 10.6) (Munro et al. 2013, 2023), and experienced increased levels of protein carbonyls after approximately a century of age (Sosnowska et al. 2014). However, there are also many notable exceptions, casting doubt on the generalisability of the free radical hypothesis of ageing and its derivatives, such as the case of the naked mole-rat illustrated in the previous paragraph.

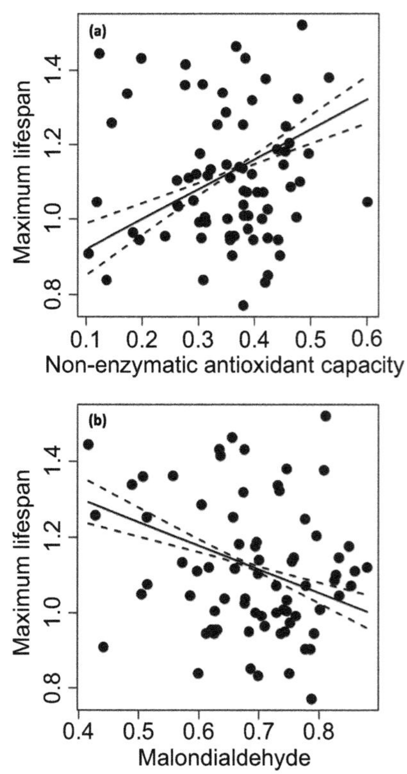


**Fig. 10.5** Bird species that live longer (maximum lifespan) have (**a**) higher levels of plasma non-enzymatic antioxidant capacity and (**b**) lower levels of oxidative damage to lipids (malondialdehyde). Model fit and standard error are shown as continuous and dashed lines, respectively. All the variables were log10-transformed. Reprinted with minor modifications from Vágási and collaborators (2019) with permission of John Wiley and Sons

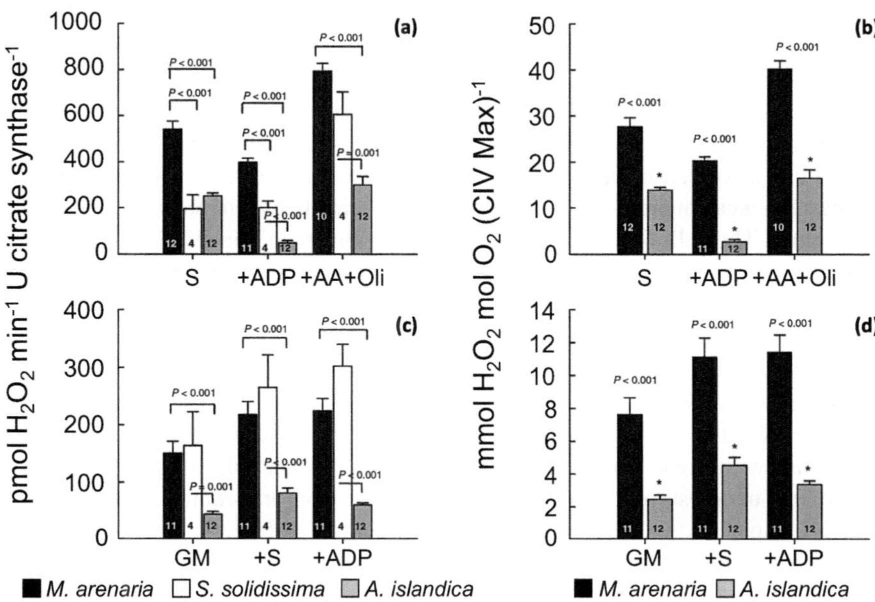


**Fig. 10.6** Hydrogen peroxide production of isolated mantle mitochondria. Results for the sequential addition of substrates and inhibitors (from left to right) are normalised with citrate synthase activity (panels **a** and **c**) and with respiratory capacity (CIV) activity (panels **b** and **d**). Substrates: G, glutamate; M, malate; S, succinate; inhibitors: AA, antimycin A; Oli, oligomycin. Species are presented in order of increasing maximum reported longevity from left to right: *M. arenaria* = 28 years, *S. solidissima* = 37 years, and *A. islandica* = 507 years. Significant differences are for Tukey's HSD (left panels) and t-test (right panels). Values are shown as mean and standard error; sample size is indicated within bars. Reprinted with minor modifications from Munro and collaborators (2013) with permission of John Wiley and Sons

There are conserved pathways that regulate the lifespan of organisms; however, many organisms appear not to age at all (the so-called *negligible senescence*), and there is substantial variation in ageing mechanisms and patterns, sometimes even between closely related species. For example, although oxidative stress is a significant factor in cellular senescence, its manifestation together with changes in antioxidant levels with age differ largely amongst species of social insects, making it difficult to find general patterns (Kramer et al. 2021). Some species of fish, *Hydra*, or sea urchins show extraordinary longevity and negligible senescence (e.g. Martinez 1998; Du et al. 2013; Dańko et al. 2015; Klimovich et al. 2018; Ebert 2019; Stenvinkel and Shiels 2019). Du and collaborators (2013) found that cellular oxidative damage, antioxidant levels, and proteasome did not vary with age in most sea urchin tissues. It is, however, unclear if negligible senescence is supportive of the role of reactive oxygen species as drivers of ageing.

Some animal groups have evolved specific adaptations to cope with oxidative stress. The transcription factor NRF2 (NF-E2-related factor 2) is a major cellular pathway for regulating the Metazoan antioxidant response (see Chap. 1). Castigilione and collaborators (2020) suggested that the remarkable longevity of birds might be due to the constitutive activation of the NRF2-mediated antioxidant response. They identified a KEAP1 mutation in the Neoavian ancestor that probably disrupted the repression of NRF2 by KEAP1, increased NRF2-driven antioxidant enzyme expression, leading to constitutive NRF2 activity, and decreased oxidative stress.

We can also identify methodological limitations in several earlier comparative studies that weaken previous inferences about the link between longevity and oxidative status. These limitations involve (i) simple bivariate correlations, (ii) limited sample sizes, (iii) lack of control for variation in body mass and phylogenetic history (closely related species share an evolutionary history, thus they are not independent from each other), (iv) use of baseline values of oxidative status markers, whilst studies relying on stress-induced values or testing resistance to stress are rare (Torres-Barceló et al. 2013), (v) neglection of the evidence showing that the link between oxidative stress and ageing might emerge more strongly in older individuals facing with high metabolic demands (e.g. Herborn et al. 2016; Zagkle et al. 2020), (vi) underestimation of the relevance of the environmental conditions under which lifespan and oxidative status are being quantified (e.g. water temperature is an important modulator of senescence in marine ectotherms, with species living at lower water temperatures tending to live longer than those species with similar life histories living at higher temperatures; Philipp et al. 2012), and (vii) estimates of lifespan that included both the pre- and post-reproductive phases, inflating the lifespan of longer-lived species with the addition of a period of life over which individuals are under a waning force of selection.

This issue with how to estimate lifespan is very relevant. Many studies showed that the post-reproductive lifespan may be pronounced (Dagg 2009). For example, guppies *Poecilia reticulata* from high-predation environments live longer than guppies from low-predation environments because they have a longer reproductive lifespan, which is the component of the life-history that makes a direct contribution to individual fitness (Reznick et al. 2006). There were no differences amongst populations in post-reproductive lifespan, which is predicted since there can be no contribution of this segment of the life-history to an individual's evolutionary fitness (Reznick et al. 2006). In a study on wild *Drosophila melanogaster*, females had a post-reproductive lifespan that, on average, made up 40% of total lifespan (Klepsatel et al. 2013). Importantly, post-reproductive lifespan was not correlated with reproductive fitness components, supporting the hypothesis that this period is a highly variable, random "add-on" at the end of reproductive life rather than a correlate of selection on reproductive fitness (Reznick et al. 2006; Klepsatel et al. 2013). Selection should favour the evolution of an extended post-reproductive lifespan only whether post-reproductive individuals contribute to their inclusive fitness in some way, either by caring for their own young, other kin, or grand-offspring, as for example appears to occur in primates or elephants (Dagg 2009). Thus, lifespan may not adequately reflect the amongst species variation in ageing, since lifespan depends on both ageing rate and intrinsic vulnerability (Kowald 2002). Maximum lifespan is routinely used as an estimate of longevity because it is supposed to be proportional to genetic limitations on longevity potential and it might reflect the species' rate of physiological ageing (Finch and Pike 1996; de Magalhães et al. 2005). However, maximum lifespan might also be influenced by mortality rates, which are independent of ageing-related mortality rates, and may show different results compared with correlations with known aged individuals (e.g. Jimenez et al. 2019b). It is also influenced by the sample size: the bigger it is, the higher the chance of having a very old individual.

The discoveries of centenarian and supercentenarian species have added a new level of uncertainty to the universality of the free radical hypothesis of ageing. The olm *Proteus anguinus*, a small cave salamander, has evolved an extreme life-history strategy with a predicted maximum lifespan of over 100 years, an adult average lifespan of 68.5 years, and an age at sexual maturity of over 15 years (Voituron et al. 2011). If we look at the relationship between the body size of amphibians and their longevity, the olm lies as an outlier in the distribution. However, neither its basal metabolism nor antioxidant activities explain this result (Issartel et al. 2009; Voituron et al. 2011). Similarly, freshwater pearl mussels *Margaritifera margaritifera*, which can live up to 200 years, did not show any enhanced resistance to oxidative stress compared to shorter-lived freshwater mussel species (Ridgway et al. 2014). The Greenland shark *Somniosus microcephalus* is one of the world's largest predatory sharks, mainly distributed in the Arctic and subarctic regions of the North Atlantic (Bigelow and Schroeder 1948; Nielsen et al. 2014). These sharks are exceptional supercentenarians, being able to live until 300 years of age (Nielsen et al. 2016). Costantini and collaborators (2017b) found that, compared to other vertebrate species, Greenland sharks had particularly high levels of glutathione peroxidase in muscle and low levels of protein carbonyls in red blood cells, respectively. However, the values of these two particular components of the oxidative status were not probably the reason for their exceptional longevity. Rather, they were probably the result of the physiological adaptations of Greenland sharks to living in cold waters and deep dives.

Studies on cell cultures do not help to shed light on the role of reactive oxygen species as drivers of ageing. Harper and collaborators (2011) assessed the stress resistance of primary fibroblasts derived from 35 species of free-living bird species. Cell lines from longer-lived species were more resistant to death caused by cadmium, paraquat, hydrogen peroxide, methyl methanesulfonate, or metabolic inhibition induced in low-glucose medium than those from shorter-live species and, overall, all avian fibroblasts were more resistant to stress than rodent fibroblasts (Harper et al. 2011). Cell lines from longer-lived species also proliferated more rapidly than cells from shorter-lived species, although there was no correlation between proliferation and cell resistance. Importantly, these results persisted after accounting for the influence of body mass and phylogeny (Harper et al. 2011). Similarly, Ogburn and collaborators (2001) investigated the response of primary embryonic fibroblast-like cells to an oxidative challenge in long- and short-lived avian species (budgerigar *Melopsittacus undulatus* vs Japanese quail *Coturnix japonica*) and short- and long-lived mammalian species (house mouse *Mus musculus* vs humans). They found that cell resistance against oxidative damage was associated with longevity in birds, but not in mammals. Pickering and collaborators (2015) found (i) significant associations between species longevity and resistance of proteins to oxidative stress after exposure of skin-derived fibroblasts to hydrogen peroxide or paraquat, and (ii) higher baseline levels of protein carbonyls in fibroblasts from shorter-lived mammals compared with longer-lived mammals. Taken together, these results supported the hypothesis that resistance to oxidative stress underlies variation between species in longevity. However, other studies on cell cultures provided weak support for this link. Strecker and collaborators (2010) suggested that maintenance of active quality control mechanisms is more important than the generation of reactive oxygen species for cellular senescence. When cultivated until replicative senescence, pigeon embryonic fibroblasts proliferated about one-third longer compared to chicken embryonic fibroblasts. However, both senescent cell populations did not show any increased levels of reactive oxygen species nor accumulation of oxidative damage to mitochondrial DNA or proteins (Strecker et al. 2010). The autophagosomal/lysosomal pathway was not downregulated in old avian cells, enhancing the resistance against oxidative stress, and prolonged the lifespan of chicken embryonic fibroblasts. Oxidative stress-mediated apoptosis also induced a dose-dependent cell proliferation in both pigeon and chicken embryonic fibroblasts. Thus, autophagy and compensatory proliferation were more important than reactive oxygen species in driving the senescence of avian fibroblasts (Strecker et al. 2010).

It might be that the molecular composition of the organism is more relevant for lifespan than the generation of reactive oxygen species because molecules differ widely in their propensity to react with reactive oxygen species. Several studies suggested that the resistance of mitochondrial DNA to oxidative damage is a critical determinant of lifespan. We know that some nucleotides are more prone to oxidation than others (e.g. guanine is most easily oxidised, Bjelland and Seeberg 2003), thus, the sequence of nucleotides might be a target of selection. The free energy is a physical property of the double-stranded DNA molecule that indicates the binding energy between the two DNA strands: the more negative the free energy is, the less likely the spontaneous separation of the two strands through thermal fluctuations (Samuels 2005). Relying on a dataset including 76 mammalian species, 14 bird species, and 4 invertebrate species, Samuels (2005) showed that shorter-lived species had higher abundances of adenine and thymine and lower abundance of cytosine than longer-lived species in mitochondrial DNA, whilst the abundance of guanine was practically constant across species. Samuels also found that this variation between taxa in relative abundances of nucleotides resulted in lower free energy in longer- than in shorter-lived species. This result would indicate that longer-lived species are equipped with a mitochondrial DNA less susceptible to mutations because the low free energy would limit the opening of the double strands and their exposure to reactive oxygen species as a result (Samuels 2005). There is some indication that the mutation rate of mitochondrial DNA is linked to longevity. Nabholz and collaborators (2008) examined the lineage-specific mitochondrial mutation rate (cytochrome *b* sequence) across 1696 mammalian species. They found that rates of mutation of mitochondrial DNA were lower in longer- than in shorter-lived species. Importantly, variation between taxa in metabolic rate did not explain this association between longevity and substitution rate, suggesting that factors like somatic damage might play a role as a constraint on somatic mutation rate of mitochondrial DNA (Nabholz et al. 2008).

By comparing the nonsynonymous and synonymous evolution of codon sites across 25 species of mammals, Jobson and collaborators (2010) found that genes involved in lipid composition, especially in terms of fatty acid biosynthesis, and vitamin C binding have collectively undergone increased selective pressure in longer-lived species, whereas genes involved in DNA replication/repair or antioxidation have not. Overall, these results suggested that selection of genes controlling for fatty acid composition and resistance of connective tissues to damage might be more relevant than that on genes directly influencing the generation of reactive oxygen species and defence mechanisms. It might be that antioxidant genes are not primary targets of selection because they can be upregulated to tackle any increase in reactive oxygen species generation (Hulbert et al. 2007). This suggestion is partly supported by (i) a comparative genomic analysis that identified unique sequence changes in DNA repair genes in the supercentenarian bowhead whale *Balaena mysticetus* genome compared to other shorter-lived cetaceans (Keane et al. 2015), (ii) a comparison of liver transcriptomes between mice, naked mole-rats, and humans which found that longer-lived species have upregulated DNA repair genes and DNA damage signalling pathways compared to shorter-lived species (Fig. 10.7) (MacRae et al. 2015), (iii) the higher capacity of skin fibroblasts of longer-lived mammals to upregulate genes involved in DNA repair compared to shorter-lived mammals (Ma et al. 2016), and (iv) a comparative genomic and transcriptomic study indicating that the long-lived bat species *Myotis myotis* exhibited a transcriptomic profile of enhanced DNA repair and autophagy pathways, compared to short-lived bat species *Molossus molossus* (Huang et al. 2020). However, several studies indicated that (i) not all DNA repair pathways contribute equally to differences between species in lifespan (e.g. DNA double-strand break is more relevant than nucleotide excision repair) and (ii) the threshold of DNA damage for activation of cell death or cell cycle arrest also differs between species (Hart and Setlow 1974; Kato et al. 1980; Woodhead et al. 1980; Lorenzini et al. 2009; Page and Stuart 2012; Freitas and de Magalhaes 2011; Tian et al. 2017, 2019).

Protein composition of tissues is also important for the organismal vulnerability to oxidative stress because amino acids differ in their propensity to oxidation; cysteine and methionine are the amino acids most sensitive to oxidation by reactive oxygen species (Chap. 1). This is very relevant because even a single change in the amino acid sequence can lead to a modification of the function and/or the stability of a protein, which could influence the mitochondrial generation of reactive oxygen species and the lifespan as a consequence (Mariadassou and Pellay 2014). Resistance of proteins to oxidative damage might explain the surprising long lifespan of naked mole-rats (Pérez et al. 2009b) and bats (Salmon et al. 2009). Moosmann and Behl (2008) compared the genome sequences across 218 animal species, including 95 mammals, 14 birds, 10 reptiles, 12 amphibians, 39 fish, and 48 invertebrates. Their

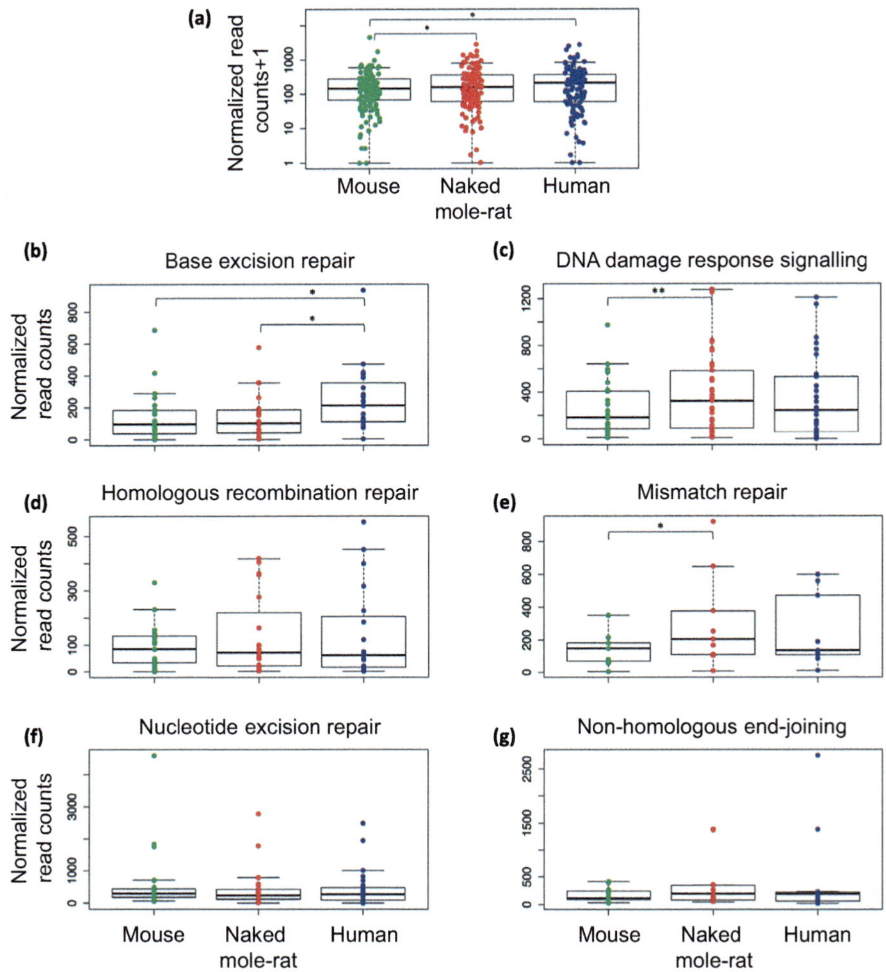


**Fig. 10.7** Higher expression of essential DNA repair genes in long-lived species. (**a**) Expression of DNA repair genes is significantly higher in livers of naked mole-rats ($P = 0.0167$) and humans ($P = 0.0429$) than in mice. For individual DNA repair pathways, (**b**) base excision repair, (**c**) DNA damage response signalling, (**d**) homologous recombination repair, (**e**) mismatch repair, (**f**) nucleotide excision repair, and (**g**) non-homologous end-joining, humans and naked mole-rats had higher expression of many essential repair proteins in the liver. Asterisks indicate significant differences amongst species: humans had significantly higher expression than mice ($P = 0.0010$) and naked mole-rats ($P = 0.0039$) of those genes involved in the base excision repair pathway; naked mole-rats had significantly higher expression than mice ($P = 0.0496$) of genes involved in mismatch repair pathway and DNA damage response signalling pathway ($P = 0.0011$), respectively. Reprinted with minor modifications from MacRae and collaborators (2015). © 2015 MacRae et al.; distributed under the terms of the Creative Commons Attribution License (CC BY)

analysis revealed that longer-lived species synthesise respiratory chain complexes that are depleted of cysteine. This relationship between lifespan and depletion in cysteine content of mitochondrial proteins persisted after correction for body mass and phylogenetic history. Moosmann and Behl (2008) also found that (i) aerobic species had a decreased content of genomic cysteine compared to anaerobic species and (ii) a significant relationship between maximum lifespan and cysteine in anaerobic species (e.g. some nematodes and parasitic platyhelminthes) did not emerge. Interestingly, in nematodes with a biphasic aerobic–anaerobic lifestyle, cysteine appeared to predict the duration of the aerobic phase of life. Overall, these results suggested that there might have been a selection against cysteine presence in mitochondrial DNA-encoded proteins of longer-lived species because cysteine thiyl radicals would promote irreversible protein cross-linking. It might be that natural selection has increased the amino acid substitution rates in mitochondrially encoded proteins (e.g. cytochrome b, which affects the rate of reactive oxygen species generation). Prior work found that these amino acid substitution rates were higher in longer- than in shorter-lived mammals (Rottenberg 2006, 2007a), and in longer- than in shorter-lived songbirds (but not in other avian taxa; Rottenberg 2007b). However, a later study on the mitochondrial genomes of 168 mammalian species did not provide support for the link between longevity and cysteine (Aledo et al. 2011).

Methionine is also a relevant amino acid because of its propensity to oxidation. By examining the mitochondrial genomes of 168 mammalian species, representing 24 different taxonomical orders, Aledo and collaborators (2011) found that mitochondrially encoded proteins from shorter-lived species contained more methionine than proteins from longer-lived species, and this result persisted after controlling for body mass and phylogeny. However, this relationship between longevity and methionine content was not significant across all orders. Their analyses also suggested that this variation in the abundance of methionine in proteins was due to the addition of methionine to the proteins in shorter-lived species, and not to the loss of methionine from proteins in longer-lived species (Aledo et al. 2011). Importantly, the methionine content was higher in mitochondrial proteins located in the inner mitochondrial membrane, where most reactive oxygen species are generated. These results were interpreted as supportive of a possible antioxidant role of methionine in shorter-lived mammals: species that live shorter suffer more oxidative stress, so that they selectively accumulate methionine in their mitochondrial proteins to protect mitochondrial DNA. This sacrifice of mitochondrial proteins would enable the organism to preserve the integrity of mitochondrial DNA, and would also be temporary because the organism is equipped with enzymes called methionine sulfoxide reductases that convert oxidised methionine to reduced methionine (Stadtman et al. 2002; Bender et al. 2008). However, another interpretation for this link between methionine content and longevity is also possible: the high accumulation of methionine in mitochondrial proteins would increase the vulnerability to oxidative stress, resulting in shorter lifespans. If so, what is the compensation for shorter-lived mammals for accumulating so much methionine? Which is the adaptive advantage of having mitochondria rich in methionine? An explanation for this hypothesis might lie in the role of methionine in sustaining the high reproductive activity typical of shorter-lived species. A study on hispid cotton rats *Sigmodon hispidus* showed that methionine supplementation enhanced population density, recruitment, and survival, and resulted in an earlier and longer reproductive season (Webb et al. 2005).

Beyond the relative importance of amino acids, maintenance of protein integrity is certainly important for longevity. Heat shock proteins provide good grounds for thinking this. Heat shock proteins function as cellular chaperones for other proteins, playing a number of roles in response to oxidative stress, such as folding and assisting in the establishment of proper protein conformation and prevention of unwanted protein aggregation, stabilising partially unfolded proteins. The quantification of constitutive levels of these proteins in brain, heart, and liver of 12 mammalian and one avian species ranging in lifespan from 3 to 30 years showed that their expression was highly correlated with maximum lifespan (whilst controlling for body mass and phylogeny), indicating that higher basal levels of heat shock protein expression are characteristic of longer-lived species (Salway et al. 2011).

## 10.4 Age-Related Changes of Oxidative Status and Links to Survival

The disposable soma hypothesis of ageing predicts that, whilst natural selection optimises trade-offs in the allocation of limited resources amongst self-maintenance and other activities, they inevitably come at a cost for the soma that would emerge later in life in the form of accumulated damage (Kirkwood 1977; Kirkwood and Rose 1991). Overall, current evidence on free-living animals provides equivocal support for this hypothesis.

We know little about how oxidative status markers change with age and whether oxidative damage accumulates over time in free-living animals. We have also a limited understanding of the links between organismal oxidative status and survival probability. In plants, Munné-Bosch and Lalueza (2007) found that leaves of oldest shrubs of *Cistus clusii* had higher oxidative stress in both chloroplasts and mitochondria than those of young shrubs when exposed to adverse climatic conditions and, even in the absence of environmental stress, newly emerged leaves accumulated higher amounts of abscisic acid as plants aged. By contrast, Morales and collaborators (2013) found that plants of the extraordinarily long-lived *Borderea pyrenaica* did not show any age-dependent signs of oxidative stress, suggesting that age-induced oxidative stress is not a universal feature of ageing in perennial plants. What about animals? Evidence tells us that even in animals the predicted age-related changes of oxidative status markers do not always occur.

Saiz and collaborators (2015) found that the activities of glutathione S-transferase and catalase were higher in older than younger calanoid copepods *Paracartia grani*. Levels of lipid peroxidation also tended to be higher in older than younger copepods, particularly shortly before death. Similarly, Lucas and collaborators (2017) found that DNA damage (comet assay) increased with age to the same degree in both queens and workers of the ant *Lasius niger* over the first year of their life, despite previous results showing that queens showed higher expression of DNA repair genes than workers (Lucas et al. 2016). A possible explanation for this finding is that queens might be subjected to higher rates of damage than workers because of increased investment into reproduction, and their increased levels of repair serve to compensate for this. Marasco and collaborators (2017) examined within-individual changes in four blood-based oxidative status markers and compared them between early- and mid-adulthood (i.e. 5- vs. 12-months old) in captive zebra finches. Birds were living under control (ad libitum food) or more challenging environmental conditions as unpredictable food availability, having previously found that the latter increased corticosterone levels when food was absent but improved survival over a period of three years. Levels of 8-OHdG and protein carbonyls increased with age in both experimental environments, with a higher increase of 8-OHdG in the challenging environment. Levels of superoxide dismutase increased with age in the controls but not in the challenged birds, whilst the opposite was documented for the plasma non-enzymatic antioxidant capacity (Marasco et al. 2017). The links between markers and survival were rather weak and varied across groups, indicating that there might be non-linear relationships between environmentally-mediated oxidative stress and its associated benefits and costs at the organismal level (Marasco et al. 2017). This is particularly important because studies on the link between oxidative status and survival metrics in free-living animals show quite variable outcomes (Table 10.1). Similarly, by means of a common garden experiment, Liparoto and collaborators (2020) examined within-individual changes of blood-based oxidative status markers in wild-caught Tyrrhenian tree frogs *Hyla sarda* maintained under common garden conditions. This species underwent a range expansion from northern Sardinia (source) up to Corsica (newly founded) during the Late Pleistocene, and then the two populations became geographically isolated. Thus, animals were collected from both populations in order to assess whether the past biogeographical history left a signature of phenotypic differentiation for ageing. At the beginning of the experiment, Sardinian and Corsican frogs had similar concentrations of all oxidative status markers analysed. One year later, Corsican frogs had higher levels of reactive oxygen metabolites and lower levels of glutathione peroxidase and serum non-enzymatic antioxidant capacity, and suffered higher mortality than Sardinian frogs, respectively (Fig. 10.8). The results of this study indicated that age-related changes in red blood cell superoxide dismutase might be a shared (or conserved) mechanism of physiological ageing. By contrast, age-related changes in the other oxidative status markers might be private (i.e. peculiar to particular evolutionary lineages) because these changes were expressed only in Corsican frogs (Liparoto et al. 2020). The possible conserved level in the expression of the superoxide dismutase across the two islands is also supported by the lack of genetic variation between Sardinian and Corsican frogs in the gene SOD-1 (Bisconti et al. 2011).

Studies on free-living animals also provided interesting evidence for possible age-related changes in oxidative status markers. A cross-sectional study on free-living alpine marmots *Marmota marmota* found that older individuals had higher

**Table 10.1** Summary of studies that examined the relationships between markers of oxidative damage or antioxidant protection and survival metrics in free-living vertebrates

| Class | Species | Age | Tissue | Marker | Effect | Study |
|---|---|---|---|---|---|---|
| Reptiles | *Niveoscincus ocellatus* | Adult | Plasma | OX—reactive oxygen metabolites | There was no link with survival to the following season | Isaksson et al. (2011) |
| Reptiles | *Niveoscincus ocellatus* | Adult | Plasma | ANT—non-enzymatic antioxidant capacity | Female skinks with higher antioxidant capacity showed a decreased probability of survival to the following season | Isaksson et al. (2011) |
| Birds | *Apus melba* | Adult | Whole blood | ANT—resistance of red blood cells to haemolysis induced by an in vitro exposure to a pro-oxidant molecule | It was higher in males, but not in females, which returned the next year | Bize et al. (2008) |
| Birds | *Parus major* | Nestling/ Adult | Red blood cells | ANT—glutathione peroxidase | Higher survival to first winter and recruitment of nestlings with intermediate levels | Norte et al. (2008) |
| Birds | *Pygoscelis adeliae* | Adult | Plasma | OX—reactive oxygen metabolites | There was no difference between birds that returned to the colony and those that did not return | Beaulieu et al. (2011) |
| Birds | *Pygoscelis adeliae* | Adult | Plasma | ANT—non-enzymatic antioxidant capacity | There was no difference between birds that returned to the colony and those that did not return | Beaulieu et al. (2011) |
| Birds | *Geothlypis trichas* | Adult | Red blood cells | OX—DNA damage | It was lower in males that returned the next year | Freeman-Gallant et al. (2011) |
| Birds | *Parus major* | Nestling | Red blood cells | ANT—glutathione peroxidase | It was lower in nestlings that successfully fledged | Koivula et al. (2011) |
| Birds | *Parus major* | Nestling | Red blood cells | ANT—glutathione reductase | It was not associated with fledging probability | Koivula et al. (2011) |
| Birds | *Parus major* | Nestling | Red blood cells | ANT—glutathione-S-transferase | It was not associated with fledging probability | Koivula et al. (2011) |
| Birds | *Parus major* | Nestling | Red blood cells | ANT—catalase | It was not associated with fledging probability | Koivula et al. (2011) |

| | | | | | | |
|---|---|---|---|---|---|---|
| Birds | *Parus major* | Nestling | Red blood cells | ANT—superoxide dismutase | It was not associated with fledging probability | Koivula et al. (2011) |
| Birds | *Phalacrocorax aristotelis* | Nestling/Adult | Plasma | OX—malondialdehyde | Lower in nestlings that had higher recruitment probability as adults | Noguera et al. (2011) |
| Birds | *Hirundo rustica* | Adult | Plasma | ANT—non-enzymatic antioxidant capacity | High levels positively predicted long-term survival | Saino et al. (2011) |
| Birds | *Acrocephalus sechellensis* | Adult | Plasma | OX—reactive oxygen metabolites | It did not predict the probability of an individual being alive in the next year | van de Crommenacker (2011) |
| Birds | *Acrocephalus sechellensis* | Adult | Plasma | ANT—non-enzymatic antioxidant capacity | It did not predict the probability of an individual being alive in the next year | van de Crommenacker (2011) |
| Birds | *Parus major* | Nestling/Adult | Plasma | OX—reactive oxygen metabolites | It did not predict post-fledging survival | De Coster et al. (2012) |
| Birds | *Parus major* | Nestling/Adult | Plasma | ANT—non-enzymatic antioxidant capacity | It did not predict post-fledging survival | De Coster et al. (2012) |
| Birds | *Aptenodytes patagonicus* | Nestling | Plasma | OX—reactive oxygen metabolites | Higher levels in those individuals that died before leaving the nest | Geiger et al. (2012) |
| Birds | *Aptenodytes patagonicus* | Nestling | Plasma | ANT—non-enzymatic antioxidant capacity | Lower levels in those individuals that died before leaving the nest | Geiger et al. (2012) |
| Birds | *Parus major* | Nestling | Whole blood | ANT—resistance of red blood cells to haemolysis induced by an in vitro exposure to a pro-oxidant molecule | Higher in nestlings that successfully fledged | Losdat et al. (2013) |
| Birds | *Parus major* | Nestling | Whole blood | ANT—resistance of red blood cells to haemolysis induced by an in vitro exposure to a pro-oxidant molecule | It did not predict local recruitment | Losdat et al. (2013) |
| Birds | *Calonectris diomedea* | Adult | Plasma | OX—reactive oxygen metabolites | Higher levels indicated lower resight probability in the following seasons | Costantini and Dell'Omo (2015) |
| Birds | *Calonectris diomedea* | Adult | Plasma | ANT—non-enzymatic antioxidant capacity | There was no link with the resight probability | Costantini and Dell'Omo (2015) |

(continued)

**Table 10.1** (continued)

| Class | Species | Age | Tissue | Marker | Effect | Study |
|---|---|---|---|---|---|---|
| Birds | *Calonectris diomedea* | Adult | Plasma | ANT—total thiols | There was no link with the resight probability | Costantini and Dell'Omo (2015) |
| Birds | *Diomedea exulans* | Adult | Plasma | OX—thiobarbituric acid reactive substances | Individuals having higher levels were less likely to return the next years | Costantini et al. (2015) |
| Birds | *Phalacrocorax aristotelis* | Nestling | Plasma | OX—reactive oxygen metabolites | Higher damage in individuals that were less likely to be resighted in the breeding colony two years later | Herborn et al. (2016) |
| Birds | *Phalacrocorax aristotelis* | Adults | Plasma | OX—reactive oxygen metabolites | Higher levels at the end of the breeding season predicted reduced return rates, irrespective of the individual age | Herborn et al. (2016) |
| Birds | *Phalacrocorax aristotelis* | Nestlings | Plasma | ANT—non-enzymatic antioxidant capacity | It did not predict the return rate | Herborn et al. (2016) |
| Birds | *Phalacrocorax aristotelis* | Adults | Plasma | ANT—non-enzymatic antioxidant capacity | It did not predict the return rate | Herborn et al. (2016) |
| Birds | *Tachycineta bicolor* | Adult | Plasma | OX—reactive oxygen metabolites | It was lower in individuals that were more likely to return the next year | Ouyang et al. (2016) |
| Birds | *Tachycineta bicolor* | Adult | Plasma | ANT—non-enzymatic antioxidant capacity | It was higher in individuals that were more likely to return the next year | Ouyang et al. (2016) |
| Birds | *Tachycineta bicolor* | Adult | Plasma | OX—reactive oxygen metabolites | It was lower in individuals that were more likely to return the next year | Stanton et al. (2017) |
| Birds | *Tachycineta bicolor* | Adult | Plasma | ANT—non-enzymatic antioxidant capacity | It was lower in individuals that were more likely to return the next year | Stanton et al. (2017) |
| Birds | *Coloeus monedula* | Nestlings | Plasma | OX—reactive oxygen metabolites | There was no relationship with post-fledging survival | Boonekamp et al. (2018) |
| Birds | *Coloeus monedula* | Nestlings | Plasma | OX—thiobarbituric acid reactive substances | There was no relationship with post-fledging survival | Boonekamp et al. (2018) |
| Birds | *Coloeus monedula* | Nestlings | Plasma | OX—oxidised glutathione | There was no relationship with post-fledging survival | Boonekamp et al. (2018) |

| | | | | | | |
|---|---|---|---|---|---|---|
| Birds | *Coloeus monedula* | Nestlings | Plasma | ANT—reduced glutathione | Individuals with higher reduced glutathione had higher post-fledging survival probability | Boonekamp et al. (2018) |
| Birds | *Parus major* | Adult | Plasma | OX—reactive oxygen metabolites | It did not predict the chance to return the next year | Casagrande and Hau (2018) |
| Birds | *Parus major* | Adult | Plasma | OX—protein carbonyls | It did not predict the chance to return the next year | Casagrande and Hau (2018) |
| Birds | *Parus major* | Adult | Plasma | ANT—non-enzymatic antioxidant capacity | It did not predict the chance to return the next year | Casagrande and Hau (2018) |
| Birds | *Parus major* | Adult | Red blood cells | ANT—glutathione peroxidase | It was lower in individuals that returned the next year | Casagrande and Hau (2018) |
| Birds | *Hirundo rustica* | Adult | Plasma | OX—malondialdehyde | It was lower in birds with higher return annual rates when measured on the first day after hatching | Pap et al. (2018) |
| Birds | *Hirundo rustica* | Adult | Plasma | OX—malondialdehyde | It did not predict the return rate when measured at day 13 after hatching | Pap et al. (2018) |
| Birds | *Hirundo rustica* | Adult | Erythrocytes | ANT—thiols | It was higher in birds with higher return annual rates when measured on the first day after hatching | Pap et al. (2018) |
| Birds | *Hirundo rustica* | Adult | Plasma | OX—malondialdehyde | It was lower in birds with higher return annual rates when measured on the first day after hatching | Pap et al. (2018) |
| Birds | *Branta bernicla* | Adult | Plasma | OX—malondialdehyde | Survival probability was lower in individuals having lower plasma levels of malondialdehyde | Bodey et al. (2020) |
| Birds | *Branta bernicla* | Adult | Plasma | ANT—non-enzymatic antioxidant capacity | The survival probability was not associated with the plasma non-enzymatic antioxidant capacity | Bodey et al. (2020) |
| Birds | *Branta bernicla* | Adult | Plasma | ANT—superoxide dismutase | The survival probability was not associated with the plasma superoxide dismutase | Bodey et al. (2020) |

(continued)

**Table 10.1** (continued)

| Class | Species | Age | Tissue | Marker | Effect | Study |
|---|---|---|---|---|---|---|
| Birds | *Sturnus vulgaris* | Nestling | Plasma | OX—reactive oxygen metabolites | It was lower in birds that successfully fledged | Costantini, unpublished data |
| Birds | *Sturnus vulgaris* | Nestling | Plasma | ANT—non-enzymatic antioxidant capacity | It did not predict the fledging success | Costantini, unpublished data |
| Birds | *Sturnus vulgaris* | Nestling | Plasma | ANT—thiols | It did not predict the fledging success | Costantini, unpublished data |
| Mammals | *Ovis aries* | Young | Plasma | ANT—superoxide dismutase | Lambs with higher activity of superoxide dismutase in August were more likely to survive their first winter | Christensen et al. (2016) |
| Mammals | *Ovis aries* | Young | Plasma | OX—protein carbonyls | Male lambs with higher protein carbonyl concentrations were less likely to survive their first winter | Christensen et al. (2016) |
| Mammals | *Ovis aries* | Young | Plasma | ANT—non-enzymatic antioxidant capacity | The probability of lamb survival did not vary with antioxidant capacity | Christensen et al. (2016) |
| Mammals | *Ovis aries* | Young | Plasma | OX—malondialdehyde | The probability of lamb survival did not vary with malondialdehyde concentration | Christensen et al. (2016) |
| Mammals | *Mungos mungo* | Adult | Plasma | OX—malondialdehyde | Survival probability was higher in individuals having lower plasma levels of malondialdehyde | Vitikainen et al. (2016) |
| Mammals | *Mungos mungo* | Adult | Plasma | OX—protein carbonyls | Survival probability was not associated with plasma levels of protein carbonyls | Vitikainen et al. (2016) |
| Mammals | *Mungos mungo* | Adult | Whole blood | ANT—superoxide dismutase | Survival probability was not associated with levels of superoxide dismutase | Vitikainen et al. (2016) |

*OX* compound derived from oxidation; *ANT* antioxidant

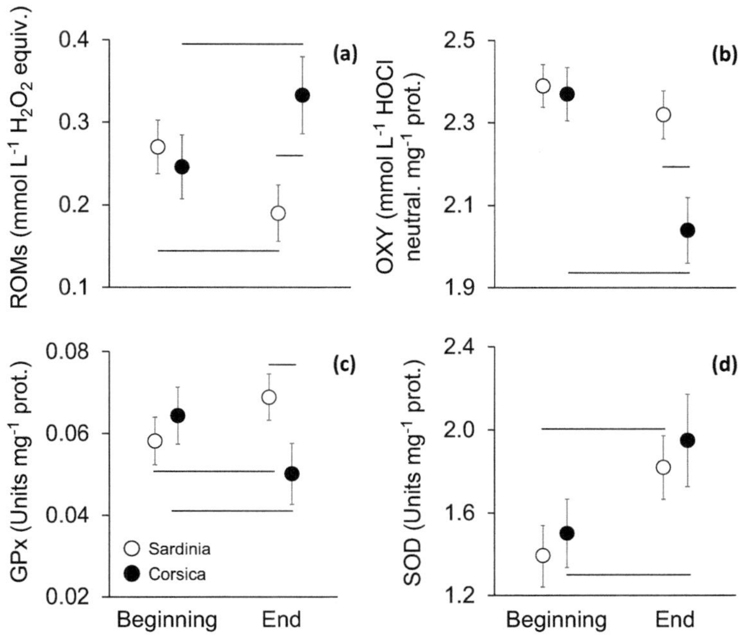


**Fig. 10.8** Differentiation between two populations (Sardinia and Corsica) of the Tyrrhenian tree frog *Hyla sarda* along a historical range expansion in age-related changes of the oxidative status markers (**a**) ROMs, reactive oxygen metabolites in serum, (**b**) OXY, serum non-enzymatic antioxidant capacity, (**c**) GPx, glutathione peroxidase in red blood cells, and (**d**) SOD, superoxide dismutase in red blood cells measured at the beginning of the common garden experiment and a year later, respectively. Significant contrasts are indicated with bars. Values are shown as estimated marginal mean and standard errors. equiv = equivalents; neutral = neutralised. Reprinted with minor modifications from Liparoto and collaborators (2020)

levels of two markers of oxidative damage (reactive oxygen metabolites and the protein adduct between 4-hydroxynonenal and histidine) than younger individuals (Fig. 10.9a, b) (Costantini et al. 2012). By contrast, a cross-sectional study on free-living rhesus macaques *Macaca mulatta* found that older individuals had lower levels of 8-OHdG than younger ones, suggesting a decline of oxidative stress in parallel with known age-related declines in reproductive investment (Georgiev et al. 2015). Similarly, a study on free-living cheetahs *Acinonyx jubatus* found that older individuals had lower levels of reactive oxygen metabolites and protein carbonyls than younger individuals (Fig. 10.9c–f) (Costantini et al. 2017a). In addition, older cheetahs had lower serum non-enzymatic micro-molecular antioxidant capacity and higher activity of superoxide dismutase in erythrocytes than younger cheetahs. Although the number of cheetahs of which there were repeated measurements of markers over time was low, the data showed within-individual changes over time similar to those illustrated by the cross-sectional analyses. Other cross-sectional studies on free-living smallmouth bass *Micropterus dolomieu* (Elmer et al. 2017), captive grey mouse lemurs *Microcebus murinus* (Noiret et al. 2020), and free-living or captive chimpanzees *Pan troglodytes* (Thompson González et al. 2020; Costantini et al. 2021) also found that older individuals had higher plasma non-enzymatic antioxidant capacity, and lower levels of lipid peroxides or 8-OHdG than younger individuals. By contrast, a study on thick-billed murres *Uria lomvia* did not find any significant correlation between individual age and oxidative status markers (Jimenez et al. 2019a). The results of these cross-sectional studies may be confounded by selective disappearance, if only high-quality individuals with low levels of oxidative damage or high antioxidant protection survive to an age where ageing takes effect.

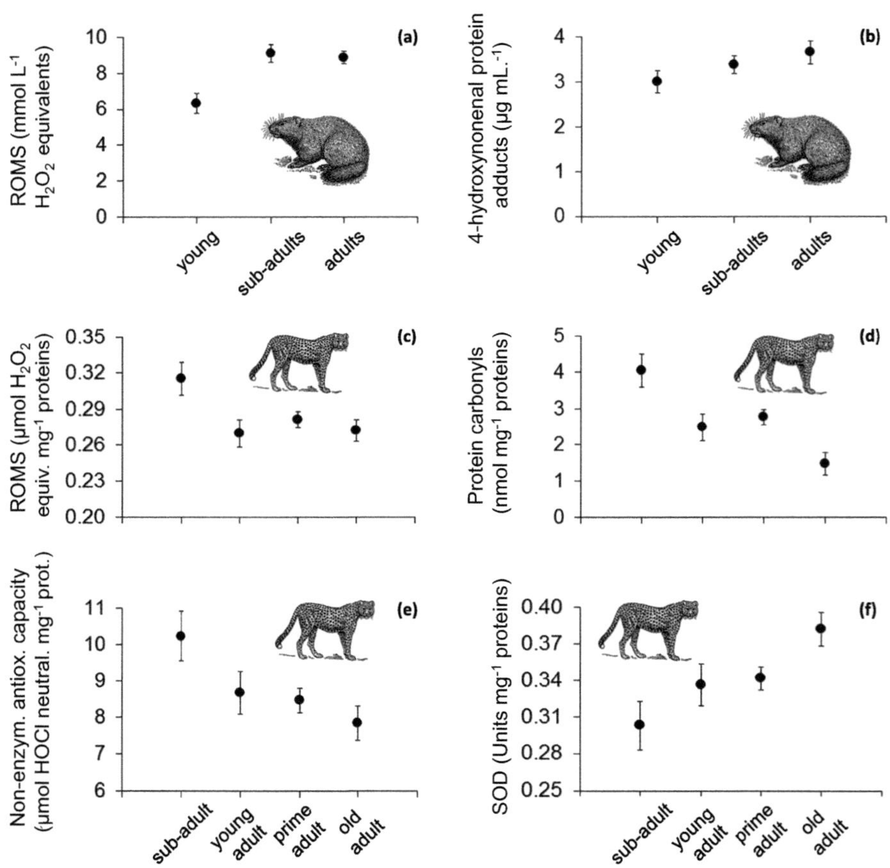


**Fig. 10.9** Age-related changes of blood-based markers of oxidative status in (**a**–**b**) free-living marmots *Marmota marmota* and (**c**–**f**) free-living cheetahs *Acinonyx jubatus*. Values in panels (**c**) and (**f**) were square-root and log-transformed, respectively. Age categories of cheetahs: sub-adults, 13–23 months of age; young adults, 24–42 months of age; prime adults >3.5–7.0 years of age; old adults >7.0 years of age. Values are shown as mean and standard error. ROMs, reactive oxygen metabolites; SOD, superoxide dismutase; equiv., equivalents; neutral., neutralised. Data illustrated in panels (a) and (b) were collected from Costantini and collaborators (2012). Panels (**c**–**f**) were reprinted with minor modifications from Costantini and collaborators (2017a); © The Author 2017, published by Oxford University Press and the Society for Experimental Biology; distributed under the terms of the Creative Commons Attribution License (CC BY)

Thus, the low levels of oxidative damage documented in older cheetahs or rhesus macaques might be due to the disappearance of individuals with high oxidative damage over time, meaning that older classes include only individuals with low oxidative damage. Although this may be true, a few longitudinal studies carried out on free-living animals found similar results. For example, Vitikainen and collaborators (2016) found a within-individual decline of plasma levels of malondialdehyde and α-tocopherol with age in free-living banded mongooses *Mungos mungo*.

Some studies suggested that individual attributes might confound correlations between individual age and markers of oxidative status or the within-individual changes of oxidative status over time. For example, when testing the correlation between individual age and thiobarbituric acid reactive substances in wandering albatrosses *Diomedea exulans*, it turned out to be significant only for birds that were not breeding when blood samples were collected (Fig. 10.10) (Costantini et al. 2014b). As animals enter the reproductive phase, they experience strong changes at multiple levels of their physiological functions. Thus, it might be that these physiological changes influenced the oxidative status independently from age, probably hiding the correlation between age and concentration of oxidative damage.

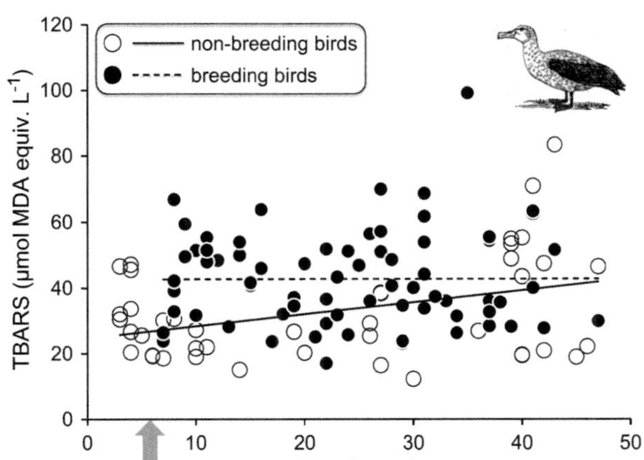


**Fig. 10.10** Correlation between individual age and a marker of lipid peroxidation. Plasma levels of thiobarbituric acid reactive substances increased significantly with individual chronological age in wandering albatrosses *Diomedea exulans* that were out of the breeding season, but not in those that were sampled during the breeding season. This increase is particularly notable when birds are around 40 years old. The correlation between oxidative damage and chronological age in non-breeding birds becomes stronger if immature birds are removed from analyses. The dashed line indicates the breeding birds; the solid line indicates the non-breeding birds. The arrow indicates the age of first reproduction (six years). Reprinted with minor modifications from Costantini and collaborators (2014b) with permission of Springer Nature

## 10.5 Does Oxidative Stress Make Life Longer or Shorter? Links to Survival and Hormesis

Actuarial senescence is the increasing risk of mortality with advancing age. A reason for this increase in mortality might lie with the deterioration of physical or physiological functioning with advancing age. If oxidative stress is an important proximate mechanism driving cellular senescence, we should expect a link with survival probability. Do we have evidence that variation in oxidative status markers between individuals predicts their survival perspectives? We have some empirical mixed evidence in support of the link between oxidative status and survival (Table 10.1). For example, Costantini and Dell'Omo (2015) found that Scopoli's shearwaters *Calonectris diomedea* having higher plasma levels of reactive oxygen metabolites during a given breeding season had a lower resight probability in the next years. Similarly, Costantini and collaborators (2015) found that wandering albatrosses *Diomedea exulans* having higher plasma levels of thiobarbituric acid reactive substances were less likely to return in the next years. In contrast to predictions, Bodey and collaborators (2020) found that light-bellied brent geese *Branta bernicla hrota* having lower plasma levels of malondialdehyde were less likely to survive. The interpretation of these contrasting results remains unsettled because we can identify a number of potential theoretical and methodological caveats.

First, the age of individuals is often unknown when tissue samples are collected. Thus, the sample involves individuals of different ages and histories of wear and tear. As stated in the previous paragraph, cross-sectional studies may be confounded by selective disappearance, if early mortality is associated with certain concentrations of oxidative status markers. For example, if younger individuals with higher levels of oxidative damage are more likely to die, we would observe individuals with lower levels of oxidative damage in older age classes. Moreover, individuals of different ages would differ in their residual lifespan. Another issue might lie in the assumption that '*two individuals of the same age showing, e.g. similar levels of a given marker of oxidative damage at time t, should have a similar probability of survival per unit of time*'. However, individuals of the same age can differ substantially in the degree to which they accumulate tissue damage, akin to bodily wear and tear, from past experiences. By exposing hand-raised adult Eurasian blackbird *Turdus merula* to a combination of repeated immune and disturbance stressors for over one year, Hau and collaborators (2015) found that stress-exposed birds showed greater decreases in telomere length and maintained higher circulating levels of oxidative damage compared with control birds.

These inconsistencies between studies on the link between survival and oxidative status might also stem from a wide array of other factors that can confound the results, making difficult their interpretation and comparability. First, the literature is dominated by studies on birds. Second, studies differ in how they address the question. For example, we have studies that (i) compared oxidative status markers between adults that returned the next breeding season and those that did not return, or (ii) focussed on the probability of nestlings to successfully leave the nest or to recruit into the breeding population the next breeding season. Third, studies used different proxies for survival that might be biased by dispersal, predation of particular phenotypes, age, or season. Fourth, most studies relied on blood-based markers of oxidative status. Although there is evidence that certain markers correlate well across different tissues (Argüelles et al. 2004; Veskoukis et al. 2009), and that the blood is a sensitive tissue to oxidative challenges (Isaksson 2010; Costantini et al. 2011), we also know that cellular ageing occurs heterogeneously across cells and tissues (e.g. Herndon et al. 2002; Passos et al. 2007). Fifth, an in-depth analysis of studies reported in Table 10.1 shows that a given marker predicts survival in some species but not in others. We do not know if this result is circumstantial or indicates that oxidative stress affects longevity only in certain species. Sixth, many studies did not take individual sex into account. This is unfortunate because females and males differ in lifespan, ageing rate, and oxidative status across a large number of animals (e.g. Trivers 1985; Carranza et al. 2004; Tower 2006; Clutton-Brock and Isvaran 2007; Maklakov et al. 2007; Bize et al. 2008; Costantini et al. 2009; Kamper et al. 2009; Nussey et al. 2009; Costantini 2010; Criscuolo et al. 2010; Casagrande et al. 2011; Vina et al. 2011; Rubolini et al. 2012; Archer et al. 2013; Maklakov and Lummaa 2013; Costantini 2018). Seventh, it is difficult, if not impossible, to record the individual history with accuracy. We know from laboratory-based work that past experiences of stress, particularly those experienced early in life, mould the individual resistance against stress. Thus, when possible, recording age-related changes in oxidative status from cradle to death might provide more insight into the role of oxidative stress as a driver of ageing.

Finally, we should not forget that exposure to oxidative stress is not necessarily detrimental in terms of fitness. There is much evidence in a range of invertebrates and vertebrates that exposure to short episodes of mild stress (e.g. radiation, caloric restriction, thermal stress, and physical effort) at some point in life may increase longevity (Rattan 2008; Calabrese 2013; see Chap. 7). Importantly, many of these studies exposed organisms to stressors that increase the generation of reactive oxygen species, implying that oxidative stress is somehow involved in driving lifespan through hormetic mechanisms (e.g. Yang and Hekimi 2010; Bazopoulou et al. 2010). For example, mice exposed to very low doses of radiation early in life lived longer than controls (Caratero et al. 1998). Similarly, exposure of nematodes *Caenorhabditis elegans* early in life to heat stress, hyperbaric oxygen, or juglone (a chemical that triggers the generation of reactive oxygen species) significantly increased subsequent resistance to the same challenge, resulting in a longer lifespan (Cypser and Johnson 2002). Research on various insect species demonstrated that exposure to mild levels of ionising radiation stimulated the extension of lifespan (Calabrese 2013). However, the extent to which early life exposure to mild stress carries benefits for longevity also depends on how the developmental environment is predictive of the adult environment (Costantini et al. 2014a; Costantini and Borremans 2019; see Chap. 7). Environmentally-primed stress resistance in early life may therefore carry survival costs if the subsequent environment does not match that to which the adjustment was made (Costantini et al. 2014a; Costantini and

Borremans 2019). This cost also seems to emerge when the environment of offspring is not as much stressful as it was that of their parents (Schade et al. 2014). Moreover, we have some indications that hormetic responses differ between females and males, reiterating the relevance of taking sex into account (Salmon et al. 2001; Le Bourg 2005; Sørensen et al. 2007; Costantini et al. 2014a). Overall, these results illustrate that the prediction '*high oxidative damage at time x is linked to lower survival probability*' is oversimplistic.

## 10.6 Conclusions and Future Directions

We have much evidence showing that longer-lived species tend to have lower generation of reactive oxygen species or oxidative damage, lower concentrations of endogenous antioxidants, or cells made of molecules that are more resistant to oxidative damage than shorter-lived species. However, we have a growing number of studies reporting data on the link between lifespan and oxidative status markers that are seemingly contradictory and confused. We have also conflicting results between studies about the expected increase of oxidative damage levels with age, and the link between oxidative status markers and metrics of survival probability in free-living animals.

In this chapter, I have identified a number of methodological and biological issues that might affect results and make comparisons complicated. For example, a short-lived species is expected to invest much less in the defence against oxidative stress than a long-lived species. Thus, it might be erroneous to expect a linear relationship between lifespan and a given marker of oxidative damage. Comparative studies also need to be more integrative, merging more molecular pathways and life-history data in multivariate and network analyses in order to elucidate the drivers of ageing (e.g. Csermely and Soti 2006; Xue et al. 2007; Brink et al. 2009; Southworth et al. 2009; Soltow et al. 2010; Cohen et al. 2013). In 1996, Kowald and Kirkwood called for a synthesis of formerly distinct hypotheses into a *network theory of ageing*. It integrates the contribution of, for example, defective mitochondria, aberrant proteins, free radicals, and antioxidant enzymes, stating that it is the collective set of maintenance processes that together determines the ability of the organism to preserve homeostasis by protecting against damage (Kowald and Kirkwood 1996). Studies on single species also need to integrate physiological data with individual attributes, ranging from age to the life-history stage and environmental conditions under which the organism lives (e.g. Costantini et al. 2015). We also need testable alternative hypotheses regarding the link between oxidative status and age. It is often assumed that oxidative damage should increase with age, but we have longitudinal studies showing that certain markers of oxidative damage decline with age. This result is not in line with predictions of the disposable soma hypothesis of ageing. Why is this? For example, a reduction with age of 8-OHdG levels might indicate a reduced capability of base excision mechanisms to repair DNA. However, studies on laboratory rodents did not provide support for this explanation (e.g. Fraga et al. 1990; Hamilton et al. 2001; Gorbunova et al. 2007). Another reason for this age-related decline of oxidative damage might lie with a general loss in physiological performance with age (e.g. reduced basal metabolic rate or energy expenditure, hormonal profile, and muscle performance; e.g. Henry 2000; Broggi et al. 2007; Briga and Verhulst 2021; Pontzer et al. 2021) that results in a reduced generation of reactive oxygen species, determining lower levels of oxidative damage as a result. Finally, we need to (i) move beyond the oversimplistic and unsupported predictions '*high oxidative damage at time x is linked to lower survival probability*' and '*two individuals of the same age showing, e.g. similar levels of a given marker of oxidative damage at time t, should have a similar probability of survival per unit of time*', and (ii) abandon a reductionist approach in favour of a more holistic view, would we aim to unravel the secrets of ageing, and to assess the extent to which oxidative stress is a relevant driver of individual or species lifespan.

## References


Abele D, Brey T, Philipp E (2009) Bivalve models of aging and the determination of molluscan lifespans. Exp Geront 44:307–315

Aledo JC, Li Y, De Magalhães JP, Ruíz-Camacho M, Pérez-Claros JA (2011) Mitochondrially encoded methionine is inversely related to longevity in mammals. Aging Cell 10:198–207

Andrews AB, Horvath TL (2009) Uncoupling protein-2 regulates lifespan in mice. Am J Physiol Endocrinol Metab 296:E621–E627

Andziak B, O'Connor TB, Buffenstein R (2005) Antioxidants do not explain the disparate longevity between mice and the longest-living rodent, the naked mole-rat. Mech Ageing Dev 126:1206–1212

Andziak B, O'Connor TP, Qi W, DeWaal EM, Pierce A, Chaudhuri AR, Van Remmen H, Buffenstein R (2006) High oxidative damage levels in the longest-living rodent, the naked mole-rat. Aging Cell 5:463–471

Angelier F, Vleck CM, Holberton RL, Marra PP (2013) Telomere length, non-breeding habitat and return rate in male American redstarts. Funct Ecol 27:342–350

Angelier F, Costantini D, Blévin P, Chastel O (2018) Do glucocorticoids mediate the link between environmental conditions and telomere dynamics in wild vertebrates? A review. Gen Comp Endocrinol 256:99–111

Archer CR, Sakaluk SK, Selman C, Royle NJ, Hunt J (2013) Oxidative stress and the evolution of sex differences in life span and ageing in the decorated cricket, *Gryllodes sigillatus*. Evolution 67:620–634

Argüelles S, García S, Maldonado M, Machado A, Ayala A (2004) Do the serum oxidative stress biomarkers provide a reasonable index of the general oxidative stress status? Biochim Biophys Acta 1674:251–259

Armstrong E, Boonekamp J (2023) Does oxidative stress shorten telomeres in vivo? A meta-analysis. Ageing Res Rev 85:101854

Austad SN (1993) Retarded senescence in an insular population of Virginia opossums (*Didelphis virginiana*). J Zool 229:695–708

Barja G (1998) Mitochondrial free radical production and aging in mammals and birds. Ann N Y Acad Sci 854:224–238

Barja G (2013) Updating the mitochondrial free radical theory of aging: an integrated view, key aspects and confounding concepts. Antioxid Redox Signal 19:1420–1445

Barja G, Cadenas S, Rojas C, Pérez-Campo R, López-Torres M (1994) Low mitochondrial free radical production per unit O2 consumption can explain the simultaneous presence of high longevity and high aerobic metabolic rate in birds. Free Radic Res 21:317–327

Bazopoulou D, Knoefler D, Zheng Y, Ulrich K, Oleson BJ, Xie L, Kim M, Kaufmann A, Lee YT, Dou Y, Chen Y, Quan S, Jakob U (2010) Developmental ROS individualizes organismal stress resistance and lifespan. Nature 576:301–305

Beaulieu M, Reichert S, Le Maho Y, Ancel A, Criscuolo F (2011) Oxidative status and telomere length in a long-lived bird facing a costly reproductive event. Funct Ecol 25:577–585

Beckman KB, Ames BN (1998) The free radical theory of aging matures. Physiol Rev 78:547–581

Begum S, Basova L, Strahl J, Sukhotin A, Heilmayer O, Philipp E, Brey T, Abele D (2009) A metabolic model for the ocean quahog *Arctica islandica* – effects of animal mass and age, temperature, salinity and geography on respiration rate. J Shellfish Res 28:1–7

Bender A, Hajieva P, Moosmann B (2008) Adaptive antioxidant methionine accumulation in respiratory chain complexes explains the use of a deviant genetic code in mitochondria. Proc Natl Acad Sci USA 105:16496–16501

Bigelow H, Schroeder W (1948) Sharks. In: Tee-Van J, Breder CM, Hildebrand SF, Parr AE, Schroeder WC (eds) Fishes of the Western North Atlantic, Part 1. Yale University, Sears Foundation for Marine Research, Yale, CT, pp 59–546

Bisconti R, Canestrelli D, Nascetti G (2011) Genetic diversity and evolutionary history of the Tyrrhenian treefrog Hyla sarda (Anura: Hylidae): adding pieces to the puzzle of Corsica–Sardinia biota. Biol J Linn Soc 103:159–167

Bize P, Devevey G, Monaghan P, Doligez B, Christe P (2008) Fecundity and survival in relation to resistance to oxidative stress in a free-living bird. Ecology 89:2584–2593

Bize P, Criscuolo F, Metcalfe NB, Nasir L, Monaghan P (2009) Telomere dynamics rather than age predict life expectancy in the wild. Proc R Soc Lond B 276:1679–1683

Bjelland S, Seeberg E (2003) Mutagenicity, toxicity and repair of DNA base damage induced by oxidation. Mutat Res 531:37–80

Blanco MA, Sherman PW (2005) Maximum longevities of chemically protected and non-protected fishes, reptiles, and amphibians support evolutionary hypotheses of aging. Mech Ageing Devel 126:794–803

Bodey TW, Cleasby IR, Blount JD, McElwaine G, Vigfusdottir F, Bearhop S (2020) Consistent measures of oxidative balance predict survival but not reproduction in a long-distance migrant. J Anim Ecol 89:872–1882

Boonekamp JJ, Mulder E, Verhulst S (2018) Canalisation in the wild: effects of developmental conditions on physiological traits are inversely linked to their association with fitness. Ecol Lett 21:857–864

Bouwhuis S, Charmantier A, Verhulst S, Sheldon BC (2010) Individual variation in rates of senescence: natal origin effects and disposable soma in a wild bird population. J Anim Ecol 79:1251–1261

Brand MD (2000) Uncoupling to survive? The role of mitochondrial inefficiency in ageing. Exp Geront 35:811–820

Brand MD, Turner N, Ocloo A, Else PL, Hulbert AJ (2003) Proton conductance and fatty acyl composition of liver mitochondria correlates with body mass in birds. Biochem J 376:741–748

Brenner RR (1984) Effect of unsaturated fatty acids on membrane structure and enzyme kinetics. Progr Lipid Res 23:69–96

Briga M, Verhulst S (2021) Mosaic metabolic ageing: Basal and standard metabolic rates age in opposite directions and independent of environmental quality, sex and life span in a passerine. Funct Ecol 35:1055–1068

Brink TC, Demetrius L, Lehrach H, Adjaye J (2009) Age-related transcriptional changes in gene expression in different organs of mice support the metabolic stability theory of aging. Biogerontology 10:549–564

Broggi J, Hohtola E, Koivula K, Orell M, Thomson RL, Nilsson J-A (2007) Sources of variation in winter basal metabolic rate in the great tit. Funct Ecol 21:528–533

Brunet-Rossinni AK (2004) Reduced free-radical production and extreme longevity in the little brown bat (*Myotis lucifugus*) versus two non-flying mammals. Mech Ageing Devel 125:11–20
Buffenstein R (2005) The naked mole-rat: a new long-living model for human aging research? J Geront Series A 60:1369–1377
Butler PG Jr, Wanamaker AD, Scourse JD, Richardson CA, Reynolds DJ (2013) Variability of marine climate on the North Icelandic Shelf in a 1357-year proxy archived based on growth increments in the bivalve *Arctica islandica*. Palaeogeogr Palaeocl 373:141–151
Buttemer W, Battam H, Hulbert AJ (2008) Fowl play and the price of petrel: long-living Procellariformes have peroxidation resistant membrane composition compared with short-living Galliformes. Biol Lett 4:351–354
Buttemer W, Abele D, Costantini D (2010) From bivalves to birds: oxidative stress and longevity. Funct Ecol 24:971–983
Calabrese EJ (2013) Low doses of radiation can enhance insect lifespans. Biogerontology 14:365–381
Canestrelli D, Bisconti R, Liparoto A, Angelier F, Ribout C, Carere C, Costantini D (2021) Biogeography of telomere dynamics in a vertebrate. Ecography 44:453–455
Caratero A, Courtade M, Bonnet L, Planel H, Caratero C (1998) Effect of continuous gamma irradiation at a very low dose on the life span of mice. Gerontology 44:272–276
Carranza J, Alarcos S, Sánchez-Prieto CB, Valencia J, Mateos C (2004) Disposable-soma senescence mediated by sexual selection in an ungulate. Nature 432:215–218
Casagrande S, Hau M (2018) Enzymatic antioxidants but not baseline glucocorticoids mediate the reproduction-survival trade-off in a wild bird. Proc R Soc Lond B 285:20182141
Casagrande S, Dell'Omo G, Costantini D, Tagliavini J, Groothuis T (2011) Variation of a carotenoid-based trait in relation to oxidative stress and endocrine status during the breeding season in the Eurasian kestrel: a multi-factorial study. Comp Biochem Physiol Part A 160:16–26
Castiglione GM, Xu Z, Zhou L, Duh EJ (2020) Adaptation of the master antioxidant response connects metabolism, lifespan and feather development pathways in birds. Nat Commun 11:2476
Chaudhuri AR, de Waal EM, Pierce A, van Remmen H, Ward WF, Richardson A (2006) Detection of protein carbonyls in aging liver tissue: a fluorescence-based proteomic approach. Mech Ageing Devel 127:849–861
Cherubini A, Martin A, Andres-Lacueva C, Di Iorio A, Lamponi M, Mecocci P, Bartali B, Corsi A, Senin U, Ferrucci L (2005) Vitamin E levels, cognitive impairment and dementia in older persons: the InCHIANTI study. Neurobiol Aging 26:987–994
Christensen LL, Selman C, Blount JD, Pilkington JG, Watt KA, Pemberton JM, Reid JM, Nussey DH (2016) Marker-dependent associations among oxidative stress, growth and survival during early life in a wild mammal. Proc R Soc Lond B 283:20161407
Clutton-Brock TH, Isvaran K (2007) Sex differences in ageing in natural populations of vertebrates. Proc R Soc Lond B 274:3097–3104
Cohen AA (2018) Aging across the tree of life: the importance of a comparative perspective for the use of animal models in aging. Biochim Biophys Acta Mol Basis Dis 1864:2680–2689
Cohen AA, McGraw KJ, Wiersma P, Williams JB, Robinson WD, Robinson TR, Brawn JD, Ricklefs RE (2008) Interspecific associations between circulating antioxidant levels and life-history variation in birds. Am Nat 172:178–193
Cohen AA, Milot E, Yong J, Seplaki CL, Fülöp T, Bandeen-Roche K, Fried LP (2013) A novel statistical approach shows evidence for multi-system physiological dysregulation during aging. Mech Ageing Dev 134:110–117
Costantini D (2008) Oxidative stress in ecology and evolution: lessons from avian studies. Ecol Lett 11:1238–1251
Costantini D (2010) Effects of diet quality on serum oxidative status and body mass in male and female pigeons during reproduction. Comp Biochem Physiol Part A 156:294–299
Costantini D (2018) Meta-analysis reveals that reproductive strategies are associated with sexual differences in oxidative balance across vertebrates. Curr Zool 64:1–11

Costantini D, Borremans B (2019) The linear no-threshold model is less realistic than threshold or hormesis-based models: an evolutionary perspective. Chem Biol Inter 301:26–33
Costantini D, Dell'Omo G (2015) Oxidative stress predicts long-term resight probability and reproductive success in Scopoli's shearwater (Calonectris diomedea). Conserv Physiol 3:cov024
Costantini D, Dell'Omo G, De Filippis PS, Marquez C, Snell H, Snell H, Tapia W, Brambilla G, Gentile G (2009) Temporal and spatial covariation of gender and oxidative stress in the Galápagos land iguana *Conolophus subcristatus*. Physiol Biochem Zool 82:430–437
Costantini D, Marasco V, Moller AP (2011) A meta-analysis of glucocorticoids as modulators of oxidative stress in vertebrates. J Comp Physiol B 181:447–456
Costantini D, Ferrari C, Pasquaretta C, Cavallone E, Carere C, von Hardenberg A, Réale D (2012) Interplay between plasma oxidative status, cortisol and coping styles in wild alpine marmots, *Marmota marmota*. J Exp Biol 215:374–383
Costantini D, Monaghan P, Metcalfe N (2014a) Prior hormetic priming is costly under environmental mismatch. Biol Lett 10:20131010
Costantini D, Meillere A, Carravieri A, Lecomte V, Sorci G, Faivre B, Weimerskirch H, Bustamante P, Labadie P, Budzinski H, Chastel O (2014b) Oxidative stress in relation to reproduction, contaminants, gender and age in a long-lived seabird. Oecologia 175:1107–1116
Costantini D, Goutte A, Barbraud C, Faivre B, Sorci G, Weimerskirch H, Delord K, Chastel O (2015) Demographic responses to oxidative stress and inflammation in the wandering albatross (Diomedea exulans). PLoS One 10:e0133967
Costantini D, Wachter B, Melzheimer J, Czirják GÁ (2017a) Socioecological and environmental predictors of physiological stress markers in a threatened feline species. Conserv Physiol 5:cox069
Costantini D, Smith S, Killen SS, Nielsen J, Steffensen JF (2017b) The Greenland shark: a new challenge for the oxidative stress theory of ageing? Comp Biochem Physiol Part A 203:227–232
Costantini D, Masi S, Rachid L, Beltrame M, Rohmer M, Krief S (2021) Mind the food: rapid changes in antioxidant content of diet affect oxidative status of chimpanzees. Am J Physiol 320:728–734
Criscuolo F, Font-Sala C, Bouillaud F, Poulin N, Trabalon M (2010) Increased ROS production: a component of the longevity equation in the male Mygalomorph, *Brachypelma albopilosa*. PLoS One 5:e13104
Csermely P, Soti C (2006) Cellular networks and the aging process. Arch Physiol Biochem 112:60–64
Cypser JR, Johnson TE (2002) Multiple stressors in *Caenorhabditis elegans* induce stress hormesis and extended longevity. J Gerontol A 57:B109–B114
Dagg AI (2009) The social behavior of older animals. The Johns Hopkins University Press
Dańko MJ, Kozłowski J, Schaible R (2015) Unraveling the non-senescence phenomenon in Hydra. J Theor Biol 382:137–149
de Costa J, Barja G, Almaida-Pagan PF (2020) Membrane peroxidation index and maximum lifespan are negatively correlated in fish of the genus Nothobranchius. J Exp Biol 223:jeb224063
De Coster G, De Neve L, Verhulst S, Lens L (2012) Maternal effects reduce oxidative stress in female nestlings under high parasite load. J Avian Biol 43:177–185
de Haan JB, Bladier C, Griffiths P, Kelner M, O'Shea RD, Cheung NS, Bronson RT, Silvestro MJ, Wild S, Zheng SS, Beart PM, Hertzog PJ, Kola I (1998) Mice with a homozygous null mutation for the most abundant glutathione peroxidase, Gpx1, show increased susceptibility to the oxidative stress-inducing agents paraquat and hydrogen peroxide. J Biol Chem 273:22528–22536
de Magalhães JP, Costa J (2009) A database of vertebrate longevity records and their relation to other life-history traits. J Evol Biol 22:1770–1774
de Magalhães JP, Cabral JA, Magalhães D (2005) The influence of genes on the aging process of mice: a statistical assessment of the genetics of aging. Genetics 169:265–274
de Magalhães JP, Costa J, Church GM (2007) An analysis of the relationship between metabolism, developmental schedules, and longevity using phylogenetic independent contrasts. J Geront A 62:149–160

De Waal EM, Liang H, Pierce A, Hamilton RT, Buffenstein R, Chaudhuri AR (2013) Elevated protein carbonylation and oxidative stress do not affect protein structure and function in the long-living naked-mole rat: a proteomic approach. Biochem Biophys Res Commun 434:815–819

Delhaye J, Salamin N, Roulin A, Criscuolo F, Bize P, Christe P (2016) Interspecific correlation between red blood cell mitochondrial ROS production, cardiolipin content and longevity in birds. Age 38:433–443

Descamps S, Boutin S, Berteaux D, Gaillard JM (2006) Best squirrels trade a long life for an early reproduction. Proc R Soc Lond B 273:2369–3237

Domínguez-de-Barros A, Sifaoui I, Borecka Z, Dorta-Guerra R, Lorenzo-Morales J, Castro-Fuentes R, Córdoba-Lanús E (2023) An approach to the effects of longevity, sexual maturity, and reproduction on telomere length and oxidative stress in different Psittacidae species. Front Genet 14:1156730

Du C, Anderson A, Lortie M, Parsons R, Bodnar A (2013) Oxidative damage and cellular defense mechanisms in sea urchin models of aging. Free Radic Biol Med 63C:254–263

Ebert TA (2019) Negative senescence in sea urchins. Exp Gerontol 122:92–98

Elmer LK, O'Connor CM, Philipp DP, Van Der Kraak G, Gilmour KM, Willmore WG, Barthel BL, Cooke SJ (2017) Oxidative ecology of paternal care in wild smallmouth bass, Micropterus dolomieu. J Exp Biol 220:1905–1914

Epel ES, Blackburn EH, Lin J, Dhabhar FS, Adler NE, Morrow JD, Cawthon RM (2004) Accelerated telomere shortening in response to life stress. Proc Natl Acad Sci USA 101:17312–17315

Filho DW, Althoff SL, Dafré AL, Boveris A (2007) Antioxidant defenses, longevity and ecophysiology of South American bats. Comp Biochem Physiol Part C 146:214–220

Finch CE, Pike MC (1996) Maximum life span predictions from the Gompertz mortality model. J Gerontol A 51:183–194

Finkel T, Holbrook NJ (2000) Oxidants, oxidative stress and the biology of ageing. Nature 408:239–247

Fraga CG, Shigenaga MK, Park JW, Degan P, Ames BN (1990) Oxidative damage to DNA during aging: 8-hydroxy-2'-deoxyguanosine in rat organ DNA and urine. Proc Natl Acad Sci USA 87:4533–4537

Freeman-Gallant CR, Amidon J, Berdy B, Wein S, Taff CC, Haussmann MF (2011) Oxidative damage to DNA related to survivorship and carotenoid-based sexual ornamentation in the common yellowthroat. Biol Lett 7:429–432

Freitas AA, de Magalhaes JP (2011) A review and appraisal of the DNA damage theory of ageing. Mutat Res 728:12–22

Furness LJ, Speakman JR (2008) Energetics and longevity in birds. Age 30:75–87

Galván I, Naudí A, Erritzøe J, Møller AP, Barja G, Pamplona R (2015) Long lifespans have evolved with long and monounsaturated fatty acids in birds. Evolution 69:2776–2784

Geiger S, Le Vaillant M, Lebard T, Reichert S, Stier A, Le Maho Y, Criscuolo F (2012) Catching-up but telomere loss: half-opening the black box of growth and ageing trade-off in wild king penguin chicks. Mol Ecol 21:1500–1510

Georgiev AV, Thompson ME, Mandalaywala TM, Maestripieri D (2015) Oxidative stress as an indicator of the costs of reproduction among free-ranging rhesus macaques. J Exp Biol 218:1981–1985

Gershman R, Gilbert DL, Nye F, Dwyer P, Fenw MV (1954) Oxygen poisoning and X-irradiation: a mechanism in common. Science 119:623–626

Golden TR, Hinerfeld DA, Melov S (2002) Oxidative stress and aging: beyond correlation. Aging Cell 1:117–123

Gorbunova V, Seluanov A, Mao Z, Hine C (2007) Changes in DNA repair during aging. Nucl Acids Res 35:7466–7474

Hall ME, Nasir L, Daunt F, Gault EA, Croxall JP, Wanless S, Monaghan P (2004) Telomere loss in relation to age and early environment in long-lived birds. Proc R Soc Lond B 271:1571–1576

Hamilton ML, Van Remmen H, Drake JA, Yang H, Guo ZM, Kewitt K, Walter CA, Richardson A (2001) Does oxidative damage to DNA increase with age? Proc Natl Acad Sci USA 98:10469–10474

Han ES, Muller FL, Pérez VI, Qi W, Liang H, Xi L, Fu C, Doyle E, Hickey M, Cornell J, Epstein CJ, Roberts LJ, Van Remmen H, Richardson A (2008) The in vivo gene expression signature of oxidative stress. Physiol Genom 34:112–126

Harley CB, Futcher AB, Greider CW (1990) Telomeres shorten during ageing of human fibroblasts. Nature 345:458–460

Harman D (1956) Aging: a theory based on free radical and radiation chemistry. J Gerontol A 11:298–300

Harman D (1972) The biologic clock: the mitochondria? J Am Ger Soc 20:145–147

Harper JM, Wang M, Galecki AT, Ro J, Williams JB, Miller RA (2011) Fibroblasts from long-lived bird species are resistant to multiple forms of stress. J Exp Biol 214:1902–1910

Hart RW, Setlow RB (1974) Correlation between deoxyribonucleic acid excision-repair and lifespan in a number of mammalian species. Proc Natl Acad Sci USA 71:2169–2173

Hau M, Haussmann MF, Greives TJ, Matlack C, Costantini D, Quetting M, Adelman JS, Miranda AC, Partecke J (2015) Repeated stressors in adulthood increase the rate of biological ageing. Front Zool 12:4

Haussmann MF, Marchetto NM (2010) Telomeres: linking stress and survival, ecology and evolution. Curr Zool 56:714–727

Haussmann MF, Winkler DW, Vleck CM (2005) Longer telomeres associated with higher survival in birds. Biol Lett 1:212–214

Hayflick L (1965) The limited in vitro lifetime of human diploid cell strains. Exp Cell Res 37:614–636

Hayflick L, Moorhead PS (1961) The serial cultivation of human diploid cell strains. Exp Cell Res 25:585–621

Heidinger BJ, Blount JD, Boner W, Griffiths K, Metcalfe NB, Monaghan P (2012) Telomere length in early life predicts lifespan. Proc Natl Acad Sci USA 109:1743–1748

Henry CJK (2000) Mechanisms of changes in basal metabolism during Ageing. Eur J Clin Nutr 54:S77–S91

Herborn KA, Daunt F, Heidinger BJ, Granroth-Wilding HMV, Burthe SJ, Newell MA, Monaghan P (2016) Age, oxidative stress exposure and fitness in a long-lived seabird. Funct Ecol 30:913–921

Herndon LA, Schmeissner PJ, Dudaronek JM, Brown PA, Listner KM, Sakano Y, Paupard MC, Hall DH, Driscoll M (2002) Stochastic and genetic factors influence tissue-specific decline in ageing *C. elegans*. Nature 419:808–814

Herrero A, Barja G (1997a) ADP-regulation of mitochondrial free radical production is different with complex I- and complex II-linked substrates: implications for the exercise paradox and brain hypermetabolism. J Bioener Biomembr 29:241–249

Herrero A, Barja G (1997b) Sites and mechanisms responsible for the low rate of free radical production of heart mitochondria in the long-lived pigeon. Mech Ageing Dev 98:95–111

Herrero A, Barja G (1998) $H_2O_2$ production of heart mitochondria and aging rate are slower in canaries and parakeets than in mice: sites of free radical generation and mechanisms involved. Mech Ageing Dev 103:133–146

Holman RT (1954) Autoxidation of fats and related substances. In: Holman RT, Lundberg WO, Malkin T (eds) Progress in chemistry of fats and other lipids, vol 2. Pergamon Press, London, pp 51–98

Hornsby PJ (2003) Replicative senescence of human and mouse cells in culture: significance for aging research. Mech Ageing Devel 124:853–855

Houben JMJ, Moonen HJJ, van Schooten FJ, Hageman GJ (2008) Telomere length assessment: biomarker of chronic oxidative stress? Free Radic Biol Med 44:235–246

Huang Z, Whelan CV, Dechmann D, Teeling EC (2020) Genetic variation between long-lived versus short-lived bats illuminates the molecular signatures of longevity. Aging 12:15962–15977

Hulbert AJ (2005) On the importance of fatty acid composition of membranes for aging. J Theor Biol 234:277–288

Hulbert AJ (2008) Explaining longevity of different animals: is membrane fatty acid composition the missing link? Age 30:89–97

Hulbert AJ, Rana T, Couture P (2002a) The acyl composition of mammalian phospholipids: an allometric analysis. Comp Biochem Physiol Part B 132:515–527

Hulbert AJ, Faulks S, Buttemer WA, Else PL (2002b) Acyl composition of muscle membranes varies with body size in birds. J Exp Biol 205:3561–3569

Hulbert AJ, Faulks SC, Buffenstein R (2006) Oxidation-resistant membrane phospholipids can explain longevity differences among the longest-living rodents and similarly-sized mice. J Gerontol 61:1009–1018

Hulbert AJ, Pamplona R, Buffenstein R, Buttemer WA (2007) Life and death: metabolic rate, membrane composition, and life span of animals. Physiol Rev 87:1175–1213

Hulbert AJ, Beard LA, Grigg GC (2008) The exceptional longevity of an egg-laying mammal, the short-beaked echidna (*Tachyglossus aculeatus*) is associated with peroxidation-resistant membrane composition. Exp Geront 43:729–733

Ilmonen P, Kotrschal A, Penn DJ (2008) Telomere attrition due to infection. PLoS One 3:e2143

Isaksson C (2010) Pollution and its impact on wild animals: a meta-analysis on oxidative stress. EcoHealth 7:342–350

Isaksson C, While GM, Olsson M, Komdeur J, Wapstra E (2011) Oxidative stress physiology in relation to life history traits of a free-living vertebrate: the spotted snow skink, *Niveoscincus ocellatus*. Integr Zool 6:140–149

Issartel J, Hervant F, de Fraipont M, Clobert J, Voituron Y (2009) High anoxia tolerance in the subterranean salamander *Proteus anguinus* without oxidative stress nor activation of antioxidant defenses during reoxygenation. J Comp Physiol B 179:543–551

Istomina AA, Zhukovskaya AF, Mazeika AN, Barsova EA, Chelomin VP, Mazur MA, Elovskaya OA, Mazur AA, Dovzhenko NV, Fedorets YV, Karpenko AA (2023) The relationship between lifespan of marine bivalves and their fatty acids of mitochondria lipids. Biology 12:837

Jennings BJ, Ozanne SE, Dorling MW, Hales CN (1999) Early growth determines longevity in male rats and may be related to telomere shortening in the kidney. FEBS Lett 448:4–8

Jimenez AG, O'Connor ES, Elliott KH (2019a) Muscle myonuclear domain, but not oxidative stress, decreases with age in a long-lived seabird with high activity costs. J Exp Biol 222:jeb211185

Jimenez AG, O'Connor ES, Tobin KJ, Anderson KN, Winward JD, Fleming A, Winner C, Chinchilli E, Maya A, Carlson K, Downs CJ (2019b) Does cellular metabolism from primary fibroblasts and oxidative stress in blood differ between mammals and birds? The (lack-thereof) scaling of oxidative stress. Integr Comp Biol 59:953–969

Jobson RW, Nabholz B, Galtier N (2010) An evolutionary genome scan for longevity-related natural selection in mammals. Mol Biol Evol 27:840–847

Jones DP (2006) Redefining oxidative stress. Antiox Redox. Signal 8:1865–1879

Jones OR, Gaillard JM, Tuljapurkar S, Alho JS, Armitage KB, Becker PH, Bize P, Brommer J, Charmantier A, Charpentier M, Clutton-Brock T, Dobson FS, Festa-Bianchet M, Gustafsson L, Jensen H, Jones CG, Lillandt BG, McCleery R, Merilä J, Neuhaus P, Nicoll MA, Norris K, Oli MK, Pemberton J, Pietiäinen H, Ringsby TH, Roulin A, Saether BE, Setchell JM, Sheldon BC, Thompson PM, Weimerskirch H, Jean Wickings E, Coulson T (2008) Senescence rates are determined by ranking on the fast-slow life-history continuum. Ecol Lett 11:664–673

Jones OR, Scheuerlein A, Salguero-Gómez R, Camarda CG, Schaible R, Casper BB, Dahlgren JP, Ehrlén J, García MB, Menges ES, Quintana-Ascencio PF, Caswell H, Baudisch A, Vaupel JW (2014) Diversity of ageing across the tree of life. Nature 505:169–173

Jurk D, Wilson C, Passos JF, Oakley F, Correia-Melo C, Greaves L, Saretzki G, Fox C, Lawless C, Anderson R, Hewitt G, Pender SL, Fullard N, Nelson G, Mann J, van de Sluis B, Mann DA, von Zglinicki T (2014) Chronic inflammation induces telomere dysfunction and accelerates ageing in mice. Nat Commun 2:4172

Kamper EF, Chatzigeorgiou A, Tsimpoukidi O, Kamper M, Dalla C, Pitychoutis PM, Papadopoulou-Daifoti Z (2009) Sex differences in oxidant/antioxidant balance under a chronic mild stress regime. Physiol Behav 98:215–222

Kato H, Harada M, Tsuchiya K, Moriwaki K (1980) Absence of correlation between DNA-repair in ultraviolet irradiated mammalian-cells and life-span of the donor species. Jpn J Genet 55:99–108

Keane M, Semeiks J, Webb AE, Li YI, Quesada V, Craig T, Madsen LB, van Dam S, Brawand D, Marques PI, Michalak P, Kang L, Bhak J, Yim HS, Grishin NV, Nielsen NH, Heide-Jørgensen MP, Oziolor EM, Matson CW, Church GM, Stuart GW, Patton JC, George JC, Suydam R, Larsen K, López-Otín C, O'Connell MJ, Bickham JW, Thomsen B, de Magalhães JP (2015) Insights into the evolution of longevity from the bowhead whale genome. Cell Rep 10:112–122

Keipert S, Voigt A, Klaus S (2011) Dietary effects on body composition, glucose metabolism, and longevity are modulated by skeletal muscle mitochondrial uncoupling in mice. Aging Cell 10:122–136

Keipert S, Ost M, Chadt A, Voigt A, Ayala V, Portero-Otin M, Pamplona R, Al-Hasani H, Klaus S (2013) Skeletal muscle uncoupling-induced longevity in mice is linked to increased substrate metabolism and induction of the endogenous antioxidant defense system. Am J Physiol Endocrinol Metab 304:495–506

Kern B, Ivanina AV, Piontkivska H, Sokolov EP, Sokolova IM (2009) Molecular characterization and expression of a novel homolog of uncoupling protein 5 (UCP5) from the eastern oyster *Crassostrea virginica* (Bivalvia: Ostreidae). Comp Biochem Physiol Part D 4:121–127

Kirkwood TBL (1977) Evolution of ageing. Nature 270:301–304

Kirkwood TBL, Rose MR (1991) Evolution of senescence: late survival sacrificed for reproduction. Phil Trans R Soc Lond B 332:15–24

Kirkwood TBL, Feder M, Finch CE, Franceschi C, Globerson A, Klingenberg CP, LaMarco K, Omholt S, Westendorp RGJ (2005) What accounts for the wide variation in life span of genetically identical organisms reared in a constant environment? Mech Ageing Devel 126:439–443

Klepsatel P, Gáliková M, De Maio N, Ricci S, Schlötterer C, Flatt T (2013) Reproductive and post-reproductive life history of wild-caught *Drosophila melanogaster* under laboratory conditions. J Evol Biol 26:1508–1520

Klimovich A, Rehm A, Wittlieb J, Herbst EM, Benavente R, Bosch TCG (2018) Non-senescent Hydra tolerates severe disturbances in the nuclear lamina. Aging 10:951–972

Koivula MJ, Kanerva M, Salminen JP, Nikinmaa M, Eeva T (2011) Metal pollution indirectly increases oxidative stress in great tit (*Parus major*) nestlings. Environ Res 111:362–370

Kotrschal A, Ilmonen P, Penn DJ (2007) Stress impacts telomere dynamics. Biol Lett 3:128–130

Kowald A (2002) Lifespan does not measure ageing. Biogerontol 3:187–190

Kowald A, Kirkwood TBL (1996) A network theory of ageing: the interactions of defective mitochondria, aberrant proteins, free radicals and scavengers in the ageing process. Mutat Res 316:209–236

Kramer BH, Nehring V, Buttstedt A, Heinze J, Korb J, Libbrecht R, Meusemann K, Paxton RJ, Séguret A, Schaub F, Bernadou A (2021) Oxidative stress and senescence in social insects: a significant but inconsistent link? Phil Trans R Soc Lond B 376:20190732

Kumar SA, Albrecht T, Kauzál O, Tomášek O (2021) No evidence for trade-offs between lifespan, fecundity, and basal metabolic rate mediated by liver fatty acid composition in birds. Front Cell Dev Biol 9:638501

Lambert AJ, Boysen HM, Buckingham JA, Yang T, Podlutsky A, Austad SN, Kunz TH, Buffenstein R, Brand MD (2007) Low rates of hydrogen peroxide production by isolated heart mitochondria associate with long maximum lifespan in vertebrate homeotherms. Aging Cell 6:607–618

Le Bourg E (2005) Hormetic protection of *Drosophila melanogaster* middle-aged male flies from heat stress by mildly stressing them at young age. Naturwissenschaften 92:293–296

Lebovitz RM, Zhang H, Vogel H, Cartwright J, Dionne L, Lu N, Huang S, Matzuk MM (1996) Neurodegradation, myocardial injury, and perinatal death in mitochondrial superoxide dismutase-deficient mice. Proc Natl Acad Sci USA 93:9782–9787

Lei XG, Zhu JH, Cheng WH, Bao Y, Ho YS, Reddi AR, Holmgren A, Arnér ES (2016) Paradoxical roles of antioxidant enzymes: basic mechanisms and health implications. Physiol Rev 96:307–364

Liparoto A, Canestrelli D, Bisconti R, Carere C, Costantini D (2020) Biogeographic history moulds population differentiation in ageing of oxidative status in an amphibian. J Exp Biol 223:jeb235002

Lopez-Torres M, Perez-Campo R, Rojas C, Cadenas S, Barja G (1993) Maximum lifespan in vertebrates: relationship with liver antioxidant enzymes, glutathione system, ascorbate, urate, sensitivity to peroxidation, true malondialdehyde, *in vivo* $H_2O_2$, and basal and maximum aerobic capacity. Mech Ageing Dev 70:177–199

Lorenzini A, Johnson FB, Oliver A, Tresini M, Smith JS, Hdeib M, Sell C, Cristofalo VJ, Stamato TD (2009) Significant correlation of species longevity with DNA double strand break recognition but not with telomere length. Mech Ageing Dev 130:784–792

Loschen G, Flohé L, Chance B (1971) Respiratory chain linked $H_2O_2$ production in pigeon heart mitochondria. FEBS Lett 18:261–264

Losdat S, Helfenstein F, Blount JD, Marri V, Maronde L, Richner H (2013) Nestling erythrocyte resistance to oxidative stress predicts fledging success but not local recruitment in a wild bird. Biol Lett 9:20120888

Lucas ER, Privman E, Keller L (2016) Higher expression of somatic repair genes in long-lived ant queens than workers. Aging 8:1941–1951

Lucas ER, Augustyniak M, Kędziorski A, Keller L (2017) Lifespan differences between queens and workers are not explained by rates of molecular damage. Exp Gerontol 92:1–6

Ma S, Upneja A, Galecki A, Tsai YM, Burant CF, Raskind S, Zhang Q, Zhang ZD, Seluanov A, Gorbunova V, Clish CB, Miller RA, Gladyshev VN (2016) Cell culture-based profiling across mammals reveals DNA repair and metabolism as determinants of species longevity. eLife 5:e19130

MacRae SL, Croken MM, Calder RB, Aliper A, Milholland B, White RR, Zhavoronkov A, Gladyshev VN, Seluanov A, Gorbunova V, Zhang ZD, Vijg J (2015) DNA repair in species with extreme lifespan differences. Aging 7:1171–1184

Maklakov AA, Lummaa V (2013) Evolution of sex differences in lifespan and aging: causes and constraints. Bioessays 35:717–724

Maklakov AA, Fricke C, Arnqvist G (2007) Sexual selection affects lifespan and aging in the seed beetle. Aging Cell 6:739–744

Marasco V, Stier A, Boner W, Griffiths K, Heidinger B, Monaghan P (2017) Environmental conditions can modulate the links among oxidative stress, age, and longevity. Mech Ageing Dev 164:100–107

Mariadassou M, Pellay FX (2014) Identification of amino acids in mitochondrially encoded proteins that correlate with lifespan. Exp Gerontol 56:53–58

Martin JG, Festa-Bianchet M (2011) Age-independent and age-dependent decreases in reproduction of females. Ecol Lett 14:576–581

Martinez DE (1998) Mortality patterns suggest lack of senescence in hydra. Exp Geront 33:217–225

Martínez-Lendech N, Osorio-Beristain M, Franco B, Pedraza-Reyes M, Obregón A, Contreras-Garduño J (2019) Does juvenile hormone prompt oxidative stress in male damselflies? J Exp Biol 222:jeb19453

Medawar PB (1952) An unsolved problem of biology. HK Lewis and Co

Melov S, Schneider JA, Day BJ, Hinerfeld D, Coskun P, Mirra SS, Crapo JD, Wallace DC (1998) A novel neurological phenotype in mice lacking mitochondrial manganese superoxide dismutase. Nat Genet 18:159–163

Melov S, Coskun P, Patel M, Tuinstra R, Cottrell B, Jun AS, Zastawny TH, Dizdaroglu M, Goodman SI, Huang TT, Miziorko H, Epstein CJ, Wallace DC (1999) Mitochondrial disease in superoxide dismutase 2 mutant mice. Proc Natl Acad Sci USA 96:846–851

Miquel J, Economos AC, Fleming J, Johnson JE Jr (1980) Mitochondrial role in cell aging. Exp Geront 15:579–591

Moe B, Rønning B, Verhulst S, Bech C (2009) Metabolic ageing in individual zebra finches. Biol Lett 5:86–89

Møller AP (2006) Sociality, age at first reproduction and senescence: comparative analyses of birds. J Evol Biol 19:682–689

Møller AP, de Lope F, Saino N (2005) Reproduction and migration in relation to senescence in the barn swallow *Hirundo rustica*: a study of avian 'centenarians'. Age 27:307–318

Monaghan P (2010) Telomeres and life histories: the long and the short of it. Ann N Y Acad Sci 1206:130–142

Monaghan P, Haussmann MF (2006) Do telomere dynamics link lifestyle and lifespan? Trends Ecol Evol 21:47–53

Monaghan P, Charmantier A, Nussey DH, Ricklefs RE (2008) The evolutionary ecology of senescence. Funct Ecol 22:371–378

Montgomery MK, Hulbert AJ, Buttemer WA (2011) The long life of birds: the rat-pigeon comparison revisited. PLoS One 6:e24138

Montgomery MK, Hulbert AJ, Buttemer WA (2012) Does the oxidative stress theory of aging explain longevity differences in birds? I. Mitochondrial ROS production. Exp Geront 47:203–210

Moosmann B, Behl C (2008) Mitochondrially encoded cysteine predicts animal lifespan. Aging Cell 7:32–46

Morales M, Oñate M, García MB, Munné-Bosch S (2013) Photooxidative stress markers reveal absence of physiological deterioration with ageing in *Borderea pyrenaica*, an extraordinarily long-lived herb. J Ecol 101:555–565

Munné-Bosch S, Lalueza P (2007) Age-related changes in oxidative stress markers and abscisic acid levels in a drought-tolerant shrub, *Cistus clusii* grown under Mediterranean field conditions. Planta 225:1039–1049

Munro D, Pichaud N, Paquin F, Kemeid V, Blier PU (2013) Low hydrogen peroxide production in mitochondria of the long-lived *Arctica islandica*: underlying mechanisms for slow aging. Aging Cell 12:584–592

Munro D, Rodríguez E, Blier PU (2023) The longest-lived metazoan, Arctica islandica, exhibits high mitochondrial $H_2O_2$ removal capacities. Mitochondrion 68:81–86

Munshi-South J, Wilkinson GS (2006) Diet influences life span in parrots. Auk 123:108–118

Mylonas A, O'Loghlen A (2022) Cellular senescence and ageing: mechanisms and interventions. Front Aging 3:866718

Nabholz B, Glémin S, Galtier N (2008) Strong variations of mitochondrial mutation rate across mammals - the longevity hypothesis. Mol Biol Evol 25:120–130

Nakagawa S, Gemmell NJ, Burke T (2004) Measuring vertebrate telomeres: applications and limitations. Mol Ecol 13:2523–2533

Nielsen J, Hedeholm RB, Simon M, Steffensen JF (2014) Distribution and feeding ecology of the Greenland shark (Somniosus microcephalus) in Greenland waters. Polar Biol 37:37–46

Nielsen J, Hedeholm RB, Heinemeier J, Bushnell PG, Christiansen JS, Olsen J, Bronk Ramsey C, Brill RW, Simon M, Steffensen KF, Steffensen JF (2016) Eye lens radiocarbon reveals centuries of longevity in Greenland shark (Somniosus microcephalus). Science 353:702–704

Noguera JC, Kim S-Y, Velando A (2011) Pre-fledgling oxidative damage predicts recruitment in a long-lived bird. Biol Lett 8:61–63

Noiret A, Puch L, Riffaud C, Costantini D, Riou J-F, Aujard F, Terrien J (2020) Sex-specific response to caloric restriction after reproductive investment in Microcebus murinus: an integrative approach. Front Physiol 11:506

Norte AC, Ramos JA, Araújo PM, Sousa JP, Sheldon BC (2008) Health-state variables and enzymatic biomarkers as survival predictors in nestling great tits (*Parus major*): effects of environmental conditions. The Auk 125:1–11

Nussey DH, Coulson T, Festa-Bianchet M, Gaillard JM (2008) Measuring senescence in wild animal populations: towards a longitudinal approach. Funct Ecol 22:393–406

Nussey DH, Kruuk LE, Morris A, Clements MN, Pemberton JM, Clutton-Brock TH (2009) Inter- and intrasexual variation in aging patterns across reproductive traits in a wild red deer population. Am Nat 174:342–357

Nussey DH, Froy H, Lemaitre J-F, Gaillard J-M, Austad SN (2013) Senescence in natural populations of animals: widespread evidence and its implications for bio-gerontology. Ageing Res Rev 12:214–225

O'Connor TP, Lee A, Jarvis JU, Buffenstein R (2002) Prolonged longevity in naked mole-rats: age-related changes in metabolism, body composition and gastrointestinal function. Comp Biochem Physiol Part A 133:835–842

Ogburn CE, Carlberg K, Ottinger MA, Holmes DJ, Martin GM, Austad SN (2001) Exceptional cellular resistance to oxidative damage in long-lived birds requires active gene expression. J Gerontol A 56:468–474

Orr WC, Sohal RS (1994) Extension of life-span by overexpression of superoxide dismutase and catalase in Drosophila melanogaster. Science 263:1128–1130

Ouyang JQ, Lendvai ÁZ, Moore IT, Bonier F, Haussmann MF (2016) Do hormones, telomere lengths, and oxidative stress form an integrated phenotype? A case study in free-living tree swallows. Integr Comp Biol 56:138–145

Page MM, Stuart JA (2012) Activities of DNA base excision repair enzymes in liver and brain correlate with body mass, but not lifespan. Age 34:1195–1209

Pamplona R (2008) Membrane phospholipids, lipoxidative damage and molecular integrity: a causal role in aging and longevity. Biochim Biophys Acta - Bioenerget 1777:1249–1262

Pamplona R, Barja G (2003) Aging rate, free radical production, and constitutive sensitivity to lipid peroxidation: insights from comparative studies. In: Van Zglinicki T (ed) Biology of aging and its modulation series. Aging at the molecular level, vol 1. Kluwer Academic Publisher, New York, pp 47–64

Pamplona R, Costantini D (2011) Molecular and structural antioxidant defences against oxidative stress in animals. Am J Physiol 301:R843–R863

Pamplona R, Prat J, Cadenas S, Rojas C, Perez-Campo R, Lopez-Torres M, Barja G (1996) Low fatty acid unsaturation protects against lipid peroxidation in liver mitochondria from long-lived species: the pigeon and the human case. Mech Ageing Devel 86:53–66

Pamplona R, Barja G, Portero-Otin M (2002) Membrane fatty acid unsaturation, protection against oxidative stress, and maximum life span: a homeoviscous-longevity adaptation? Ann N Y Acad Sci 959:475–490

Pap PL, Vincze O, Fülöp A, Székely-Béres O, Pătraş L, Pénzes J, Vágási CI (2018) Oxidative physiology of reproduction in a passerine bird: a field experiment. Behav Ecol Sociobiol 72:18

Parry AJ, Narita M (2016) Old cells, new tricks: chromatin structure in senescence. Mamm Genome 27:320–331

Passos JF, Saretzki G, Ahmed S, Nelson G, Richter T, Peters H, Wappler I, Birket MJ, Harold G, Schaeuble K, Birch-Machin MA, Kirkwood TB, von Zglinicki T (2007) Mitochondrial dysfunction accounts for the stochastic heterogeneity in telomere-dependent senescence. PLoS Biol 5:e110

Pauliny A, Wagner RH, Augustin J, Szep T, Blomqvist D (2006) Age-independent telomere length predicts fitness in two bird species. Mol Ecol 15:1681–1687

Pearl R (1928) The rate of living. Alfred A. Knopf, New York

Pérez VI, Van Remmen H, Bokov A, Epstein CJ, Vijg J, Richardson A (2009a) The overexpression of major antioxidant enzymes does not extend the lifespan of mice. Aging Cell 8:73–75

Pérez VI, Buffenstein R, Masamsetti V, Leonard S, Salmon AB, Mele J, Andziak B, Yang T, Edrey Y, Friguet B, Ward W, Richardson A, Chaudhuri A (2009b) Protein stability and resistance to oxidative stress are determinants of longevity in the longest-living rodent, the naked mole-rat. Proc Natl Acad Sci USA 106:3059–3064

Perez-Campo R, Lopez-Torres M, Rojas C, Cadenas S, Barja G (1994) Longevity and antioxidant enzymes, non-enzymatic antioxidants and oxidative stress in the vertebrate lung: a comparative study. J Comp Physiol B 163:682–689

Pérez-Estrada JR, Hernández-García D, Leyva-Castro F, Ramos-León J, Cuevas-Benítez O, Díaz-Muñoz M, Castro-Obregón S, Ramírez-Solís R, García C, Covarrubias L (2019) Reduced lifespan of mice lacking catalase correlates with altered lipid metabolism without oxidative damage or premature aging. Free Radic Biol Med 135:102–115

Peterson RO, Vucetich JA, Fenton G, Drummer TD, Spencer Larsen C (2010) Ecology of arthritis. Ecol Lett 13:1124–1128

Philipp E, Brey T, Heilmayer O, Abele D, Pörtner HO (2006) Physiological ageing in a polar and a temperate swimming scallop. Mar Ecol Progr Ser 307:187–198

Philipp EER, Strahl J, Sukhotin AA (2012) Aging in marine animals. In: Abele D, Vázquez-Medina JP, Zenteno-Savín T (eds) Oxidative stress in aquatic ecosystems. Wiley-Blackwell, pp 193–207

Pickering AM, Lehr M, Kohler WJ, Han ML, Miller RA (2015) Fibroblasts from longer-lived species of primates, rodents, bats, carnivores, and birds resist protein damage. J Gerontol A 70:791–799

Pike TW, Blount JD, Metcalfe NB, Lindström J (2010) Dietary carotenoid availability and reproductive effort influence the age-related decline in performance. Behav Ecol 21:1048–1053

Pontzer H, Yamada Y, Sagayama H, Ainslie PN, Andersen LF, Anderson LJ, Arab L, Baddou I, Bedu-Addo K, Blaak EE, Blanc S, Bonomi AG, Bouten CVC, Bovet P, Buchowski MS, Butte NF, Camps SG, Close GL, Cooper JA, Cooper R, Das SK, Dugas LR, Ekelund U, Entringer S, Forrester T, Fudge BW, Goris AH, Gurven M, Hambly C, El Hamdouchi A, Hoos MB, Hu S, Joonas N, Joosen AM, Katzmarzyk P, Kempen KP, Kimura M, Kraus WE, Kushner RF, Lambert EV, Leonard WR, Lessan N, Martin C, Medin AC, Meijer EP, Morehen JC, Morton JP, Neuhouser ML, Nicklas TA, Ojiambo RM, Pietiläinen KH, Pitsiladis YP, Plange-Rhule J, Plasqui G, Prentice RL, Rabinovich RA, Racette SB, Raichlen DA, Ravussin E, Reynolds RM, Roberts SB, Schuit AJ, Sjödin AM, Stice E, Urlacher SS, Valenti G, Van Etten LM, Van Mil EA, Wells JCK, Wilson G, Wood BM, Yanovski J, Yoshida T, Zhang X, Murphy-Alford AJ, Loechl C, Luke AH, Rood J, Schoeller DA, Westerterp KR, Wong WW, Speakman JR, IAEA DLW Database Consortium (2021) Daily energy expenditure through the human life course. Science:808–812

Porter RK, Brand MD (1993) Body mass dependence of $H^+$ leak in mitochondria and its relevance to metabolic rate. Nature 362:628–630

Rattan SIS (2008) Hormesis in aging. Ageing Res Rev 7:63–78

Reed T, Kruuk LEB, Wanless S, Frederiksen M, Cunningham EJM, Harris MP (2008) Reproductive senescence in a long-lived seabird: rates of decline in late life performance are associated with varying costs of early reproduction. Am Nat 171:E89–E101

Reichert S, Stier A (2017) Does oxidative stress shorten telomeres in vivo? A review. Biol Lett 13:20170463

Rey B, Roussel D, Romestaing C, Belouze M, Rouanet JL, Desplanches D, Sibille B, Servais S, Duchamp C (2010) Up-regulation of avian uncoupling protein in cold-acclimated and hyperthyroid ducklings prevents reactive oxygen species production by skeletal muscle mitochondria. BMC Physiol 10:5

Reznick D, Bryant M, Holmes D (2006) The evolution of senescence and post-reproductive lifespan in guppies (*Poecilia reticulata*). PLoS Biol 4:e7

Ricklefs RE (2010) Life-history connections to rates of aging in terrestrial vertebrates. Proc Natl Acad Sci USA 107:10314–10319

Ridgway I, Bowden TJ, Roman-Gonzalez A, Richardson CA (2014) Resistance to oxidative stress is not associated with the exceptional longevity of the freshwater pearl mussel, Margaritifera margaritifera nor three unionid species. Aquat Sci 76:259–267

Robert KA, Brunet-Rossinni A, Bronikowski AM (2007) Testing the 'free radical theory of aging' hypothesis: physiological differences in long-lived and short-lived colubrid snakes. Aging Cell 6:395–404

Rottenberg H (2006) Longevity and the evolution of the mitochondrial DNA-coded proteins in mammals. Mech Ageing Dev 127:748–760

Rottenberg H (2007a) Coevolution of exceptional longevity, exceptionally high metabolic rates, and mitochondrial DNA-coded proteins in mammals. Exp Geront 42:364–373
Rottenberg H (2007b) Exceptional longevity in songbirds is associated with high rates of evolution of cytochrome b, suggesting selection for reduced generation of free radicals. J Exp Biol 210:2170–2180
Rubner M (1908) Das problem der lebensdauer und seine beziehunger zum wachstum und ernaibrung. Oldenburg, Munchen
Rubolini D, Colombo G, Ambrosini R, Caprioli M, Clerici M, Colombo R, Dalle-Donne I, Milzani A, Romano A, Romano M, Saino N (2012) Sex-related effects of reproduction on biomarkers of oxidative damage in free-living barn swallows (Hirundo rustica). PLoS ONE 7:e48955
Ruggiero C, Metter EJ, Melenovsky V, Cherubini A, Najjar SS, Ble A, Senin U, Longo DL, Ferrucci L (2008) High basal metabolic rate is a risk factor for mortality: the Baltimore longitudinal study of aging. J Geront A 63:698–706
Saino N, Caprioli M, Romano M, Boncoraglio G, Rubolini D, Ambrosini R, Bonisoli-Alquati A, Romano A (2011) Antioxidant defenses predict long-term survival in a passerine bird. PLoS ONE 6:e19593
Saiz E, Calbet A, Griffell K, Bersano JG, Isari S, Solé M, Peters J, Alcaraz M (2015) Ageing and caloric restriction in a marine planktonic copepod. Sci Rep 5:14962
Salmon AB, Marx DB, Harshman LG (2001) A cost of reproduction in *Drosophila melanogaster*: stress susceptibility. Evolution 55:1600–1608
Salmon AB, Leonard S, Masamsetti V, Pierce A, Podlutsky AJ, Podlutskaya N, Richardson A, Austad SN, Chaudhuri A (2009) The long lifespan of two bat species is correlated with resistance to protein oxidation and enhanced protein homeostasis. FASEB J 23:2317–2326
Salmon AB, Richardson A, Pérez VI (2010) Update on the oxidative stress theory of aging: does oxidative stress play a role in aging or healthy aging? Free Radic Biol Medic 48:642–655
Salomons HM, Mulder GA, van de Zande L, Haussmann MF, Linskens MH, Verhulst S (2009) Telomere shortening and survival in free-living corvids. Proc R Soc Lond B 276:3157–3165
Salway KD, Gallagher EJ, Page MM, Stuart JA (2011) Higher levels of heat shock proteins in longer-lived mammals and birds. Mech Ageing Dev 132:287–297
Samuels DC (2005) Life span is related to the free energy of mitochondrial DNA. Mech Ageing Dev 126:1123–1129
Sanz A, Pamplona R, Barja G (2006) Is the mitochondrial free radical theory of aging intact? Antiox Redox Sign 8:582–599
Sasaki T, Unno K, Tahara S, Shimada A, Chiba Y, Hoshino M, Kaneko T (2008) Age-related increase of superoxide generation in the brains of mammals and birds. Aging Cell 7:459–469
Schade FM, Clemmesen C, Mathias Wegner K (2014) Within- and transgenerational effects of ocean acidification on life history of marine three-spined stickleback (Gasterosteus aculeatus). Mar Biol 161:1667–1676
Scholer MN, Arcese P, Puterman ML, Londoño GA, Jankowski JE (2019) Survival is negatively related to basal metabolic rate in tropical Andean birds. Funct Ecol 33:1436–1445
Schriner SE, Linford NJ, Martin GM, Treuting P, Ogburn CE, Emond M, Coskun PE, Ladiges W, Wolf N, Van Remmen H, Wallace DC, Rabinovitch PS (2005) Extension of murine life span by overexpression of catalase targeted to mitochondria. Science 308:1909–1911
Selman C, McLaren JS, Collins AR, Duthie GG, Speakman JR (2008) The impact of experimentally elevated energy expenditure on oxidative stress and lifespan in the short-tailed field vole *Microtus agrestis*. Proc R Soc Lond B 275:1907–1916
Sharpless NE, DePinho RA (2007) How stem cells age and why this makes us grow old. Nat Rev Mol Cell Biol 8:703–713
Shefferson R, Jones O, Salguero-Gómez R (2017) The evolution of senescence in the tree of life. Cambridge University Press, Cambridge
Simons MJ (2015) Questioning causal involvement of telomeres in aging. Ageing Res Rev 24:191–196
Sohal RS, Orr WC (2012) The redox stress hypothesis of aging. Free Rad Biol Medic 52:539–555

Sohal RS, Mockett RJ, Orr WC (2002) Mechanisms of aging: an appraisal of the oxidative stress hypothesis. Free Radic Biol Med 33:575–586
Soltow QA, Jones DP, Promislow DEL (2010) A network perspective on metabolism and aging. Integr Comp Biol 50:844–854
Sørensen JG, Kristensen TN, Kristensen KV, Loeschcke V (2007) Sex specific effects of heat induced in Hsf-deficient *Drosophila melanogaster*. Exp Gerontol 42:1123–1129
Sosnowska D, Richardson C, Sonntag WE, Csiszar A, Ungvari Z, Ridgway I (2014) A heart that beats for 500 years: age-related changes in cardiac proteasome activity, oxidative protein damage and expression of heat shock proteins, inflammatory factors, and mitochondrial complexes in *Arctica islandica*, the longest-living noncolonial animal. J Gerontol A 69:1448–1461
Southworth LK, Owen AB, Kim SK (2009) Aging mice show a decreasing correlation of gene expression within genetic modules. PLoS Genet 5:e1000776
Speakman JR (2005a) Body size, energy metabolism and lifespan. J Exp Biol 208:1717–1730
Speakman JR (2005b) Correlations between physiology and lifespan–two widely ignored problems with comparative studies. Aging Cell 4:167–175
Speakman JR, Talbot DA, Selman C, Snart S, McLaren JS, Redman P, Krol E, Jackson DM, Johnson MS, Brand MD (2004) Uncoupled and surviving: individual mice with high metabolism have greater mitochondrial uncoupling and live longer. Aging Cell 3:87–95
Stadtman ER, Moskovitz J, Berlett B, Levine RL (2002) Cyclic oxidation and reduction of protein methionine residues is an important antioxidant mechanism. Mol Cell Biochem 234-235:3–9
Stanton R, Clark RG, Morrissey CA (2017) Intensive agriculture and insect prey availability influence oxidative status and return rates of an aerial insectivore. Ecosphere 8:e01746
Stenvinkel P, Shiels PG (2019) Long-lived animals with negligible senescence: clues for ageing research. Biochem Soc Trans 47:1157–1164
Stier A, Bize P, Roussel D, Schull Q, Massemin S, Criscuolo F (2014) Mitochondrial uncoupling as a regulator of life-history trajectories in birds: an experimental study in the zebra finch. J Exp Biol 217:3579–3589
Storer JB (1967) Relation of lifespan to brain weight, body weight and metabolic rate among inbred mouse strains. Exp Geront 2:173–182
Strecker V, Mai S, Muster B, Beneke S, Bürkle A, Bereiter-Hahn J, Jendrach M (2010) Aging of different avian cultured cells: lack of ROS-induced damage and quality control mechanisms. Mech Ageing Develop 131:48–59
Sun J, Tower J (1999) FLP recombinase-mediated induction of Cu/Zn-superoxide dismutase transgene expression can extend the lifespan of adult Drosophila melanogaster flies. Mol Cell Biol 19:216–228
Tacutu R, Craig T, Budovsky A, Wuttke D, Lehmann G, Taranukha D, Costa J, Fraifeld VE, de Magalhaes JP (2013) Human ageing genomic resources: integrated databases and tools for the biology and genetics of ageing. Nucleic Acids Res 41:D1027–D1033
Talbot DA, Lambert AJ, Brand MD (2004) Production of endogenous matrix superoxide from mitochondrial complex I leads to activation of uncoupling protein 3. FEBS Lett 556:111–115
Thompson González NT, Otali E, Machanda Z, Muller MN, Wrangham R, Thompson ME (2020) Urinary markers of oxidative stress respond to infection and late-life in wild chimpanzees. PLoS One 15:e0238066
Tian X, Seluanov A, Gorbunova V (2017) Molecular mechanisms determining lifespan in short- and long-lived species. Trends Endocrinol Metab 28:722–734
Tian X, Firsanov D, Zhang Z, Cheng Y, Luo L, Tombline G, Tan R, Simon M, Henderson S, Steffan J, Goldfarb A, Tam J, Zheng K, Cornwell A, Johnson A, Yang JN, Mao Z, Manta B, Dang W, Zhang Z, Vijg J, Wolfe A, Moody K, Kennedy BK, Bohmann D, Gladyshev VN, Seluanov A, Gorbunova V (2019) SIRT6 is responsible for more efficient DNA double-strand break repair in long-lived species. Cell 177:622–638
Tobler M, Gómez-Blanco D, Hegemann A, Lapa M, Neto JM, Tarka M, Xiong Y, Hasselquist D (2022) Telomeres in ecology and evolution: a review and classification of hypotheses. Mol Ecol 31:5946–5965

Torres-Barceló C, Cabot G, Oliver A, Buckling A, MacLean RC (2013) A trade-off between oxidative stress resistance and DNA repair plays a role in the evolution of elevated mutation rates in bacteria. Proc R Soc Lond B 280:20130007

Tower J (2006) Sex-specific regulation of aging and apoptosis. Mech Ageing Dev 127:705–718

Tricola GM, Simons MJP, Atema E, Boughton RK, Brown JL, Dearborn DC, Divoky G, Eimes JA, Huntington CE, Kitaysky AS, Juola FA, Lank DB, Litwa HP, Mulder EGA, Nisbet ICT, Okanoya K, Safran RJ, Schoech SJ, Schreiber EA, Thompson PM, Verhulst S, Wheelwright NT, Winkler DW, Young R, Vleck CM, Haussmann MF (2018) The rate of telomere loss is related to maximum lifespan in birds. Phil Trans R Soc Lond B 373:20160445

Trivers R (1985) Social evolution. Menlo Park, Benjamin/Cummings Publishing, California

Ungvari Z, Ridgway I, Philipp EE, Campbell CM, McQuary P, Chow T, Coelho M, Didier ES, Gelino S, Holmbeck MA, Kim I, Levy E, Sosnowska D, Sonntag WE, Austad SN, Csiszar A (2011) Extreme longevity is associated with increased resistance to oxidative stress in Arctica islandica, the longest-living non-colonial animal. J Gerontol A 66:741–750

Vaanholt LM, Daan S, Schubert KA, Visser GH (2009) Metabolism and aging: Effects of cold exposure on metabolic rate, body composition, and longevity in mice. Physiol Biochem Zool 82:314–324

Vaanholt LM, Daan S, Garland T Jr, Visser GH (2010) Exercising for life? Energy metabolism, body composition, and longevity in mice exercising at different intensities. Physiol Biochem Zool 83:239–251

Vágási CI, Vincze O, Pătraş L, Osváth G, Pénzes J, Haussmann MF, Barta Z, Pap PL (2019) Longevity and life history coevolve with oxidative stress in birds. Funct Ecol 33:152–161

Valencak TG, Ruf T (2007) N-3 polyunsaturated fatty acids impair lifespan but have no role for metabolism. Aging Cell 6:15–25

van de Crommenacker J (2011) Hard times in paradise? Oxidative status, physiology and fitness in the tropical Seychelles warbler. PhD Thesis, University of Groningen, The Netherlands

van Deursen JM (2014) The role of senescent cells in ageing. Nature 509:439–446

Van Remmen H, Ikeno Y, Hamilton M, Pahlavani M, Wolf N, Thorpe SR, Alderson NL, Baynes JW, Epstein CJ, Huang TT, Nelson J, Strong R, Richardson A (2003) Life-long reduction in MnSOD activity results in increased DNA damage and higher incidence of cancer but does not accelerate aging. Physiol Genom 16:29–37

Veskoukis AS, Nikolaidis MG, Kyparos A, Kouretas D (2009) Blood reflects tissue oxidative stress depending on biomarker and tissue studied. Free Rad Biol Med 47:1371–1374

Vina J, Gambini J, Lopez-Grueso R, Abdelaziz KM, Jove M, Borras C (2011) Females live longer than males: role of oxidative stress. Curr Pharm Des 17:3959–3965

Vitikainen EIK, Cant MA, Sanderson JL, Mitchell C, Nichols HJ, Marshall HH, Thompson FJ, Gilchrist JS, Hodge SJ, Johnstone RA, Blount JD (2016) Evidence of oxidative shielding of offspring in a wild mammal. Front Ecol Evol 4:58

Voituron Y, de Fraipont M, Issartel J, Guillaume O, Clobert J (2011) Extreme lifespan of the human fish (*Proteus anguinus*): a challenge for ageing mechanisms. Biol Lett 7:105–107

von Zglinicki T (2002) Oxidative stress shortens telomeres. Trends Biochem Sci 27:339–344

Vyas CM, Manson JE, Sesso HD, Cook NR, Rist PM, Weinberg A, Moorthy MV, Baker LD, Espeland MA, Yeung LK, Brickman AM, Okereke OI (2024) Effect of multivitamin-mineral supplementation versus placebo on cognitive function: results from the clinic sub-cohort of the COSMOS randomized clinical trial and meta-analysis of three cognitive studies within COSMOS. Am J Clin Nutr. in press

Wasser DE, Sherman PW (2010) Avian longevities and their interpretation under evolutionary theories of senescence. J Zool 280:103–155

Webb RE, Leslie DM Jr, Lochmiller RL, Masters RE (2005) Impact of food supplementation and methionine on high densities of cotton rats: support of the amino-acid-quality hypothesis? J Mammal 86:46–55

Weinert BT, Timiras PS (2003) Theories of aging. J Appl Physiol 95:1706–1716

Weirsma P, Nowak B, Williams JB (2012) Small organ size contributes to the slow pace of life in tropical birds. J Exp Biol 215:1662–1669

Williams GC (1957) Pleiotropy, natural selection, and the evolution of senescence. Evolution 11:398–411

Woodhead AD, Satlow RB, Grist E (1980) DNA repair and longevity in three species of cold-blooded vertebrates. Exp Gerontol 15:301–304

Xia C, Møller AP (2018) Long-lived birds suffer less from oxidative stress. Avian Res 9:41

Xue H, Xian B, Dong D, Xia K, Zhu S, Zhang Z, Hou L, Zhang Q, Zhang Y, Han J-DJ (2007) A modular network model of aging. Mol Syst Biol 3:147

Yang W, Hekimi S (2010) A mitochondrial superoxide signal triggers increased longevity in Caenorhabditis elegans. PLoS Biol 8:e1000556

Yarian CS, Rebrin I, Sohal RS (2005) Aconitase and ATP synthase are targets of malondialdeyde modification and undergo an age-related decrease in activity in mouse heart mitochondria. Biochem Biophys Res Commun 330:151–156

Young AJ (2018) The role of telomeres in the mechanisms and evolution of life-history trade-offs and ageing. Phil Trans R Soc Lond B 373:20160452

Yu BP, Yang R (1996) Critical evaluation of the free radical theory of aging. A proposal for the oxidative stress hypothesis. Ann N Y Acad Sci 786:1–11

Zagkle E, Grosiak M, Bauchinger U, Sadowska ET (2020) Rest-phase hypothermia reveals a link between aging and oxidative stress: a novel hypothesis. Front Physiol 11:575060

Zhang L, Yang F, Zhu WL (2021) Evidence for the 'rate-of-living' hypothesis between mammals and lizards, but not in birds, with field metabolic rate. Comp Biochem Physiol Part A 253:110867

GPSR Compliance

*The European Union's (EU) General Product Safety Regulation (GPSR) is a set of rules that requires consumer products to be safe and our obligations to ensure this.*

*If you have any concerns about our products, you can contact us on ProductSafety@springernature.com*

In case Publisher is established outside the EU, the EU authorized representative is:

Springer Nature Customer Service Center GmbH
Europaplatz 3
69115 Heidelberg, Germany

**Batch number: 10732076**

Printed by Printforce, the Netherlands